UNDERSTANDING DIGITAL INDUSTRY

PROCEEDINGS OF THE CONFERENCE ON MANAGING DIGITAL INDUSTRY, TECHNOLOGY AND ENTREPRENEURSHIP (COMDITE 2019), BANDUNG, INDONESIA, 10-11 JULY 2019

Understanding Digital Industry

Editors

Siska Noviaristanti
Telkom University, Indonesia

Hasni Mohd Hanafi
Multimedia University, Malaysia

Donny Trihanondo
Telkom University, Indonesia

LONDON AND NEW YORK

Routledge is an imprint of the Taylor & Francis Group, an informa business

© 2020 Taylor & Francis Group, London, UK

Typeset by Integra Software Services Pvt. Ltd., Pondicherry, India

All rights reserved. No part of this publication or the information contained herein may be reproduced, stored in a retrieval system, or transmitted in any form or by any means, electronic, mechanical, by photocopying, recording or otherwise, without written prior permission from the publisher.

Although all care is taken to ensure integrity and the quality of this publication and the information herein, no responsibility is assumed by the publishers nor the author for any damage to the property or persons as a result of operation or use of this publication and/or the information contained herein.

Publisher's Note
The publisher has gone to great lengths to ensure the quality of this reprint but points out that some imperfections in the original copies may be apparent.

Library of Congress Cataloging-in-Publication Data

Applied for

Published by: CRC Press/Balkema
                Schipholweg 107C, 2316XC Leiden, The Netherlands
                e-mail: Pub.NL@taylorandfrancis.com
                www.crcpress.com – www.taylorandfrancis.com

First issued in paperback 2021

ISBN 13: 978-1-03-224088-6 (pbk)
ISBN 13: 978-0-367-41076-6 (hbk)

DOI: https://doi.org/10.1201/9780367814557

Table of contents

Preface	xi
Organizing Committee	xiii
Scientific Committee	xv
Acknowledgements	xvii

Digital know-how and FinTech readiness 1
S. Muthaiyah & T.O.K. Zaw

An overarching architectural framework for smart cities 5
S. Muthaiyah, T.O.K. Zaw & Indrawati

Using 3D sculpting machine to link art, nature, and community – case study from Thailand Biennale, Krabi 2018 9
V. Mukdamanee

Green product advantage: Mediating effects on green SMEs performance 13
S. Nuryakin & T. Maryati

Bringing Telkom Indonesia to be a global digital player 17
R.D Pasaribu

Corporate innovation: Democratizing decision-making 18
F. Feisal

The three giants in the IT industry and their valuation 20
F.R. Kurnia & P.M.T. Sitorus

Analysis of the role of brand orientation in brand performance in B2B firms 24
S.L. Retno

The influence of e-service quality on the data package buying experience in Telkomsel 27
P. Kusdinar & M. Ariyanti

Develop a perceivable financial statement: A qualitative study at SME four coffee 31
E. Wahyudi & S.K. Widhaningrat

The effect of internal, external, and interactive marketing on customer decisions 35
A. Wulandari & B. Suryawardani

Analysis of how to increase purchases through the Telkomsel Virtual Assistant 40
Ismawaty & G. Ramantoko

Optimizing Instagram as a digital marketing channel: A qualitative study at Larnis 45
Y. Shigeno & S.K. Widhaningrat

The development of marketing channels and inventory management systems for SMEs 49
Edbert & S.K. Widhaningrat

Valuation analysis of skill based asset company: Acquisition by PT Elnusa, Tbk *D. Christian & E. Tanudjaya*	54
Synergy effect analysis of PT XL Axiata Tbk after acquisition, 2011–2013 *R.A. Gitasari & R. Rokhim*	58
The main determinants in the distress / failure probability of the public-private partnership financing scheme in infrastructure development *M.B. Octaviani & Z.A. Husodo*	63
Bandung smart technology readiness index *R. Kurniawan & Indrawati*	67
Leverage up predictive analytics to increase prepaid credit purchasers *R. Arnaz & G. Ramantoko*	71
Users' continuance intention toward hospitality service applications *Indrawati, S.A. Hasana & S.K.B. Pillai*	76
Customers' continuance acceptance toward automated guided transportation *Indrawati & P.A. Nidya*	81
Corporate transformation, digital culture, and employee performance in a digital company *R. Bachtiar & A.I. Susanty*	85
How internal and organizational factors influence employees' innovative behavior *Adityawarman & R. Rachmawati*	90
Effect of work placement and workloads on performance of the PKH companion *N. Maulida & R. Wahyuningtyas*	94
Designing KPIs for measurement of operations in PERUM Jaminan Kredit Indonesia *R.E. Banjarnahor, U. Cariawan & Arviansyah*	99
Customer continuance intention toward digital banking applications *Indrawati, N.P. Pradhina & S.K.B. Pillai*	103
Self-regulated learning for smart learning in a university at Cyberjaya *L. Anthonysamy, K.A. Choo & H.S. Hin*	108
Analysis of factors influencing consumer intention toward Indonesia QR mobile payment *H.A. Baskoro & A. Amini*	112
Patterning consumer behavior to school innovations from two different perspectives *E. Princes & J. Setiawan*	117
Analysis of purchase decision-making in different surroundings *E. Princes & Sasmoko*	121
Is the relationship between entrepreneurial competency and business success contingent upon the business environment? *M.A.S. Ismiraini & L. Yuldinawati*	125
A proposed model for measuring cloud accounting adoption among SMEs in Indonesia *E. Widaryanti & Indrawati*	129
Digital literacy deficiencies in digital learning among undergraduates *L. Anthonysamy*	133
Smart city concept based on Nusantara culture *D. Trihanondo & D. Endriawan*	137

Credit scoring model for SME customer assessment in a telco company *L.A. Baranti & A. Lutfi*	141
Go Beyond training and its impact on the performance of employees in PT Telkomsel *A. Dharmiko & N. Dudija*	145
Exploring jewelry design for adult women by developing the pineapple skin *A.S.M. Atamtajani & S.A. Putri*	150
Identifying factors that influence interest using credit scoring by implementing customer journey mapping *L. Nadeak & D. Tricahyono*	154
Assessing Indonesia's textile company listed in IDX market: The valuation case *P.W. Satriawana & R. Hendrawan*	158
Influence of perceived quality of mobile payment application towards loyalty *T.N. Sandyopasana & M.G. Alif*	163
Implementation of marketing innovation at PT. Pegadaian in the revolutionary 4.0 *M.A. Sugiat & K. Sudiana*	168
Analysis of customer intention in adopting automated parcel station services *P.M.T. Sitorus & M.O. Alexandra*	172
Predicting VR adoption on e-commerce platforms using TAM and porter five forces *C.S. Mon*	176
Transportation mathematical model teaching using inquiry-based learning *R. Aurachman & Nopendri*	180
Associative learning method for teaching operation research *R. Aurachman*	184
Optimization model of the number of one-way servers to minimize cost and constraint *R. Aurachman*	188
Employee behavior towards big data analytics: A research framework *W. Ahmed, S.M. Hizam, H. Akter & I. Sentosa*	192
Indicators for measuring green energy: An Indonesian perspective *Indrawati, C. Januarizka, D. Tricahyono & S. Muthaiyah*	196
Indicators for measuring green waste: An Indonesian perspective *Indrawati, F. Andriawan, D. Tricahyono & S. Muthaiyah*	201
Essence and implementation of ERP in culinary industry: Critical success factors *R. Suryani, R.W. Witjaksono & M. Lubis*	205
Interpersonal skills in project management *A. Said, H. Prabowo, M. Hamsal & B. Simatupang*	210
Employees' post-adoption behavior towards collaboration technology *K. Rapiz & Arviansyah*	214
Designing RecyclerApp: A mobile application to manage recyclable waste *A. Jalil, A.A.A. Hussin & R. Sham*	219
Stock market prediction using multivariate neural network backpropagation *T. Kristian & F.T. Kristanti*	223
IDR/USD forecasting: Classical time series and artificial neural network method *M.F.Q. Alam & B. Rikumahu*	227

Development of Stock Opname application with integration to SAP Business One using Scrum 232
Leonardo, M. Lubis & W. Puspitasari

How to reduce the risk of fintech application through cooperative organizations? 236
Sugiyanto

Factors affecting adoption of mobile banking in Bank Mandiri customers 240
I. Fitri & T. Widodo

EVA approach in measuring distribution company performance listed in IDX 244
P.E. Sembiring & J. Jahja

Analysis of Sharia banking health with the risk based bank rating and bankometer 248
M.I. Hidayat & F.T. Kristanti

Implementation of a decision supporting system against satisfaction of Tower Provider 253
Sarman & A.M.A. Suyanto

Cosmopolitanism in the fashion industry in Bali as an impact of the tourism sector 258
A. Arumsari, A. Sachari & A.R. Kusmara

Pulp and paper companies and their fair value: Indonesian stock market evidence 262
M.I. Miala & F.T. Kristanti

Sunda culture values at Sunda restaurant design in Bandung 267
T. Cardiah, R. Wulandari & T. Sarihati

Post acquisition analysis and evaluation with regard to diversification 271
K. Rosby & D.A. Chalid

Financial efficiency of metal and mineral mining company in Indonesia 275
N.F. Sembung & F.T. Kristanti

Identification of tourism destination preferences based on geotag feature on Instagram using data analytics and topic modeling 280
H. Irawan, R.S. Widyawati & A. Alamsyah

Social network analysis of the information dissemination patterns and stakeholders' roles at superpriority tourism destinations in Indonesia 285
H. Irawan, D.A. Digpasari & A. Alamsyah

The effect analysis of financial performance of companies before and after M&As 289
A.R. Bionda, I. Gandakusuma & Z. Dalimunthe

Effect of transformational leadership and culture on creativity and innovation 293
A. Sulthoni & J. Sadeli

Essence and implementation of enterprise resource planning in the textile industry: Critical success factor 297
A.R. Muafah, R.W. Witjaksono & M. Lubis

Innovation of motif design for traditional batik craftsmen 302
F. Ciptandi

Design documentation of software specifications for a maternal and child health system 307
G.S. Hana, T.L.R. Mengko & B. Rahardjo

Design strategies in the market competition of capsule hostels 311
S. Rahardjo

The link of e-servqual & perceived justice of e-recovery to satisfaction & loyalty *S. Syafrizal & S.L. Geni*	316
Technical efficiency of information and communications technology companies in East Asia and Southeast Asia *Y. Yuliansyah & P.M. Sitorus*	320
Examining factors influencing internal audit effectiveness *F.D. Izzuddin & H.M. Hanafi*	324
Implementation of cognitive multimedia learning theory on mobile apps interactive story Kisah Lutung Kasarung *D. Hidayat, M.I.P. Koesoemadinata & M.A.B.M. Desa*	328
Efficiency of legal and regulatory framework in combating cybercrime in Malaysia *S. Khan, N. Khan & O. Tan*	333
The case of women entrepreneurs: Comparison study between two cities in Indonesia and Zimbabwe *R.L. Nugroho & S. Mapfumo*	337
The compliance to the statement of risk management and internal control in Malaysia *H. Johari & N. Jaffar*	342
Toward a model of social entrepreneurship using the soft system methodology approach *G. Anggadwita & G.C.W. Pratami*	346
Motivation to become agropreneurs among youths in Malaysia *M.F.A. Rahim, K.W. Chew, M.N. Zainuddin & A.S. Bujang*	350
Customer journey maps of Muslim young agropreneurs *M.F.A. Rahim, J.W. Ong, N.M. Yatim, H.A. Yanan & M.N.M. Nizat*	354
New culinary trends based on the most popular Instagram accounts *G. Anggadwita, E. Yuliana, D.T. Alamanda, A. Ramdhani & A. Permatasari*	358
Sentiment analysis of social media engagement to purchasing intention *A. Wiliam, Sasmoko, W. Kosasih & Y. Indrianti*	362
User satisfaction among Malaysian music streamers *A.H. Kaur & S. Gopinathan*	366
Effective tax rate and reporting quality for Malaysian manufacturing companies *S.M. Ali, M. Norhashim & N. Jaafar*	370
The effects of viral marketing on users' attitudes toward JOOX Indonesia *I. Nilasari & D. Tricahyono*	374
Political connection, internal audit and audit fees in Malaysia *A.A. Saprudin*	379
Cyber-entrepreneurial intentions of the Malaysian university students *N. Ismail, N. Jaffar & T.S. Hooi*	383
Financial literacy of the younger generation in Malaysia *Z. Selamat, N. Jaffar, H. Hamzah & I.S. Awaludin*	387
Financial performance evaluation for network facility and service providers *S. Segaran & L.T.P. Nguyen*	391

Nonlinear impact of institutional quality on economic performance within the
comprehensive and progressive agreement for trans-pacific partnership 396
C.Y. Chong

External knowledge acquisition and innovation in small and medium-sized enterprises 400
I.S. Rosdi, A.M. Noor & N. Fauzi

Overcoming math anxiety and developing mathematical resilience via e-learning 404
R.A. Razak

Relational social capital, innovation capability, and small and medium-sized enterprise
performance 407
R. Widayanti, R. Damayanti, Nuryakin & Susanto

Author index 411

Preface

As a World-Class University, Telkom University is prepared to contribute to knowledge development by conducting a conference with all papers published in the proceedings. This proceedings compiles papers from presenters at the Conference on Managing Digital Industry, Technology and Entrepreneurship 2019 (CoMDITE 2019) which was held on July 10-11, 2019.

This conference had two main sessions, i.e. a plenary session and a parallel session with 142 presenters. The plenary session consisted of a keynote speaker that was delivered by Ridwan Kamil, Governor of West Java Province, and a panel discussion which featured some experts as the invited speakers, i.e. Associate Professor Indrawati, Ph.D., from Telkom University Indonesia; Professor Dr. Saravanan from Multimedia University, Malaysia; Dr. Vichaya Mukdamanee from Silpakorn University, Thailand; Dr. Rina D Pasaribu, M.Sc., CPM, Senior General Manager Telkom Corporate University from Telkom Indonesia; Fauzan Feisal, MIB, as CEO Digital Amoeba Program from Telkom Indonesia; Dr. Ratri Wahyuningtyas, Vice Dean School of Economic and Business Telkom University as the moderator of the panel discussion session 1; and Dr. Maya Ariyanti, Lecturer Master of Management Program as the moderator of the panel discussion session 2.

The 122 papers are from various universities and higher educational institutions from Indonesia and Malaysia.

CoMDITE 2019 was successfully held in collaboration between Magister of Management Program Telkom University (MM Tel-U) and Multimedia University (MMU). This event is supported by Telkom Corporate University.

On behalf of the committee, I would like to express our gratitude to all distinguished speakers, authors, presenters, participants and sponsors for contributing to the successful event of CoMDITE 2019. I hope this proceeding will contribute to the development and improvement of digital industry knowledge & practices.

Siska Noviaristanti, PhD

Organizing Committee

Chair
Siska Noviaristanti, Ph.D
Telkom University, Indonesia

Co-chair
Dr. Hasni binti Mohd Hanafi
Multimedia University, Malaysia

PR
Donny Trihanando, M.Ds
Telkom University, Indonesia

Seminar
Dr Chong Choy Yoke
Multimedia University, Malaysia

Parallel Session
Risris Rismayani, M.M
Telkom University, Indonesia

Administration
Hamsatulazura Hamzah
Multimedia University, Malaysia

Proceeding
Mohammad Tyas Pawitra
Telkom University, Indonesia

Documentation
Zarehan Selamat
Multimedia University, Malaysia

Treasure
Astri Angelina
Telkom University, Indonesia

Scientific Committee

Prof. Nuran Acur
University of Glasgow, UK

Prof. Elisabetta Lazzaro
Hogeschool Kunst Utrecht, The Netherlands

Prof. Hiro Tugiman
Telkom University, Indonesia

Assoc. Prof. Indrawati, Ph.D
Telkom University, Indonesia

Assoc. Prof. Dr. Nahariah Jaffar
Multimedia University, Malaysia

Dr. Vichaya Mukdamanee
Silpakorn University, Thailand

Dr. Rinyaphat Nithipataraahnan
Silpakorn University, Thailand

Assoc. Prof. Dr. Maya Arianti
Telkom University, Indonesia

Dr. Mariati Norhasim
Multimedia University, Malaysia

Dr. Hasni binti Mohd Hanafi
Multimedia University, Malaysia

Dr. Siska Noviaristanti
Telkom University, Indonesia

Dr. Gadang Ramantoko
Telkom University, Indonesia

Dr. Palti Sitorus
Telkom University, Indonesia

Dr. Astri Gina
Telkom University, Indonesia

Dr. Ratna L Nugroho
Telkom University, Indonesia

Dr. Ratri Wahyuningtyas
Telkom University, Indonesia

Dodie Tricahyono, Ph.D
Telkom University, Indonesia

Acknowledgements

We would first like to express our gratitude to the Dean of the Faculty of Economics and Business at Telkom University and the Dean of the Management Faculty, Multimedia University, Malaysia for all their support and advice during the preparation and the successful organisation of CoMDITE 2019. Further, thanks to all authors who have submitted a paper and came to Bandung to present it. In particular, we thank the seminar keynote speakers who shared their valuable knowledge and experiences.

We would like to address a warm appreciation to the members of the scientific committee for their participation and expertise in the preparation of the conference. We also thank all people who agreed to play the role of moderator and session chair.

Finally, thanks also to all organising committee members who already put their best efforts to make CoMDITE 2019 happen.

Bandung, September 2019
CoMDITE 2019 Organising Committee

Digital know-how and FinTech readiness

Saravanan Muthaiyah & Thein Oak Kyaw Zaw
Faculty of Management, Multimedia University, Cyberjaya, Selangor, Malaysia

ABSTRACT: The finance and accounting domains today require institutions of higher learning to produce competent graduates who are familiar with digital soft skills. These include Blockchain, e-audit, big data, cryptocurrencies, auditing on the cloud, softbot counseling, robotics in business process automation, peer-to-peer transactions, application of artificial intelligence in treasury-related work, and crowdfunding. The voguishness of technology usage in the digital economy today has become an impelling force to the accounting and finance professions. Industry experts argue that the lack of FinTech skills will adversely impact recruitment. FinTech-enabled platforms create disruption to the accounting workforce as newer emerging business models and processes are slowly rising across banking, finance, audit, and integrated reporting. In this article, we present the top 10 skills needed in terms of FinTech readiness that should be included in existing curriculums offered at the tertiary level from expert opinion gathered by relevant stakeholders.

1 INTRODUCTION

A cross-sector research program concluded by Capgemi with collaboration with LinkedIn in 2017 reveals that the worldwide digital gap is widening. According to their report, 54% of companies agreed that the digital talent gap in the market is hampering their progress, causing them to lose their competitive advantage in their respective industries. The study uncovers potential attrition problems with regard to the widening digital gap. About 29% of employees feel that their knowledge level is already redundant, and 38% believe that they will be redundant in four to five years. Compared to hard coding skills, the report suggested that soft digital skills are more indicative of a well-rounded digital professional. Although the study mentions customer centricity and passion for learning as soft digital skills, other studies also include creative thinking, collaborative skills, problem solving, and productivity as main attributes for digital skills needed today. In essence, learning how technology can be used to achieve higher working standards is of great value to employers. The Robert Half survey in 2014 included 2,100 chief financial officers (CFOs). A total of 61% of the participants agreed that business analytics, business intelligence, and data mining were deemed mandatory for accounting and finance employees.

2 DESCRIPTIVE ACCOUNTING TO PRESCRIPTIVE ACCOUNTING

Accounting education has always propagated descriptive accounting until the twentieth century, but today's digital ecosystem has created a need for more prescriptive accounting systems. Figure 1 shows specific practices that can be used for prescriptive and predictive accounting, which are illustrated in Table 1. In general, four types of soft digital accounting skills are relevant to the accounting and finance professions, and they are descriptive, diagnostic, predictive, and prescriptive.

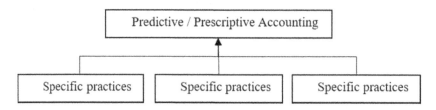

Figure 1. Descriptive approach.

Table 1. Digital skills and practices.

Digital Skills	Explanation	Specific Practices
Descriptive	What happened?	Profit and loss (P&L) and balance sheet analysis
Diagnostic	Why did it happen?	Forecasting, planning and budgeting
Predictive	When and what will happen?	Modeling or simulation (Monte Carlo/What if?)
Prescriptive	What action do we take?	Optimization decision

To illustrate this, imagine a CFO in a budget airline company. By applying predictive and prescriptive skills, he or she can decide how to appropriately optimize tariff pricing rates for additional cargo that is loaded onto the airplane and renegotiate landing rates for the best profit margins. This sort of future outlook in decision-making for better value creation will now be possible with big data and other technologies.

3 COMPETENCIES REQUIRED TO RESOLVE THE DIGITAL GAP

Emerging digital tools are enabling advanced analytical thinking and understanding that can benefit internal and external stakeholders. After speaking to a group of CFOs from sectors like finance, banking, service, and consultancy, skills gaps were mostly prevalent in areas such as data visualization, data mining, analytics, and Blockchain. These experienced CFOs indicated that knowledge and skills in FinTech technologies will create a better finance workforce and that education providers must focus on these areas to eventually close the skills gaps that exist today. This section presents the core competencies and know-how that are most significant for digital skills that are relevant for the accounting curriculum. We summarize seven core competencies from the frequency of votes provided by our CFOs during the interview session. Competencies required are in the areas of: 1) blockchain, 2) analytics, 3) crypto assets, 4) data portability, 5) data visualization, 6) e-audit, and 7) cloud services.

4 RESEARCH OBJECTIVES AND QUESTIONS

The objective of this study was to ascertain attributes that are mostly prevalent for producing finance and accounting students who can truly demonstrate predictive and prescriptive digital skills in the FinTech digital ecosystem. From the interviews conducted with several CFOs, we found that institutions of higher learning should concentrate on the seven aforementioned competencies. As such, RQ1 has already been answered in Section 3. Additional research questions (RQs 2, 3, and 4) that we explored in this study are listed next.

RQ 1 – What core competencies are required for the FinTech domain?
RQ 2 – How many public and private tertiary institutions teach the seven competencies?
RQ 3 – What is the extent of FinTech inclusion in the existing curriculum?
RQ 4 – What challenges are associated with the deployment of FinTech competencies?

5 METHODOLOGY

The approach of this research was to replicate the essence of Robert Half (2014) and EY's Global Finance Performance Improvement studies in Malaysia. A total of 18 CFOs were interviewed in the first part of the study to understand the most needed competencies in FinTech. In Section 3, we have discussed all seven competencies that were listed as significant based on frequency of votes obtained. In the second part of our study, and particularly to answer RQs 2, 3, and 4, we interviewed academicians, counselors, and program coordinators from a total of 15 universities, both public and private, with well-established accounting programs.

5.1 *How many public and private tertiary institutions teach the seven core competencies?*

Out of 15 institutions, the inclusion of the seven competencies discussed earlier is less than 25%. This means that an undergraduate program that has 40 subjects will only have a maximum of 10 subjects. This seemed logical at first, but that was only for institution 8 (INST 8). The other institutions do not show much promise. Overall, the exposure of students to the competencies mentioned in Section 3 is disappointing.

5.2 *What is the extent of FinTech inclusion in the existing curriculum?*

From discussions, we gathered that inclusion is very slim. Academics realize that they need to include these competencies; however, they feel that students cannot be exposed to tools that they themselves aren't too familiar with, and they argue that these tools will be obsolete anyway by the time the students graduate.

5.3 *What challenges are associated with the deployment of FinTech competencies?*

Figure 2 summarizes the top six reasons academicians highlighted when asked why inclusion would be poor. Some instructors mentioned that these skills can be learned on the job and do not have to be taught. Counselors were afraid that the costs for acquiring technological resources such as proprietary software would outweigh the benefits. Program coordinators also felt that it is best that FinTech be a whole program instead of embedding parts of it into an existing curriculum.

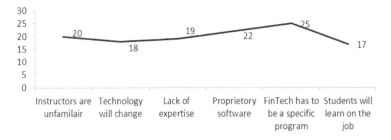

Figure 2. The extent of FinTech inclusion.

6 CONCLUSION

The Capgemi report concluded that the digital talent gap is widening faster and is directly causing companies to lose their competitive advantage. The Robert Half survey concluded that core competencies for digital skills are necessary, and this confirms our study as well. On top of the three competencies listed in the Robert Half study, our research shows a more comprehensive list of core FinTech skills and competencies that can be used as a baseline for academic course planning and update. Overall, we conclude that the gap is widening fast and that tertiary education providers can take a more serious role in providing more inclusiveness when teaching digital skill sets and eventually narrowing this gap. We also feel that instructors must be trained accordingly to produce an integrated curriculum. The curriculum has to be built on solid technology architecture with the focus of creating a skill portfolio that includes critical thinking, regulation, and compliance. Financial engineering specialization in the past provided a niche for finance graduates with skills in derivatives, liability management, and sophisticated technical analysis. Nevertheless, these skills alone did not prepare students to understand the economical ecosystem completely; as a result, financial engineers could not foresee the subprime crisis waiting to happen. Size of leverage, fraud, and misrepresentation must be embedded into courses like ethics to produce a wholesome curriculum. In terms of digital know-how for FinTech readiness, our study clearly shows that we are not ready and that what may be further lacking is a holistic curriculum that can create thinkers who have the ability to think critically.

REFERENCES

Arner, D. W., Barberis, J. N., & Buckley, R. P. 2015. The evolution of Fintech: A new post-crisis paradigm? *SSRN Electronic Journal*. doi: 10.2139/ssrn.2676553

Bradford, M. 2018, February 18. How Blockchain will disrupt the financial sector – Raconteur. www.raconteur.net/sponsored/blockchain-will-disrupt-financial-sector.

Calderon, T. G., & Cheh, J. J. 2002. A roadmap for future neural research in auditing and risk assessment. *International Journal of Accounting Information Systems 3*, 203–236.

Capgemi. 2017. https://reports.capgemini.com/2017/wpcontent/uploads/2018/03/CapG_RA17_UK-2.pdf. Retrieved January 2019.

Clarke, P. 2017, March 20. How artificial intelligence will eat your finance job – and how to survive. https://news.efinancialcareers.com/my-en/277602/how-machine-learning-will-eliminate-230000-finance-jobs.

EY's Global Finance Performance Improvement. www.ey.com/en_gl/performance-improvement. Retrieved March 2019.

Lam, M. 2004. Neural network techniques for financial performance prediction: Integrating fundamental and technical analysis. *Decision Support Systems 37*(4), 567–581.

Moffitt, S. 2018, April 30. The top 30 emerging technologies (2018–2028). https://medium.com/@seanmoffitt/the-top-30-emerging-technologies-2018-2028-eca0dfb0f43c.

PWC. 2016, June 6. Financial services technology 2020 and beyond: Embracing disruption. www.pwc.com/gx/en/industries/financial-services/publications/financial-services-technology-2020-and-beyond-embracing-disruption.html.

Robert Half. 2014. www.annualreports.com/HostedData/AnnualReportArchive/r/NYSE_RHI_2014.pdf, retrieved January 2019.

Tuncay, E. 2010. Effective use of cloud computing in educational institutions. *Proscenia Social and Behavioral Sciences, 2*, 938–942.

An overarching architectural framework for smart cities

Saravanan Muthaiyah & Thein Oak Kyaw Zaw
Multimedia University, Cyberjaya, Malaysia

Indrawati
Faculty of Economics and Business, Telkom University, Bandung, Indonesia

ABSTRACT: The journey toward realizing a "smart city" leads back to one path that many learned people argue is "the best solution for a smart city." This solution embraces the Internet of Things (IoT) in every aspect of the city as much as possible. The debates on IoT as a solution have been ongoing, and not many have come up with an answer that differs from the mainstream. This research intends to find out whether IoT is a viable solution and if other potential options are better. This study was based on a qualitative approach that began with 1) a survey of literature, 2) preliminary results highlighting that IoT is not a viable solution, and 3) with IoT alone, cities are deemed partially smart. Findings indicate that we require two crucial components, i.e., being autonomous and self-learning in order to achieve a smart city.

1 INTRODUCTION

The Internet of Things (IoT) has been drawing in enormous volumes of research as the world progresses toward making everything connected. It is a future vision that everyday devices will be equipped with microcontrollers, transceivers, and protocol stacks that will communicate with each other, making them integrated with the Internet (Atzori, Iera, & Morabito, 2010). Connected devices will enable easier access to sensors, actuators, and so on, which will create many applications that will make life easier for people. Currently, a wide range of IoT applications is available on the market in different areas such as medical aids, traffic management, home automation, and many more (Bellavista, Cardone, Corradi, & Foschini, 2013). However, each IoT system has its own mechanisms and methods in reaching its objectives. This makes integration for all ranges of IoT applications into one main system of control almost impossible. This complexity, however, does not stop organizations and small and medium-sized enterprises (SMEs) from adopting IoT, as the push factor comes from the users and the government (Schaffers, Komninos, Pallot, Trousse, Nilsson, & Oliveira, 2011). The immense demand for a smart city causes researchers and enthusiasts to produce volumes of IoT research applications in which organizations try to realize the dream of becoming the first movers. Almost all of the solutions suggested by research will point back to IoT, which shows that the pure adoption of IoT has become a mainstream concept. However, with no clear definition of a smart city and no solid research on how to build it, the path toward a smart city remains vague.

As no comprehensive definition exists for a smart city, one could argue that a city may not reach the status of smart even though most of the devices, if not all, are connected. One could also argue that IoT may not be the answer in creating a smart city. Therefore, elements to achieve in a smart city must have solid guidelines. From a literature review and the scenario presented earlier, we know a gap exists between a unified vision of a smart city and the method of designing it. This research is intended to narrow this gap by proposing the architecture layers needed in constructing a smart city.

2 OVERARCHING FRAMEWORK FOR SMART CITIES

Currently, only Dameri (2012) has been able to propose a comprehensive definition for smart cities. Dameri lists three crucial aspects: terminology, components, and boundaries and scope. *Terminology* refers to land, citizens, technology, and governance. *Components* means that the city must be intelligent, wired, sustainable, inclusive, and democratic. *Boundaries* signifies the area of a smart city, while *scope* includes well-defined, measurable goals such as environmental sustainability and citizen participation.

Summing up, Dameri states that "a smart city is a well-defined geographical area, in which high technologies such as ICT, logistics, energy production, and so on, cooperate to create benefits for citizens in terms of well-being, inclusion and participation, environmental quality, [and] intelligent development; it is governed by a well-defined pool of subjects, able to state the rules and policy for the city government and development." While we agree, our contribution is an overarching framework that includes specific layers, i.e., 1) acquisition, 2) data, 3) business, and 4) application architectures. Therefore, after connecting all the dots, we propose a framework (see Figure 1) with an emphasis on autonomy and self-learning at the Business Architecture layer. These two items were proposed because *smart* should consist of the ability to reason with very little or no human input (i.e., independent or autonomous).

The ability to think independently requires learning algorithms, and actions must be autonomous. Therefore, it is clear that in order to be called smart, a system or device needs to be able to achieve two things: (1) the ability to think independently in difficult situations and (2) the ability to act independently. With the current IoT as the solution, both of these elements will remain unfulfilled, as IoT does not possess learning algorithms for it to think independently in difficult situations; IoT only possesses the first main component, complete autonomy with minimum human interaction. Even so, it still requires human intervention even though the interactions are minuscule. It will be made possible by the combination of two technologies – IoT and artificial intelligence (AI). Smart technology implies automatic computing principles like self-configuration, self-healing, self-protection, and self-optimization (Sairamesh, Lee, & Anania, 2004). Therefore, in order to reach smart city status, IoT alone is not adequate and a combination of IoT and AI is needed.

Table 1 illustrates each layer's capabilities within the proposed framework. Artificial intelligence for self-learning must perform tasks that require machines to learn reasoning capabilities. Douglas (2018) states that AI has four main abilities: sense, converse, act, and learn. Although AI and IoT are not mainstream solutions, the concept of combining them in realizing a smart city is catching on. Huawei in 2018 unveiled its artificial intelligence smart city

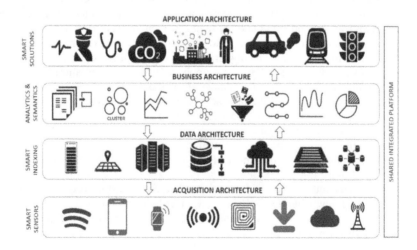

Figure 1. Overarching autonomous learning city (smart city) framework.

Table 1. Layer functionality.

Layers	Functionality
Application Architecture	Present data for service allocation Respond to waste, traffic, carbon emission, health care, and surveillance Mitigation responses and decision-making
Business Architecture	Data analytics – prescriptive and predictive Semantic analysis with aided reasoning capabilities Agent learning capabilities supported by autonomous features
Data Architecture	Data acquisition – multiple source Data optimization and gleaning – from structured as well as unstructured data Learning and training capabilities
Acquisition Architecture	Smart readers – feeder for data architecture Smart sensors – feeder for data architecture IoT sensors – feeder for data architecture Smart contracts for participative user network

platform, which utilizes IoT capabilities (Reichert, 2018). Roven (2018) and Bailey and Coleman (2019) highlight solutions with both of the elements mentioned earlier.

3 CONCLUSION

Our world's population will hit the 10 billion mark by 2050, and this will add enormous pressure on our water, sanitation and waste management, carbon footprint, noise pollution, homelessness, urban health, and security. Most of these urban pressures and negative externalities will require an agile problem-solving mechanism via social justice and an intelligent urban governance system. More important, cities have to be sustainable economically, environmentally, and socially. This should be the value proposition for an autonomous learning city (smart city) administration.

A broad range of urban policy papers have been presented; this includes a national urban strategy, national and local government integration, and the concept of smart cities. As the World Urban Forum no. 9 (WUF 9) declaration concluded, sustainability is the core driving force for a smart city. In this context, "urban governance" must also be incorporated into the planning, design, and management of cities. Our contribution in this study is threefold: we propose 1) an overarching autonomous learning city framework (OALCF), 2) transition strategies for the OALCF, and 3) a more realistic definition for smart cities, i.e., an autonomous learning city, keeping in mind the value proposition that is needed for resolving urban pressures, good governance, and sustainability.

REFERENCES

Atzori, L., Iera, A., & Morabito, G. 2010. The Internet of Things: A survey. *Computer Networks*, *54*(15), 2787–2805. doi:10.1016/j.comnet.2010.05.010

Bailey, D. J., & Coleman, Y. 2019, January 8. Urban IoT and AI: How can cities successfully leverage this synergy? Retrieved from https://aibusiness.com/future-cities-iot-aio/.

Bellavista, P., Cardone, G., Corradi, A., & Foschini, L. 2013. Convergence of MANET and WSN in IoT urban scenarios. *IEEE Sensors Journal*, *13*(10), 3558–3567. doi:10.1109/jsen.2013.2272099

Dameri, R. P. 2012. Searching for smart city definition: A comprehensive proposal. *International Journal of Computers and Technology*, *11*(5), 2544–2551. doi:10.24297/ijct.v11i5.1142

Douglas, E. 2018, February 8. Artificial intelligence: Abilities & expectations. Retrieved from https://orzota.com/2018/02/08/artificial-intelligence-abilities/.

Reichert, C. 2018, November 14. Huawei unveils artificial intelligence smart cities platform. Retrieved from www.zdnet.com/article/huawei-unveils-artificial-intelligence-smart-cities-platform/.

Roven, A. 2018, November 27. AI in smart cities. Retrieved from https://medium.com/neuromation-blog/ai-in-smart-cities-dfe2fa7d2829.

Sairamesh, J., Lee, A., & Anania, L. 2004. Information cities. *Communications of the ACM, 47*(2), 28–31.

Schaffers, H., Komninos, N., Pallot, M., Trousse, B., Nilsson, M., & Oliveira, A. 2011. Smart cities and the future Internet: Towards cooperation frameworks for open innovation. *The Future Internet, 6656*, 431–446. doi:10.1007/978-3-642-20898-0_31

Using 3D sculpting machine to link art, nature, and community – case study from Thailand Biennale, Krabi 2018

V. Mukdamanee
Silpakorn University, Bangkok, Thailand

ABSTRACT: 3D Scanning, 3D Printing, and 3D Sculpting machines have been widely used by contemporary artists. The technology functions through the process of collecting, transferring, and analyzing digital data. As computer software and machines become more updated and affordable, artists, nowadays, can create artificial objects with much more complex details than what could have been imagined in the past. There is no difference between artworks done by hand and by machine. One is able to create artificial objects that look identical to nature, with various materials. By using four artworks exhibited in Thailand Biennale, Krabi 2018, I examine some practical approaches to how artists collaborate with these technologies to create site-specific artworks which are inspired by nature and the local community. The artworks mentioned are examples of how technology and art, together, can remind or teach people about the way to develop our society into a better one.

1 INTRODUCTION

1.1 *Background of study*

Digital technology has, unquestionably, changed the way humans live and think. It exists in everything and is used in all professions, including the fields of art and design. For almost half a decade, digital technology has become one of the most important tools and medium for contemporary artists, existing as part of the creative or exhibition processes. Digital images and information are essential sources of inspiration. Conventional art practices, such as drawing, painting, sculpture, and music have found ways to collaborate with digital technology, while digital technology has also developed into new forms of art, such as computer art, net art, digital installation art, virtual reality, etc. Digital photography and videos commonly appear in many exhibitions, from the level of students' shows in art colleges to international art exhibitions in famous art museums. Artists learn about today's art world mainly from digital media and, at the same time, express themselves and their art processes through the same channel. In the end, we no longer deal with digital technology only as "form" or "medium", but also have embedded it into the "methods" of thinking and making. We are looking at and reflecting the world digitally.

Most people may be more familiar with digital technology in the forms of 2-dimensional and moving images. However, in the past decade, the technologies of 3-dimensional scanner, CNC sculpting machine, and 3D printing have become more commonly used in the art world. Initially developed for industrial purposes, these technologies allow the user to create objects from computer-generated digital files with precision and accuracy to detail according to the virtual models. The objects can be re-printed or re-sculpted as many times as one wants, and as quickly as the machine's capacity. Digital information that the machine uses as reference can be drawn and generated from a computer software, or it can also come directly from the 3D scanner. This opens up great possibilities for artists to create objects anywhere in the world as long as there are machines in the area and a way to send digital information to a technician who knows how to operate the machines. The materials used for the printed or sculpted objects are various, such as foam, plastic, plaster, wood, ceramic, glass, and metal.

When these technologies were first developed, due to the limitation of softwares and oversized files, it was very difficult to use machines to sculpt or print objects with complex detail. Like the difference between hand-drawn animation and 3D digital modelling software in the past, the handmade sculptures acquired a more "natural" look than the machine-made ones. As new software gets updated and these machines becomes more affordable, new printed materials have been introduced, including the development of 3D scanning technology that allowed us to sculpt or print objects that look exactly like nature. The difference between handmade or computer-generated artworks is no longer created by technological limitations, but rather the skills of the artists or technicians. It is true that there are some sculptures that cannot be done by machine as well as some sculptures that cannot be made by human; this may not be for technical reasons, but rather a conceptual decision made by the artists. Artists are now capable of choosing how they want to sculpt, by hand or machine. The outcome of the objects may be the same, but the conceptual aspects are definitely very different.

The use of 3D sculpting and printing techniques in contemporary art is not only about how to make the work, but also how to talk, discuss, and collaborate with others, especially the technicians. It is very clear that these technologies become the greatest tools for the contemporary art community, where artists might not be able to be in the studio for the production process, as they are busy traveling between countries, exhibiting in international art events happening all over the world. The artist talks with one group of technicians in order to work with the 3D scanner or draw the model of their artwork in a computer software. Then, the artist has to manage how to transfer all of the digital information to another group of technicians to do 3D sculpting or printing. During this process, the artist may check the quality or the development of artwork through images, text, and video clips sent via email. A lot of online discussions and conversations emerge from this process. After the work is produced, the artist needs to contact another group of technicians to pack and transport the artwork from the foundry or factory to the installing site, then contact another group of technicians to install the work, which is a process that also requires a lot of online communication. Digital technology and machine have changed the role of contemporary artists from the maker to organizer, manager, and collaborator. In most cases, artists who deal with 3D sculpting and printing technique have never been to the foundry. The presence of the artist, whose name is eventually stated as the creator of objects, seems to appear more in the digital world rather than in the actuality.

2 METHOD

To explore the use of 3D sculpting and printing technique in contemporary art, I focus on four artworks exhibited in Thailand Biennale, Krabi 2018, including *How to Carve a Sculpture* by Japanese female artist Aki Inomata, *Voyage in Time* by LUXURYLOGICO the artists group from Taipei, *View-Review* by Alicja Kwade from Berlin, and *Apocalypse* by Beijing-based artist Wang Sishun. The reason for choosing Thailand Biennale, Krabi 2018 as a case study is because of the exhibition's concept, which had many aspects that may seem in conflict with the digital world. The most visible aspect is the fact that the exhibition was planned to be completely outdoor in the natural sites, so there was no electricity to power any digital equipments. Another aspect comes from the essential idea of this international art event that would like to create the platform to connect art, nature, and community together, which many may think of this combination as opposite to the aggressive development of digital and technology. However, after looking at these examples of artworks, it appears to me that we can no longer separate nature from technology. The digital world is already embedded within our lives.

3 DISCUSSION

For *How to Carve a Sculpture*, Aki Inomata used digital information from the 3D Scanner and collaborated with stone sculptors in Thailand to create multiple pieces of sandstone sculptures, which were installed around the pond located in the city of Krabi. For years, the artist's main concept was to create comparisons of living activity between human and animals. For this particular case of Thailand Biennale 2018, she focused on the activity of constructing the living environment via process of sculpting. The artist started the project by observing the behavior of beavers in Japanese zoos. *"Beavers create artificial ponds by cutting down the surrounding trees and assembling them into dams to catch and hold streams. These beaver ponds provide a distinctive environment that attracts moose, herons, and many other creatures."* Such environment of giving and sharing is totally different from the idea of how human build their house and design the surroundings. Inomata's sculptures were the enlarged version of these gnawed trees. The artist installed the work in a way that one may see them as remnants of the construction process, and, at the same time, something very similar to a Japanese stone garden. *How to Carve a Sculpture* sits between the mixed feelings of natural or artificial, human or non-human, objects to be seen or objects to be hidden.

Figure 1. *How to Carve a Sculpture*, [comparing the CNC sculpted foam and the carved sandstone in the studio], Thailand Biennale, Krabi 2018.

To start the production of *How to Carve a Sculpture*, Inomata provided several wood pieces for these beavers in the zoo to bite and chew. After several weeks, Inomata collected and 3D scanned the wood pieces that had been transformed and covered with tooth mark of the beavers. All information and series of digital images were sent to the technicians in Thailand, who used it to produce enlarged foam models by using CNC sculpting machine. These models were later used as the reference for handmade sandstone carving. The same set of digital information had been transferred and gone through many processes in order to complete the final artwork. Although Inomata flew to meet with the technicians in Thailand many times to oversee the process of stone carving and installation - she even took part in carving several pieces by herself - we could still say that the artist was not there with the technicians during most of the production process. Technology of 3D scanner, CNC sculpting machine, and online networking became the essential tools that link artist, technicians, animals, and materials together, running the whole production from the beginning to the end.

Voyage in Time by LUXURYLOGICO was another example of how the artist used 3D printed model as the reference to create a new handmade objects. Not only was *Voyage in Time* a collaboration with a group of technicians, but the collaboration rather extended to the entire village. LUXURYLOGICO proposed a project of commissioning the local boat builder to construct a new twenty-six meters long wooden boat based on the model made from computer software and 3D printing machine. The artist group used a computer program to slice the image of local long-tailed boat into multiple pieces. Then, they reformatted and combined the fragments to create a new, unexpected shape. This unusual boat structure became a great challenge for the local technicians, both in terms of the technical and the conceptual aspects. The boat making process indeed gave rise to the question of "what is art?" to the entire village of the boat builders. After a few months of trial-and-errors, the building process was finished. The boat was pushed into the water and it floated. On this last day of production, almost everyone in the village came together. They helped pulling and pushing the boat. They came to eat together, share their thoughts, and took pride in the production of an artwork that they all took part in.

For the last two cases of *View-Review* by Alicja Kwade and *Apocalypse* by Wang Sishun, both artists approached the 3D printing technology in a very similar way. Kwade and Sishun were both interested in making the aluminium sculptures that looked exactly identical to the actual stones found in Krabi. On one hand, Kwade installed the aluminium stone next to the actual one on the beach, with mirrors in between them to create the illusion that the natural and artificial objects are a reflection of each other. On the other, Sishun decided to install his stone-shaped aluminium as a monumental sculpture standing in the center of the local national park. The production processes of two artists were very similar, as they started by searching for the stones in the right shape and used 3D scanner to gather all information of the found stones. The digital information of stones was sent to the foundry outside of Thailand to fabricate the identical objects with aluminum as material. Then the objects would be shipped to be installed in Krabi. The concept of both artists focused on the same issue of how today's technology allowed us to reproduce the mirror copy of nature. "*It seems like the real and the unreal objects perfectly overlap. Whenever a viewer passes by, the objects seem to be transforming into the opposing materiality, illustrating an alchemistic process, another version of reality.*"

4 CONCLUSION

These four artworks from Thailand Biennale, Krabi 2018, are just examples of how today's artists collaborate with the 3D scanning, 3D printing, and 3D sculpting machines. Due to the nature of the site-specific exhibition in the natural space, these artworks may approach the technique in quite similar ways. However, it must be noted here that there are many other artists who deal with the same technology in many different directions. Nevertheless, by looking at the work of Aki Inomata, LUXURYLOGICO, Alicja Kwade, and Wang Sishun, we see the possibility of collaboration between nature, community, and technology. In creating artworks to reflect on nature and build relationships in the community, one does not have to use only natural materials and deny using digital media to connect people together. Digital and technology are not just something appearing on computer screen, but they are already implanted in our minds.

REFERENCES

Elsom, Cook, 2001. Principles of Interactive Multimedia. New York: MacGrawHill.
Fred T. Hofstetter, 2001, Multimedia Literacy, Third Edition. New York: MacGrawHill.
Gjedde, Lisa, 2005. Designing for Learning in Narrative Multimedia Environment. London: Idea Group Inc.
Mayer, R., 2009. Multimedia Learning, Second Edition. New York: Cambridge University Press.
Office of Contemporary Art and Culture, Ministry of Culture, Edge of the Wonderland; Catalogue of Thailand Biennale, Krabi 2018, Bangkok (2019), Page 104.
Sweller, J. 1999. Instructional Design in Technical Areas. Camberwell, Australia: ACER Press.

Green product advantage: Mediating effects on green SMEs performance

Susanto Nuryakin & Tri Maryati
Universitas Muhammadiyah Yogyakarta

ABSTRACT: This research aims to analyze the effect of green product innovation and green product advantage on green *batik* small and medium-sized enterprises' (SMEs) performance in Bantul Regency, Yogyakarta Province, Indonesia. The researchers collected information from 200 managers or owners of *batik* SMEs in Bantul. This research used purposive sampling methods to collecting the data. The results of this study showed that green product innovation had a significant positive effect on green product advantage and green SMEs performance. Furthermore, the green product advantage had a mediating effect on green product innovation and green SMEs performance.

1 INTRODUCTION

Internal resources are a source of competitive advantage that can create a company's distinctiveness compared to its competitors. Companies that adapt and orient to the market have advantages in their performance. Companies also need to maintain business sustainability by taking care of the environmental aspects (Barbier, 2016). The role of company owners and managers becomes interesting when they strive to create distinctive products and can adapt to customer desires by utilizing eco-friendly production materials (Cao & Chen, 2018).

Previous research on the issue of green innovation has been carried out by (Ar, 2012; Chang, & Wu, 2012; Dangelico & Pujari, 2010). The studies revealed that the current focus of research on green innovation in reaching customers was done not only at the organizational level but also at the customer level. Other studies explained the importance of business organizations orienting their care toward maintaining aspects of the business environment in achieving a competitive advantage (Chang, 2011). Researchers have also found a strong relationship between innovation and environmental issues (Lin, Chen, & Huang, 2014), economic activities, and long-term sustainability of business organizations (Lavrinenko, Ignatjeva, Ohotina, Rybalkin, & Lazdans, 2019).

The researchers conducted the study at *batik* SMEs in Yogyakarta by considering natural resources in their production processes. In the current global warming situation, many businesses run operational processes that use production machinery and modern technology. On the contrary, the *batik* SMEs in Yogyakarta choose to use human capital and natural resources in the production process. This study empirically examines the crucial mediating role of green product advantages in the relationship of green product innovation to green SMEs performance. The green product advantage in this study is the mediating construct in the relationship between green product innovation and green SMEs performance.

2 THEORETICAL REVIEW AND HYPOTHESIS DEVELOPMENT

2.1 *Green product innovation and green product advantages*

Research conducted by Huang, Yang, and Wong (2016) found that green product innovation developed by companies is a determinant of green capabilities. This requires a deep understanding of internal factors in developing green product innovation. It is also essential for organizations to build green product innovations and green process innovations in achieving excellence and success in the market (Kam & Wong, 2012). Based on the previous study, the researchers propose the hypothesized model in this study:

H1: Green product innovation has a positive effect on green product advantages.

2.2 Green product innovation and green SMEs performance

Products with eco-friendly characteristics bring additional value for companies; then the companies have an advantage over their competitors. This study examines the critical role of green product innovation in the scope of SMEs. The study of Zhang and Zheng (2018) emphasized green innovation activities consisting of input processes with proportional ingredients in terms of easily decomposed materials. Based on the previous study, the researchers propose the hypothesized model in this study:

H2: Green product innovation has a positive effect on green SMEs performances.

2.3 Green product advantages and green SMEs performance

Core competence is key for SMEs businessmen in achieving a degree of uniqueness of their products in the market by utilizing internal resources through innovation capability, market capability, and production capability.

Organizations will have a competitive advantage and reputation if they carry out green process innovation. A study conducted by Enzing, Batterink, Janszen, and Omta (2011) found that innovation affects the success of the product in the market. Other literature studies, such as the study of Bakar and Ahmad (2010), also figured that innovation dramatically helps an organization in achieving optimal product position in the market. Based on the previous studies, the researchers propose the hypothesized model in this study:

H3: Green product advantages have a positive effect on green SMEs performance.

3 RESEARCH METHODS

3.1 Research design

This study proved empirically three research hypotheses. The researchers conducted the study by surveying the *batik* SMEs in Yogyakarta. The survey approach was carried out using a questionnaire to explore information related to the research sample. The quantitative approach used was testing the research hypotheses. The sample used in this study was *batik* SMEs, which use natural materials and have developed a production system using eco-friendly materials. The sample included 223 respondents consisting of owners/managers of *batik* SMEs in Yogyakarta. The data collection process involved distributing questionnaires to the research sample. The sampling technique used a purposive sampling method by considering the scope of the SMEs and the use of natural ingredients in the production process. This research used a quantitative research design to analyze the third hypothesis, and an empirical research model.

4 RESULT

4.1 Hypotheses test

This research utilizes structural equation modeling to analyze the research design and to test the hypotheses using AMOS software. Table 1 indicates the results of the structural model test in this research. Overall, model fit measurement with structural equation modeling indicates that the model fit values are good (GFI = 0.930, AGFI = 0.904, RMSEA = 0.035, CFI = 0.986, TLI

= 0.983). Overall, the paths estimated test shows that all of the hypotheses significantly show results that support the hypotheses test. The three hypotheses developed in this research, after the test, show positive and significant results, that green product innovation positively significantly influences green product advantages and green SMEs performance. Green product advantages significantly influence green SMEs performance. The results of each hypothesis test are presented in the table that follows.

Table 1 explains the standardized path coefficient value in the test of each exogenous construct with the endogenous construct. The test results empirically prove what forms the three hypotheses.

The statistic test H1 finds the path coefficient value of 7.137 and significance value of 0.000, showing that green product innovation positively, significantly influences green product advantages. The statistic test H2 finds the path coefficient value of 3.914 and significance value of 0.000, showing that green product advantages positively, significantly influence green SMEs performance. The statistic test H3 finds the path coefficient value of 4.218 and significance value of 0.000, showing that green product innovation positively, significantly influences green SMEs performance.

5 DISCUSSION, IMPLICATIONS, AND LIMITATIONS

Interesting findings in this study empirically proved all of the tested hypotheses. Empirical studies conducted by previous researchers support the results of this study.

The first hypothesis testing found that green product innovation has a positive effect on green product advantages, which proved significant. The results of this study are in line with previous studies conducted by Huang et al. (2016), which found that green product innovation became an essential part of the company's strategy. Other studies are also in line with the results of this research. They found green product innovation and green process innovations helped in achieving excellence and success for their products in the market (Kam & Wong, 2012).

The second hypothesis testing found that green product innovation has a positive effect on green SMEs performance, which the researchers significantly proved. This study is also in line with the research conducted by Zhang and Zheng (2018), which found that green innovation activities consist of input processes in product creation activities.

The third hypothesis testing found that green product advantages have a positive effect on green SMEs performance, which the researchers significantly proved. The results of this study are in line with previous studies that state that competitive advantage and company reputation can be enhanced through green innovation.

Another interesting finding in this study explains that green innovation is also an important strategy when companies develop new products and production processes. Even green innovation is believed by some researchers to be a breakthrough for the long-term orientation of a company while maintaining a competitive advantage and environmental sustainability (Ar, 2012; Li, Lei, & Han, 2018). In enhancing the company's reputation and the access of customers in broader new markets, green innovation is important in a company's strategy, specifically adopting environmentally friendly technologies and resources (Low & Shang Gao, 2015).

Table 1. The testing of the hypothesis model.

Hypothesis	Proposed Effect	Path Coefficient	Sig	Result
H1	Positive	7.137**	0.000	H1 is supported
H2	Positive	3.914*	0.000	H2 is supported
H3	Positive	4.218**	0.000	H3 is supported

* $p < 0.05$, ** $p < 0.01$.

This research makes an essential contribution to the body of knowledge about how SMEs optimize internal organizational capabilities (internal resources capabilities), especially related to innovation capability.

ACKNOWLEDGMENTS

The authors sincerely thank the Ministry of Research and Technology of Indonesia for endorsing this research project, and all the respondents who participated in the survey.

REFERENCES

Ar, I. M. 2012. The impact of green product innovation on firm performance and competitive capability: The moderating role of managerial environmental concern. *Procedia – Social and Behavioral Sciences*, *62*, 854–864. doi: 10.1016/j.sbspro.2012.09.144

Bakar, L. J. A., & Ahmad, H. 2010. Assessing the relationship between firm resources and product innovation performance. *Business Process Management Journal*, *16*(3), 420–435. doi: 10.1108/14637151011049430

Barbier, E. B. 2016. Building the green economy. *Canadian Public Policy*, *42*(S1), S1–S9. doi: 10.3138/cpp.2015-017

Cao, H., & Chen, Z. 2018. The driving effect of internal and external environment on green innovation strategy: The moderating role of top management's environmental awareness. *Nankai Business Review International*, *13*(1), 38–52. doi: 10.1108/NBRI-05-2018-0028

Chang, C.-H. 2011. The influence of corporate environmental ethics on competitive advantage: The mediation role of green innovation. *Journal of Business Ethics*, *104*(3), pp. 361–370.

Chen, Y. S., Chang, C. H., & Wu, F. S. 2012. Origins of green innovations: The differences between proactive and reactive green innovations. *Management Decision*, *50*(3), 368–398. doi: 10.1108/00251741211216197

Dangelico, R. M., & Pujari, D. 2010. Mainstreaming green product innovation: Why and how companies integrate environmental sustainability. *Journal of Business Ethics*, *95*, 471–486. doi: 10.1007/s10551-010-0434-0

Enzing, C. M., Batterink, M. H., Janszen, F. H. A., & Omta, S. W. F. O. 2011. Where innovation processes make a difference in products' short- and long-term market success. *British Food Journal*, *113*(7), 812–837. doi: 10.1108/00070701111148379

Huang, Y.-C., Yang, M.-L., & Wong, Y.-J. 2016. The effect of internal factors and family influence on firms' adoption of green product innovation. *Management Research Review*, *39*(10), 1167–1198. doi: 10.1108/MRR-02-2015-0031

Kam, S., & Wong, S. 2012. The influence of green product competitiveness on the success of green product innovation. *European Journal of Innovation Management Decision*, *15*(4), 468–490. doi: 10.1108/14601061211272385

Lavrinenko, O., Ignatjeva, S., Ohotina, A., Rybalkin, O., & Lazdans, D. 2019. The role of green economy in sustainable development (case study: the EU states). *Entrepreneurship and Sustainability Issues*, *6*(3), 1113. doi: http://doi.org/10.9770/jesi.2019.6.3(4)

Li, L., Lei, L., & Han, D. 2018. Regional green innovation efficiency in high-end manufacturing. *Journal of Coastal Research*, *82*, 280–287. doi: 10.2112/si82-040.1

Lin, R.-J., Chen, R.-H., & Huang, F.-H. 2014. Green innovation in the automobile industry. *Industrial Management & Data Systems*, *11*(6). doi: 10.1108/IMDS-11-2013-0482

Low, S. P., & Shang Gao, W. L. T. 2015. Comparative study of project management and critical success factors of greening new and existing buildings in Singapore. *Structural Survey*, *32*(5), 413–433. doi: 10.1108/SS-12-2013-0040

Zhang, Z., & Zheng, X. 2018. Global supply chain integration, financing restrictions, and green innovation: Analysis based on 222,773 samples. *Kybernetes*, *29*(2), 539–554. doi: 10.1108/K-06-2018-0339

Bringing Telkom Indonesia to be a global digital player

Rina D. Pasaribu
PT. Telkom Indonesia, Tbk (Telkom), Indonesia

In this Industrial Revolution 4.0 era, computers are connected and communicate with one another to ultimately make decisions without human involvement. A combination of cyber-physical systems, the Internet of Things, and the Internet of Systems make Industry 4.0 possible and the smart factory a reality. As a result of the support of smart machines that keep getting smarter as they get access to more data, our factories will become more efficient and productive and less wasteful. This disruptive era makes it necessary for Telkom Indonesia to move as quickly as possible to follow rapid technological advancements.

Hootsuite, a social media management company, has researched and found that Indonesia's digital growth is rising rapidly in some sectors, especially in Internet users and active social media users. This growth requires a company to turn into a digital company in order to survive in this global competition. Therefore, Telkom Indonesia must change from a telecommunication (TELCO) company to a telecommunication digital company (TELCO DICO). One thing that must be considered is rapid shifting into DICO competencies, so Telkom CorpU has a focus on recruiting people with digital talent. To find these people, Telkom CorpU is implementing a human resources development mechanism with a three pronged-concept that supports Telkom's digital capability development.

This three-pronged concept, also called competency management, is a holistic approach to developing great digital staff. It consists of technical/functional competencies, professional competencies, and digital leadership competencies. These three competencies offer full support to enhance digital technical and leadership capabilities through a domestic and global development program. In another effort to drive digital culture, Telkom Indonesia has been accelerating massive digital innovation with the Hack Idea and Digital Amoeba programs, promoting flexible working arrangements, enhancing knowledge of management systems, and encouraging creative space. To strengthen corporate culture, Telkom Indonesia is digitizing employee activity with mobile applications such as Diarium and Cognitium.

In terms of learning technology, more and more organizations are considering updated and better ways to educate their workforces. Learning technology trends, including Micro Learning, artificial intelligence (AI), and AR/VR, are set to dominate this year. Learning technology trends will provide insights that are cutting-edge, engaging, and effective, and that drive employee growth and improve the organization's learning culture. To adapt to this phenomenon, Telkom has the Learning Value Chain for human resources development, and the Telkom Integrated Learning Cycle for delivering education.

Corporate innovation: Democratizing decision-making

Fauzan Feisal
PT. Telkom Indonesia, Tbk (Telkom), Indonesia

One key trait of the new digital economy is disruption. The latest so-called digital technologies open up opportunities for almost everyone to join the digital movement. Goods and services, and access to them, are made easy, thus lowering costs of consumerism as well as production. Many costs have become nearly zero, e.g. the cost of communication (from SMS to instant messaging), the cost of showcasing merchandise (from physical outlets to online shopping), the cost of content production (from TV studios to personal YouTube channels), etc. All of these have made existing corporate processes, products, and their prices irrelevant for many kinds of industries. Telkom, basically overall Telco industries, is included among the top industries being disrupted by the digital economy.

We have learned that the key to managing digital industry is managing people. Just like the past Industrial Revolution, this fourth (digital) revolution is triggered by technological change, i.e. the Internet. When technology changes, the economy will adapt, and people are the first responders to such shifts. Those who are brought up with technology, those who respond to technology, those who buy technology-related products and services, and those who create and sell them are all people. When people change their behavior, the economy changes. So, in a corporation, everything about managing people will change as well: how employees react to technology, how employees react to customers, how employees react to product and services, how employees react to flows of work and money, how the labor market reacts to company recruitment, how employees react to the managerial style of the company. Telkom's capabilities in managing people will be tested during this era, far more intensely than ever before in Telkom's history.

Through the efforts of intrapreneurs, a flock of employees who take additional burden and risk for their company's success, Telkom has been innovating since day one, from adopting telephony technologies, satellite technologies, and cellular technologies up to IPTV technologies. Digital technologies are different. The aforementioned technologies require a massive infrastructure. Telkom can run these products' life cycle only on a massive scale. Digital technologies come in very small denominators so that a start-up company can create a tech-based business with just three persons in a garage with capital close to zero. But, even in the era of massive infrastructure, the de facto influencer of a company's decision are employees with strong character and exceptional skills. They are the ones who prescribe how Telkom's management should choose, design, and operate the adoption of technologies. These people, intrapreneurs, will be required more than ever before by Telkom, not only because of the miniature nature of digital technologies, but also due to their miniature and diverse market.

Telkom has introduced the Digital Amoeba Program, a playground for intrapreneurs. Within this program, Telkom's employees will learn to adopt updated mind-sets and concepts, to bring confidence to their work, and to adopt the chosen digital technologies and build related business on it.

In this program, Telkom distributes technology-related decision-making to the participants. This is very different from the previous organizational structure, where decision-making was made down the structure of research and development divisions. However, intrapreneurs are there already; the program is only gathering them and scaling up their influence. Contextually, the program is an innovation playground. Digital technologies and their related businesses are being developed as innovation. Employees compete for their ideas, and the top chosen ideas then incubated and accelerated in the program.

Founded in January 2017, the program has been failing. "Fail fast, succeed faster" has been the core principle of Amoeba's incubation processes, learning from the start-up industry. Currently, Amoeba has more than 40 active teams validating their innovations, and it is still looking for new ideas. At least three new products and two new processes are expected to be triggered from these teams within 2019. Each team consists only of two to five employees, working within an agile environment. They validate digital technology adoption ideas, and they are attracting users and prospective customers in only seven months with costs less than Rp 0.5 (far below typical product development schemes in corporations, especially Telcos).

The program is now examining how it provides impact. *Innovation accounting* is a collective and multivariable approach to assessing whether investment in innovation can solve business sustainability issues. The impact of the Digital Amoeba Program can be calculated from its human impact (technical and leadership capabilities improvement, employee happiness improvement, etc.), costs impact (reduced costs, faster asset development, etc.), revenues (new revenues, customer experience improvement, etc.), and market capitalization (investor confidence level). A concluding model has not yet been established, but the program is providing positive significant impacts. When the power of decision-making is utilized to revive dying businesses, many unexpected responses should be anticipated, some of which can point to a strong future.

The three giants in the IT industry and their valuation

Fajar Ramdhani Kurnia & Palti M.T. Sitorus
Master of Management Department, Telkom University, Bandung, Indonesia

ABSTRACT: This study aims to estimate the fair value of stock prices from technology-based companies listed on the National Association of Securities Dealers Automated Quotations (NASDAQ), i.e., Apple Inc. (AAPL), Alphabet Inc. (GOOG), and Microsoft Corp. (MSFT). This study uses the discounted cash flow (DCF) method with the free cash flow to firm (FCFF) approach and the relative valuation (RV) method with the price-to-book value (PBV) and price-to-earnings ratio (PER) approaches. Using the DCF–FCFF method, in the pessimistic scenario, all the shares are overvalued. In the moderate scenario, MSFT shares are overvalued while AAPL and GOOG shares are undervalued. In the optimistic scenario, all shares are undervalued. Using the RV–PER approach, in all scenarios, investors are advised to buy AAPL shares. Meanwhile, the RV–PBV approach, specifically in the pessimistic scenario, recommends buying MSFT shares and GOOG shares in two other scenarios.

1 INTRODUCTION

The National Association of Securities Dealers Automated Quotations (NASDAQ), founded in 1971, was the first stock exchange to operate electronically. More than 3,000 shares are listed on the NASDAQ, most of which are shares of technology-based companies, so that issues regarding the latest technology greatly affect the value of the NASDAQ Composite Index. In this study, three technology-based companies that had almost the same business fields and were the richest companies in the field of technology were examined; they are Microsoft Corp. (MSFT), Alphabet Inc. (GOOG), and Apple Inc. (AAPL).

The main problem in this study is how to conduct an analysis of the intrinsic value of the share prices of three technology-sector companies on the NASDAQ stock market (Microsoft Corporation, Alphabet Inc. and Apple Inc.) in 2018. Further analysis is needed regarding the intrinsic value of the stock prices of these three companies using the discounted cash flow (DCF) approach, the free cash flow to firm (FCFF) method, and relative valuation (RV) using the price-to-earnings ratio (PER) and price-to-book value (PBV) approaches in three conditions/scenarios, namely optimistic, moderate, and pessimistic.

The purpose of this study is to answer research questions that have been formulated as follows: to identify the intrinsic value of AAPL, GOOG, and MSFT shares in three scenarios, i.e., pessimistic, moderate, and optimistic; to provide recommendations for investors about the three shares; and to provide theoretical and practical benefits for interested parties.

2 LITERATURE REVIEW

Stock prices of companies included in the NASDAQ-100 over the past 10 years (2009–2018) have a volatile tendency when viewed on a year-on-year basis. The percentage change in stock prices (increase or decrease) is also not constant. When investors only focus on low prices, it can lead to inappropriate decision-making.

Neaxie and Hendrawan (2017) used the DCF method with the FCFF and RV approaches to calculate the fair price of telecommunications company shares listed on the Indonesia

Stock Exchange (IDX). The results of this study indicate that the DCF method with the FCFF approach is optimistic, resulting in fair value of TLKM in an undervalued condition, fair value of ISAT under overvalued conditions, and fair value of EXCL under undervalued conditions.

Alexandre (2016) evaluated AAPL shares on NASDAQ as of October 2015 with DCF, the dividend discount model (DDM), and RV. From this study, it was concluded that the market price of AAPL's shares at that time was undervalued.

3 METHODOLOGY

3.1 Database

A market share price sample was taken from nonfinancial companies listed on the NASDAQ-100 from five years back until September 2018. Historical data for calculating DCF–FCFF and RV are taken from publicly available yearly company financial reports submitted to the US Securities and Exchange Commission (SEC) website.

3.2 Free cash flow to firm

Free cash flow to firm (FCFF) is net cash flow for a company that is used as a basis for finding company value based on equity or liability. The FCFF method is usually used for valuing companies that have equity and debt in order to maximize company value; FCFF is based on earnings before interest and tax (EBIT). According to Damodaran (2006), FCFF is the sum of all cash flow for all company owners.

$$FCFF = EBIT * (1 - Tax) - CAPEX + DA \pm \Delta WC \quad (1)$$

Where: EBIT = earnings before interest and tax; Tax = tax; CAPEX = capital expenditure for long-term investment; DA = depreciation and amortization; and WC = short-term working capital.

3.3 Relative valuation

Relative valuation (RV) is a method for finding the value of an asset by comparing a company with other companies of the same type or within the industry where the company is located. The tool used to do an RV is multiples. One form of multiples is price multiples, where market prices become the main component in price multiples. Examples of price multiples include price/earnings per share (P/ER), price/book value per share (P/BV), price/sales per share (P/S), and price/cash flow per share.

3.4 Framework

We calculate the price per share of three companies (AAPL, GOOG, and MSFT) using FCFF, PER, and PBV and then compare them with the value of shares in the market. After knowing the condition of the stock price, it can be used as an input for investors in optimizing investment in the shares of the three companies that are the object of research.

4 RESULT AND DISCUSSION

4.1 Discounted cash flow–free cash flow to firm

The results of the calculation of the intrinsic value of shares of AAPL, GOOG, and MSFT using DCF–FCFF can be seen in Table 4.1.

Table 4.1. Intrinsic value of shares based on DCF–FCFF.

Share's code	Scenario	Intrinsic value	*	**	Share's price condition
AAPL	Pessimistic	138.70	157.17	11.75%	overvalued
	Moderate	213.29	157.17	35.70%	undervalued
	Optimistic	357.02	157.17	127.16%	undervalued
GOOG	Pessimistic	1,013.55	1,039.46	2.49%	overvalued
	Moderate	1,132.05	1,039.46	8.91%	undervalued
	Optimistic	1,283.02	1,039.46	23.43%	undervalued
MSFT	Pessimistic	45.31	100.56	54.94%	overvalued
	Moderate	92.25	100.56	8.27%	overvalued
	Optimistic	297.47	100.56	195.81%	undervalued

* Share's market value as per December 26, 2018, ** Deviations from market value

Based on Table 4.1, in the pessimistic scenario, all three companies' stocks are overvalued. In the moderate scenario, only MSFT is overvalued while AAPL and GOOG are undervalued. In the optimistic scenario, all three companies' stocks are undervalued.

As we can see from Table 4.1, if we check the minimal value of the deviations of the calculated intrinsic value from its market value for each share, AAPL is close to the pessimistic scenario, GOOG is close to the pessimistic scenario, and MSFT is close to the moderate scenario. Our intrinsic value calculation can predict the market value with up to 88% accuracy for AAPL, 97% accuracy for GOOG, and 91% accuracy for MSFT. This shows that our data source and method are reliable.

4.2 *Relative valuation–PER and PBV*

In addition to using the DCF–FCFF approach, we also use the RV, PER, and PBV methods. The PER is obtained from the calculation of equity value divided by earnings after tax (EAT). The PBV is obtained by dividing the calculation of equity value by the book value. Relative valuation is divided into three scenarios: pessimistic, moderate, and optimistic.

The results of the PER and PBV calculations for the three companies can be seen in Table 4.2.

Quarterly (Q3) NASDAQ data show that the average PER value for the technology industry is 40.56 times, with the lowest value at HP Inc. (HPQ) of 6.50 times and the highest PER for Cisco Systems Inc. (CSCO) of 158.89. This shows that the research results for all scenarios are in the PER range of the market. Also, quarterly (Q3) NASDAQ data show that the average PBV value for the technology industry is 12.06 times, with the lowest value at Intel Corp. (INTC) of 2.92 times and the highest PBV of Oracle Corp. (ORCL) of 82.57. This shows that the results of the study are in the PBV range in the market.

Table 4.2. PER and PBV for each scenario.

Share's code	Scenario	PER	PBV
AAPL	Pessimistic	11.00	6.38
	Moderate	16.69	9.69
	Optimistic	27.65	16.05
GOOG	Pessimistic	56.25	4.69
	Moderate	62.72	5.23
	Optimistic	70.98	5.92
MSFT	Pessimistic	20.21	4.35
	Moderate	40.45	8.71
	Optimistic	128.95	27.75

5 CONCLUSION

Based on the DCF–FCFF method, the intrinsic value of shares from AAPL, GOOG, and MSFT are in various conditions depending on the scenario chosen. In the pessimistic scenario, all of them are overvalued. In the moderate scenario, the intrinsic value of the shares of MSFT is overvalued, while the AAPL and GOOG shares are undervalued. In the optimistic scenario, the intrinsic value of shares of AAPL, GOOG, and MSFT is undervalued. Investors are advised to choose the undervalued shares in each scenario. If there's no undervalued condition, then investor should hold or sell their shares.

Using the PBV approach, in the pessimistic scenario, investors are advised to choose MSFT shares because they have the lowest PBV among the three. In the moderate and optimistic scenarios, investors are advised to choose GOOG shares because they have the lowest PBV.

Using the RV–PER approach, in all scenarios, investors are advised to prefer AAPL shares because they have the lowest PER value among the three.

In carrying out valuations, subjective assumptions are very influential on the final results of the calculation. For this reason, it is better to look to valid data sources so that there is a strong basis in determining the value of a calculation variable. Future research should use more valuation methods as a comparison of results so as to provide more information to investors and companies in making decisions both in investment and in policy making.

REFERENCES

Bogdan, R., & Taylor, S. J. 1975. *Introduction to Qualitative Research Methods: A Phenomenological Approach to the Social Sciences*. Hoboken, NJ: Wiley.

Brealey Myers, M. 2001. *Fundamentals of Corporate Finance*. Upper Saddle River, NJ: Prentice Hall.

Damodaran, A. 2006. *Damodaran on Valuation*. Second Edition. Hoboken, NJ: Wiley.

Djaja, I. 2018. *All about Corporate Valuation*. Second Edition. Jakarta: PT. Elex Media Komputindo.

Gonçalo Lopez, A. 2016. *Apple Inc. Equity Valuation*. Lisbon: Fakultas Bisnis dan Ekonomi Universidade Católica Portuguesa.

Ivanovska, N., Ivanovski, Z., & Narasanov, Z. 2014. Fundamental analysis and discounted free cash flow valuation of stock at MSE. *UTMS Journal of Economics*, 5(1), 11–24.

Megginson, W. L., Lucey, B. M., & Smart, S. B. 2008. *Introduction to Corporate Finance*. Andover, Hampshire: Cengage Learning EMEA.

Neaxie, L. V., & Hendrawan, R. 2017. Stock valuations in telecommunication firms: Evidence from the Indonesia Stock Exchange. *Journal of Economic & Management Perspectives*, 11(3), 455.

Zemba, S., & Hendrawan, R. 2018. Does rapidly growing revenues always produce an excellent company value? DCF & P/E valuation assessment of the hospital industry. *e-Proceeding of Management*, 5(2), 2045.

Analysis of the role of brand orientation in brand performance in B2B firms

S.L. Retno
University of Indonesia, Jakarta, Indonesia

ABSTRACT: This research aims to examine the relationship between brand orientation and brand performance in business-to-business (B2B) firms in Indonesia, especially for the healthcare industry. This research highlights the role of brand-oriented strategy in transforming internal organizational resources into brand performance through customer value co-creation and brand credibility. A questionnaire-based survey was conducted to collect data from 145 B2B healthcare firms in Indonesia. The results indicate that the brand orientation strategy is significantly related to brand performance. The results also show that the brand orientation strategy is supported by organizational resources, which is entrepreneurial orientation, but this is proven otherwise by marketing capability. It could be concluded that the B2B healthcare firms in Indonesia fully understand the benefit of B2B branding for their brand performance, but they consider external factors when they decide to allocate their resources to develop brand orientation.

1 INTRODUCTION

Branding is one of the marketing instruments that is proven to give success to a product. But branding is generally used more in the business-to-consumer (B2C) market than in the business-to-business (B2B) market, and this is related to the notion that branding can only be used well in the B2C market. Most managers make the assumption that brand loyalty can be applied only to the B2C market; this is because the B2B customers tend to be rational and objective in purchase-making decisions, where they are more focused on things such as product features, prices, etc. compared to the product brand itself. However, people who are responsible for the purchase-making decisions process are humans with emotions (Kotler, Pfoertsch, & Michi, 2006), and this will make them try to find a trusted brand among the available choices.

Many firms have not understood the value of B2B branding strategies (Leek & Christodoulides, 2011; Urde, Baumgarth, & Merrilees, 2013), which ultimately can lead to failure to adopt a brand orientation strategy. Such firms are still not entirely confident in turning their marketing strategies into brand-oriented strategies because they believe that this strategy cannot have a direct effect on the firm's financial performance (Chang, Wang, & Arnett, 2018). Financial performance is considered very important for B2B firms because they reckon B2B branding to be a long-term investment and the branding needs to provide them with financial benefits. But, in reality, B2B firms must understand that B2B branding needs to be done to keep their firms competitive in an ever-changing business environment (Kotler et al., 2006). Especially with the increasing number of competitors emerging in the market, branding could have more impact on the value of products offered by the firm, where product values are no longer unique so consumers will tend to make their choices based on prices and product availability, which will ultimately lead to very competitive market conditions (Urde, 1994).

Based on data held by the National Drug and Food Control, currently the number of pharmaceutical firms in Indonesia is estimated to have reached more than 200, while based on data held by the Republic of Indonesia Ministry of Health, the number of medical device firms in Indonesia is estimated to have reached more than 500. This certainly increases the competition between healthcare firms in Indonesia. The existence of the universal coverage program, known as Jaminan Kesehatan Nasional (JKN), carried out by the government since 2014, has also had an impact on the growth of the healthcare industry in Indonesia, causing it to decline. Under these conditions, healthcare firms in Indonesia must create competitive advantages, especially for products intended for use in hospitals. As generally happens in the B2B market, B2B marketing generally prioritizes personal relationships. The purchase decision-making by the hospital does not fully look at the quality or features of the product but also considers other aspects such as price. But the universal coverage program means that the prices offered by healthcare firms to the hospitals tend to remain the same. Hence, the firms should start doing brand-oriented marketing in order to make their products more competitive.

Our research contributes to the existing literature in several ways and also provides guidance for firms that would like to adopt a brand-oriented strategy concerning the factors that will affect the implementation of brand orientation. This research also provides additional insights for practitioners, especially for firms in the healthcare industry, to understand the role of B2B branding in brand performance.

2 METHODS

The data were collected from 145 B2B healthcare firms in Indonesia. The respondents to our questionnaire were selected from among senior managers/directors in sales or marketing who had worked for more than five years. The questionnaires were distributed through email or message applications asking them to participate in the study by completing a web-based questionnaire. The number of responses from pharmaceutical and medical device firms were almost equal, with 48.27% coming from medical device firms, 46.89% from pharmaceutical firms, and 4.82% from firms that deal in both product types. All of the respondents held senior managerial positions: managing directors (12.41%), sales managers (67.59%), marketing managers (17.24%), and product managers (2.75%). The sample included both multinational and local firms – about 22.07% of the firms are multinational and 77.93% of the firms are local.

3 RESULTS AND DISCUSSION

The data were analyzed using partial least square (PLS) analysis. The internal reliabilities were measured using composite reliabilities that are greater or equal to 0.864, which is above the 0.70 level recommended (Nunnally, 1978). The convergent validity was examined using the factor loadings and average variance extracted (AVE). All factor loadings were greater or equal to 0.689, which is above the 0.50 level recommended. Meanwhile, the AVE value for each construct was greater or equal to 0.614, which is above the 0.50 level recommended. Based on these results, the measurement model demonstrates acceptable internal reliability and convergent validity.. The results suggest that the measurement model demonstrates acceptable discriminant validity.

Eight hypotheses are supported and one is not. Entrepreneurial orientation is positively related to brand orientation (t = 4.159). Thus H1 is supported. Marketing capability is not positively related to brand orientation (t = 0.285) Thus H2 is not supported. Brand orientation is positively related to customer value co-creation (t = 6.016). Thus H3 is supported. Customer value co-creation is positively related to brand performance (t = 3.025). Thus H4 is supported. Brand orientation is positively related to brand credibility (t = 6.244). Thus H5 is supported. Brand credibility is positively related to brand performance (t = 3.243). Thus H6 is supported. Brand orientation is positively related to brand performance (t = 1.786). Thus H7 is supported. Customer value co-creation and brand credibility are partially mediated brand orientation to brand performance (t = 2.984 and t = 2.952). Thus H8 and H9 are supported.

Our study provides evidence that brand orientation can influence brand performance through customer value co-creation and brand credibility. But even though brand orientation has been proven to impact brand performance, many healthcare firms in Indonesia are still slow to adopt brand orientation. This study examines two factors that could influence the development of brand orientation: entrepreneurial orientation and marketing capability. The result suggests that only entrepreneurial orientation can influence the adoption of brand orientation. This means that when firms have higher levels of entrepreneurial orientation, they are more innovative, more proactive, and more likely to take risks, hence they could become more brand-oriented. Marketing capability proved otherwise because the firms also consider external factors, and that could affect their decision to adopt brand orientation. In this case, the healthcare market situation in Indonesia greatly influences the healthcare firms in allocating their marketing capability to brand orientation. Since B2B customers in Indonesia are more price-oriented, healthcare firms focus their resources accordingly.

Our findings suggest that firms that desire to develop brand orientations should promote strategic entrepreneurship among their top managers. As defined by Low and Fullerton (2006), brand orientation consists of entrepreneurial activities that involve a firm's resources intensively. It is also recommended for B2B firms in emerging market to adopt entrepreneur-oriented strategies instead of focusing on building the brand, because it could make the firms more flexible in responding to customers' needs (Reijonen et al., 2015). Our findings also suggest that firms who adopt brand orientation should consider building brand credibility and customer value co-creation to improve their brand performance.

REFERENCES

Chang, Y., Wang, X., & Arnett, D. B. 2018. Enhancing firm performance: The role of brand orientation in business-to-business marketing. *Industrial Marketing Management, 72*(17), 17–25. https://doi.org/10.1016/j.indmarman.2018.01.031

Kotler, P., Pfoertsch, W., & Michi, I. 2006. *B2B Brand Management*. https://doi.org/10.1007/978-3-540-44729-0

Leek, S., & Christodoulides, G. 2011. A literature review and future agenda for B2B branding: Challenges of branding in a B2B context. *Industrial Marketing Management, 40*(6), 830–837. https://doi.org/10.1016/j.indmarman.2011.06.006

Low, G. S., & Fullerton, R. A. 2006. Brands, brand management, and the brand manager system: A critical-historical evaluation. *Journal of Marketing Research*. https://doi.org/10.2307/3152192

Nunnally, J. 1978. *Psychometric Methods*. New York: McGraw-Hill.

Reijonen, H., Hirvonen, S., Nagy, G., Laukkanen, T., & Gabrielsson, M. 2015. The impact of entrepreneurial orientation on B2B branding and business growth in emerging markets. *Industrial Marketing Management, 51*, 35–46. https://doi.org/10.1016/j.indmarman.2015.04.016

Urde, M. 1994. Brand orientation: A strategy for survival. *Journal of Consumer Marketing, 11*(3), 18–32.

Urde, M., Baumgarth, C., & Merrilees, B. 2013. Brand orientation and market orientation: From alternatives to synergy. *Journal of Business Research, 66*(1), 13–20. https://doi.org/10.1016/j.jbusres.2011.07.018

The influence of e-service quality on the data package buying experience in Telkomsel

P. Kusdinar & M. Ariyanti
Telkom University, Bandung, Indonesia

ABSTRACT: This study aims to analyze the influence of e-service quality and its subdimensions on the experience of buying data service packages as a prepaid customer through the MyTelkomsel purchase channel. The dimensions of e-service quality to be analyzed here include electronic service quality (ES-Qual), consisting of efficiency, fulfillment, system availability, and privacy; and electronic recovery service quality (E-RecSQUAL), consisting of responsiveness, compensation and contact. This research employed a quantitative research method with descriptive analysis techniques, along with a data collection method using secondary data from Telkomsel's internal big data. Purposive sampling techniques were used. Based on specified criteria, 164,394 samples of data package purchase transactions were analyzed. This study also utilized the logistic regression analysis method. Based on the overall fit test with omnibus tests of model coefficients, it can be concluded that, simultaneously and partially, ES-Qual has a significant effect on the buying experience. Nagelkerke R squares values show that this model can predict the value of its observations.

1 INTRODUCTION

PT Telkomsel is one of the largest operators in Indonesia, with 157 million prepaid subscribers in 2018. Data service users have increased in Indonesia every year and have become one of the biggest contributors to Telkomsel's revenue after voice service. In 2018, prepaid customers carried out around 200 million data service package purchase transactions through MyTelkomsel. The high transaction value of PT Telkomsel's data service package proves that customers are willing to buy a data package in order to enjoy the service immediately. However, there are times when customers cannot enjoy the service immediately due to various reasons, such as disruption of information technology (IT) systems, business rules/restrictions, etc.

Although there have been many improvements, based on Telkomsel's internal big data, the number of failed transactions to purchase data packages by prepaid customers via MyTelkomsel is still high – more than 2 million per month in 2018. Therefore in this study the focus is to use Telkomsel's internal big data to examine the service quality when purchasing data service packages because it is more accurate and the facts of the buying experience are recorded in the system.

One of the best ways to measure quality of service is by implementing e-service quality (ES-Qual). E-service quality is a continuation of service quality that measures the quality of services by considering IT factors. Many studies related to the influence of e-service quality have shown that it has an effect on the customer experience, customer satisfaction, and customer loyalty; most of them use primary data taken through surveys. Therefore, this study focused on using secondary data from Telkomsel's internal big data, so the analysis is more accurate.

There are various opinions regarding the dimensions of e-service quality used in previous studies, but in this study, the authors used the seven e-service quality dimensions tested by Parasuraman, Zeithaml, and Malhotra (2005): efficiency, fulfillment, system availability, privacy, responsiveness, compensation, and contact.

2 RELATED WORK

Most of the previous studies related to the influence of e-service quality on customer satisfaction used secondary data sourced from surveys or questionnaire; at the time when the authors conducted this study, no studies had used secondary data sourced from big data. Therefore, the focus of this study was to analyze the effect of e-service quality on the experience of buying data service packages as recorded in Telkomsel's internal big data. Following is a list of some of the researchers' literature related to e-service quality using secondary data.

3 METHODOLOGY

This research was descriptive and employed quantitative methods; data collection was performed using research instruments. The author here describes the value of the research variable to find out the relationship between the independent variable (e-service quality) and the dependent variable (buying experience)

Table 1. Research characteristics.

Research Characteristics	Research Type
Method	Quantitative
Purpose	Descriptive
Type	Causal
Research setting	Non-contrived setting
Unit analysis	Organization
Time Frame	Cross-sectional

In this research variables that were used are as shown:

Table 2. Research variable.

No	Variable	Sub-variable	Indicator	Scale
1	E-Service Quality (X) (Parasuraman et al.,)	Efficiency (X)	Response Time Application (0:>10 sec and 1: <10 sec)	Nominal
		Fulfilment (X2)	Stock Availability (0: not available 1: available)	Nominal
		System Availability (X3)	System Availability (0: not running and 1: running)	Nominal
		Privacy (X4)	Customer Authentication (0: invalid and 1:valid)	Nominal
		Responsiveness (X5)	Complaint ticket (0: No complaintand 1: Complaint)	Nominal
		Compensation (X6)	Refund Status (0: no refund and 1: refund)	Nominal
		Contact (X7)	Customer Contact (0: No interaction and 1: Interaction)	Nominal
2	Buying Experience (Y) (Nasermoadeli, Choon-Ling, & Maghnati, 2012, p. 128)		Purchase Status, 1: Success and 0: Failed	Nominal

Prepaid customer purchase transactions were sourced from Telkomsel's internal big data via several tools such as Splunk Query, SQL Query, and Shell Script for the process of extracting data, distinct data, aggregation data, and loading data (ETL Process). This study also utilized logistics regression to determine the effect of dimensions of e-service quality both simultaneously and partially on the experience of buying data service packages via MyTelkomsel.

4 RESULT AND DISCUSSION

4.1 Characteristics of research data

From PT Telkomsel's internal big data that was processed, the total transactions were 164,394.

Table 3. Total transactions processed.

Total Purchase Transactions	Total Successful Transactions	Total Failed Transactions
164,394	147,334	17,060

4.2 Data cleansing

The composition of research data after the data-cleansing process was conducted to detect inconsistent and dirty data is shown in Table 4.

4.3 Hypothesis

The hypothesis was examined using logistic regression because the dependent variable (buying experience) is dichotomous, where successful buying is marked with a value = 1 and failed buying is marked with a value = 0. Following are the steps to test hypotheses using logistic regression (Ghozali, 2016).

Assessing the overall model (fit overall model), the significance of the model is <0.05; it can be concluded that efficiency, fulfillment, system availability, privacy, responsiveness, compensation, and contact simultaneously influence the buying experience (purchase status).

1. Wald Test: Partial testing of each variable obtained a significance value of <0.05; then each of the subdimensions of e-service quality partially affects the buying experience.
2. The Coefficient of Determination, Nagelkerke R square value is 0.941, which suggests that the variability of the dependent variable that can be explained by the independent variable is 94.1%, while the remaining 5.9% is explained by other variables outside the research model.
3. Feasibility Test Regression Model (Hosmer and Lemeshow's test): With a significance value of 0.231 (> 0.05), the model can be concluded to predict the value of observation. The regression model equations are as follows:

$$Ln\ p(1-p) = -15.366 + 3.781\ Efficiency + 5.967\ Fulfillment + 5.138\ System\ Availability \\ + 6.179\ Privacy + 0.160\ Contact - 0.190\ Responsiveness - 0.577\ Compensation + e \quad (1)$$

Table 4. After the data-cleansing process.

Total Purchase Transactions	Total Successful Transactions	Total Failed Transactions
164,367	147,307	17,060

4.4 *Discussion of research results*

1. Simultaneous e-service quality has a significant effect on the experience of buying data service packages as a prepaid customer through MyTelkomsel. The value of the Sig. model is 0.000 (<0.05) and the value of Nagelkerke's R square is 0.941. Cox and Snell's R square is 0.458, which has similarities with previous research, such as research by Hansel Jonathan (2013) and by Vadivelu Tharanikaran, Sutha Sritharan, and Vadivelu Thusyanthy (2017).
2. Partially each subdimension of e-service quality (efficiency, fulfillment, system availability, privacy, responsiveness, compensation, and contact) has a significant effect on the experience of buying data service packages as a prepaid customer through MyTelkomsel. The value of the Sig. model is 0.000 (<0.05).
3. The results showed that the e-service quality variable explains 94.1% of the buying experience, while the remaining 5.9% is another factor outside the model.

5 CONCLUSION

This study has provided empirical evidence about the effect of e-service quality partially and simultaneously on the buying experience when Telkomsel prepaid customers purchased data service packages through MyTelkomsel. This study used a sample of 164,367 transactions originating from Telkomsel's internal big data for one year in 2018.

The variables used consist of dependent variables, namely buying experience; of independent variables of e-service quality consisting of E-S-QUAL variables, namely efficiency, fulfillment, system availability, privacy; and E-Recs-Qual Variables, namely responsiveness, compensation, and contact.

REFERENCES

Dumbill, E. 2012. *Big Data Now: 2012 Edition. What Is Big Data?*: O'Reilly Media, Inc.
Ghozali, I. 2016. *Aplikasi Analisis Multivariete Dengan Program IBM SPSS 23, Cetakan ke VIII*. Semarang: Badan Penerbit Universitas Diponegoro.
Indrawati, P. 2015. *Metode Penelitian Manajemen dan Bisnis*. Bandung: PT Refika Aditama.
Kotler, P., & Armstrong, G. 2018. *Principles of Marketing*, 17th edition. Upper Saddle River, NJ: Pearson Prentice Hall.
Meyer, C., & Schwager, A. 2007. Understanding customer experience. *Harvard Business Review*, 85(2), 116–126.
Nasermoadeli, A., Choon-Ling, K., &Maghnati, F. 2012. Evaluating the impact of customer experience on repurchase intention. *International Journal of Business and Management*, 8(6), 2013.
Parasuraman, A., Zeithaml, V. A., & Malhotra, A. 2005. E-S-QUAL: A multiple item scale for assessing electronic service quality. *Journal of Service Research*, 7(3), 213–233.
Sugiyono. 2014. *Metode penelitian kuantitatif, kualitatif dan R&D*. Bandung: Alfabeta.
Sun, H., & Heller, P. 2012. *Oracle Information Architecture*. Oracle Information Architecture.
Zeithaml, V.A., Bitner, M.J., & Gremler, D.D. 2013. *Services Marketing. Integrating Customer Focus across the Firm*, 6th edition. New York: McGraw-Hill Irwin.

Develop a perceivable financial statement: A qualitative study at SME four coffee

Eko Wahyudi & Sisdjiatmo K. Widhaningrat
Master of Management, Faculty of Economics and Business, Universitas Indonesia, Kampus UI Salemba, Jakarta, Indonesia

ABSTRACT: This study aims to develop a perceivable financial statement format at small and medium-sized enterprise (SME) Four Coffee. The research was conducted using a qualitative method. The data analysis technique employed descriptive analysis with a qualitative approach. The researcher describes, interprets, and explains the situation at hand, together with suggesting improvements to the research object. Four Coffee has the correct financial statement format, which includes internal decision-making materials and reference data for prospective creditors or investors. Web-based accounting software can be utilized by doing optimization and customization of the output produced in order to get a desirable financial statement. The use of accounting software does not always provide an output that meets the expectation of the SME. However, by accepting technological advances such as accounting software technology, SMEs can speed up the process and reduce human error to build a perceivable financial statement.

1 INTRODUCTION

1.1 *SMEs in Indonesia*

In small and medium-sized enterprises (SMEs), creating financial statements is often overlooked and is usually considered a trivial thing. On the other hand, small-scale business owners sometimes do not have clear information on whether their business is in a state of profit or loss, so they often take a late or inappropriate policy decision. The object of this research was an SME faced with the challenge of doing business on a tiny scale in Indonesia called Four Coffee. Four Coffee serves food and beverages with a specialization in different types of coffee drinks.

The initial analysis was carried out using data collection techniques to collect relevant data through in-depth interviews and field observations on research subjects. The next step was to interpret the data using analytical tools such as segmenting, targeting, positioning (STP), analysis of marketing mix (7P), strength, weakness, opportunity, and threat (SWOT) analysis, analysis of business model canvas (BMC), and analysis of Porter's 5 Forces. The conclusion and interpretation of the data is the gap between the actual conditions and the ideal conditions for Four Coffee.

1.2 *Objective*

The analysis examined in Point 1.1 resulted in some of the main problems experienced by Four Coffee. In this research, researchers chose one of these problems in order to define the objectives of this study: developing a micro, small, and medium-sized (MSME) financial report format for Four Coffee that is easy to understand so that it can be used as reference material in internal decision-making and as reference financial data for creditors and investors.

Table 1. Pareto analysis result.

No	Code	Value	Weight	Contribution	Distribution	Accumulation
1	Financial Statement	8	9	72	47%	47%
2	Raw Material Inventories	7	7	49	32%	80%
3	SOP	5	4	20	13%	93%
4	Channel Distribution	3	3	9	6%	99%
5	Certification	2	1	2	1%	100%
Total				198	100%	

2 LITERATURE REVIEW

2.1 *Financial statement*

Financial statements are a structured presentation of the financial position and performance of an entity. Financial statements aim to provide information about the financial situation, financial performance, and cash flow of a company that is beneficial for most users of the report in making economic decisions.

According to Anthony, Hawkins, and Merchant (2011), the preparation of financial statements can be carried out in the following stages:

1. Analyze transactions
2. Record data into journals
3. Post a journal entry in the ledger
4. Adjust journal entries
5. Journalize and post closing journal entries

2.2 *Management accounting*

Management accounting is "the process of identifying, measuring, accumulating, analyzing, compiling, interpreting, and communicating information used by management to plan, evaluate and control within an entity and to ensure appropriate and accountable use of these resources. Management accounting also includes the preparation of finance reposts for non-management groups such as shareholders, creditors, regulatory bodies and tax authorities" (Chartered Institute of Management Accountants)

2.3 *Previous research*

The preparation of financial statements based on Microsoft Excel is familiar to SMEs. It requires several development features, for example adding the cross-check feature of the cost of goods sold, which is calculated based on cost sheet and sales volume (Faisal, 2018). Creating SMEs financial statements begins with the identification of evidence. At this stage, SMEs are one step ahead because the owner has documented transaction evidence in one container and given that evidence a reference code. Based on these documents, we can prepare a simple financial statement that includes only a balance sheet and an income statement (Maulana, 2016).

3 RESEARCH METHODOLOGY

This qualitative research used descriptive methods. Qualitative research produces written or oral data from observing behavior or people. The research objective of this qualitative approach is to understand social phenomena from participants' point of view, looking at motivation, behavior, and other characteristics.

The techniques used in collecting data included interviews, observation, and documentation. The data-collection process utilizing these three methods was carried out for seven months, with 11 face-to-face meetings. Compiling this research resulted in two types of data sources, namely primary data sources and secondary data sources. General stages used in analyzing qualitative data in this study included:

1. Data reduction
2. Data visualization
3. Data conclusion and verification

4 RESULT AND DISCUSSION

4.1 *Result*

This study focused on resolving the problems of financial statements, namely the development of financial report formats, by standardizing financial recording procedures up to the preparation of financial statements required by MSMEs. Four Coffee already has financial records in the forms of a general journal and Microsoft Excel files. The financial records show transaction history from October 2017 until the last month of this research activity. For data for 2017, the earliest transaction date is October 18, 2017, while the most recent data show transactions that occurred on December 29, 2018.

The financial reports owned by Four Coffee are in the form of a simple Microsoft Excel format in the way of a trial balance. For Four Coffee as an MSME, manual recording like this is easier than using accounting software. Researchers analyzed the financial statements and found several issues:

a) In the general journal
 • Transaction numbers were missing.
 • The account name was not complete.
 • Transactions were not yet chronologically arranged.

b) On the balance sheet
 • There is no apparent balance sheet or statement of financial position.
 • There is no calculation of accumulated depreciation of fixed assets.
 • There is no calculation of the end-of-period inventory so that the cost of goods sold will be the same as the cost of the inventory purchased throughout the year. The recording as carried out by Four Coffee is not correct, and if these records are desired as material for internal decision-making and reference data for potential creditors or investors, they are far from perfect. Therefore, researchers and MSMEs simulated all of these transaction data to be entered into accounting software to see the comparison and to solve the problems faced by Four Coffee. We also did optimization and adjusted the resulting output to provide the desired financial statement presentation. In this research, we used Jurnal.id accounting software as the simulator.

4.2 *Discussion*

The advantages of financial records using web-based accounting software (here Jurnal.id) are as follows:

1. Journals, ledgers, and financial statements are integrated.
2. Data can be accessed anywhere using a computer or smartphone.
3. Users do not need to understand accounting in detail.
4. Added value is produced in the form of speed to process data.

The disadvantages of recording using web-based accounting software such as Jurnal.id are as follows:

1. It requires the cost of a software license per month, quarter, or year.
2. The server stores the data of the software owner (there is a risk of misuse of data).

The value added can be processed using the output of accounting software as follows:

1. A trend can be established from the historical data of an account (e.g., sales, operating costs).
2. The data needed for financial ratios are known.
3. Financial projections can be made and used to determine company policy.

5 CONCLUSIONS

- The use of accounting software does not always provide an expected output, so it requires customization.
- Accepting technological advances such as accounting software technology is advantageous for SMEs because such technologies speed up the process and reduce human error when building a perceivable financial statement.

REFERENCES

Anthony, R., Hawkins, D., & Merchant, K. 2011. *Accounting: Text and Cases*. 13th edition. New York: McGraw-Hill.
Cooper, D. R., & Scindler, P. S. 2014. *Business Research Methods*. 12th edition. New York: McGraw-Hill.
Faisal. 2018. *Pengembangan Format Laporan Keuangan dan Analisis Skenario atas Skema Pre-Order Pada CV Azka Syahrani*. Jakarta: Tesis Universitas Indonesia.
Laporan Tahunan Kementerian KUKM Tahun 2015. Jakarta: Departemen Koperasi dan UKM.
Laporan Tahunan Kementerian KUKM Tahun 2016. Jakarta: Departemen Koperasi dan UKM.
Maulana, A. R. 2016. *Penetapan harga jual produk dan penyusunan laporan keuangan pada UKM Kusuma*. Jakarta: Tesis Universitas Indonesia.
Mullins, J. W. & Walker, O. C. 2013. *Marketing Management: A Strategic Decision-Making Approach*. 8th edition. New York: McGraw-Hill.
Profil Bisnis Usaha Mikro, Kecil Dan Menengah (UMKM). 2015. Jakarta: Bank Indonesia.
UU No. 20 Tahun 2008. *Tentang Usaha Mikro, Kecil dan Menengah*.
Waluyo. 2012. *Akuntansi Pajak*. Jakarta: Salemba Empat.

The effect of internal, external, and interactive marketing on customer decisions

A. Wulandari & B. Suryawardani
Telkom University, Bandung, Indonesia

ABSTRACT: Success or failure in the sales process is determined by marketing activities, thus the role of marketing activities in a company is crucial. Three types of marketing occur in the sales process, namely internal marketing, external marketing, and interactive marketing. Internal marketing connects employees and banks, external marketing is the relationship between customers and banks, and interactive marketing is the relationship between employees and customers. This research aimed to examine the influence of internal marketing, external marketing, and interactive marketing on the decision-making of customers of Bank Mandiri in Bandung. A qualitative research method was implemented with multiple linear regression. Purposive sampling was employed, with the number of respondents reaching 400 people. The results of the study showed that internal marketing, external marketing, and interactive marketing variables have a simultaneous influence on the customer decision variable of 89.4%.

1 INTRODUCTION

Marketing activities play a crucial role in a company, as they determine the success or failure of the product sales process. The more successful the marketing process is, the higher its impact on the sale of products will be, which can mean profits for a company. In Indonesia, a number of banks have become favorite places to save; of course they have been chosen for different reasons. The following table lists the top brands award in 2017 in the phase 1 savings product category:

Table 1 shows that Mandiri Bank's savings product placed last in the 2017 phase 1 savings product category.

Three types of marketing occur in the product sales process, namely internal marketing, external marketing, and interactive marketing. Internal marketing connects employees and banks. In order to market a bank's products well, employees should be involved in the process. They must be given all information about the bank's products and services that can be marketed to customers. They should comprehend all the products and services their bank offers, and then help to provide information to customers when needed.

External marketing connects customers and banks. Direct relationships between customers and banks are generally through front office officers or customer service. Here the front office officer will try to provide an explanation of the bank's products and services in detail. The success or failure of customers to purchase the bank's products and services will be greatly influenced by the service results of officers in the front office.

Interactive marketing connects employees and customers. Here the employees must comprehend the products and services their bank offers, in order to get involved and help marketing programs, and to explain the products and services interactively and accurately to customers. If a customer wants to purchase the bank's products and services, and asks the employees, but the employees have a negative impression, of course

Table 1. Top brand award Tabungan 2017.

Brand	TBI	TOP
Tahapan BCA	25.0%	TOP
BRI Simpedas	15.5%	TOP
BRI BritAma	13.2%	TOP
BNI Taplus	9.2%	
Tabungan Mandiri	7.6%	

the customer will not make a purchase. Thus, these three marketing concepts are interrelated and constitute a single entity.

2 METHODOLOGY

This study included quantitative descriptive research, which, according to Zikmund (2013, p. 49), is designed to obtain data that describe the characteristics of topics of interest in research. The type of data examined in this study were quantitative. Quantitative analysis is systematic scientific research on parts and phenomena and their relationships (Zikmund, 2010, p. 136).

2.1 *Population and sample*

Population is a generalization area consisting of objects/subjects that have certain qualities and characteristics set by researchers to be studied and from which conclusions can be drawn (Sugiyono, 2017, p. 61). The population in this study comprised customers of Mandiri Bank in Bandung. The sample was taken by using purposive sampling with 400 respondents.

2.2 *Data analysis*

The data analysis used in this study was to determine the effect of internal marketing (IM), external marketing (EM), and interactive marketing (ITM) on deciding whether to become a customer of Bank Mandiri in Bandung. In this study, the measurements were carried out using multiple regression analysis. According to Sugiyono (2017, p. 260), regression analysis is used to predict how far the value of the dependent variable changes, if the value of the independent variable is manipulated/changed. The benefit of regression analysis is the ability to decide whether the dependent variable can be manipulated through increasing the independent variable. According to Wibisono (2013, p. 277), multiple regression analysis describes the relationship between one dependent variable (bound) and several independent variables (free).

3 THEORETICAL FRAMEWORK

3.1 *Internal marketing (IM), external marketing (EM), and interactive marketing (ITM)*

Internal marketing connects employees and banks. In order to be able to market a bank's products well, employees should be involved in the process. They must be given all information about bank products and services that can be marketed to customers. They should comprehend all the products and services their bank offers, and then help to provide information to customers when needed.

External marketing connects customers and banks. Direct relationships between customers and banks are generally through front office officers or customer service. Here the front office officer will try to explain bank products and services in detail. The success or failure of customers to purchase bank products and services will be greatly influenced by the service results of officers in the front office.

Interactive marketing connects employees and customers. Here employees must comprehend the products and services of their bank, in order to get involved and help marketing programs, and to explain the bank's products and services interactively and accurately to customers. If a customer wants to purchase the bank's products and services, and asks the employees, but the employees have a negative impression, of course the customer will not make a purchase. These three concepts constitute a single interrelated entity.

3.2 *Customer decision*

According to Anoraga (2000), the decision-making process has five stages:

a. Need for introduction
   A customer might need to be introduced to products and services because of receiving new information about a product, economic conditions, advertising, or an accident. In addition, lifestyle, demographics, and personal characteristics can also influence purchasing decisions.
b. Consumer information process
   This stage covers consumers' search for information sources. The information process is carried out selectively. Consumers choose the information that is the most relevant to the benefits sought and in accordance with their beliefs and attitudes. The information process includes searching activities, paying attention, understanding, storing in memory, and seeking additional information.
c. Product/brand evaluation
   Consumers evaluate the characteristics of various products/brands and choose the one that best meets their needs.
d. Purchasing
   In purchasing, several other activities are performed, such as shopping, selecting, determining when to buy, and weighing the financial implications of the purchase. After the customer finds the right place and time and is supported by purchasing power, then finally the purchasing activity is carried out.
e. Post-purchase Decision
   Once a consumer makes a purchase, then a post-purchase evaluation occurs. If the product performance is in accordance with consumer expectations, the consumer will be satisfied. If not, then the possibility of purchase will decrease.

The framework of this study can be seen in Figure 1:

4 RESULTS AND DISCUSSION

4.1 *Hypothesis test/partial effect (t-Test)*

The testing of partial effect was conducted by comparing the t-value and t-table or level of significance and alpha. In this study, researchers used an alpha level of 5%. Thus, using t-table ($\alpha/2$) and df (n−k), the value of the t-table = 5% (two-way test), and df (400−4) = df (396) = 1.96597. The testing result shows that internal marketing, external marketing, and interactive marketing have positive and significant influences on customer decisions: 3.8% for internal marketing, 43.4% for external marketing, and 54.9% for interactive marketing.

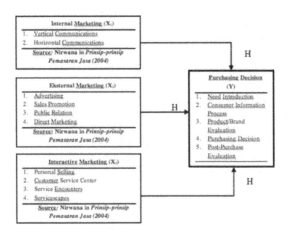

Figure 1. Research framework.

4.2 Simultaneous test (F-Test)

The F-test value was 23,403,289 with Sig. 0.000 < alpha 0.5. The calculation results obtained from data processing were F count (23,403,289) > F table (2.6274), so internal marketing, external marketing, and interactive marketing have a simultaneous and significant influence on the customer decision variable.

4.3 Determination of coefficient test

The independent variables of internal marketing, external marketing, and interactive marketing impact the customer decision variable by a coefficient of determination indicated by an R square of 89.4%. The R square shows that all the independent variables studied increased the customer decision variable by 89.4%.

5 CONCLUSION AND SUGGESTIONS

From the results of the study, the contribution of internal marketing, external marketing, and interactive marketing variables partially affected the customer decision variable at Mandiri Bank in Bandung. The effect of the internal marketing variable was 0.109 points, the effect of the external marketing variable was 0.529 points, and the effect of the interactive marketing variable was 0.685 points. The results show that the contribution of internal marketing, external marketing, and interactive marketing simultaneously influenced the customer decision variable by 89.4%.

Based on these conclusions, Mandiri Bank should pay attention and improve its internal marketing, external marketing, and interactive marketing because these three variables influence people to become customers of Mandiri Bank in Bandung. Scholars should continue this study by adding other variables related to other marketing activities.

REFERENCES

Anoraga, P. 2000. *Manajemen Bisnis*, Jakarta: PT Rineka Cipta, p. 226.
Ari Luhur Sasangka. 2010. *Analisis Faktor-Faktor yang Mempengaruhi Keptusan dalam Pembelian Minuman Energi (Studi Kasus pada Extra Joss di PT. Bintang Toedjoe Cabang Semarang)*.
Arikunto. 2010. *Prosedur Penelitian: Suatu Pendekatan Praktek*. Jakarta: Rineka Cipta.
Husein, U. 2009. *Metode Penelitian Untuk Skripsi dan Tesis Bisnis*. Jakarta: Rajawali Persada.

Kotler, P., & Keller, K. L. 2009. *Manajemen Pemasaran Edisi Ketiga Belas Jilid I*, Jakarta: Erlangga, p. 166.

Malhotra, N. K., & Birks, D. F. 2016. *Marketing Research: An Applied Approach*. Updated Second European Edition. Harlow, UK: Pearson Education Limited.

Sugiyono. 2017. *Statistika Untuk Penelitian, Cetakan ke-28*. Bandung: CV. ALFABETA (Penerbit Afabeta).

Wibisono, D. 2013. *Panduan Penyusunan Skripsi, Tesis & Disertasi*. Yogyakarta: CV. ANDI OFFSET (Penerbit ANDI).

Zikmund, W. G., Babin, B. J., Carr, J. C., & Griffin, M. (2013). Business research methods (9th ed.). Mason, OH: South-Western.

Zikmund, W. G., Babin, B. J., Carr, J. C., & Griffin, M. (2010). *Business research methods* (8th ed.). Mason, HO: Cengage Learning.

Nirwana (2004), Prinsip-prinsip Pemasaran Jasa, Penerbit Dioma, Malang.

Analysis of how to increase purchases through the Telkomsel Virtual Assistant

Ismawaty & Gadang Ramantoko
Telkom University, Bandung, Indonesia

ABSTRACT: Telkomsel as a digital telecommunication company has released a digital-based virtual assistant (VA) application, which is used to facilitate customer transactions and to serve as an information channel. The numbers of purchasing transactions through VA are very low compared to the total active VA users, so it is necessary to analyze how to increase purchasing via VA. This research seeks to provide insight into customer characteristics by using prediction models based on historical data employing behavioral segmentation (occasion, benefits sought, user status, usage rate, and loyalty status). Data were analyzed using a random forest algorithm that utilizes per-segment data and all-segment data. Theoretical approaches used included digital marketing, behavioral segmentation, and customer behavior. Random forest analysis with a total of 22 input variables and 130,388 records resulted in an all-segment data model that produced the highest accuracy, 96%, with 49,948 predicted customers and the predictor importance variables of loyalty point, recharge, data payload, short message service (SMS), and voice transactions.

1 INTRODUCTION

PT. Telekomunikasi Selular (Telkomsel) is the largest cellular telecommunication service operator in Indonesia, which has four main products: Kartu Halo (postpaid), simPATI, Kartu As, and LOOP (prepaid). In order to achieve its vision of becoming a world-class trusted provider in mobile digital lifestyle services, Telkomsel launched its virtual assistant (VA) application. The virtual assistant is a digital customer service solution that combines artificial intelligence, customer analytics, and human interaction to provide a better, faster, and more precise self-service customer experience. The virtual assistant can be accessed via LINE @Telkomsel, Facebook Messenger "Telkomsel," and Telegram @Telkomsel_official_bot (www.telkomsel.com/). Based on Asosiasi Penyelenggara Jasa Internet Indonesia (APJII) survey data in 2017, internet users grew by 7.9% from 2016 to 2017 and have grown more than 600% in the past 10 years. The growth of internet users in Indonesia amounts to 54.69% of the total population. The most widely used internet services are chat (89.35%), social media (87.13%), and search engines (74.84%) (https://databoks.katadata.co.id/datapublish/2018/02/20/).

Along with the shifting customer behavior, revenues from legacy short message service (SMS) and voice have decreased. In 2017, broadband services increased 5% monthly on average. Beginning in November 2017, broadband services brought higher revenue with an 8% gap with legacy services. These phenomena made Telkomsel continue to innovate in creating digital products, such as the virtual assistant (VA). The virtual assistant is a digital chat-based service with a chatbot model. Based on the results of these researches, it can be summarized that the chat service with the chatbot model has excellent capabilities in terms of customer care and virtual assistance.

Meanwhile, the productivity of Telkomsel's VA is not optimal yet. From September 2017 through May 2018, VA interactions totaled 310,605, with 170,479 unique active users and 10,196 purchase transactions. Based on these data, purchase transactions contributed only 3% of the total interactions. In addition, the unique users who had made purchases totaled 7,415 customers, only 4% of the total VA active users. There is a big gap between interaction data and purchasing data; it is important to analyze the behavior of Telkomsel customers, especially those who have made interaction and purchase transactions through VA. One strategy to increase sales through VA is to create a customer segmentation so that the target of a 50% increase in purchases through VA can be achieved. Customer segmentation is the first step in the customer analysis process and can help companies make strategic plans (Lu & Furukawa, 2012). This research aims to predict customers who will make purchases through VA and to find important predictor variables that influence purchasing decisions by using the random forest model in the behavioral segmentation method.

2 LITERATURE REVIEW

2.1 *Literature review*

Digital marketing is the application of digital technology, ranging from identification and anticipation, to providing value to customers, to marketing activities aimed at increasing consumer knowledge (Chaffey & Smith, 2013). According to Chaffey and Smith (2013), in digital marketing, strategy development and implementation strategy must be interrelated in order to compete because the world of technology changes so fast and dynamically. The following planning framework can be applied in digital marketing strategy development, such as segmentation, targeting and positioning, online value proposition, sequence, integration and database, and tools (web functionality, email, IPTV, etc.). Segmentation is one of the important digital marketing strategies to analyze market models, and it consists of demographic, psychographic, geographic, and behavioral aspects. In this research, the segmentation that will be analyzed is based on customer behavior. Segmentation based on customer behavior is the right initial analysis process to form a market segment model (Armstrong & Kotler, 2015). Behavioral segmentation is a grouping of consumers based on knowledge, attitudes, uses, and responses to the products offered, and it consists of the occasion, benefit sought, user status, usage rate, and loyalty status. The occasion, benefit sought, user status, usage rate, and loyalty status have significant relationships to predicting the customer who will make purchases through VA.

2.2 *Research framework*

The research framework for this research is:

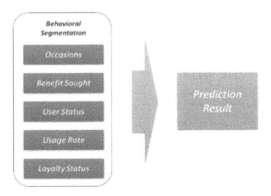

Figure 1. Research framework.

3 RESEARCH METHODOLOGY

In this research, the population studied included Telkomsel active prepaid customers (sim-PATI, Kartu As and LOOP) who interacted and purchased through VA during the period from January to May 2018. Random sampling was conducted on a total of 146,549 customers divided into training (70%) and testing (30%) data sets. Validity and reliability are used in a confusion matrix with two parameters: sensitivity (TP/(TP+FN)) and accuracy ((TP+TN)/(TP+FP+FN+TN). Operational variables are occasion (total SMS transactions on working days and weekends, total voice transactions on working days and weekends, total transaction data payload on working days and weekends), benefit sought (total SMS transactions on-net and off-net, total duration of voice transactions on-net and off-net, total data payload, total VA points redeemed), user status (total VA interactions, total recharge transactions, tier point), usage rate (SMS revenue, voice revenue, data payload revenue, total recharge), and loyalty status (length of stay, loyalty points).

4 RESEARCH RESULTS AND DISCUSSION

4.1 *Research characteristics and cleaning data*

This research used Telkomsel's big data, consisting of historical VA for the period from January to May 2018. In the initial stages, the authors conducted a data audit to see the missing values and outlier data for the population of 146,549 customers. The data audit process was carried out using SPSS Modeler version 18 to see the characteristics and statistics of the 22 variables under examination. Based on the data-cleansing process, data that were processed included 130,388 records with a composition of 3,407 purchase records (3%) and 126,981 non-purchase records (97%).

4.2 *Model and prediction of per-segment and all-segment data behavioral segmentation*

The per-data segment model was grouped based on occasion, benefit sought, user status, usage rate, and loyalty status. The per-data segment grouping was intended to focus the research on each of the characteristics of the behavioral segmentation in order to study which data segment is the most optimal in terms of accuracy and prediction. Based on the processing result, the accuracy in training and testing data was occasion (77%, 71%), benefit sought (78%, 75%), user status (89%, 88%), usage rate (90%, 87%), and loyalty status (93%, 92%). Sensitivity in training and testing data was occasion (83%, 74%), benefit sought (83%, 75%), user status (85%, 85%), usage rate (89%, 85%), and loyalty status (89%, 90%). Predictor importance variables were occasion (SMS transactions on weekdays and weekends, voice transactions on weekdays and weekends, data transactions on weekends and weekdays), benefit sought (SMS transactions on-net and off-net, voice transaction duration on-net and off-net, data payload), user status (tier point), usage rate (SMS revenue, voice revenue, recharge revenue, and data revenue) and loyalty points (loyalty points). The all-segment data model was processed as a comparison to the previous per-segment data model. In this model, the data segments of occasion, benefit sought, user status, usage rate, and loyalty status were processed as a whole group in order to produce the most optimal segments in terms of accuracy and prediction. Based on the processing result, accuracy in training and testing data was 97% and 96%. Sensitivity in training and testing data were 96% and 94%. Predictor importance variables were loyalty points, recharge, data revenue, SMS transactions on weekdays, and voice revenue.

4.3 *Testing model and prediction of per-segment and all-segment data behavioral segmentation*

After the model was formed, the testing prediction model was implemented for all data sets after cleansing the 126,981 non-purchase VA records using the SPSS modeler stream version

18. The predicted results were: occasion 65,710 customers; benefit sought 60,716 customers; user status 40,000 customers; usage rate 18,565 customers; and loyalty status 49,323 customers. The testing prediction model for all segment data using data sets after data cleansing resulted in 49,948 customers.

4.4 *Discussion*

Based on these processing data, the random forest algorithm can be used to make predictions with the behavioral segmentation method. According to previous research, the random forest method produces excellent performance in predicting customer buying behavior and can be used for modeling because of its ease of implementation and fast calculation time (Liao & Du, 2017; Peker, Kocyigit, & Eren, 2017).

Table 1. Behavioral segmentation.

Behavioral Segmentation	Accuracy Training	Testing	Sensitivity Training	Testing	Prediction
Occasion	77%	71%	83%	74%	65,710
Benefit Sought	78%	75%	83%	75%	60,716
User Status	89%	88%	85%	85%	40,000
Usage Rate	90%	87%	89%	85%	18,565
Loyalty Status	93%	92%	89%	90%	49,323
All Segment Data	97%	96%	96%	94%	49,948

The results obtained the highest accuracy produced in the all-segment data model, with an accuracy rate of 97%, and the dominant predictor importance variable was loyalty points. This shows that loyalty status has a very big influence on the prediction model. The results of this study are consistent with previous research that states that customer loyalty and satisfaction are the most influential indicators of the positive effect on purchasing decisions of products and services (Dah, Chen, & Prempreh, 2015; Susanto, 2013.

5 CONCLUSIONS

a. The all-segment behavioral data segmentation produced the best results, with an accuracy level of 97%, and customer prediction was 49,948 (37%), which was closest to the target.
b. The top five variables of predictor importance generated based on random forest algorithm results in the all-segment data model were loyalty points, recharge data, data revenue, SMS transactions on weekdays, and voice revenue.

REFERENCES

Armstrong, G., & Kotler, P. 2015. *Marketing: An Introduction*. 12th Edition. Harlow, UK: Pearson Education.
Chaffey, D., & Smith, P. 2013. *Emarketing Excellence*. London: Routledge.
Dah, H. M., Chen, W., & Prempreh, V. M. 2015. Assessing the impact of loyalty program on consumer purchasing behavior in fine-dining restaurant. *European Journal of Business and Management*, 7(30).
Heo, M., & Lee, K. J. 2017. Chatbot as a new business communication tool: The case of Naver TalkTalk. *Business Communication Research and Practice*, 1(1), 41–45.
Liao, Y., & Du, W. E. 2017. *Prepaid or Postpaid? That Is the Question*. Research Gate.

Lu, K., & Furukawa, T. 2017. *Similarity of Transactions for Customer Segmentation.* Fukuoka: Kyushu University.

Peker, S., Kocyigit, A., & Eren, P. E. 2017. A hybrid approach for predicting customers' individual purchase behavior. *Kybernetes, 46*(10), 1614–1631.

Susanto, A. H. (2013). The influence of customer purchase decision on customer satisfaction and its impact on customer loyalty. *The EMBO Journal*, 1(4) (December), 1659–1666. https://databoks.katadata.co.id/datapublish/2018/02 accessed May 12, 2018. www.telkomsel.com accessed May 12, 2018.

Optimizing Instagram as a digital marketing channel: A qualitative study at Larnis

Y. Shigeno & S.K. Widhaningrat
University of Indonesia, Jakarta, Indonesia

ABSTRACT: Business opportunities have grown widely available over the past few years in the Muslim clothing industry, especially hijab products. With the high demand for hijab, an effort is needed to support small and medium-sized enterprises (SMEs) in taking advantage of these opportunities. Larnis is an SME that sells hijab products made of knitting material, which is less popular in Indonesia. This study aims to optimize the use of Instagram as a digital marketing channel for Larnis. The data were obtained through qualitative methods that included in-depth interviews, observations, surveys, and focus group discussions. Based on the data and analysis, there has been a lack of promotional activities and a less than optimal use of online marketing channels owned by Larnis. The result of this research is a set of steps to take in order to optimize the use of Larnis's Instagram account for business purposes, which leads to increased exposure for Larnis and its products.

1 INTRODUCTION

1.1 Background

One business that is developing and becoming a trend in recent years is the Muslim clothing business, on large, medium, and small scales. According to a bulletin from the director-general of national export development (2015), the development of the Muslim fashion world has experienced a significant increase in the past few decades. Indonesia is a country where the majority of the population is Muslim; the increasing number of Indonesian women who wear the hijab certainly presents a lucrative business opportunity.

Larnis is one of the small and medium-sized enterprises (SMEs) that sells hijab products. Larnis was founded by Mrs. Firda Hafsoh in July 2014. The business issue Larnis faces at present is its underdeveloped sales and marketing. During this time Larnis carries out no strategy or any special promotional activities to attract customers. Its online marketing channel, Instagram, is not used to its full potential in communicating and marketing its products. This is unfortunate as the Internet is currently increasing its role in daily life.

Mrs. Firda's expectation as Larnis's owner is that her products will be widely recognized by the Indonesian people. Therefore, this research in the form of business coaching was conducted in order to create remedial steps that can help Larnis develop and apply strategies to achieve its goals.

2 LITERATURE REVIEW

2.1 Social media marketing

Kaufman and Horton (2015) state that social media can help companies to build their digital presence, namely by:

a. increasing brand awareness where social media can provide a very broad reach to increase the efficiency of disseminating content and messages compared to traditional marketing channels;
b. building a reputation where social media can help brands in telling their background;
c. driving visitors to a company's website; and
d. improving relations with the audience.

Instagram, one of the social media platforms with significant growth in its overall user base, can be used as a medium for promoting product and connecting a company with its customers because it can reach millions of people at once without incurring any costs. Visual content allows companies to post pictures and videos about products, events, and other information related to a particular company or brand to increase exposure in creative ways (Belch & Belch, 2018).

More specifically, Mia Nummila conducted research related to Instagram. In her research, entitled "Successful Social Media Marketing on Instagram. Case: @minoshoes," Nummila (2015) states that for any company, Instagram can be very useful when used correctly. When successfully adapting Instagram as part of a social media strategy and learning how to use it most efficiently, companies can gain a competitive advantage over other brands. Marketing on Instagram can also quickly increase company sales and brand awareness (Nummila, 2015).

3 METHODOLOGY

3.1 *Research approach*

The business coaching research used qualitative methods by gathering information from sources concerned with research.

3.2 *Object/subject of the research*

The object of this research was digital marketing channels, namely social media platform Instagram, which is used to increase micro and small to medium-sized enterprise (MSME) exposure. The subject of this study was Larnis.

3.3 *Research data collection and sources*

The data collected in this business coaching activity used interviews, observations, focus group discussions, and surveys conducted with related parties. According to the source, the data in the study were divided into two, namely primary data and secondary data. Primary data were obtained from interviews, observations, focus group discussions, and surveys conducted with related parties. Secondary data were obtained from the study of literature, including internal documents, previous research, scientific articles, books, and the Internet, as well as data from related parties. In this study, the secondary data used were Larnis's internal data such as customer and sales data, as well as data from other similar studies that can be used for comparison, such as journals and previous theses, and literature as a reference in solving problems.

3.4 *Data analysis method*

In analyzing these data, three steps were followed according to Malhotra (2010):
a. Data selection
b. Data presentation
c. Conclusion and verification

These three steps were carried out using content analysis and triangulation. The results of interviews and observations were identified and codified using key phrases and sentences under the questions given, words with similar concepts, and experience repetition. In this study, the authors triangulated information and data obtained from the results of interviews, observations, focus group discussions, and surveys to obtain a comprehensive picture and confirmation regarding Larnis's conditions.

4 RESULTS AND DISCUSSION

From the results of the conducted analysis, the problems summarized in the gap analysis were given a weight based on the urgency and ability of MSMEs to resolve the problem, which is called Pareto analysis. Gap analysis is a comparison between ideal conditions (the conditions that SME owners want to achieve) with the real conditions that are currently being experienced. Gap analysis can also identify actions that need to be taken to reduce gaps and achieve expected performance. The results of the Pareto analysis can be seen in Table 1.

This research focused on the first Pareto analysis, which examined digital marketing channels. Through business coaching activities, the writer tried to resolve the problem by optimizing digital marketing channels on Instagram. This step is expected to increase exposure and can also educate the market about the products Larnis offers.

The Instagram platform was chosen for several reasons, one of which is the ease of access for Larnis owners and employees to anytime and anywhere communicate with followers, either by uploading new photos or by replying to messages and comments from followers. The limitation of Larnis's owner in accessing a computer requires that the proposed improvement can overcome the existing problem without placing an additional burden on her. Mrs. Firda, as Larnis's owner, has an Instagram account for Larnis and has no difficulty in accessing the account.

Second, the additional features of an Instagram business account can provide insight into demographic data and follower activity as well as the extent of the uploaded content. This, of course, can assist Larnis in evaluating and comparing which content is the most interesting and liked, as well as being one of the measuring tools in seeing the performance of each uploaded image.

Third, Meggs and Purvis (2006) state that information delivery will be more effective when using visual communication because humans can more quickly capture messages in visual form when compared to text. The optimization process carried out in this study implements steps or stages that must be completed in developing strategies for digital marketing, according to Gilmore et al. (cited in Hollensen, Kotler, & Opresnik 2017):

1. Create social media marketing objectives. Establish the objectives and goals that the company hopes to achieve.
2. Conduct a social media audit (where are we today?). Assess current social media use and how it is working.
3. Choose the most relevant social media platforms to work with. Choose which networks best meet the company's social media objectives.

Table 1. Pareto urgency table.

No	Problems	Value	Ability Weight	Contribution	Distribution	Accumulation
1	Digital Marketing Channel	4	10	40	48%	48%
2	Signage	3	9	27	32%	80%
3	Human Resources	3	4	12	14%	94%
4	Products	1	5	5	6%	100%
	Total			84	100%	100%

Source: Research Data Analysis (2018).

4. Get social media inspiration from industry leaders, competitors, and key opinion leaders in the online community.
5. Create a time plan for the company's social media efforts. The social media marketing plan should also include a content plan.
6. Test, evaluate, and adjust the social media marketing plan. The entire plan should be constantly tested.

The result after the Instagram development was that the number of user interactions with Larnis increased by almost five times, from 23 to 109 interactions in one week, where eight users clicked on the website and one user on the directions to Larnis's location. Improvements were also seen in the discovery section, which is the total number of users who saw uploaded content, which increased from 32 times to 8,032 times. Besides, the impression gained by Larnis also increased, from 80 in the period January 5–11, 2019, to 12,129 impressions in the period May 11–17, 2019. Accounts following Larnis also increased, from 1,966 accounts to 2,134.

5 CONCLUSION

The conclusion from this series of business coaching activities is that Larnis's Instagram platform has been optimized to support Larnis's marketing activities. These achievements – resulting from optimizing online marketing channels through Instagram – included an increase in exposure and in the number of user interactions with Larnis that can be measured by looking at Instagram, as well as an increase in the number of followers who follow Larnis's Instagram account. Therefore, we need to continue to optimize the use of Instagram on an ongoing basis to maintain the engagement of Larnis's followers.

REFERENCES

Belch, G. & Belch, M. 2018. *Advertising and Promotion: An Integrated Marketing Communications Perspective*. New York: McGraw-Hill Education.
Direktorat Jenderal Pengembangan Ekspor Nasional Kementrian Perdagangan. 2015. *Warta Ekspor*. Retrieved on March 30, 2019 from http://djpen.kemendag.go.id/app_frontend/admin/docs/publication/9871447132408.pdf.
Hollensen, S., Kotler, P., & Opresnik, M. 2017. *Social Media Marketing: A Practitioner Guide*. Second Edition. Opresnik Management Guides.
Kaufman, I., & Horton, C. 2015. *Digital Marketing: Integrating Strategy and Tactics with Values*. New York: Routledge.
Kotler, P., & Keller, K. L. 2016. *Marketing Management*. 15th Edition (Global Edition). Edinburgh: Pearson Education Limited.
Malhotra, N. 2010. *Marketing Research: An Applied Orientation*. Sixth Edition. Upper Saddle River, NJ: Pearson Education.
Meggs, P. B., & Purvis, A. W. 2006. *Meggs' History of Graphic Design*. Hoboken, NJ: Wiley.
Nummila, M. 2015. Successful Social Media Marketing on Instagram. Bachelor's Thesis Degree Programme in International Business, Haaga-Helia University of Applied Sciences.

The development of marketing channels and inventory management systems for SMEs

Edbert & S.K. Widhaningrat
Universitas Indonesia, Jakarta, Indonesia

ABSTRACT: The emergence of technology offers several opportunities for business owners to expand their markets. The main purpose of this article is to help the owner of Liliana Collection, a retail children's clothing store, to gain better understanding of how to improve its business operation, specifically in marketing, distribution, and inventory management systems. This research used a business coaching approach. Data were collected from in-depth interviews, observations, and secondary data, and then were analyzed using political, economic, social, and technological (PEST) analysis; marketing mix; segmentation, targeting, and position (STP) analysis; a business model canvas; and strengths, weaknesses, opportunities, and threats (SWOT) analysis. The findings were analyzed using gap and Pareto analyses. Then, an e-commerce channel was adopted for marketing and distribution, and an inventory management application was used to help the owner in recording inventory transactions. The implementations then carried out along with the coaching process educated the owner about the new system in order to make the execution successful.

1 INTRODUCTION

Small to medium-sized enterprises (SMEs) make an important contribution to national economic development through their role in opening up employment opportunities for the community, where at the same time it can reduce social problems and increase the level of the community's consumption. Liliana Collection is a manufacturer and distributor of children's clothing in Jakarta. The children's clothing industry is already packed with competitors; therefore, good operational management, innovation, and strong cash flow and capital are needed. However, Liliana Collection was still managed through a traditional system. In this article, we aim to improve Liliana Collection's business through the development of marketing channels and a new inventory management system.

2 LITERATURE REVIEW

2.1 *Digital marketing*

Ryan and Jones (2009) state that marketing in the digital era is not about technology but rather about how technology can better connect to consumers. Digital marketing is a way for marketers to connect with consumers with varied and innovative content (Ryan & Jones, 2009).

2.2 *Inventory management system*

Wild (2004) mentions that a good inventory management system is a recording system using a computer where transactions can be directly inputted into the computer without the need for manual recording; data updates are quick and simple and computerized automatically; the choice of data is immediately formed in the system; each transaction is integrated in various business activities; and the information is quickly updated, including information about exceptions and errors in the data.

2.3 *Past research*

Theresia Pradiani (2017) has shown that shifting from traditional marketing systems to digital marketing has a positive impact on certain businesses, which in this case is increasing sales volume. Hendra Agusvianto (2017) finds that the application of technology to inventory systems will be beneficial in terms of efficiency and accuracy.

3 RESEARCH METHODOLOGY

This research belongs to the qualitative research method. Data obtained from the business coaching activities were analyzed using qualitative analysis methods. The data were collected through interviews, observation, and documentation. These methods were carried out for seven months with face-to-face meetings with representatives of the SME. The objective of this approach was to understand the participants' perspective, behavior, and motivation. Primary and secondary data mining were also performed for deeper analysis. The data were then analyzed through three stages of qualitative data analysis: familiarizing and organizing, coding and reducing, and interpreting and representing (Ary, Jacobs, Sorensen, & Razavieh, 2010).

4 RESULTS

The collected data were analyzed through market opportunity analysis (STP analysis); marketing mix; product, place, promotion, and price (4 Ps) analysis; business model canvas analysis; strengths, weaknesses, opportunities, and threats (SWOT) analysis; and political, economic, social, and technological (PEST) analysis. The results were identified through codification using gap analysis.

Table 1. Gap analysis of Liliana Collection.

No	Analysis	Ideal Condition	Actual Condition	Gap	Condition
1	BMC	Good inventory management system Strong cash flow Reach customers all over Indonesia Online and offline marketing channels Good accounts receivable management system	Traditional inventory management system Tight cash flow Customers are mainly on Java Offline marketing channels only Traditional account receivable management system	Inventory management system not optimal Tight cash flow Reach customers all over Indonesia Marketing channels not optimal Accounts receivable management system not optimal	Inventory managemen system Cash flow Marketing channels Marketing channels Account receivable
2	PEST	Forex hedging system Online marketing system	No forex hedging system Traditional marketing methods	No forex hedging system Offline and online marketing channels	Forex Eksposure marketing channel

(*Continued*)

Table 1. (Continued)

3	Marketing Mix	Online marketing Modern and integrated inventory management system	Traditional marketing methods Traditional inventory management system	Offline and online marketing channels Inventory management system not optimal	Marketing channels Inventory management system
4	STP	Reach customers all over Indonesia	Customers are mainly on Java	Reach customers all over Indonesia	Marketing channels

These problems were analyzed through Pareto analysis to see which are the most important and which are not too important (from an urgency standpoint).

Table 2. Pareto analysis of Liliana Collection.

No.	Gap	Score	Weight	Contribution (S xW)	Distribution	Accumulation
1	Marketing Channels	9	8	72	40%	40%
2	Inventory Management System	9	8	72	40%	80%
3	Accounts Receivable	5	3	15	8%	88%
4	Cash Flow	5	3	15	8%	96%
5	Forex Exposure	3	2	6	4%	100%
	Total			180	100%	

5 IMPLEMENTATION

5.1 Building an online presence for liliana through e-commerce website

Chapin (2010) mentions that the easiest way to sell goods online is through a marketplace website. Chapin also added several criteria that must be considered: the method of payment, website regulations, and customer service. When considering these criteria, Tokopedia is the most suitable website for Liliana Collection. According to Chapin, several steps must be taken before entering an online marketplace – setting up an online shop, creating product photos, writing product descriptions, establishing customer service, and providing methods of shipment and payment.

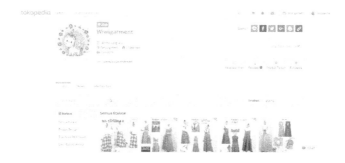

Figure 1. Liliana Collection's new online presence.

5.2 Developing a new inventory management system

The next solution implemented was to improve the current traditional inventory management system for Liliana Collection. Previously, Liliana Collection used a traditional ledger recording system to record its inventory. This method was inefficient because a lot of stock went undetected and miscommunication often occurred between the store and the warehouse.

Figure 2. Manual inventory management system.

Based on theoretical references, a good inventory management system must be integrated, centralized, and updated with every transaction or movement of stock or goods (Waters, 2003). The new inventory management system proposed for this SME was an integrated recording system using the Ipos 5.0 application and the addition of recording the codification of goods as well as the use of travel documents for distribution to stores.

Figure 3. New inventory management system.

6 CONCLUSION

Based on the analysis of business coaching activities for SME Liliana Collection, the problem with Liliana Collection was that the marketing channels have not been maximized and there is no good inventory management system. Technology must be maximized for business owners in order to improve business and survive the current digital disruption.

6.1 *Online marketing channels for liliana collection through e-commerce website*

One of the problems was that Liliana Collection's marketing channel system was still traditional, so this caused limitations in its development. The creation of an online store on the e-commerce site was intended to expand its market and form a foundation so that it can adapt to the technological disruption that is happening in the garment industry. The development of marketing channels in the e-commerce store for Liliana Collection has reached 92.50%.

6.2 *New inventory management system development*

An inventory management system has been created for and implemented in the SME Liliana Collection. The addition of the recording system through coding, input to the application, and the use of travel documents is expected to improve the SME's inventory management system, which was previously still recorded manually in ledgers. Implementation was also done by using the Ipos application system, which is intended to improve the recording system and to integrate the warehouse and the store so that the information obtained is more accurate and in real time. Implementation of the new inventory management system has reached 100%.

REFERENCES

Agusvianto, H. 2017. Sistem Informasi Inventori Gudang Untuk Mengontrol Persediaan Barang Pada Gudang Studi Kasus: PT. Alaisys Sidoarjo. *Journal of Information Engineering and Educational Technology* [online], *1* (01). Available at https://journal.unesa.ac.id/index.php/jieet/article/view/679/-545.

Ary, D., Jacobs, L. C., Sorensen, C., & Razavieh, A. 2010. *Introduction to Research in Education.* Eighth Edition. Belmont, CA: Wadsworth.

Bidgoli, H. 2002. *Electronic Commerce: Principles and Practice.* San Diego, CA: Academic Press.

Chapin, K. 2010. *The Handmade Marketplace: How to Sell Your Crafts Locally, Globally and Online.* North Adams, MA: Storey Publishing.

Muller, M. 2003. *Essentials of Inventory Management.* New York: Amacom.

Pradiani, T. 2017. Pengaruh Sistem Pemasaran Digital Marketing Terhadap Peningkatan Volume Penjualan Hasil Industri Rumahan. *Jurnal Ilmiah Bisnis dan Ekonomi Asia (Jibeka)* [online], *11*(2), 46–53. Available at https://lp2m.asia.ac.id/wp-content/uploads/2018/04/7.-JURNAL-THERESIA-JIBEKA-VOL-11-NO-2-FEB-2017.pdf

Ryan, D., & Jones, C. 2009. *Understanding Digital Marketing: Marketing Strategies for Engaging the Digital Generation.* London: Kogan Page.

Schneider, G. P. 2011. *Electronic Commerce.* Ninth Edition. Boston, MA: Course Technology Cengage Learning.

Waters, D. 2003. *Inventory Control and Management.* First Edition. West Sussex: Wiley.

Wild, T. 2004. Improving Inventory Record Accuracy: Getting Your Stock Information Right. First Edition. Oxford: Elsevier.

Valuation analysis of skill based asset company: Acquisition by PT Elnusa, Tbk

Deny Christian & Edward Tanudjaya
University of Indonesia, Indonesia

ABSTRACT: In 2017, PT Elnusa acquired PT CALE with the aim of increasing the company's capability in the Engineering, Procurement, Construction (EPC) business. The business diversification carried out by PT Elnusa in the EPC field encountered obstacles in terms of Engineering because the company did not have these capabilities despite having mastered the Procurement and Construction side. The acquisition is expected to not only provide added value in terms of capability but also cost savings. The acquisition of Engineering companies is basically a service-based company, so that the company's valuation cannot only use valuation based on the assessment of fixed tangible assets but also the valuation of intangible assets.

1 INTRODUCTION

1.1 Background

Over the past few years, the oil and gas business has shown poor results. The decline has a significant impact on almost all oil and gas exploitation activities in Indonesia. In the national oil and gas condition, there is a change in profit sharing scheme between contractors with the government through Ministerial Regulation (Permen) No. 08 of 2017 concerning Gross Split production sharing contracts from those previously using the Cost Recovery system in which all exploration costs borne by the government are entirely the responsibility of the contractor to force contractors to evaluate the costs of oil and gas production. Service providers in the oil and natural gas sector were also affected, with these companies being forced to wage price wars.

PT Elnusa Tbk is a service provider company in the energy sector, one of which is oil and gas services. Since 2016, PT Elnusa has been developing business in the areas of Engineering, Procurement, Construction (EPC) and Operation and Maintenance (O&M) services.

PT CALE is an Engineering services provider company that was founded in 2008 and provides project management services, process and safety engineering, civil/structure engineering, mechanical engineering, package engineering, piping & pipeline engineering, instrument engineering, and electrical engineering. At the end of 2018, PT CALE's shareholders offered ownership of 99.9% of the outstanding shares to be acquired by PT Elnusa.

This Research will analyze the valuation of company with intangible asset (in this case PT CALE- engineering skill based company), which different to company with tangible asset (Property, Plant, Equipment).

1.2 Research objectives

This research will conduct an analysis of Pricing determined by PT CALE in the acquisition process, mainly because PT CALE is a company engaged in services and its main assets are Intangible assets.

2 LITERATURE REVIEW

2.1 *Company acquisition*

Form of Acquisition

Theoretically it is divided into stock acquisition and asset acquisition. Share acquisition is the purchase of all or part of a company's stock, so that the buying company has management control over the company that is purchased. While the acquisition of assets is the purchase of all or part of assets related to the company's business to be purchased.

2.2 *Assessment of target acquisition companies*

The approach for evaluating target companies in general can be divided into four namely (Partner, 1991), namely:

a. Liquidation value approach
   This valuation assumes that the company is liquidated, then assets and liabilities are valued at fair prices
b. A transaction value approach that can be compared
   This approach is based on the average price of transactions of several similar companies on the market.
c. Market value approach
   This approach is simply stated as the current price per share of the company multiplied by the number of outstanding shares
d. Discounted cash flow approach
   DCF analysis conducted with data and discount rates that are appropriate and supportive is one of the acceptable valuation methods in the income approach.

Calculation of FCFF over Enterprise Value can use the following formula (Damodaran, 2006):

Value of the Firm

$$\frac{FCFF}{WACC - g} \quad (1)$$

FCFF = Free Cash Flow to the Firm
WACC = Weighted Average Cost of Capital
g = Growth Rate

2.3 *Valuation of intangible asset*

1. *Relief from Royalty Method (RRM)*
   RRM calculates the value based on the payment of a hypothetical royalty that will be saved by owning assets instead of licensing them. The reason behind RRM is quite intuitive: Having intangible assets means the underlying entity does not have to pay for the privilege of using those assets.
2. *Multiperiod Excess Earnings Method (MPEEM)*
   MPEEM is a variation of discounted cash flow analysis. Instead of focusing on all entities, MPEEM isolates cash flows that can be associated with an intangible asset and measures fair value by discounting it to present value
3. *With and Without Method (WWM)*
   WWM estimates the value of intangible assets by calculating the difference between two discounted cash flow models: one that represents the status quo for a business enterprise with existing assets, and the other without it.

4. *Real Option Pricing*
   As Aswath Damoradan noted, “the most intangible assets that are difficult to assess are assets that have the potential to create cash flows in the future but not now.” These assets have option characteristics that make them suitable for valuation using the option pricing model and include patents undeveloped and undeveloped natural resource options
5. *Replacement Cost Method Less Obsolescence*
   This method requires valuation of replacement costs for new intangible assets, i.e. “costs to build, at current prices at the date of analysis, intangible assets with utility equivalent to intangible subjects, using modern materials, standard production, design, layout, and quality of workmanship.

3 COMPANY OVERVIEW

3.1 *General overview of the engineering, procurement, construction industry*

EPC is an abbreviation of Engineering, Procurement and Construction, a term for work that starts with the process of design/system design to be built, procurement/purchase of goods and continued with building/construction of what has been designed.

4 QUANTITATIVE ANALYSIS, DETERMINATION OF VALUE AND PRICE OF TARGET COMPANIES

4.1 *Analysis of PT CALE's financial ratios*

The data used for analysis are the financial statements of independent auditors from 2010 to 2017. The following is a balance sheet table of PT CALE:

	2010	2011	2012	2013	2014	2015	2016	2017
Pendapatan Usaha	24.313	31.515	52.444	79.147	105.840	112.442	73.660	76.190
	-	-	-	-	-	-	-	-
Biaya Teknik & Engineering					59.517	44.034	28.119	48.844
Transportasi & Perjalanan Dinas	Tidak Terdefinisikan di Laporan Keuangan				3.455	3.813	4.822	3.639
Biaya Peralatan					348	450	591	542
Biaya Pokok Lain					212	1.103	401	2.695
Beban Pokok Pendapatan	7.048	16.873	30.926	47.499	63.218	49.400	33.933	55.720
	-	-	-	-	-	-	-	-
Beban Gaji Karyawan	6.066	6.093	9.570	8.274	12.889	42.525	26.090	7.431
Beban Kantor	717	1.628	3.678	8.775	9.269	6.374	5.460	2.651
Transportasi					503	435	1.041	197
Perawatan & Perbaikan					379	1.180	626	139
Perlengkapan & ATK					1.576	911	431	574
Legal & Perijinan	Tidak Terdefinisikan di Laporan Keuangan				90	273	258	144
Beban Rumah Tangga					474	456	313	239
Asuransi					4.679	1.362	1.704	530
Biaya Sewa Gedung					1.060	569	723	703
Biaya Lain Lain					507	1.189	364	125
Beban umum dan Administrasi	6.783	7.722	13.248	17.049	22.159	48.899	31.550	10.082
Depresiasi	502	241	606	874	1.318	1.634	1.200	1.017
Laba Bruto	9.980	6.679	7.664	13.725	18.638	12.489	6.977	9.370
Beban Lain lain	- 23	- 249	- 304	- 226	-	-	-	-
total Pendapatan/ Beban Lainlain	- 1	- 173	241	549	839	2.183	5.172	5.750
Beban Bunga	-	-	-	-	- 87	- 1.114	- 1.839	- 3.147
Laba Bersih	9.981	6.852	7.423	13.175	17.799	10.306	1.804	3.621

Figure 1. Profit and Loss Report 2010-2017.

1. Projected income (S)
   The average growth over the past 5 years is an average of 12%, taking into account the risk factors for oil and gas price fluctuations in the 2012-2014 period to decrease by more than 50%, then the assumption of revenue growth is set at 6%
2. Operating Cost (OE) and Overhead (OH) Costs
   In making a simple projection of operating costs, the average percentage of income is used during the period of 2013-2017, which is 55% (OE = 55% * S) and overhead costs by an

average percentage of net lease income of 59% (OH = 59 % * (S-OE)) during the same time period in which net lease income is income minus operating costs.
3. Income Tax
The tax rate that is the burden of the company until 2018 uses a 25% tariff base.
4. Depreciation Costs
Through calculations, the average percentage depreciation of the value of fixed assets is 8.3%. Furthermore, the value of fixed assets is assumed to grow every year with a growth of 17%
5. Residual Value
The EPC industry, especially in the oil and gas sector, if withdrawn since 2010 has experienced quite fluctuating dynamics, the industry has decreased by -12% growth, so that the figure is used as a reference for growth during a conservative review but management is also preparing another scenario which is -1% based on historical data during the years 2010-2017 and also an optimistic scenario which obtained the growth rate of the last 2 years where the growth of the oil and gas industry experienced growth of 6% which is expected to occur constantly
6. Capital Cost Estimates

$$WACC = \frac{E}{V} * Re + \frac{D}{V} * Rd * (1 - Tc) \qquad (2)$$

5 CONCLUSION

Through calculations with the liquidation method, we get a book value from PT CALE of Rp. 71.56 billion. For the intrinsic value of PT CALE, using the assumption of income growth of 12% per year, if using three scenarios, namely pessimism, normal and aggressive using the assumption of residual growth of -12% for the pessimistic scenario, -1% for the normal scenario, and 6% for an aggressive scheme, then we get the company's intrinsic value of Rp. 62.7 billion for the pessimistic scenario, Rp. 75.3 billion for the normal scenario, and Rp. 110 billion for the aggressive scenario.

REFERENCES

Brigham, E.F., & Louis C., G. (1994). Financial Management,Seventh Edition. Philadelhia: The Dryden Press.
Clark, J.J. (1985). Business Merger and Acquisition Strategies. New Jersey: Prentice Hall, Inc.
Cooke, T.E. (1986). Mergers and Aqcuisition Strategies. Oxford: Basil Blackwell.

Synergy effect analysis of PT XL Axiata Tbk after acquisition, 2011–2013

R.A. Gitasari & R. Rokhim
University of Indonesia, Jakarta, Indonesia

ABSTRACT: Telecommunications companies have reached a critical point. A high level of investment, competition from technology companies, and dynamic customer behavior mean companies must strategize in order to continue to grow. PT XL Axiata Tbk (XL) conducted a merger and acquisition with PT Axis Telekom Indonesia (AXIS) in 2014. This research connects it with two companies engaged in the analysis field after the acquisition. Evaluations carried out in this study were published in the XL period before acquisition (2011–2013) up to five years after acquisition (2014–2018). Based on the results of the synergy effects, the value of the acquisition transaction was higher than the fair market price, and the results of revenue synergies and the synergy of cost for several aspects showed positive values, while financial synergy found a negative value.

1 INTRODUCTION

World telecommunications company executives say that in recent years the telecommunications industry has reached a tipping point. Based on the analysis of Davenport, Loucks, and Schatsky (2017), telecommunications companies continue to invest in updating networks and developing technology. To monetize networks, telecommunications companies provide a variety of data services to meet consumer needs, but have not yet succeeded in excelling in these services. The tight competition is not only from telecommunications companies but also from over-the-top (OTT) technology, which offers applications and streams content directly to consumers via the Internet and takes its main services from telecommunications, namely voice and messaging applications (El-Darwiche, Peladeau, Rupp, & Groene, 2017). The current generation on average spends 315 minutes on OTT applications every day. WhatsApp, Viber, and Apple's iMessage represent more than 80% of short messages, and Skype controls one-third of international telephone traffic (in minutes). The result is that many operators have experienced a significant decline in their main business services. Short message sales have decreased by 30%, international telephone services by 20%, and roaming services by 20% (El-Darwiche et al., 2017).

In Indonesia, PT XL Axiata Tbk signed an acquisition agreement with PT Axis Telecom Indonesia (XL Axiata, 2014). The acquisition agreement was stated in a conditional sales purchase agreement (CSPA) with the Saudi Telecom Company (STC) and Teleglobal Investment B.V. STC, which is the largest shareholder of Axis – 95% of its shares were sold to XL Axiata Tbk (KPPU, 2014). Based on the agreement, it was stated that Axis Corporation had a value of US $865 million. Merger and acquisition (M&A) transactions continue to increase, with the highest record in world M&A transactions reached in 2017. A total of 3,389 transactions took place with a total nominal value of US $498.2 trillion (Krikhaar, Loucks, & Squazzin, 2018). Many studies have cited failures in M&A. Some have said that M&A provides a negative synergy or does not have an impact on a company. Therefore, the objective of this research was to find out the revenue synergy, financial costs, and the opportunity for a company to grow after the occurrence of M&A.

2 LITERATURE REVIEW

A merger is defined if all assets and liabilities of one company are transferred to a recipient company with consideration of payment in the form of equity shares from the transfer recipient company, debt securities, cash, or a mix of these (Kumar, 2009). Another study says that mergers involve a combination of some or all the assets of organizations, people, processes, and technologies that produce "wholeness" (Mehta & Hirschheim, 2007).

This synergy – namely, revenue and cost synergy – can be achieved by increasing revenue or reducing costs. According to Gaughan (2015), increasing income is more difficult to achieve. A survey conducted by McKinsey estimates that 70% of mergers fail to achieve revenue synergy (Christofferson, McNish, & Sims, 2004). The results from a company's M&A can create more revenue than when the company was independent. An increase in income can be obtained through various sources such as strength in prices, increase in marketing, new strategies, a combination of functional strength, and market power (Ross, Westerfield, Jaffe, & Jordan, 2016).

The combination of two companies in the same business industry can lead to high selling prices or high purchase prices. A previous study found that the benefits of horizontal mergers can be attributed to increased efficiency rather than an increase in market power (Fee & Thomas, 2007). The research examines whether the stock market reacts to competitors, consumers, and suppliers. The increase in income can also come from the integration of functional forces. If one company has strong research and development (R&D) and the other company has good marketing and distribution, then the merger process can increase the capacity or potential of the company. A 2016 survey conducted by Deloitte (n = 528) found that 50% of the expected synergy was an increase in income. But few of the acquirers managed to reach more than 80% of the target (Bamford, Chickermane, & Nandy, 2017).

Through M&A, companies can increase operational efficiency. According to an analysis by Ryu, Yang, and Kim (2018), 83% of merger agreements do not increase shareholder returns, but the level of performance after a telecommunications company M&A has decreased significantly, to 10%, from a long-term perspective.

Generally, M&A will strengthen a business by making operations more synergistic. Studies in India (Gupta & Banerjee, 2017) find no improvement in the financial performance of acquisition companies; indicators of profitability and post-merger liquidation are deteriorating. Other research says that post-merger size and company growth have a positive impact on company profitability while debt capital reduces corporate profitability (Long, 2015). A 2006 survey conducted by the Economist Intelligence Unit said that only 45% of companies had achieved synergies from M&A activities, which confirmed that the expected synergy had been achieved.

3 METHODOLOGY

This research was a historical case and was a type of descriptive research used to evaluate the impact of activities on the financial aspects of a company; this research used quantitative methods with the time dimension used in the period 2011–2018. The object of this research was to examine PT XL Axiata Tbk as an acquirer that has conducted M&A for PT Axis Telekom Indonesia in the telecommunications industry. Therefore, this article explains the steps of research using a fishbone diagram. The research diagram gives an idea of how the research was carried out.

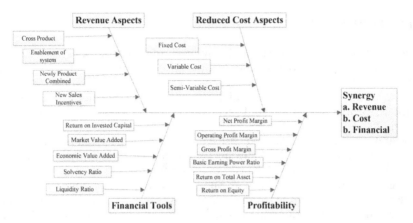

Figure 1. Fishbone diagram.

To achieve revenue synergy, cost synergy and financial synergy are discussed through four aspects: revenue, cost reduction, financial analysis, and profitability. This study analyzes the increase in company income after the acquisition. Components in increasing company income are included in the analysis. This study evaluates the two variables plus the semi-variable costs in the period before and after the acquisition.

Furthermore, an analysis is conducted based on the company's profitability. The profitability variables used are net profit margin, operating profit margin, gross profit margin, basic earning power margin, return on assets, and return on equity. Using these variables, the company profitability is compared from 2011 to 2018. The added value of a company from M&A is calculated based on return on capital (RoIC), market value adjustment (MVA), economic value added (EVA), solvency ratios, and liquidity ratios. These variables measure how efficiently the company utilized the capital used for investment.

The four aspects that are examined will result in whether XL has successfully acquired with a synergy of revenue, costs, and financial analysis. The results are used as material for analysis and interpretation. At the end of the research, conclusions and recommendations related to the synergy that has been made before are given.

4 RESULTS AND DISCUSSIONS

In this study, there is a difference in the data on the valuation of AXIS's fixed assets from the value in the 2014 financial statements, which results in goodwill with the value of fixed assets in the initial valuation of the company conducted by Y&R. Market valuation of fixed assets carried out by Y&R was Rp5,640 trillion, while in 2014 financial statements the fixed assets owned were worth Rp3,705 trillion. The initial asset valuation was Rp1,935 trillion higher than the valuation of AXIS assets after the acquisition integration process was completed.

According to Ross and colleagues (2016), additional revenue can use new strategies, product combinations, and competitive prices. The joining of AXIS into a new business unit in XL adds to XL's revenue volume through a new strategy (dual brand); the combination of bundling products and competitive prices is a factor of increasing revenue. So the company managed to increase revenue in the five-year period after the acquisition. Even though the company failed to achieve profit projection five years after acquisition, in this study revenue synergy can be achieved by XL in the five-year period after the acquisition. XL's total revenue after the acquisition was higher than the total revenue from the combined XL and AXIS after the acquisition. Only a few acquirers achieved 80% of the revenue target from the acquisition (Bamford et al., 2017).

The total operating expenses of XL and AXIS prior to the acquisition were quite high. XL achieved efficiency in operating expenses. This includes infrastructure costs, employee salaries and benefits, sales, and marketing expenses. The total fixed cost of XL after the acquisition increased, and there was no expected cost synergy. The increasing infrastructure burden is a factor in the fixed cost. In addition, XL had to spend quite a lot in 2017 on program transformation. Interconnection services have decreased because variable costs can be reduced. Semi-variable expense – namely sales and marketing – increased after the acquisition. XL achieved efficiency in operating expenses by reducing operational expenses in infrastructure, which is the biggest burden in operating expenses.

NPM, OPM, BEP, ROA, and ROE after the acquisition do not show financial synergy from after the acquisition. Only GPM has increased. The high level of corporate investment and net income that are affected by depreciation also affect the company's profitability ratio. The high depreciation and infrastructure costs borne by XL after the acquisition became an invoice to reduce the company's profitability.

5 CONCLUSIONS

The company's revenue increased 10.8% after the acquisition. Although the revenue increase is not significant enough, the company managed to increase its sales volume. Cost synergy shows positive results. After the acquisition of XL, cost efficiency came up to 8.8% from the combined cost of the previous XL and Axis. Financial synergy using financial performance ratios – namely RoIC, MVA, EVA, solvency ratio, and liquidity ratio – failed to show positive results after the acquisition. XL's financial performance before the acquisition was better than its financial performance after the acquisition using this ratio. In terms of profitability, the average shows negative from NPM, OPM, BEP, ROA, and ROE. The increase occurred at GPM XL after the acquisition. The high depreciation expense after the acquisition that affected the company's net profit caused XL's profitability ratio after the acquisition to show a negative average.

In the future, the acquiring company needs to consider operational expenses and network investment levels after the acquisition. The speed of technology that influences customer behavior makes the market dynamic. This can be a special focus for shareholders and managers when acquiring. Further research can develop additional qualitative analyses, including human resources. The acquisition not only affects the company's performance but also affects all the stockholders in the company.

REFERENCES

Bamford, I., Chickermane, N., & Nandy, R. 2017. Revenue synergies in acquisitions: In search of the Holy Grail. Deloitte Development, LLC.

Christofferson, S. A., McNish, R. S., & Sims, D. L. 2004. Where mergers go wrong. *McKensey Quarterly*, 2, 92–99.

Davenport, T. H., Loucks, J., & Schatsky, D. 2017. Cognitive technologies survey. Deloitte Development, LLC.

El-Darwiche, B., Peladeau, P., Rupp, C., & Groene, D. F. 2017. Aspiring to digital simplicity and clarity in strategic identity. *2017 Telecommunications Trends*.

Fee, C. E., & Thomas, S. 2007. Source of gains in horizontal acquisitions: Evidence from consumer, supplier and rival firms. *Journal of Financial Economics* (74), 423–460.

Gaughan, P. A. 2015. *Mergers, Acquisitions and Corporate Restructurings*. Sixth Edition. Hoboken, NJ: Wiley.

Gupta, B., & Banerjee, D. P. 2017. Impact of merger and acquisitions on financial performance: Evidence from selected companies in India. *International Journal of Commerce and Management Research*, 1, 14–19.

Krikhaar, J., Loucks, J., & Squazzin, M. 2018. Charting a well-defined integration strategy. In *Mergers and Acquisitions in Tech, Media and Telecom*. Deloitte Development, LLC p. 1.

Kumar, R. 2009. Post-merger corporate performance: An Indian perspective. *Management Research News*, *32*(2), 145–157.

Long, P. H. 2015. Merger and acquisitions in the Czech banking sector: Impact of bank mergers on the efficiency of banks. *Journal of Advanced Management Science*, *3*(2), 1.

Mehta, M., & Hirschheim, R. 2007. Strategic alignment in mergers and acquisitions: Theorizing is integration decision making. *Journal of the Association for Information Systems*, *8*(3), 143–174.

Ross, S. A., Westerfield, R. W., Jaffe, J., & Jordan, B. D. 2016. *Corporate Finance*. 11th Edition. New York: McGraw-Hill Education.

Ryu, M. H., Yang, S., & Kim, S. 2018. Do telecom carrier takeovers create value? Longitudinal analysis of U.S. telecom carrier takeovers from 1996 to 2005. *Telecommunication Policy*, 406.

XL Axiata. 2014. *Annual Report 2014*. Jakarta: XL Axiata Tbk.

The main determinants in the distress / failure probability of the public-private partnership financing scheme in infrastructure development

Mira Budi Octaviani & Zaafri A Husodo
Magister of Management, Faculty of Economics and Business, University of Indonesia, Jakarta, Indonesia

ABSTRACT: This study analyzes the Main Determinants in the Distress / Failure Probability of the Public-Private Partnership Financing Scheme in Infrastructure Development. Based on the processed data, namely the Private Participation in Infrastructure Project Database (PPI) World Bank Project Database from 1990 to 2017 that there were 364 projects out of 7050 projects that were canceled or depressed from 128 countries. In terms of data processing, this empirical study uses probit regression with instrumental variable on types of projects that are canceled / distressed. Furthermore, this study also determines instrumental variables where project results are influenced by independent variables which also have variations between independent variables. Another advantage of identifying instrument variables is to minimize the effect of bias on independent variables on the dependent variable, namely PPP failure. The purpose of this research is to determine the main determinant in the performance of the PPP Project. From the results of the study it was found that independence in Government and Government quality significantly affected the failure of PPP projects and the level of government corruption had a significant effect in increasing the potential failure of PPP projects. In addition, macroeconomic stability also has an effect by showing that GDP growth has a significant effect in reducing the potential failure of PPP projects. The involvement of debt providers has an effect on reducing PPP project failures but is not significant due to other factors that have an effect before a debt provider chooses the project or not, such as project bankability.

1 INTRODUCTION

Infrastructure is an absolute necessity in a country to develop, especially in the economic sector. With the support of a proper infrastructure, it can increase the progress of community economic activities. Especially in countries that have territories in the form of islands, so that they need infrastructure, in this case roads / bridges, as a means of equitable distribution of economic conditions. Based on the importance of infrastructure, efforts to develop infrastructure are a priority in several countries. Infrastructure is public goods which, based on Indonesian law, is goods managed by the government. There are budget limitations owned by the Government that require alternative financing schemes that are non-APBN/APBD or that do not use government funds. This is because fiscal alternatives are not the best choice for funding infrastructure development because it will potentially increase the amount of foreign debt of the State. As for the monetary alternative, it is also not a solution for funding infrastructure development because it will affect the stability of the domestic inflation rate. Therefore an efficient funding scheme is needed to support the implementation of infrastructure projects. One of the infrastructure financing schemes that can be an alternative to limited government budget is the Private-Public Partnership (PPP), the private sector can participate in the provision of infrastructure starting from the aspects of funding, design, construction, operation, to maintenance. PPP can be a financing solution for infrastructure development where if you take fiscal steps the State will tend to increase the amount of foreign debt so that it is considered not the best choice. Meanwhile, if you take monetary steps it has the potential to disrupt the stability of the domestic inflation rate. The government in the PPP scheme has the goal of minimizing costs related to project risk by transferring risks during construction, operating period, and demand risk to the private sector (Yescombe, 2007). PPP, which is an innovation in financing infrastructure development, does not always succeed in accordance with its expectations. There are several projects with PPP schemes that have been canceled or distressed. The causes of the failure of PPP implementation in the project were macroeconomic conditions, government conditions, project type, and several other factors. The purposes of this research are (a) know the main factors that cause the PPP project failure; (b) Determine the influence of the involvement of debt providers in PPP projects that use debt funding schemes.

2 LITERATURE REVIEW

2.1 *Infrastructure development*

Investment in infrastructure is crucial for economic growth, quality of life, poverty reduction, access to education and health care, and achieving many goals of a strong economy (Delmon, 2009). Understanding infrastructure according to Kodoatie (2005) is a system that supports the social system and economic system which is also a liaison with the environmental system, where this system can be used as a basis for making policies. According to Stone (1974) infrastructure is a variety of physical facilities that are needed and developed by public agencies that aim to meet social and economic goals and government functions in terms of transportation, electricity, water supply, waste disposal and other services that similar. Three main drivers of increasing demand for infrastructure can be identified, namely demographic development, environmental challenges and drivers, technical change, and economic growth and reliability.

2.2 *Private-public partnership scheme*

PPP is a funding scheme for infrastructure development which is an alternative to the limited availability of the Government budget. Infrastructure, which is public goods, is the responsibility of the government in terms of procurement, so that by using the PPP financing scheme, the Government can work with the private sector (investors) in an infrastructure development agreement. In this PPP financing scheme, the Government can share risks with the private sector. This is because in the design, construction, operation to maintenance phases are the responsibility of the investor.

3 RESEARCH METHOD

The design of this study aims to test the hypothesis to determine the effect of several variables, especially the effect of the condition of the debt provider to the distressing PPP Project. The object of this study takes global coverage, namely in countries included in the World Bank list with research objects, the biggest factor influencing the occurrence of a failed PPP project or experiencing difficulties (distress), especially in times of crisis. This research uses two types of research, namely explanatory research and descriptive research. The data used in this study were sourced from the World Bank's Private Participation in Infrastructure (PPI) Project Database which records infrastructure projects in various countries in the form of PPP since 1990. The database consists of 7050 projects from 1990 to 2017 which have accumulated for 128 countries with variable variables. recorded for each project, including project status (active, distressed, canceled, or concluded), project sectors and sub-sectors, committed investments, contract duration, contract form, multilateral support, and project sponsors, among other variables. The regression model used in this study is probit regression, which is based on conditions where the dependent variable is quantitative and uses a normal cumulative distribution. In this study, probit regression was used in the 7050 observations that were included.

The purpose of this probit regression is to see which variables are the main factors causing failure of a project (distress/cancel).

Of the several independent variables, a regression is carried out to get the most optimal equation so that the factors that most influence the distress / failure of the project with the PPP financing scheme are obtained. Because the relationship between risk and project outcomes is quite complex, as described above, this study uses the equation as follows:

$$y_i = a1 + \beta . X_i + u_i \tag{1}$$

$$y*1_i = \beta y21 + \gamma .x + u_i \tag{2a}$$

$$y2_i = x1_i \Pi_1 + x2_i \Pi_2 + v_i \tag{2b}$$

where y = Dependent Variable; xi = Independent Variable; u = *error*; Π_1 Π_2 = matrix of parameter.

The equation is translated into the model as below:

Project outcome (fail or not fail) = f (various endogenous and exogenous factors), with endogenous variables functioning of instrument variables.

The endogenous variables are variables that have a strong correlation with the error value. Also in this correlation analysis can also be seen instrumental variables that are variables that have a strong correlation with endogenous variables but do not have a correlation with errors.

3.1 *Variables*

3.1.1 *Independent variables/IV*

The independent variables used are as shown in the table below:

Table 1. Table of variable used.

Category	Variable Independent	Category	Variable Independent
Legal and Institutional Framework	Government effectiveness Control of Corruption Political Stability Voice and accountability	Contract	Government fiscal support (direct/indirect) Contracted with federal or local or state/province government
Macro Economics	GDP Growth annual Real effective exchange rate Inflation Export Import	Country Fiscal Capacity	Fiscal position annually (surplus or deficit) Short term debt to exports ratio Debt to GDP ratio Tax revenue
Structure of Transaction	Type Project (brownfield or divesture or greenfield or management and lease contract)	Debt Providers	Debt Provider involved Bank involved

3.1.2 *Dependent variables (DV)*

The dependent variable used is a dummy variable which means 1 for the PPP failure project (using canceled and distressed project status data) and 0 is the other status (using project data with active and concluded status).

4 RESULT AND CONCLUSION

Dominant factors that influence the success or failure of PPP projects are government effectiveness, annual GDP growth and control of corruption. These three variables are variables that have a significant influence on project status, both increasing or decreasing project failure. (i) Government effectiveness reduces the failure rate of PPP projects. This is in line with the better level of civil service capacity and independence from political pressure and the quality of policy formulations, the lower the potential for failure of a PPP project. (ii) Annual GDP growth reflects a country's economic quality, so the higher the annual GDP growth, the lower the potential for failure of PPP projects. This can be seen from the regression results which show that annual GDP growth has a significant negative relationship to the failure of PPP projects. (iii) Control over corruption is a factor in increasing the failure of PPP projects. This is in accordance with the understanding of the control variable over corruption which is the level that represents the possibility of a governmental authority exercised for personal gain, so the higher the value of control over corruption, the higher the failure of the PPP project.

A significant factor in influencing project success is governance conditions. This is due to the existence of government policies that affect the implementation of the project with the PPP scheme, so that the government conditions during the PPP project are very influential on the success of the PPP project.

From the results of the above study it is known that the debt provider has an influence on the failure of a PPP project although it is not significantly related. Debt providers have a negative relationship with project status. This shows that if a project has a debt provider involvement, the lower the potential for failure of a PPP project. The insignificance of the influence contributed by the debt provider to the status of the project, due to other factors that influence the involvement of the debt provider, namely the bankability of a project. Debt providers conduct studies to look at project bankability before becoming involved as debt providers, so that if there is a debt provider involved in the project, the more bankable the project will be.

REFERENCES

Delmon, Jeffrey. (2005). Project Finance, BOT Projects and Risk. Kluwer Law International. The Hague, The Netherlands.
Delmon, Jeffrey. (2011). Public-private partnership projects in infrastructure: an essential guide for policy makers. Cambridge University Press. United States of America.
Independent Evaluation Group. (2012). World Bank Group Support to Public-Private Partnerships: Lessons from Experience in Client Countries, FY02–12.
Kodoatie, R.J., (2003). Manajemen dan Rekayasa Infrastruktur. Pustaka Pelajar, Yogyakarta.
Peraturan Presiden Republik Indonesia No. 38 Tahun 2015. Kerjasama Pemerintah dengan Badan Usaha dalam Penyediaan Infrastruktur.
Walker C, Mulcahy J, Smith A, Lam PTI, Cochrane R. 1995. Privatized infrastructure. London, UK: Thomas Telford.
Yescombe, E.R (2007). Public-Private Partnerships Principles of Policy and Finance. Great Britain: Elsevier Ltd.

Bandung smart technology readiness index

Redima Kurniawan & Indrawati
Telkom University Bandung, Indonesia

ABSTRACT: The United Nations (2014) has projected that by 2050, 66% of the world's population will become city residents. This situation may escalate into problems for cities caught unaware. Smart technology, which is one of the dimensions of the smart city concept, is a proposed solution. The research discussed in this article was based on previous research conducted by Indrawati, Setiawan, and Amani (2017). The aim was to learn the index value of smart technology readiness in Bandung. This research utilized a mixed method approach, and data collection was done through literature studies, interviews, observations, and surveys. The source of this research was based on the quadruple helix method. Based on the results of data processing, the index value of the application of smart technology in Bandung is 72.89. This shows that the level of readiness of smart technology in Bandung is quite good and satisfying, but in some cases, there are still shortcomings.

1 INTRODUCTION

Bandung, as one of the largest cities in Indonesia with a population density of 15,713 people/square kilometer, has a role as a tourist destination, a center for business activities, and a place of education (Pusat Statistik Bandung, 2015). This makes Bandung a densely populated city with a high level of activity and all of its associated problems.

Bandung has implemented the concept of smart city management. Ten priority areas have been applied to the smart city of Bandung, including smart technology (Kamil, 2015). Smart technology is a dimension of the smart city, which is described as the ability of a city to connect homes, offices, and cellular phones through a telecommunications system.

2 RESEARCH OBJECTIVES AND QUESTIONS

The objective of this research was to measure the readiness index of smart technology in Bandung. The research questions of this study were as follows:

1. In accordance with the experiences, feelings, and insights of respondents, as well as secondary data regarding Bandung and best practices, what are the values of the variables and indicators used to measure Bandung's smart technology readiness?
2. Based on question 1, what is the level of Bandung's smart technology readiness?

3 RESEARCH METHODOLOGY

A mixed method was used in this research, an explorative research through data retrieval using descriptive research and data analytics in order to find data related to smart technology readiness. The explorative research on this project aimed to compare selected data based on interview results.

In conducting the descriptive research, the first step was to review the variables and indicators based on research conducted by Indrawati et al. (2017). The second phase was the search for best practice data on smart technology, followed by examining the application of smart technology in Bandung. The fourth stage was to conduct structured interviews with respondents selected from quadruple helix dimensions, where this approach involves four parties, namely: the government, business players, researchers, and experts in smart technology and citizenship (Rufaidah, 2013). The respondents were directly involved in the implementation of smart technology in Bandung and could provide the desired information. The fifth stage was measuring the readiness index from the value of indicators obtained in the previous stage

The respondents of this research comprised 30 people with information technology (IT) backgrounds in Bandung, which consisted of government workers, business players, researchers and experts, and users or citizens.

4 SMART TECHNOLOGY MODEL

Based on previous research conducted by Indrawati et al. (2017), the smart technology model can be summarized as shown in Figure 1.

The validity of the proposed model was then measured using the Pearson Product Moment correlation formula. The test was conducted on 30 respondents within a threshold = 0.3 retrieved from the r-table (Indrawati, 2015; Jikmund, Babin, Carr, & Griffin, 2013).

From the calculation of the validity, it was found that one indicator, "content usage," was not valid, but the rest were valid, so the rest of the variables and indicators were included in the development of this model.

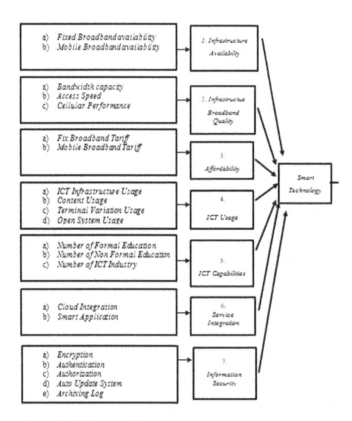

Figure 1. Proposed smart technology model.

Reliability was measured using the 0.05 significance level and the Cronbach's Alpha technique. The alpha value was 0.905, from which it can be concluded that the indicators and variables in this research were qualified and reliable.

The quantitative validation process showed that one indicator was declared not valid or qualified, so that indicator was removed from the previous model. The model was then applied to measure the readiness index of smart technology implementation in Bandung through in-depth interviews, focus group discussions, and questionnaire distribution. The new model can be used as shown in Figure 2.

5 SMART TECHNOLOGY READINESS INDEX RESULT

To measure the readiness index, secondary data from smart technology best practices and from Bandung were prepared first. Secondary data were prepared to provide information for respondents during the assessment process. Secondary data were taken from the results of literature reviews. Meanwhile, the selected city was included in the Global Cities Index 2018 released by A. T. Kearney, Singapore, then picked with ease in obtaining information, data availability, and data disclosure regarding the application of smart technology in the city.

The City of Bandung data were taken from interviews with related agencies, such as the Department of Communication and Information, to capture smart technology implementations in the said city. Any changes and differences that occurred were then updated according to the actual situation.

Based on the result of interviews, focus group discussions, and questionnaire distribution to the 30 respondents, the result showed that the index value of smart technology in Bandung is

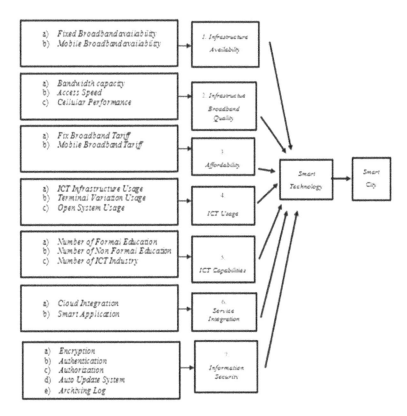

Figure 2. Smart technology model.

72.89. This shows that the level of readiness of smart technology in Bandung is satisfactory, but in some cases, there are still shortcomings. The indicator with the highest value was "Level of Formal Education," while the indicator with the lowest value was "Archiving Log."

6 CONCLUSION

The readiness index indicates that stakeholders must continuously make improvements in order to optimize the implementation of smart technology. Every indicator has various weaknesses, strengths, and problems. It is expected that the implementation of smart technology in the future can be more targeted and can be carried out based on priority level. So the improvement and development of smart technology in Bandung can also be more efficient and in accordance with needs.

ACKNOWLEDGMENT

This work was partially supported by the Ministry of Research, Technology and Higher Education of Indonesia.

REFERENCES

Indrawati. 2015. *Metode Penelitian Manajemen dan Bisnis. Bandung.* Jawa Barat, Indonesia: PT. Refika Aditama.
Indrawati, Setiawan, I. M. A., & Amani, H. 2017. *Indicators to Measure a Smart Technology: Bandung City Perspective.*
Jikmund, W. G., Babin, B. J., Carr J. C., & Griffin, M. 2013. *Business Research Methods.* 9th Edition. Mason, OH: South Western CENGAGE Learning.
Kamil, R. 2015. Statement by Mr. Ridwan Kamil, Mayor of Bandung, Indonesia. Retrieved May 10, 2018, from https://sustainabledevelopment.un.org/index.php?page=view &type=255&nr=12659&menu=35.
Pusat Statistik Bandung Badan. 2015. Bandung Dalam Angka 2016. Retrieved April 28, 2018, from: https://bandungkota.bps.go.id/dynamictable/2015/11/19/10/kepadatan-penduduk-per-km2-kota -bandung-tahun-2008-2014.html
Rufaidah, P. 2013. Branding Strategi Berbasis Ekonomi Kreatif: Triple helix vs. quadruple helix. In T. P. Barat (Ed.), *Branding Strategi Berbasis Industri Kreatif Fashion.* Bandung, Jawa Barat, Indonesia: Direktorat Jenderal Pendidikan Tinggi Kementrian dan Kebudayaan.
United Nations. 2014. World Urbanization Prospects. Retrieved May 10, 2018 Available at: https://esa.un.org/unpd/wup/publications/files/wup2014-highlights.pdf.

Leverage up predictive analytics to increase prepaid credit purchasers

Ronny Arnaz & Gadang Ramantoko
Faculty of Economics and Business, Telkom University, Bandung, Indonesia

ABSTRACT: The conversion rate of purchasing prepaid credit on e-commerce applications is still very low. The objective of this research was to gain insights into predicting customers who have a tendency to purchase prepaid credit using the clustering and logistic regression techniques. Logistic regression was used to predict customers using 17 numeric variables. After building a development model, it was then applied to all populations and identified customers with a high tendency to purchase. To increase efficiency and effectiveness in marketing programs, marketers need to know customers' priority. Segmentation was conducted with K-means clustering in order to determine which cluster gives the greatest predictive gain. The prediction results of high-prospect customers were divided into two clusters with the two-step cluster techniques, namely low and medium high-value customers (HVC). Based on the results, behavioral targeting can be done so that the marketing campaign is more targeted.

1 INTRODUCTION

During the era of digital transformation, Telkomsel wants to change customer habits in re-charging from conventional to digital channels, especially through e-commerce applications whose contribution is still very small, only 2.2%. Telkomsel has a target to make 25% of active users become prepaid credit buyers in e-commerce, which targets 4 million customers. The transactions and sales trend of e-commerce, both in Telkomsel and globally, are growing fast while the recharge conversion rate in e-commerce is still low. Thus, a digital marketing plan is needed to increase recharge users and transactions in e-commerce applications. This research is expected to solve the problem of how to increase the number of prepaid credit purchasers in e-commerce applications. The scope of this research was prepaid subscribers who have been active users in e-commerce applications but have not purchased prepaid credit yet.

2 BASIC THEORY AND RESEARCH FRAMEWORK

2.1 *Market segmentation and targeting, digital marketing and micro-targeting*

Segmenting, targeting, and positioning (STP) is a general framework that summarizes and simplifies the process of market segmentation, consisting of market segmentation, market target setting, and market positioning (Kotler, Armstrong, Saunders, & Wong, 1999).. Tom Agan (2007) states that micro-targeting is making customized winning messages, roof points, and offers, predicting their impact accurately, and delivering them directly to concerned individuals.

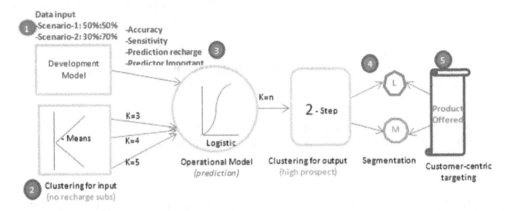

Figure 1. Research framework.

2.2 Research framework

The research framework used in this study referred to the process in Figure 1:

1. Prediction model. The model was created with Scenario 1 – when the model uses input data with a composition of 50%:50% for consumers purchasing and not purchasing prepaid credit – and Scenario 2 – when the model uses input data with a composition of 30%:70%.
2. Segmentation for non-purchaser prepaid credit. Clustering using the K-means technique was carried out in order to determine the best clusters that will give the biggest gain predictions after applying the model. The number of clusters formed was three, namely three, four, and five clusters.
3. Operationalization models. Each cluster formed with the K-means technique was applied in the prediction model so that the priority of clusters with the highest ranking and the largest gain could be obtained.
4. Segmentation for highly valuable customers. The two-step cluster was applied to high-prospect consumers so their profiles were known for marketing strategies.
5. Behavior targeting for the best cluster. The marketing strategy will be developed based on consumers' profiles and behaviors so that the products will be more easily accepted by customers.

3 DISCUSSION

3.1 Data population

This study used the population of 15.35 million active users of e-commerce applications, and the number of prepaid credit purchasers reached 297,172. Using an analytics-based table as a baseline from which to build predictive modeling, the researcher divided it into two data sets, namely a 70% training data set and a 30% testing data set, by following two scenarios as in Table 1:

Table 1. Analytics table scenario 1 (50%:50%).

User	Training data	Testing data	Total
0	207,851	89,321	297,172
1	208,097	89,075	297,172
Total	415,948	178,396	594,344

Table 2. Analytics table scenario 2 (30%:70%).

User	Training data	Testing data	Total
0	89,200	38,159	127,359
1	207,790	89,382	297,172
Total	296,990	127,541	424,531

3.2 Predictive analytics for model development

Logistic regression addresses the issue of predicting a target variable, which may be binary or binomial (such as 1 or 0, yes or no), using predictors, which may be numeric. The regression method has become an integral component in any data analysis concerned in describing the relationship between a response variable and one or more explanatory variables (Hosmer & Lemeshow, 2000). The equation follows (Peng, Lee, & Ingersoll, 2002):

$$\pi = \frac{e^{\alpha + \beta_1 X_1 + \beta_2 X_2 + \beta_3 X_3 + \beta_n X_n}}{1 + e^{\alpha + \beta_1 X_1 + \beta_2 X_2 + \beta_3 X_3 + \beta_n X_n}} \qquad (1)$$

where π = probability of the outcome; α = Y intercept; β = regression coefficient; x = variable input; and e = natural logarithm (2.71828).

In analyzing the results, the researcher created a confusion matrix as in Tables 3 and 4. By applying the logistic regression equation to 17 numeric predictors, the predictive model is as follows:

Based on the result of the confusion matrix, the researcher chose scenario 2 as the highest user prediction, accuracy, and sensitivity scores and applied it to all populations of non-prepaid credit purchaser e-commerce applications across 15,055,473 customers. The results identified 3,089,120 customers with a high tendency to purchase prepaid credit in e-commerce applications.

3.3 Customer segmentation for prioritization

The clustering process with the K-means technique was applied to 15,055,473 non-purchasers in order to prioritize and determine the best cluster by the highest rank and variable comparison. The clusters to be formed were three, namely three, four, and five clusters (K3, K4, and K5). The highest rank of each cluster showed that data packet transactions in e-commerce, the number of transactions in e-commerce, and the volume usage in e-commerce are the most influential in the best cluster result. Clusters C1K3, C1K4, and C5K5 had the highest ranking for each cluster, so they were chosen as a priority to be applied in the model.

Table 3. Confusion matrix scenario 1.

	Population	Actual Purchaser	Actual Non-purchaser	Total	
Prediction	Purchaser	137,327	9,734	147,061	Accuracy: 80.65%
	Non-purchaser	70,770	98,117	168,887	Sensitivity: 66.0%

Table 4. Confusion matrix scenario 2.

	Population	Actual Purchaser	Actual Non-purchaser	Total	
Prediction	Purchaser	163,171	18,251	181,422	Accuracy: 78.8%
	Non-purchaser	44,619	70,949	115,568	Sensitivity: 78.5%

Table 5. Operational gain of non-purchaser.

Cluster	Non-purchaser	Prospect Purchaser	Gain	Remark
All Subscribers	15,055,473	3,089,120	20.5%	
C1K3	646,904	636,848	98.4%	
C1K4	616,862	608,634	98.7%	
C5K5	346,782	345,965	99.8%	The best priority

3.4 Operationalization model

The author operationalized the development model in order to predict which high-prospect customers will buy prepaid credit from the three best clusters (C1K3, C1K4, C5K5) and all of the non-purchasers. Based on Table 5, the author chose C5K5 as the best priority for further studies.

3.5 Customer segmentation for behavioral targeting

The author then developed customer segmentation based on 345,965 customers as the top cluster priority using the two-step clustering method. Of these customers, 32.5% came from the medium-value and e-commerce-addicted segments. Deeper analysis based on demography (youth/non-youth) and length of stay (LoS) segmentation was done to get more segmented and detailed target customers. The marketing program was given according to the customers' profile with the aim of increasing the usage of low high-value customers (HVC) and maintaining the loyalty for the medium HVC.

4 CONCLUSION

This study concluded that logistic regression combined with clustering techniques addresses the issue of predicting high-prospect consumers who will purchase prepaid credit and of prioritizing segment consumers in order to make marketing programs more effective and efficient. By applying it to all populations of non-prepaid credit purchasers – as many as 15,055,473 customers – 3,089,120 customers were identified as being willing to purchase prepaid credit, with a 20.5% gain in value. It was more effective to apply a selected segment of 346,783 customers using the K-means method because 345,965 high-prospect customers were identified and also provided a big gain under this model, reaching 99.8%. Using two-step clustering for high-prospect customers, 32.5% of them came from the medium-value and e-commerce-addicted segments. In the development model, the researcher chose scenario 2, using 30% random sampling of non-purchasers and 70% random sampling of prepaid credit purchasers due to the higher accuracy and sensitivity. The top six significant variables that influenced customers to purchase prepaid credit in e-commerce applications were: *recharge non-e-commerce transactions, total recharge transactions, number of transactions using e-commerce applications, number of data packet transactions using e-commerce applications, broadband revenue, and volume usage of e-commerce applications*. At the end of study, the author created a behavioral targeting design, based on the result from the predictive model and the classification and qualitative variables (age and stay length) in order to offer an accurate marketing program.

REFERENCES

Chaffey, D. 2002. *Digital Business and e-Commerce Management*. London: Pearson.
Chapman, P., Clinton, J., Kerber, R., Khabaza, T., Reinartz, T., Shearer, C., & Wirth, R. 2000. CRISP-DM 1.0 Step-by-Step Data Mining Guide. SPSS Inc.
Gröttrup, S., & Wendler, T. 2016. *Data Mining with SPSS Theory, Exercises and Solutions*. Berlin: Springer International Publishing Switzerland. doi:10.1007/978-3-319-28709-6

Hosmer, D. W., & Lemeshow, S. 2000. *Applied Logistic Regression*. Second Edition. Hoboken, NJ: Wiley.

Kotler, P., Armstrong, G., Saunders, J., & Wong, V. 1999. *Principles of Marketing*. Second European Edition. Englewood Cliffs, NJ: Prentice Hall Europe.

Kotu, V., & Desphande, B. P. 2015. Predictive Analytics and Data Mining. Waltham, MA: Elsevier.

Peng, C.-Y., J., Lee, K. L., & Ingersoll, G. M. (2002). An Introduction to Logistic Regression Analysis and Reporting. The Journal of Educational Research.

Users' continuance intention toward hospitality service applications

Indrawati & S.A. Hasana
Faculty of Economics and Business, Telkom University, Bandung, Indonesia

S.K.B. Pillai
Goa Business School, Goa University, Goa, India

ABSTRACT: Airbnb, an online platform through which users register and rent hospitality services to tourists, is one of the biggest online marketplaces in most countries around the world, including Indonesia. This study attempted to measure factors influencing users' continuance intention (CI) of Airbnb in Indonesia by using the Modified Unified Theory of Acceptance and Use of Technology (UTAUT) 2 Model through surveying 400 valid respondents. The result revealed that five factors out of the eight factors examined influence users' CI, namely, habit (H), trust (T), hedonic motivation (HM), price-saving orientation (PSO), and facilitating conditions (FC). Only gender moderates the influence of FC and HM on CI. Airbnb management can use this model to improve their services and increase their customer base in the coming years.

1 INTRODUCTION

In 2018, Indonesia ranked as the sixth largest nation in the world in terms of Internet users with an estimated 103 million users, which is expected to increase to 123 million people (Botsvadze, 2018; Millward, 2018) as a result of various mobile applications available in the area of e-commerce, especially in the field of hospitality. Airbnb is one such hospitality service provider with the biggest online marketplace for registering and renting accommodation around the world to tourists who want to have a different travel experience (Knowledge, 2018). Airbnb has become globally successful and is now ranked first (EMarketerChart, 2018). Airbnb has expanded its operations in Indonesia also (*Jakarta Post*, 2018) as customers are happy with the services provided (Rentivo, 2018). In 2017, Airbnb's growth in Indonesia reached 72% compared to the previous year (Wicaksono, 2018), and the number of orders increased by up to 3 million (Jamaludin, 2018). The present article tries to identify what factors are responsible for influencing the continuance intention (CI) of customers toward using the Airbnb application in Indonesia and also tests for significant differences of behavior among the users in terms of age and gender.

2 THEORETICAL FRAMEWORK AND METHODOLOGY

The Unified Theory of Acceptance and Use of Technology (UTAUT) 2 Model is the latest theory in acceptance models of technology in the consumer context and was developed by Venkatesh, Morris, Davis, and Davis (2003). The UTAUT 2 Model is used in the present study, and it has also been used in previous studies (Escobar-Rodríguez & Carvajal-Trujillo, 2014; Indrawati & Marhaeni, 2015). Four modifications were carried out of the UTAUT 2 Model. First, the "Price Value" variable was replaced with the "Price Saving" variable (Escobar-Rodríguez & Carvajal-Trujillo, 2014) because users can save money using the Airbnb application while booking hotel rooms. Second, a new variable, "Trust," was added as trust influences users (Escobar-Rodríguez & Carvajal-Trujillo, 2014). Third, the "Behavioral Intention" variable was replaced with "Continuance Intention" because the present study aimed to analyze the Airbnb application's users in adopting the hospitality service application continuously (Xu, 2014). Finally, the variable of "Experience" was excluded as the moderating variable because data collection was not carried out based on a longitudinal sample, but from a cross-sectional sample.

The proposed modified UTAUT 2 Model, based on the aforementioned four modifications, consists of eight independent variables (IV), namely performance expectancy (PE), effort expectancy (EE), social influence (SI), facilitation condition (FC), hedonic motivation (HM), price-saving orientation (PSO), habit (H), and trust (T). These IVs are moderated with two demographic variables, namely age and gender. Finally, the resulting dependent variable (DV) is defined as continuance intention (CI). There were five statements each for these eight IVs and one DV, totaling 45 statements. The data so collected were tested statistically in order to see if there is any significant influence of these eight IVs on the DV and also to examine the moderating effect of age and gender.

The collected data were initially tested for content validity, expert validity, and reliability. A pilot test was carried out for making necessary modifications of the questionnaire based on expert opinions (Escobar-Rodríguez & Carvajal-Trujillo, 2014; Indrawati & Marhaeni, 2015;; Venkatesh et al., 2003). The overall validity of the 45 items on the questionnaire gave a correlation coefficient value greater than 0.30, which is in line with the suggestion given by Friedenberg and Kaplan (Indrawati & Marhaeni, 2015). Similarly, the result of the reliability test gave a Cronbach Alpha greater than 0.70, which is also considered reliable.

Eight main hypotheses and 16 sub-hypotheses were used for testing the objectives of the present study on Airbnb customers' CI. The main eight hypotheses were "*PE (H_1) / EE (H_2) / SI (H_3) / FC (H_4) / HM (H_5) / PSO (H_6) / H (H_7) / T (H_8) has a positive and significant influence toward CI.*" Similarly, the 16 sub-hypotheses that tested the moderating effect of age and gender, which influences the eight IVs toward CI, were: "*Age and gender as a moderator have an influence on PE / EE / SI / FC / HM / PSO / H / T toward CI.*" All of these hypotheses were tested at 5% one-tailed (Indrawati, 2017). The methodology used to test the hypotheses was a quantitative method using smart PLS 3.0. software.

3 ANALYSIS AND DISCUSSION

Using Google Forms, a questionnaire was circulated through email to 512 respondents using purposive sampling during the period November 16–December 21, 2018. Only 400 questionnaires were usable with a response rate of 78%; the remaining 112 respondents were eliminated as they had used Airbnb fewer than three times. The result revealed that the majority (54%) of the respondents were youngsters (17–25 years old), which is in tune with the diffusion of innovation technology theory that youngsters are inclined toward innovation and tend to be early adopters of technology (Kotler & Keller, 2012). Cronbach Alpha and composite reliability gave a reference value of 0.7, which ensured the reliability of the 45 items. Using average variance extracted indicators, the convergent validity was also tested, which gave a threshold score of 0.50 and higher. The path coefficients and t-values of each variable are shown in Table 1, which shows that of the eight IVs, five (H, T, HM, PSO, and FC) have a positive and significant influence on the DV, namely, CI, hence the formulated hypotheses H_4, H_5, H_6, H_7, and H_8 were accepted.

The remaining three IVs (PE, EE, and SI) do not have any significant influence on the DV (CI), hence the formulated hypotheses H_1, H_2, and H_3 were rejected, which is in line with the findings of earlier studies (Gou, Nikou, & Bouwman, 2013 Indrawati, 2017) that customers are happy with the use of Airbnb, they are eager to adopt new technologies, and they are not influenced by other people's opinions.

Table 1. Path, coefficients, t-values, and hypotheses status.

No.	Path	Coefficient	T-value	Hypothesis
1	H→CI	0.411***	5.764	H_4 Accepted
2	T→CI	0.164***	3.140	H_5 Accepted
3	HM→CI	0.144**	1.975	H_6 Accepted
4	PSO→CI	0.133***	2.363	H_7 Accepted
5	FC→CI	0.109**	1.913	H_8 Accepted
6	PE→CI	−0.083	1.549	H_1 Rejected
7	SI→CI	0.122	1.509	H_2 Rejected
8	EE→CI	−0.047	0.569	H_3 Rejected

*** Significant at 1%.
** Significant at 5%.

To validate the overall model, a goodness-of-fit (GoF) index was obtained based on Henseler and Sarstedt (2013) and the result gave a GoF of 0.694, indicating that the overall model is valid. Effect size was also used to validate the magnitude of the difference between groups in the model based on Cohen (Sullivan & Feinn, 2012), and the result is given in Table 2 showing all effects are very small.

Structural model (inner model) test results gave an R^2 value of 80.6%, considered "good," which confirms that the model has a strong explanatory power (Indrawati, 2017). To test the sub-hypotheses, two moderating variables, age and gender, were used in the Chin formula, which gives a t-value for comparing the paths of each group (Indrawati, 2017). Table 3 indicates that age does not affect the influence of consumers' CI, maybe because Airbnb is a gender-neutral application that can be used by anyone in any age group.

The gender variable moderates the influence of FC and HM on CI. It shows that between the male and female categories, there is a different perception in FC and HM. This means that males consider FC more than females while using Airbnb. Second, females consider HM more than males while using Airbnb.

Table 2. Effect size by DV.

Link	F-square	Size/weight
EE to CI	0.002	Small
FC to CI	0.008	Small
H to CI	0.184	Small
HM to CI	0.012	Small
PE to CI	0.015	Small
PSO to CI	0.026	Small
SI to CI	0.014	Small
T to CI	0.038	Small

Table 3. Moderation effect.

Path	T-value Age	Gender
FC→CI	0.093	6.825
H→CI	0.018	0.018
HM→CI	0.076	4.924
PSO→CI	0.025	0.020
T→CI	0.029	0.059

4 CONCLUSION AND SUGGESTION

The result reveals that the model used in this study has strong explanatory power with an R^2 value of 80.6%, indicating the model's ability to predict customers' CI. Of the eight IVs, five showed positive and significant influence on the customers' CI. Among the two moderating variables, only gender affects the influence of HM and FC on CI. Male customers consider more of FC, whereas female customers consider more of HM while using Airbnb.

Habit has become the strongest variable in predicting CI to use Airbnb, which Airbnb should give priority while developing promotional schemes that will engage customers in using Airbnb for all their hospitality needs.

Ensuring complete security while using Airbnb, with prompt and accurate confirmation of payment, ensures that trust is being developed in the minds of customers, which is critical for CI. Motivation is assured while using Airbnb as HM influences customers, else it might lead to dissatisfaction among customers. Providing recommended places with special offers might keep customers motivated. Finally, Airbnb must ensure that the hospitality services offered are reasonably priced compared to other e-commerce players without sacrificing the quality of the services offered.

REFERENCES

Botsvadze, V. 2018. Top 25 countries, ranked by Internet users, 2013–2018. Available from: https://vladimerbotsvadze.wordpress.com/2015/09/21/top-25-countries-ranked-by-internetusers-2013-2018-2/. Retrieved on April 2, 2018.

EMarketerChart. 2018. Top 10 US hotel & accommodation sites, ranked by total traffic, Q1 2017 (millions and % change vs. same period of prior year). Available from: www.emarketer.com/Chart/Top-10-US-Hotel-Accommodation-Sites-Ranked-by-Total-Traffic-Q1-2017millions-change-vs-same-period-of-prioryear/208268. Retrieved on October 21, 2018.

Escobar-Rodríguez, T., & Carvajal-Trujillo, E. 2014. Online purchasing tickets for low cost carriers: An application of the unified theory of acceptance and use of technology (UTAUT) model. *Tourism Management*, *43*, 70–88.

Guo, J., Nikou, S., & Bouwman, H. 2013. Analyzing the business model for mobile payment from banks' perspective: An empirical study. Proceedings of the 24th European Regional Conference of the International Telecommunication Society, Florence, Italy, October 20–23, 2013, 1–16.

Henseler, J., & Sarstedt, M. 2013. Goodness-of-fit indices for partial least squares path modeling. *Computational Statistics*, *28*(2), 565–580. doi: 10.1007/s00180-012-0317-1

Indrawati. 2015. *Metode Penelitian Manajemen dan Bisnis, Konvergensi Teknologi Komunikasi dan Bisnis*. Bandung, Indonesia: PT Refika Aditama.

Indrawati. 2017. *Perilaku Konsumen Individu*. Bandung, Indonesia: PT Refika Aditama.

Indrawati & Marhaeni, G. A. M. 2015. Measurement for analyzing instant messenger application adoption using the Unified Theory of Acceptance and Use of Technology 2. *International Business Management (IBM)*, 9(4), 391–396.

Jakarta Post. 2018. Airbnb eyes expansion in Indonesia. Available from: www.thejakartapost.com/news/2017/05/30/airbnb-eyes-expansion-in-indonesia.html. Retrieved on October 6, 2018.

Jamaludin, F. 2018. Bahagianya bos AirBnB, 3 juta orang memesan penginapan saat malam tahun baru. Available from: www.merdeka.com/teknologi/bahagianya-bos-airbnb-3-juta-orang-memesan-penginapan-saat-malam-tahun-baru.html. Retrieved on January 2, 2018.

Knowledge, H. W. 2018. The Airbnb effect: Cheaper rooms for travelers, less revenue for hotels. Available from: www.forbes.com/sites/hbsworkingknowledge/2018/02/27/the-airbnb-effect-cheaper-rooms-for-travelers-less-revenue-for-hotels/#3a56fd9ad672. Retrieved on September 17, 2018.

Kotler, P., & Keller, K. L. 2012. *Marketing Management*. 14th Edition. Upper Saddle River, NJ: Prentice Hall.

Millward, S. 2018. Indonesia world's fourth largest smartphone 2018. Available from: www.techinasia.com/indonesia-worlds-fourth-largest-smartphone-2018-surpass-100-million-users/. Retrieved on September 17, 2018.

Rentivo. 2018. The results are in: What owners & managers really think about OTAs (HomeAway, VRBO, Airbnb etc.). Available from: www.rentivo.com/blog/online-travel-agencies-opinion-statistics/. Retrieved on October 21, 2018.

Sullivan, M. G., & Feinn, R. 2012. Using effect size: Or why the P value is not enough. *Journal of Graduate Medical Education*, *4*(3), 279–282.

Venkatesh, V., Morris, M. G., Davis, G. B., & Davis, F. D. 2003. User acceptance of information technology: Toward a unified view. *MIS Quarterly*, *27*(3), 425–478.

Venkatesh, V., Thong, J., & Xu, X. 2012. Consumer acceptance and use of information technology: Extending the Unified Theory of Acceptance and Use of Technology. *MIS Quarterly*, *36*(1), 157–178.

Wicaksono, S. 2018. Hal yang Bisa Dipelajari Startup Pariwisata Indonesia dari Airbnb. Available from: https://phinemo.com/hal-yang-bisadipelajari-startup-wisata-dari-airbnb/. Retrieved on April 2, 2018.

Xu. X. 2014. Understanding users' continued use of online games: An application of UTAUT2 in social network games. In *Proceedings of MMEDIA 2014: The Sixth International Conference of Advanced Multimedia*. 68–86.

Customers' continuance acceptance toward automated guided transportation

Indrawati & P.A. Nidya
Faculty of Economics and Business, Telkom University, Bandung, Indonesia

ABSTRACT: In the past two years, Palembang, Indonesia, has become a city where Association of Southeast Asian Nation (ASEAN) games are held. Therefore, to facilitate transportation for ASEAN games, the government has built light rail transit (LRT) as a means of transportation in Palembang. This research aimed to examine factors according to a modified Unified Theory of Acceptance Use of Technology 2 (UTAUT) Model from Venkatesh, Thong, and Xu (2012), especially the influence of the continuance intention of customers toward LRT by surveying 400 valid respondents. The result revealed that four factors of the modified UTAUT 2 Model have positive significant influence on user's continuance intention in adopting LRT. From the highest to the lowest, respectively, these factors are habit, social influence, perceived security, and facilitating conditions. Meanwhile, the moderating variables of gender and age have no positive and significant influence.

1 INTRODUCTION

Palembang is a densely populated city in Indonesia. In the past four years, Palembang has also become a city where Association of Southeast Asian Nation (ASEAN) games are held (Rosana & Saptiyulda, 2018). Therefore, to support the ASEAN games, the government has built light rail transit (LRT) as a mode of transportation in Palembang. The Palembang LRT connects the Sultan Mahmud Badaruddin II International Airport to the Jakabaring Sports City. Construction began in 2015; the project was built to facilitate the 2018 Asian Games and was completed in the middle of 2018, a few months before the event. The system has 13 stations, 6 of which were operational as of August 2018 (Siregar, 2018). Within a month, the number of LRT passengers had reached 198,300. Public enthusiasm for LRT was high enough. Public relations representatives of LRT Palembang said that the passenger data itself had been recorded from July 23, 2018, until August 13, 2018 (Irwanto, 2018). Increasing the number of passengers until maximum capacity has been reached is important, and there is a need to analyze the factors that influence the continuance intention of customers toward using LRT. The aim of this study was to examine the factors that influence customers' continuance intention toward LRT in Palembang.

2 THEORETICAL FRAMEWORK AND METHODOLOGY

To accomplish the point of this study, the authors reviewed previous studies related to users' adoption of technology-based services. The authors used a modified UTAUT 2 Model as a basis theory for this study. The UTAUT 2 Model was introduced by Venkatesh, Thong, and Xu (2012), and it was used to measure consumer behavior in individual contexts (Venkatesh et al., 2012). This research framework adjusted and changed the UTAUT 2 Model based on the expectations for this research. This study replaced behavioral intention with continuance intention since the research needed to break down customer continuance intention toward smart mobility adoption of LRT users. This study did exclude use behavior variables since it needed to dissect just the continuance intention of users. In addition, the respondents included in this study were users who already had used LRT at least three times, and the study planned to see if the current users expect to keep utilizing LRT. Xu (2014) also replaced behavioral intention with continuance intention and eliminated use behavior.

The proposed modified UTAUT 2 Model consists of eight independent variables (IVs): performance expectancy (PE), effort expectancy (EE), social influence (SI), facilitation condition (FC), hedonic motivation (HM), price value (PV), habit (HA), and perceived security (PS). These IVs were moderated with two demographic variables, namely age and gender. Finally, the resulting dependent variable was defined as continuance intention (CI). There were five statements each for these eight independent variables and one dependent variable, totaling 45 statements. The information so gathered was tried measurably to check for any significant influence of these eight independent variables on the dependent variable and also the moderating effect of age and gender.

The gathered data were initially tested for content validity, expert validity, and reliability. Finally, a pilot test was conducted to check for necessary modifications of the questionnaire based on expert opinions (Indrawati & Marhaeni, 2015;; Venkatesh et al., 2012). The overall validity of the 45 items of the questionnaire gave a correlation coefficient value greater than 0.30, which is in line with the suggestion given by Friedenberg and Kaplan (Indrawati, 2015). Similarly, the result of the reliability test gave a Cronbach Alpha of greater than 0.70, which is also considered reliable.

Eight principle hypotheses and 16 sub-hypotheses were utilized for testing the objectives of the present study on customers' CI of LRT. The main eight hypotheses were PE (H_1) / EE (H_2) / SI (H_3) / FC (H_4) / HM (H_5) / PV (H_6) / HA (H_7) / PS (H_8) has a positive and significant influence toward CI. Additionally, the 16 sub-hypotheses that tested the impact of age and gender, which impacts the eight independent variables toward CI, were: "Age and gender as a moderator has influence on PE / EE / SI / FC / HM / PV / HA / PS toward CI." All of these hypotheses were tested at 5% one-tailed (Indrawati & Yuliansyah, 2017).

3 ANALYSIS AND DISCUSSION

To test the variable of each item, the authors distributed the questionnaire through Google Forms to collect the data. The period of collecting the data was November 19–December 22, 2018. The data received from 403 respondents with 400 respondents were valid. The authors picked consumers who were aged 17–60 years old who could answer the questions. The age category was separated into two classifications, which were young age (17–25 years old) and adult age (25–60 years old). The respondents' age was dominated by the young classification, in line with the diffusion of innovation technology theory that claims the young are more likely to be innovators and to fall into the early adopter category (Kotler & Keller, 2016). Innovators and early adopters are valiant and have no fear of going out on a limb in embracing LRT services. The respondents' gender was mostly female (54.75%). This result is in line with research conducted by Universitas Indonesia that revealed that Go-Jek users are 69% female and 31% male (Puskakom, 2017).

In this research, the authors used a significance level of 5%. Based on the result in Table 1, it can be concluded that the independent variables that have a positive and significant impact on continuance intention are HA, SI, FC, and PS. The other variables – PE, EE, PV, and HM – had no significant impact on continuance intention.

The predicting power of the model is shown by the R^2 for the dependent latent variable. The R^2 results of 0.67, 0.33, and 0.19 show where the model is good, moderate, and weak, respectively (Indrawati, 2015). The result of bootstrapping in this study showed that an R^2 of 72.1% is good. Table 2 shows that in the young age and adult age categories, gender does not affect the impact of the modified UTAUT 2 Model factor on continuance intention of consumers in the context of LRT services in Palembang. All of the t-values are less than 1.96, which implies that there is no critical distinction in the impressions of consumers as far as age and gender orientation.

Table 1. T-value for each variable.

No	Path	Coefficient	T-value		Conclusion
1	HA -> CI	0.378***	6.35	H1	Accepted
2	SI -> CI	0.195***	3.64	H1	Accepted
3	FC -> CI	0.129***	2.88	H1	Accepted
4	PS -> CI	0.142***	2.48	H1	Accepted
5	PE -> CI	0.062	1.41	H1	Rejected
6	EE -> CI	0.049	1.18	H1	Rejected
7	PV -> CI	0.037	0.72	H1	Rejected
8	HM -> CI	0.042	0.70	H1	Rejected

Table 2. The result of the moderation effect.

Paths	T-value for moderating variable	
	Age	Gender
FC -> CI	0.390	0.112
HA -> CI	0.623	0.460
PS -> CI	0.869	0.055
SI -> CI	0.144	0.871

4 CONCLUSION AND SUGGESTION

The variables with positive and significant impact on consumer continuance intention toward LRT, from the most astounding impact to the least, were habit, social influence, perceived security, and facilitating conditions. In this research, no difference was found in the impression of consumers in terms of age and gender.

The R^2 of this research was 72.1%, which means that continuance intention is 72.1% affected by social influence, facilitating conditions, habit, and perceived security while the rest (27.9%) is impacted by alternate factors that weren't contemplated in this research. This model has good predicting capacity to foresee consumer continuance intention toward LRT services adoption.

The results of this research indicate that LRT management is required to almost certainly recognize factors inside the adjusted UTAUT 2 Model influencing consumer's continuance intention of LRT services. The LRT management has to know what customers' needs are in order to provide good services.

This research has discovered that the most significant factor from the modified UTAUT 2 Model that impacts customers' continuance intention to utilize LRT is habit. This factor has become the first priority because habit has the highest t-value, which is 6.35. This means that, in order to make sure that consumers use LRT continuously, the LRT management could create LRT reward levels. The LRT management could make LRT points available for LRT users so they can used to get free entry and higher discounts. Customer will begin using LRT automatically get used to using LRT for daily transportation.

The second factor that significantly influences consumers' continuance intention to utilize LRT is social influence. The LRT management would be smart to have more associations with certain networks or with people who are vital in the networks.

The third factor that significantly influences customers' continuance intention to utilize LRT is facilitating conditions. This refers to how much an individual trusts that an authoritative and specialized foundation exists to help with the utilization of LRT services. The LRT management could improve the facilities and make customers feel more satisfied and comfortable to use LRT services.

The fourth factor that significantly impacts customers' continuance intention to use LRT is perceived security. This factor is supported by the results of the respondents' questionnaires that people still worry about using LRT. But the community believes that LRT has a security framework. The LRT management should therefore make improvements in the security sector; for example, LRTs can implement body checking before passengers enter stations just like in airports.

REFERENCES

Indrawati. 2015. *Metode Penelitian Manajemen dan Bisnis, Konvergensi Teknologi Komunikasi dan Bisnis.* Bandung: PT Refika Aditama.

Indrawati & Marhaeni, G. A. M. 2015. Measurement for analyzing instant messenger application adoption using a unified theory of acceptance and use of technology 2. *International Business Management (IBM)*, 9(4): 391–396.

Indrawati & Yuliansyah, S. 2017. Adoption factors of online-web railway ticket reservation service (a case from Indonesia). In Information and Communication Technology (ICoICT), 2017 5th International Conference on (pp. 1–6). IEEE.

Irwanto. 2018, August 15. Belum sebulan beroperasi, penumpang LRT Palembang tercatat 198.300 orang. Retrieved from www.merdeka.com/peristiwa/belum-sebulan-beroperasi-penumpang-lrt-palembang-tercatat-198300-orang.html.

Kotler, P., & Keller, K. L. 2016. *Marketing Management*. 15th Edition. Upper Saddle River, NJ: Prentice Hall.

Puskakom. 2017. *Hasil Reset Manfaat Sosial Aplikasi On- Demand: Studi Kasus Go-Jek Indonesia.* Retrieved on December 3.

Rosana, D., & Saptiyulda, E. 2018, August 18. Palembang ukir sejarah baru berkat Asian Games. Retrieved from https://asiangames.antaranews.com/berita/738742/palembang-ukir-sejarah-baru-berkat-asian-games.

Siregar, R. A. 2018, June 8. LRT Palembang Ditarget Beroperasi di 6 Stasiun Mulai 15 Juli. *Detik* (in Indonesian). Retrieved August, 14 2018.

Venkatesh, V., Thong, J., & Xu. X. 2012. Consumer acceptance and use of information technology: Extending the Unified Theory of Acceptance and Use of Technology. *MIS Quarterly*, 36(1), 157–178.

Xu, X. 2014. Understanding users' continued use of online games: An application of UTAUT2 in social network games" *MMEDIA*.

Corporate transformation, digital culture, and employee performance in a digital company

Reza Bachtiar
Faculty of Economic and Business, Telkom University, Bandung, Indonesia

Ade Irma Susanty
Faculty of Communication and Business, Telkom University, Bandung, Indonesia

ABSTRACT: In 2012, the company carried out a corporate transformation for the sake of surviving and growing in the digital age. In addition to the business model, the corporate culture has also been changed. Due to a slowdown in the growth of the company's business performance and an increase in entropy of corporate culture, the company began implementing digital culture in 2016 to improve its performance and become a digital mastery. The company defines the values of digital culture as matters relating to agility, anticipation, creativity, innovation, experimentation, open thinking, and networking. The purpose of this research was to find out and analyze the effects of corporate transformation and digital culture on employee performance. Data were collected using questionnaires from 288 respondents out of a population of 855 employees. After analyzing the data using Path Analysis, the results revealed that the two variables of directly and indirectly affected employee performance.

1 INTRODUCTION

As business competition among companies continues to get tougher and tougher these days, human resources are the key to success. A company is considered to have a good performance if it has increased profits and improved employee performance (Siren, Kohtamaki, & Kuckertz 2012). The quality and future of a company is largely determined by the quality of its human resources.

Therefore, companies must transform themselves in order to survive or win the competition. A company's competitive advantage is often the result of a continuous transformation process. Indeed, transformation is not an easy thing to do since the company will be faced with a number of obstacles and consequences during the process.

In addition, each company has different characteristics considered as their identities. These characteristics are often referred to as corporate culture. Corporate culture is a unique blend of norms, values, beliefs, and typical behavior of a group of individuals in completing certain tasks. Culture is related to how companies build commitment to realize their vision, win the hearts of customers, win competition, and build company strength (Robbins 2014).

Today's business developments require companies to carry out digital-based operations on various aspects. Therefore, it is a challenge for companies to build digital culture through digital-based work behavior.

Considering the fact that it takes a number of transformations for companies in Indonesia to survive and win competition in the digital age, the objectives of this study are prepared as follows:

1. To analyze the effect of corporate transformation on the company's digital culture.
2. To analyze the effect of company transformation on employee performance.
3. To analyze the effect of the company's digital culture on employee performance.
4. To analyze the effect of corporate transformation and digital culture on employee performance.

2 LITERATURE REVIEW

Basically, all transformations lead to an increase in the effectiveness of the company to be aligned with changes in the environment and behavior of members (Robbins 2014, p. 339). Robbins (2014, p. 484) stated that corporate transformation can be carried out with regard to 4 things namely company structure, technology, physical structure and human resources.

Robbins (2014) added that corporate culture is a perception generally agreed upon by company members in a system of shared meanings. Corporate culture is related to how workers feel incorporated in the overall characteristics of the company. Meanwhile, the company's digital culture is a consequence of the process of cultural transformation within the company influenced by technological developments and innovations that shift the behavior and habits of the company, so they can survive in facing challenges and continue to grow in an era of disruption. This is in line with what was delivered by Rudito (2017: 98) that the digital culture in a company is able to support the process of adopting and developing digital technology to generate business performance and achieve sustainable success. Digital culture that must be owned by a company includes agility, innovation, anticipation, experimentation, open mindset and networking.

Employee performance is considered as an asset for company effectiveness (Arinanye 2015, p. 1). According to Arinanye (2015), factors that influence employee performance are efficiency, quality, productivity, and timeliness.

3 RESEARCH FRAMEWORK

The framework used in this study is illustrated in the following Figure 1:

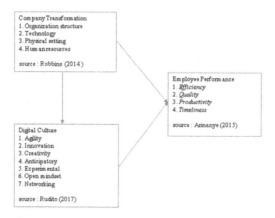

Figure 1. Research framework.

4 RESEARCH HYPOTHESES

Referring to the research framework, the research hypotheses proposed are as follows:
H1. Corporate transformation has a positive and significant effect on employee performance
H2. Corporate transformation has a positive and significant effect on digital culture.
H3. Digital culture has a positive and significant effect on employee performance.

5 RESEARCH METHOD

Data in this study were collected by surveying employees of a digital company. The total sample was 273 respondents from a total population of 855. Samples were taken based on the Slovin model. Questionnaires were used to collect data. Data were analyzed using a path analysis technique.

6 RESEARCH MODEL

The following figure shows the path diagram of the causal relationship between the variables of corporate transformation (X1), digital culture (X2) and employee performance (Y) proposed in the theory:

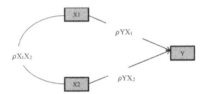

Figure 2. Path diagram model.

7 RESEARCH RESULTS

Simultaneous Path Coefficient Testing

Table 1. Simultaneous path analysis testing.

Hipothesis	F Calculate	db	F Table	Decision	Summary
X_1 dan X_2 effect Simultaneously to Y	126.693	$db_1=2$ $db_2=285$	3.027	H_0 Rejected	Significant

The simultaneous Path Analysis test results in Table 1 above show corporate transformation and digital culture have a significant and simultaneous effect on employee performance.
Partial Path coefficient Testing

Table 2. Partial path analysis testing.

Hipothesis	t Calculate	db	t table	decision	summary
$Phi_{yx1} = 0$	4.512	285	1.968	Ho rejected	Significant
$Phi_{yx2} = 0$	9.148	285	1.968	Ho rejected	Significant

Partial Path Analysis test results in Table 2 show that the variables of corporate transformation and digital culture partially have a significant effect on employee performance.

Direct and Indirect Effects of Corporate Transformation and Digital Culture on Employee Performance:

Table 3. Direct and indirect effects.

Connection	Path coefficient	Direct Effect	Indirect Effect X1	Indirect Effect X2	Total Effect
X1	0.248	0.062		0.078	0.139
X2	0.503	0.253	0.078		0.331
		Total			0.470

Based on Table 3, it is known that the direct and indirect effects are:

1. The direct effect of X1 on Y is 0.062; the indirect effect of X1 on Y through X2 is 0.078; so the total effect of X1 on Y through X2 is 0.139.
2. The direct effect of X2 on Y is 0.253; the indirect effect of X2 on Y through X1 is 0.078; so the total effect of X2 on Y through X1 is 0.331.

8 RESEARCH DISCUSSION

The first finding of this study is that digital culture variable has a significant positive effect on employee performance. This is indicated by the path coefficient value of 0.502 and the t-count value of 9.148 regarding the effect of company's digital culture variable on employee performance. Among the activities that have been able to increase awareness and implementation of digital culture in the company include organizing an employee representative program.

This study also revealed that the variables of corporate transformation and digital culture have a significant positive effect on employee performance. This is indicated by the path coefficient value of 0.471 and the t-count value of 126.693. This supports research conducted by Muliaty & Basri (2017) stating that transformation and company culture have a significant positive effect on employee performance.

Another result of this study is that the variable of corporate transformation has a significant positive effect on employee performance. This is indicated by the path coefficient value of 0.248 and the t-count value of 4.512. The effect of corporate transformation is driven by technological development, thus corporate transformation can be maximized by utilizing technology.

Finally, the last result of the study indicates that the company transformation variable has a significant positive effect on the company's digital culture. This is indicated by the path coefficient value of 0.622 and the t-count value of 13.430. This defines that the better the implementation of corporate transformation is carried out, the stronger the digital culture existence in the company will become.

9 CONCLUSION

Based on data processing and analysis, it can be concluded as follows:

1. Based on the path analysis, it is known that corporate transformation has a significant positive effect on the digital culture of the company with a path coefficient value of 0.622,
2. Based on path analysis, it is known that corporate transformation has a significant positive effect on employee performance with a path coefficient of 0.248,

3. Based on path analysis, it is known that the company's digital culture has a significant positive effect on employee performance with a path coefficient of 0.503.
4. Based on path analysis, it is known that corporate transformation and corporate digital culture have a significant positive effect on employee performance with a path coefficient of 0.471.

REFERENCES

Arinanye, R. T. (2015). Organizational Factors Affecting Employee Performance at the College of Computing and Information Sciences (Cocis). A Master's Dissertation to Makerere University: Not Published.

Indrawati, P. D. (2015). Metode Penelitian Manajemen dan Bisnis: Komvergensi Teknologi Komunikasi dan Informasi (Management and Business Research Methods: Convergence of Communication and Information Technology). Bandung: PT Refika Aditama.

Muliaty, Muliaty; Basri, Muhammad; Jasruddin, Jasruddin. (2017). Effects of Organizational Transformation and Culture on Employees Performance. Journal of Economic & Management Perspectives. Vol. 11. 1287–1292.

Noor, J. (2011). Metodologi penelitian: skripsi (Doctoral dissertation, tesis, disertasi, dan karya ilmiah) Research methodology: thesis (Doctoral dissertation, thesis, dissertation, and scientific work). Jakarta: Prenada Media.

Robbins, S. P., & Judge, T. (2014). Essentials of Organizational Behavior. London, UK:Pearson,.

Rudito, P., & Sinaga, M. F. N. (2017). Digital Mastery, Membangun Kepemimpinan Digital Untuk Memenangkan Era Disrupsi. (Digital Mastery, Building Digital Leadership to Win the Era of Disruption). Jakarta: Gramedia Pustaka Utama.

Sirén, C. A., Kohtamäki, M., & Kuckertz, A. (2012). Exploration and exploitation strategies, profit performance, and the mediating role of strategic learning: Escaping the exploitation trap. Strategic Entrepreneurship Journal, 6(1),18–41.

How internal and organizational factors influence employees' innovative behavior

Adityawarman
Indonesian Agency for Meteorology, Climatology and Geophysics, Indonsia

R. Rachmawati
University of Indonesia, Indonesia

ABSTRACT: Having employees with high innovative work behavior (IWB) is one option for an organization to innovate in order to adapt to rapid changes and maintain its performance. This article builds on previous research on how internal and organizational factors affect IWB and finds solutions to enhance IWB, especially in public service organizations. We propose servant leadership, a leadership style that promotes the well-being of employees, and learning goal orientation (LGO). We also propose the promotion of psychological capital (PsyCap) as a psychological resource to manage IWB. Using CFA to test the measurement model and SEM to test the structural model, we found that external factors only influence IWB indirectly. Both internal and organizational factors draw from employees' psychological resources to generate IWB. While servant leadership builds a safe environment and LGO provides engagement, PsyCap helps employees to take a proactive role in innovation.

1 INTRODUCTION

Innovation is an indispensable factor in enabling organizations to adapt to rapid economic changes and gain a competitive advantage (Bos-Nehles, Rankema, & Janssen, 2017). Employees with high innovative work behavior (IWB) can help organizations to innovate (Agarwal, 2014). Both organizational and internal support foster IWB (Huang & Luthans, 2015; Yoshida, Sendjaya, Hirst, & Cooper, 2014). In public service organizations, IWB is perceived as an extra-role behavior to be compensated for (Bysted & Jespersen, 2014). This means that a clear performance reward is necessary for public employees to perform IWB. Since the performance reward system in public organizations is bound to government regulations, management needs to find other practical solutions for enhancing IWB.

2 LITERATURE REVIEW

2.1 *Innovative Work Behavior (IWB)*

Innovative work behavior can be characterized as "the intentional creation, introduction and application of new ideas within a work role, group or organization, in order to benefit role performance, the group, or the organization" (Janssen, 2000). Innovation occurs only if employees engage in activities aimed at generating and implementing ideas. Employees can initiate innovations because they are in frequent contact with processes and products and can detect potential improvements and opportunities for new developments (Bos-Nehles et al., 2017).

Innovative work behavior consists of three stages: idea generation, idea promotion, and idea realization (Janssen, 2000). Idea generation refers to creativity-oriented behavior, which is preceded by the process of problem recognition, while the last two behavioral sets refer to implementation-oriented behavior (Dorenbosch, Van Engen, & Verhagen, 2005). Since innovation processes are often characterized by discontinuous activities, individuals can be expected to be involved in any combination of these behaviors at any time (Scott & Bruce, 1994).

2.2 *Servant leadership*

Servant leaders put followers' development and interests above those of the organization. This in turn leads followers to reciprocate in the form of discretionary behaviors (Newman, Schwarz, Cooper, & Sendjaya, 2015). Yoshida et al. (2014) show the importance of servant leadership in promoting follower trust and representing the collective, which in turn fosters employee creativity and team innovation.

Hypothesis 1 Servant leadership will be positively related to innovative work behavior.

2.3 *Learning Goal Orientation (LGO)*

Learning goal orientation is an individual preference to develop new skills, acquire new knowledge, and master new situations. It also has been found to significantly impact employees' skill acquisition and learning at work. Employees with high LGO usually proactively seek new information and methods to acquire role, organizational, and social learning (Tan, Au, Cooper-Thomas, & Aw, 2016). In contrast, employees with low LGO are more likely to get frustrated, exhausted, or drained in the face of setbacks and challenges (DeShon & Gillespie, 2005). They are likely to believe they lack the resources necessary to cope with challenges (Peng, Zhang, Xu, Matthews, & Jex, 2018).

Hypothesis 2 Learning goal orientation will be positively related to innovative work behavior.

2.4 *Psychological Capital (PsyCap)*

Psychological capital plays an important role in impacting the performance attitudes of employees, and may contribute to an organization's competitive advantage (Luthans, Norman, Avolio, & Avey, 2008). In a resourceful work environment where employees obtain sufficient social support, have various high-performance work practices, and observe ethical behaviors displayed by superiors, the employees will be high in PsyCap (Karatepe & Talebzadeh, 2016).

Hypothesis 3 Servant leadership will be positively related to psychological capital.

Learning goal-oriented individuals are aware of the complexity of the task environment and thus show an adaptive ability associated with their self-regulatory focus on the achievement of goals. They tend to see the positive aspects of the uncertain task environment because of their intrinsic interests in exploring opportunities and their task-related confidence in overcoming obstacles (Huang & Luthans, 2015).

Hypothesis 4 Learning goal orientation will be positively related to psychological capital.

Sweetman, Luthans, Avey, and Luthans (2011) and Abbas and Raja (2015) found that employees with higher psychological capital will more likely apply IWB in the workplace. Each component of PsyCap is needed for developing or maintaining IWB.

Hypothesis 5 Psychological capital will be positively related to innovative work behavior.

Hypothesis 6 Psychological capital will mediate the relationship between servant leadership and innovative work behavior.

Hypothesis 7 Psychological capital will mediate the relationship between learning goal orientation and innovative work behavior.

3 METHODS

3.1 Sample and procedure

Using a purposive sampling method, we collected online survey data from 407 nonmanagerial employees from one government organization based in several areas in Indonesia. The respondents included 250 males and 157 females; 76.4% of the respondents were younger than 35 years old.

Table 1. Descriptive analysis.

No	Variable	Mean	SD	1	2	3	4	5	6	7	8
1	Jenis Kelamin	1,381	0.486	1							
2	Usia	31,880	0.872	−0.071	1						
3	Pendidikan	2,877	0.651	0.039	−0.055	1					
4	Masa Kerja	2,064	0.846	−0.029	0.729**	0.0321					
5	LGO	4,739	0.855	−0.176**	0.068	0.063	0.088	1			
6	SL	4,544	1.081	−0.27	−0.016	−0.018	0.001	0.289**	1		
7	PsyCap	4,627	0.900	−0.136**	0.090	0.050	0.088	0.533**	0.340**	1	
8	IWB	4,005	1.052	−0.125*	0.146**	0.039	0.159**	0.559**	0.327**	0.734**	1

3.2 Measures

Innovative work behavior was measured using nine employee-rated items from Janssen's (2000) scale for individual innovative behavior (α = 0.929). Servant leadership was measured at the individual level using Van Dierendonck's (2017) short version of the Servant Leadership Survey (α = 0.965). The five-item LGO subscale was drawn from VandeWalle's (1997) goal orientation scale (α = 0.852). Psychological capital was measured using the short version of the Psychological Capital Questionnaire (α = 0.965). All items were measured on a six-point Likert scale.

4 RESULT

4.1 Construct validity

For addressing discriminant validity, confirmatory factor analysis and average variance extracted (AVE) were evaluated. The model showed a good fit for several indices (RMSEA = 0.049; CFI = 0.945; IFI = 0.945; TLI = 0.941; AGFI = 0.819; PGFI = 0.749). The model also showed a good construct validity with factor loading > 0.5, AVE > 0.5, and CR > 0.7.

4.2 Hypotheses testing

For hypothesis 1, the model failed to show the significant path between servant leadership and IWB (β = 0.09; P > 0.05). Hypotheses 2 to 5 were supported (P < 0.05). For indirect effect testing (hypotheses 6 and 7), we used a bias-corrected bootstrapping method with 500 bootstrap samples and a 95% confidence interval. The result supported hypotheses 6 and 7, which means that PsyCap fully mediated servant leadership and IWB and partially mediated LGO and IWB.

5 DISCUSSION

Consistent with previous work by Yoshida et al. (2014), we found that servant leadership does not directly influence IWB. Despite its weak indirect effect, servant leadership is still important in promoting follower trust and in providing an environment that fosters IWB. On the other hand, LGO positively influences IWB both directly and indirectly and strongly influences PsyCap. Learning goal orientation provides the engagement that is required in the earlier stages of IWB. This result supported the findings of Huang and Luthans (2015). Psychological capital also has a strong effect on IWB, which means that, as an extra-role behavior, IWB requires internal resources to fuel the process. Psychological capital drives employees to take proactive steps. Both organizational and internal factors are drawn from employees' psychological resources to generate IWB.

6 MANAGERIAL IMPLICATIONS

Innovative work behavior is one of the important determinants for organizational performance that are not easy to foster, especially in public service organizations. This research finding shows that internal factors (LGO and PsyCap) have more influence to IWB compared to external factors (servant leadership). Thus, rather than depending on managers to enhance their employees' IWB, organizations need to focus on selecting employees with high LGO and enhancing their PsyCap through capacity building.

REFERENCES

Abbas, M., & Raja, U. 2015. Impact of psychological capital on innovative performance and job stress. *Canadian Journal of Administrative Sciences*, 32, 128–138.

Agarwal, U. A. 2014. Linking justice, trust, and innovative work behaviour to work engagement. *Personnel Review*, 43(1), 41–73.

Bos-Nehles, A., Renkema, M., & Janssen, M. 2017. HRM and innovative work behaviour: A systematic literature review. *Personnel Review*, 46(7), 1228–1253.

Bysted, R., & Jespersen, K. R. 2014. Exploring managerial mechanisms that influence innovative work behaviour: Comparing private and public employees. *Public Management Review*, 16(2), 217–241.

DeShon, R. P., & Gillespie, J. Z. 2005. A motivated action theory account of goal orientation. *Journal of Applied Psychology*, 90(6), 1096–1127.

Dorenbosch, L., Van Engen, M. L., & Verhagen, M. 2005. On-the-job innovation: The impact of job design and human resource management through production ownership. *Creativity and Innovation Management*, 4(2), 129–141.

Huang, L., & Luthans, F. 2015. Toward better understanding of the learning goal orientation–creativity relationship: The role of positive psychological capital. *Applied Psychology*, 64(2), 444–472.

Janssen, O. 2000. Job demands, perceptions of effort–reward fairness and innovative work behaviour. *Journal of Occupational and Organizational Psychology*, 73(3), 287–302.

Karatepe, O. M., & Talebzadeh, N. 2016. An empirical investigation of psychological capital among flight attendants. *Journal of Air Transport Management*, 55, 193–202.

Newman, A., Schwarz, G., Cooper, B., & Sendjaya, S. 2017. How servant leadership influences organizational citizenship behavior: The roles of LMX, empowerment, and proactive personality. *Journal of Business Ethics*, 145(1), 49–62.

Peng, Y., Zhang, W., Xu, X., Matthews, R., & Jex, S. 2018. When do work stressors lead to innovative performance? An examination of the moderating effects of learning goal orientation and job autonomy. *International Journal of Stress Management*, 26(3), 250–260.

Scott, S. F., & Bruce, R. A. 1994. Determinants of innovative behavior: A path model of individual innovation in the workplace. *Academy of Management Journal*, 37(3), 580–607.

Sweetman, D., Luthans, F., Avey, J. B., & Luthans, B. C. 2011. Relationship between positive psychological capital and creative performance. *Canadian Journal of Administrative Sciences*, 28(1), 4–13.

Tan, K. W. T., Au, A. K. C., Cooper-Thomas, H. D., & Aw, S. S. Y. 2016. The effect of learning goal orientation and communal goal strivings on newcomer proactive behaviours and learning. *Journal of Occupational and Organizational Psychology*, 89(2), 420–445.

Yoshida, D. T., Sendjaya, S., Hirst, G., & Cooper, B. 2014. Does servant leadership foster creativity and innovation? A multi-level mediation study of identification and prototypicality. *Journal of Business Research*, 67(7), 1395–1404.

Luthans, F., Norman, S.M., Avolio, B.J. and Avey, J.B (2008) The Mediating Role of Psychological Capital in the Supportive Organizational Climate-Employee Performance Relationship. Journal of Organizational Behaviour.

Effect of work placement and workloads on performance of the PKH companion

Nova Maulida & Ratri Wahyuningtyas
Telkom University, Bandung, Indonesia

ABSTRACT: The purpose of this study was to see the extent of the effectiveness of work placements and workload to improve companion performance. The study also wanted to see the extent of the effect of workload on performance Program Keluarga Harapan companion. The research method used is a quantitative method. The sampling technique used was using saturated sampling, where all members of the population were sampled, namely 310 people. The study also used the help of SPSS 24 software to process questionnaire data. Meanwhile, the analysis of the study used descriptive analysis, multiple regression analysis, partial testing and hypothesis testing. The results showed that work placement and workload had a positive and significant influence on the performance of Program Keluarga Harapan companion in Bandung Regency. This means that the higher the work placement and the low workload will increase performance of Program Keluarga Harapan companion.

1 INTRODUCTION

Companion of program keluarga harapan is a government program implemented to alleviate poverty. In order to achieve the objectives of the Program Keluarga Harapan, a maximum performance by the companion is requested. Companion of Program Keluarga Harapan in Bandung Regency. The purpose of this study is to discuss plans for work placement and workload on the accompanying performance of the companion of Program Keluarga Harapan.

Reasons for research on work placements and workloads on Program Keluarga Harapan Companion opinions taken at Program Keluarga Harapan Bandung regency is a district based in Bandung regency, which consists of neighboring, cross, hilly, and various areas of urban coastal areas. In addition, the distance from one village in one district is quite far because Bandung Regency has a wider area of the city of Bandung with an area of 1,767.96 km2 and 167.67 km2.

2 LITERATURE REVIEW

The concept of placement includes even promotions, transfers and even demotion (Siagian, 2015: 168-169). Hasibuan (2017: 64), argues that in accordance with the principle "The right man in the right place and the right man behind the right job" or "the right place for the right place and the right placement of people for the right position".

Workloads based on the COR theory on Hobfoll (1989), Employees in companies face a heavy workload and therefore tend to lose valuable resources such as energy, time and emotions. (Mansour and Tremblay 2015: 1783). Workload is a set or number of activities that must be completed in an organizational unit and holders of positions in certain times (Dhania, 2010: 16). Gibson and Ivancevich (1993: 163), explain that pressure is a response of someone who is unable to adjust, which is influenced by individual or psychological differences, and becomes a consequence of any external action (environment, situation, events that make too many psychological demands or physical) towards someone.

Etymologically the term "performance" comes from the word "Job Performance" or "Actual Performance" which is a work achievement or a real achievement achieved by someone. As for terminologically, the definition of performance is "the results of work in quality and quantity achieved by an employee in carrying out his duties in accordance with the responsibilities given to him" (Mangkunegara, 2017: 67). According to Wibowo (2017: 7) performance comes from the notion of performance as a result of work or work performance.

2.1 *Conceptual framework*

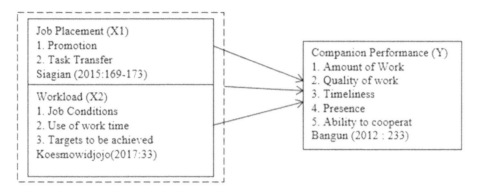

Figure 1. Conceptual framework.

Hypothesis I: Job placement has a significant effect on the performance of the Program Keluarga Harapan Companion in Bandung Regency.

Job placement can be said to have a significant influence according to the companion when there are things that the companion feels makes his job easier due to the existence of an appropriate work placement. If the inappropriate work placement in this case is transferred to a place that is further away from its domicile, the companion regarding work placement becomes less effective to performance.

Based on previous research conducted by Tobing and Zamora (2018) proving that work placements have a significant and positive effect on performance by being influenced by several factors namely work knowledge, attitude and work skills. These three factors will determine whether the work placement is appropriate or not. Inappropriate work placements will affect the results of the work done which is then assessed as performance.

Hypothesis II: Workload has a significant effect on the performance of the Program Keluarga Harapan Companion in Bandung Regency.

Workload is the amount of work carried by a person in a particular office/unit and is the result of work volume and time norms. Workload can be said to be light and heavy depending on how much volume of work there is with the time provided. If the time provided is not in accordance with the amount of work provided, the work has a heavy workload.

The research conducted by Oktaviana (2017) shows that workload is an influential thing in assessing the performance of this study comparing employee attainment seen from the type of task and time given. So that the assumption or perception of employees of workload will focus on the way to assess the work received and the time given.

Hypothesis III: Job placement and workload have a significant effect on the performance of the Program Keluarga Harapan Companion in Bandung District.

Performance which means work results in quality and quantity achieved by an employee in this case the companion in carrying out his duties in accordance with the responsibilities given to him is influenced by whether he is placed, the position he has and the workload he is responsible for. Performance is also influenced by work placement and workload as the results of the study revealed.

3 RESEARCH METHODOLOGY

Multiple linear regression analysis uses to examine the effect of independent variables (work placement and workload) on the dependent variable (companion performance), and to predict or estimate the dependent variable-bell value based on the influence of the independent variable. The regression equation is as follows:

$$Y = a + b1X1 + b2X2 \qquad (1)$$

Information
Y = PKH Companion Performance Variables
a = Number of regression constants
b = slope (regression coefficient) of the independent variable
X1 = Job placement variable
X2 = Workload variable

Based on data processing using SPSS version 24 software, the results of multiple linear regression analysis can be arranged from the coefficient values shown in the following table

Table 1. Results of multiple regression analysis.

Model	Unstandardized Coefficients		Standardized Coefficients		
	B	Std. Error	Beta	t	Sig
1 (Constant)	18.481	2.512		7.356	.000
PK	.297	.161	.125	1.840	.067
BK	.703	.103	.465	6.820	.000

Based on the results of data processing displayed in table, the regression coefficient values of each variable are obtained, so that multiple linear regression equations can be arranged as follows:

$$Y = 18,481 + (0,297)PK + (0,703)BK \qquad (2)$$

The interpretation of the multiple linear regression equation above is as follows:
  a = A constant of 18.481 means that if there is no influence of placement and workload, the performance will be worth 18.481 units.
  $\beta1$ = Regression coefficient value of variable X1 (Work Placement) is equal to 0.297 These results indicate that work placement has a positive effect on companion performance, namely the more effective job placement
  a better performance of the companion. Each increase in work placement by one unit while the constant value variable (0) will increase companion performance by 0.297.
  $\beta2$ = The regression coefficient of variable X2 (Workload) is 0.703. These results indicate that the workload has a positive effect on companion performance, namely the increasing workload will improve the performance of the companion. Each increase in workload is equal

to one unit while the workload variable is of constant value (0), will increase the Companion's performance by 0.703.

4 CONCLUSION

Based on the results of the research and discussion presented in the previous chapter, the conclusions in this study can be presented as follows:

1. The implementation of the Program Keluarga Harapan Companion's work placement is at the calculation of the percentage is 89.03%. This shows that the transfer of tasks in the organization is very effective where the transfer of tasks can increase morale. Meanwhile, according to Hariandjo (2006) work placement is the process of placing/filling in a position or reassigning a different new assignment/position.
2. Program Keluarga Harapan Companion Workload is in the low category or the percentage is 79.49% meaning that the work and tasks that must be done are not burdensome to the companion. The companion workload is low because the work target has been set by the organization (PPKH) and the time is not excessive. In line with the opinion according to Tarwaka (2011:106) is a work condition with a description of the work must be completed at a certain time limit.
3. Program Keluarga Harapan Companion Performance is in a very high category Overall the work placement variable has a percentage of 89.00%. . According to Hasibuan (2007) Performance is a result of work achieved by someone in carrying out the tasks given to him based on experience skills as well as seriousness and time.
4. Job Placement has a positive effect on Performance. Based on the processed data, the value of t is calculated for the Job Placement (PK) variable of 1,840. While the value of t table is 1,649, it is obtained t count PK> t table or 1,840> 1,649. In line with the opinion of Muaja (2017) there is a positive influence between job placement and performance.
5. Workload has a positive effect on Performance. Variables. Workload where t count Workload (BK) is 6.820. While t table is 1,649, it is obtained t count BK> t table or 6,820> 1,649. Whereas according to Paramitadewi (2017) there is a negative influence on Workload on performance.
6. Work Placement and Workload simultaneously influence the performance of Placement Variables and Workloads on Performance amounting to 0.316. The results show that the performance variable is explained by the variable work placement and workload of 31.6%, the remaining 68.4 explained the influence of other factors not examined in this study. According to Runtuwene (2016) there is a significant influence between Workload, Job Placement and Workload.

5 RECOMMENDATION

Regarding the results of the research on the effect of work placement on performance, the Bandung District Social Service and Bandung Regency PKH companion should conduct work placement evaluations related to placement effectiveness that have been determined periodically at least once a month so that PKH Facilitators can convey their constraints and obstacles in the field caused by placement of a remote working area or difficult geographical conditions. The results of this evaluation at the monthly coordination meeting will be the basis for the change in work placement through a new assignment letter.

Regarding the results of the study of the effect of workload on performance, the Bandung District Social Service and Bandung Regency PKH companion could hold capacity building aimed at increasing PKH facilitators' capacity to manage their stress so as not to overdo the burn out. In addition, the Bandung District Social Service together with the PKH Supervisor carried out the provision of motivation both through oral and with a reward system where

PKH companion who successfully performed assignments within a period of time received rewards in any form.

REFERENCES

Arifin, Johar. (2017). SPSS 24 untuk Penelitian dan Skripsi. Jakarta: PT. Elex Media Komputindo.
Arikunto, Suharsimi. (2013). Prosedur Penelitian: Suatu Pendekatan Praktik. Jakarta: Rineka Cipta.
Badan Perencanaan Daerah. (2007). Master Plan Pendidikan Kabupaten Bandung 2008 – 2025. BPD Kabupaten Bandung.
Bangun, Wilson. (2012). Manajemen Sumber Daya Manusia., Jakarta: Erlangga.
Clifton P. Campbell (1996).Job performance aids. Vol. 20 Issue: 6, pp.3–21, https://doi.org/10.1108/03090599610119269
Cooper, D.R. dan Schindler, P.S. 2014. Business Research Methods. New York: McGraw-Hill.
Croswell, John W. (2008) Research Design, Pendekatan Kualitatif, Kuantitatif, dan. Mixed, Edisi Ketiga. Bandung: Pustaka Pelajar.
Dinas Sosial Kabupaten Bandung (2016). Rencana Strategis 2016-2021. Bandung: Pemerintah Kabupaten Bandung.
Hasibuan, Malayu. (2017). Manajemen Sumber Daya Manusia. Jakarta: PT. Bumi Aksara.
Huraerah, Abu. (2008). Pengorganisasian dan Pengembangan Masyarakat. Model dan Strategi Pembangunan Berbasis Masyarakat. Bandung: Humaniora, Penerbit Buku Pendidikan – Anggota IKAPI.

Designing KPIs for measurement of operations in PERUM Jaminan Kredit Indonesia

R.E. Banjarnahor, U. Cariawan & Arviansyah
University of Indonesia, Jakarta, Indonesia

ABSTRACT: Perusahaan Umum (Perum) Jaminan Kredit Indonesia, hereinafter referred to as Perum Jamkrindo, is the only state-owned enterprise engaged in credit guarantee that has been established since 1970. Until now, Perum Jamkrindo has not had a performance measurement that integrates a financial perspective with a nonfinancial perspective. As a company with a vision to become the leader of guarantee that supports the development of the national economy, Perum Jamkrindo supposedly has a performance measurement that does not only focus on the financial perspective as reflected in aspects of the evaluation of Kriteria Penilaian Kinerja Unggul (KPKU) as mandatory by the Ministry of State-Owned Enterprises of the Republic of Indonesia. The purpose of this research was to provide suggestions and input in compiling and designing performance measurements for Perum Jamkrindo.

1 INTRODUCTION

At present the business world is facing an era of digitalization caused by technological advances. Rhenald Kasali (2017) argues that digitalization means an exponential multiplication of growth rates. Every business activity must be able to follow every change that occurs. In response to this, companies need to establish and define clear, focused visions. In implementing predetermined work goals, continuous change is needed, as Chris McChesney, Sean Covey, and Jim Huling explain in their book *The 4 Disciplines of Execution*: "When you execute a change in the behavior of other people, you are facing one of the greatest leadership challenges" (Chris, Covey, & Huling, 2012).

2 LITERATURE REVIEW

2.1 *Performance measurement*

For decades, a number of performance measurement methods and frameworks have been introduced and implemented, including performance measurement matrices, SMART pyramids, results frameworks and determinants, balanced scorecards, and performance prisms. The performance appraisal framework is created with the aim of helping companies to determine a series of performance measurements that reflect their objectives and to assess company performance (Farokhi & Roghanian, 2017). According to Neely, Gregory, and Platts (1995), performance measurement is closely related to quality, time, cost, and flexibility. Michael Armstrong (2014) states that performance management is an ongoing process to improve performance by setting individual and team goals that are aligned with the company's strategic goals, planning performance to achieve goals, reviewing and assessing progress, and developing knowledge, skills, and individual abilities.

2.2 Balanced scorecard

One method that can be used to measure the performance of a company is the balanced scorecard as introduced by Robert Kaplan. In addition to using a financial perspective (traditional approach), the balanced scorecard also measures company performance from a nonfinancial perspective such as consumers, business processes, growth, and learning. According to Kaplan and Norton (1996), the balanced scorecard must explain the company's mission and vision in various objectives and measurements. The balanced scorecard presents the company's vision and mission in performance measurements classified into four perspectives: financial, consumer, internal business process, and growth and learning.

2.3 Strategy map

According to Kaplan and Norton (2004), the strategy illustrates how a company seeks to create sustainable value for shareholders. The value created can come from tangible or intangible assets. This results in companies needing a general strategy model to use as a performance measurement system. One model that can be used is the balanced scorecard framework. The balanced scorecard only offers a corporate strategy framework in relation to value creation (Kaplan & Norton, 2004). The four perspectives in the balanced scorecard are related to each other in the form of cause-and-effect linkage. The relationship and linkages of that perspective are known as strategy maps, which are visual representations of cause-and-effect relationships between organizational strategy components.

2.4 *Kriteria Penilaian Kinerja Unggul (KPKU)*

The Ministry of State-Owned Enterprises instructs all stated-owned enterprises (SOEs) in Indonesia to harmonize in structuring a performance management system through the Letter of the State Owned Enterprises No. S-08 / S.MBU / 2013, dated January 16, 2013, regarding the Submission of Guidelines for Determining KPIs and Superior Performance Assessment Criteria of State-Owned enterprises. The Kriteria Penilaian Kinerja Unggul (KPKU) is a guideline in improving Badan Usaha Milik Negara (BUMN) performance and in measuring the performance of SOEs, adopted and adapted from the Malcolm Baldrige Criteria for Performance Excellence (MBCfPE), where the MBCfPE has proven to be the most comprehensive performance management system that can make companies superior. The KPKU BUMN criteria involve how the day-to-day business of a company is managed, so that the KPKU BUMN is designed to encourage the company's vision, mission, and objectives, and is also designed to be used as a holistic assessment tool to measure companies and show what companies need to improve performance in the long term (Indonesia, 2015). The performance measurement of superior criteria according to the KPKU consists of seven characteristics: leadership, strategic planning, customer focus, measurement, analysis and knowledge management, workforce focus, focus on operation, and focus on results.

3 RESEARCH METHODOLOGY

3.1 Research framework

This study utilized qualitative research. This research was limited to the preparation of key performance indicators (KPIs), from the preparation of a corporate strategy map, to corporate KPIs (as determined by the Ministry), to the level of division KPIs, especially divisions under the Directorate of Operasional dan Jaringan. The selection of the work unit was carried out because at this time the researcher was working in one of the divisions under the Directorate of Operasional dan Jaringan. A case study method was used in this research. Case studies were carried out on two SOEs that have also implemented KPKU in performance measurement.

4 DATA COLLECTION TECHNIQUE

Data collection was carried out using semi-structured interview methods. Semi-structured interviews generally begin with specific questions and then flow along the line of thought (Cooper & Schindler, 2014). In this research, interviews were conducted by preparing specific questions in advance. The interviews consisted of six main questions conducted with two interviewees, Mr. Tjipto Ismoyo as PT Garuda Indonesia's financial expert on February 22, 2019, and Mr. Muhartono as AVP of PT Telkom on March 11, 2019.

5 RESULTS

5.1 *Strategic map design*

Corporate strategic mapping is a management tool used to translate the vision, mission, and strategy of Perum Jamkrindo. Company results or performance categories assess performance achievements in all key areas, including product and process results, customer focus results, workforce focus results, leadership and governance results, and financial and market results. It is necessary to make adjustments in the company's strategy map to suit the KPKU and the needs of the Ministry of State-Owned Enterprises.

5.2 *KPI design for performance measurement*

In developing a strategic plan, companies must begin by determining their vision and end with the implementation of the plan by aligning the functions of each unit in the company (Kaplan & Norton, 2008). Alicia Mateos Ronco and Jose Manuel Hernandez Mezquida (2018) mention the sequence in the preparation of performance appraisal based on balanced scorecards in the following stages: determine the company's vision and mission, develop a company strategy map, and make performance measurement indicators. McChesney et al. (2012) provide a description of the lead and lag indicators as two interrelated indicators in achieving the company's vision and mission. A lag indicator is a measure of the results to be achieved while a lead indicator is a measure of achieving the goals. Thus, the lag indicator (final goal) will be achieved only if the lead indicator (the process of achieving the final goal) has been clearly determined.

Based on the result of the interviews and research and taking into account the application of PT Garuda Indonesia and PT Telkom, a reduction in the company's KPI to the KPI of the unit can be arranged with the following matriculation:

Table 1. Matriculation of cascading KPI.

Robert S. Kaplan and David P. Norton	Alicia Mateos and Jose Manuel Hernandez Mezquida	Kladogeni Anthoula Hatzigeorgiou Alexandros	PT Garuda Indonesia	PT Telkom
1. Penentuan visi	1. Penentuan visi dan misi	1. Penentuan visi dan misi	1. Penentuan visi dan misi	1. Peta strategi
2. Penentuan strategi perusahaan	2. Peta strategi	2. Lead or tag indicators	2. Peta strategii	2. KPI Korporat
3. Peta strategi	3. Menyusun indikator pengukuran kinerja	3. Kuantifikasi Strategi	3. KPI Korporat	
4. Penyusunan rencana kinerja			4. KPI Direktorat	3. Tactical
			5. Penyusunan program Gram kerja	4. Reporting Management
				5. Daily Operation

By using the stages that Alicia Mateos Ronco and Jose Manuel Hernandez Mezquida employ, and based on the organizational structure and job descriptions applicable, the stages of the preparation of performance measurements can be arranged. Preparing the performance measurement starts with a strategic map. The strategic map is then used as a basis for the preparation of corporate KPIs and then divided into KPIs for each directorate. From the corporate KPI down to the directorate KPI, the performance measurement indicators are determined in the form of lead and lag indicators.

6 CONCLUSION

Based on the analysis carried out in this research, it can be concluded that the preparation of a strategy map based on the KPKU perspective is carried out based on a balanced scorecard. In preparing the strategy map, the strategies of each perspective are interrelated and have a causal relationship. Each strategy will support the other strategies so that in one strategy map there are no stand-alone strategies. Performance measurement as mandated by the BUMN consists of KPI from the KPKU. The corporate KPI of Perum Jamkrindo are grouped according to lead or lag indicators. To reduce the corporate KPI to a unit KPI, cascading of the corporate KPI is conducted. The cascading is carried out by taking into account the organizational structure and job descriptions that apply to Perum Jamkrindo.

REFERENCES

Armstrong, M. 2014. *Armstrong's Handbook of Performance Management: An Evidence-Based Guide to Delivering High Performance*. Kogan Page.
Baldrige, M. 2015. *The Metrology of Organizational Performance: How Baldrige Standards Have Become the Common Language for Organizational Excellence around the World*. Baldrige Performance Excellence Program.
Chris, M., Covey, S., & Huling, J. 2012. *The 4 Disciplines of Execution*. New York: Free Press.
Cooper, D. R., & Schindler, P. S. 2014. *Business Research Methods*. 12th Edition. New York: McGraw-Hill/Irwin.
Gummeson, E. 2000. *Qualitative Methods in Management Research*. Thousand Oaks, CA: Sage.
Indonesia, D. B. I. B. K. B. U. M. R. 2015. *Kriteria Penilaian Kinerja Unggul BUMN 2015*. Jakarta: Kementerian Badan Usaha Milik Republik Indonesia.
Kaplan, R. S., & Norton, D. P. 1996. *The Balanced Scorecard: Translating Strategy into Action*. Cambridge, MA: President and Fellows of Harvard College.
Kaplan, R. S., & Norton, D. P. 2004. *Strategy Maps: Converting Intangible Assets into Tangible Outcomes*. Cambridge, MA: Harvard Business School Press.
Kladogeni, A., & Hatzigeorgiou, A. 2011. Designing a balanced scorecard for the evaluation of a local authority organization. *European Research Studies*, 14(2), 65–80.
McChesney, C., Covey, S., & Huling, J. 2012. *The 4 Disciplines of Execution*. New York: Free Press.
Neely, A., Gregory, M., & Platts, K. 1995. A literature review and research agenda. *International Journal of Operations & Production Management*, 15(4), 80–116. doi.org/10.1108/01443579510083622
Rhenald, K. 2017. *Disruption*. Jakarta: Gramedia Pustaka Utama.
Ronco, A. M., & Mezquida, J. M. H. 2018. Developing a performance management model for the implementation of TQM practices in public education centres. *Total Quality Management & Business Excellence*, 29(5–6), 546–579.
Sekaran, U., & Bougie, R. 2010. *Research Methods for Business: A Skill Building Approach*. Chichester, UK: Wiley.
Toto, P., & Jamkrindo, B. P. 2017. *Industri Penjaminan, Empowering UMKM dan Koperasi Naik Kelas*. Jakarta: Lembaga Manajemen FEB UI dan Jamkrindo Press.
Farokhi, S.; Roghanian, E. Determining quantitative targets for performance measures in the balanced scorecard method using response surface methodology.

Customer continuance intention toward digital banking applications

Indrawati & N.P. Pradhina
Faculty of Economics and Business, Telkom University, Bandung, Indonesia

S.K.B. Pillai
Goa Business School, Goa University, Goa, India

ABSTRACT: Internet banking in Indonesia increased from 150 million transactions in 2012 to 406 million transactions in 2016. The present article examines the digital banking application ABC of Bank ABC of Indonesia to find the factors influencing customers' continuance intention (CI) while using ABC. Results of the analysis using 423 valid respondents identified three main influencing factors, from highest to lowest, as trust (T), habit (H), and price-saving orientation (PSO). The model used can predict the CI of customers toward ABC as the R^2 gave a value of 71%. Results also revealed that male customers consider trust a more important factor than do female consumers. Bank ABC may use this information in its future marketing strategies for retaining existing customers and attracting potential consumers.

1 INTRODUCTION

In parallel with the advancement of use of the Internet in Indonesia (APJII, 2018), e-banking services increased by 27%, from 13 million users in 2012 to 54 million in 2016 – a growth rate of 169% over the four-year period mainly due to the digital banking services offered by Indonesian banks (Beritasatu, 2018). This resulted in 150 million transactions in 2012, which increased to 406 million transactions in 2016 (CNN Indonesia, 2018). Bank ABC is a foreign exchange bank with operations in Indonesia and is the result of a merger between PQR Bank and WYZ Bank. To facilitate banking services for its customers, Bank ABC launched its own banking application, ABC, which is a progressive financial application that allows its customers to carry out every budgetary action from making all payments, withdrawing and depositing cash, and managing personal finances. ABC accepts deposits, which has enabled it to attract 3 million application downloads with 700 dynamic users in Indonesia (BTPN, 2018). ABC's historical milestones include being named Best Digital Bank in Indonesia in 2018 at the Asian Banker for Indonesia Country Awards (Jenius, 2018) and also being ranked among the Top 50 Digital Only Banks by *Financial IT* magazine (Infobanknews, 2018). The main objective of the present article is to assess the factors influencing customers' continuance intention (CI) toward using ABC, and also to see the moderating effect of two demographic variables, age and gender. The result may give insights into what factors make ABC so popular among Indonesian customers.

2 THEORETICAL FRAMEWORK AND METHODOLOGY

Scholars have presented well-developed theories and models of technology adoption for assessing what factors influence users in adopting and using new technology (Sharma & Mishra, 2014). These include *Diffusion of Innovation Theory* (1960) by Roger, *Theory of Reasoned Action* (TRA) (1975) by Fishbein and Ajzen, *Theory of Planned Behavior* (TPB)

(1985) by Ajzen, *Social Cognitive Theory* (SCT) (1986), by Bandura, *Technical Adaption Model* (TAM) (1989) by Davis, *The Model of PC Utilization* (MPCU) (1991) by Thompson et al., *The Motivation Model* (MM) (1992) by Davis et al., *Extended TAM2 Model* (2000) by Venkatesh and Davis, *Unified Theory of Acceptance and Use of Technology* (UTAUT) (2003) by Venkatesh, and *Model of Acceptance with Peer Support* (MAPS) (2002) by Sykes et al.

The most commonly used model for assessing users' technology acceptance is the UTAUT 2 Model as it combines all other earlier theories and models. For the present article, the modified UTAUT 2 Model (Venkatesh, Thong, & Xu, 2012) was used as few modifications were required specific to Indonesian customers and the banking industry. First, the variable of behavioral intention (BI) was replaced with continuance intention (CI) as it was appropriate for assessing customers' use of banking applications. Second, use behavior (UB) was eliminated as the aim was to identify CI (Xu, 2014). Third, a new variable, trust (T), was added (Alawan, Dwivedi, & Rana, 2017; Farah, Hasni, & Abbas, 2017; Indrawati & Ariwiati, 2015) based on the pilot survey. Fourth, price value (PV) was replaced with price-saving orientation (PSO) as there is no cost involvement while using digital banking services. And finally, experience (E) was excluded as a moderating variable since the data used in the study were cross-sectional rather than longitudinal.

The final eight independent variables (Escobar-Rodríguez & Carvajal-Trujillo, 2014; Indrawati & Ariwiati, 2015; Venkatesh et al., 2012) were performance expectancy (PE), effort expectancy (EE), social influence (SI), facilitating conditions (FC), hedonic motivation (HM), price-saving orientation (PSO), habit (H), and trust (T). These IVs were moderated with two demographic variables, namely age and gender. Finally, the resulting dependent variable (DV) was defined as continuance intention (CI). Altogether 45 statements were included, five each for the eight IVs and one DV, and given to customers to answer. The data so collected were used to test the significance of the IVs on the DV and also the moderating effects of age and gender. To facilitate the testing of the influence of the IVs on the DV, eight main hypotheses were formulated. For the purpose of testing the moderating effects of age and gender, a set of 16 sub-hypotheses was also formulated.

The main eight hypotheses were: "*PE (H_1)/EE (H_2)/SI (H_3)/FC (H_4)/HM (H_5)/PSO (H_6)/H (H_7)/T (H_8) has a positive and significant influence toward CI.*" For testing the moderating effects of age and gender, which influence the IVs toward the DV, the 16 formulated sub-hypotheses were: "*Age (a) and Gender (b) as a moderator has an influence on PE (H_{1a} and H_{1b})/EE (H_{2a} and H_{2b})/SI (H_{3a} and H_{3b})/FC (H_{4a} and H_{4b})/HM (H_{5a} and H_{5b})/PSO (H_{6a} and H_{6b})/H (H_{7a} and H_{7b})/T (H_{8a} and H_{8b}) toward CI.*" All of these hypotheses were tested at 5% one-tailed (Indrawati, 2017).

A structured questionnaire was developed based on earlier works (Alawan et al., 2017; Escobar-Rodríguez & Carvajal-Trujillo, 2014; Indrawati & Ariwiati, 2015) and administered initially through a pilot survey. The data collected were tested for content validity, reliability, and readability. Subsequently, a pilot test was carried out on a sample of 30 respondents to ensure that the items in the questionnaire for each of the constructs were valid and reliable. Finally, using Google Forms, data were collected from 645 respondents during the period of November 23 to December 23, 2018, of which only 423 were valid (66% response rate), as the remaining 222 respondents had not used the digital banking application for more than three months. The majority of the respondents were female (57.2%) and also belonged to a students' community (60.9%). The Corrected/Adjusted Item-Total Correction (CITC) gave a correlation coefficient of more than 0.3 and was found to be valid (Indrawati, 2017). Cronbach Alpha also gave a value of more than 0.7 for all 45 statements and was also found to be valid.

3 ANALYSIS AND DISCUSSION

The data were initially tested for validity and reliability using Cronbach Alpha (CA) and composite reliability (CR). Convergent validity was also tested with the help of the average variance extracted (AVE), which is given in Table 1, and the result clearly shows that all 45 statements are valid and reliable. To determine the influence of the latent variables toward another latent variable, based on a structural model and using a bootstrap procedure, the path coefficients and t-test values were obtained, which are given in Table 2. Using a one-tailed significance level of 5%, any t-value greater than 1.65 is considered significant and the formulated hypotheses are accepted, hence it can be said that only three IVs – T, H, and PSO – have positive and significant influence on CI. H_6, H_7, and H_8 were therefore accepted and the remaining hypotheses were rejected. None of the remaining five IVs – HM, FC, SI, PE, and EE – had any influence on CI, which is in line with earlier studies (Anjarsari & Ariyanti, 2017; Arenas-Gaitan, Begoña, & Ramón-Jerónimo, 2015; Qasim & Abu-Shanab, 2016; Venkatesh et al., 2012).

Structural model (inner model) test results gave an R^2 value of CI as 71%, which clearly indicates that the proposed model has a strong predicting ability of the factors influencing CI. The goodness-of-fit (GoF) index value came to 0.71, hence the model is also considered valid. The effect of each of the eight IVs on the DV was also studied using the effect size formula and criteria, the result of which is given in Table 3, which indicates that all the values are very small and confirms again that the model has a strong predictive power. The present study also utilized a group comparison approach to test the sub-hypotheses to see the moderating effects of age and gender with the Chin Formula, which gives the t-values as presented in Table 4.

The result shows that only gender has a moderating effect on T toward CI, hence only one sub-hypothesis (H_{8b}) among the 16 was accepted. Age does not have any influence on moderating IVs toward the DV. Finally, it can be said that trust only has a positive and significant influence on CI with gender moderating. The other two IVs that also have an influence on CI were habit and price-saving orientation.

Table 1. CA, CR, and AVE.

Factor	CA	CR	AVE
PE	0.88	0.91	0.68
EE	0.90	0.93	0.73
SI	0.92	0.94	0.77
FC	0.80	0.87	0.63
HM	0.90	0.92	0.67
PSO	0.85	0.89	0.63
H	0.91	0.93	0.75
T	0.91	0.93	0.74

Table 2. PAth, coefficients, t-values, and hypotheses status.

No.	Path	Coefficient	T-value	Hypothesis
1	T→CI	0.377	6.923	H_8 Accepted
2	H→CI	0.345	5.762	H_7 Accepted
3	PSO→CI	0.111	2.096	H_6 Accepted
4	HM→CI	0.065	1.071	H_5 Rejected
5	FC→CI	−0.047	0.892	H_4 Rejected
6	SI→CI	0.040	0.819	H_3 Rejected
7	PE→CI	0.043	0.818	H_1 Rejected
8	EE→CI	0.026	0.442	H_2 Rejected

Table 3. Effect size by DV.

Link	F-square	Size/weight
EE to CI	0.001	Small
FC to CI	0.002	Small
H to CI	0.116	Small
HM to CI	0.004	Small
PE to CI	0.002	Small
PSO to CI	0.015	Small
SI to CI	0.003	Small
T to CI	0.090	Small
FC to UB	0.219	Small

Table 4. Moderation effects.

	T-value	
Path	Age	Gender
T→CI	0.860	**0.547**
H→CI	0.955	1.354
PSO→CI	0.209	1.976

4 CONCLUSION AND SUGGESTION

The result reveals that of the eight IVs, only three (trust, hedonic motivation, and price-saving orientation) have a positive and significant influence on CI. Of the two moderating variables, only gender moderates trust toward CI, and age does not have any moderating effect. Hence it can be concluded that trust is the only variable that makes ABC the most favored digital banking application in Indonesia. Bank ABC must ensure that its security system is updated periodically so that the safety and security of online banking transactions are fully protected against all categories of threat. Bank ABC must also ensure that its customers are completely satisfied with the use of ABC, making them use it frequently for all their financial needs, so that it becomes a habit among customers. Bank ABC may also think about providing a higher amount of cash back or points that can be redeemed for all the transactions customers make using ABC. Minimizing the administration fee may also influence consumers positively, as customers considered price-saving orientation as an influencing factor.

REFERENCES

Alawan, A. A., Dwivedi, Y. K., & Rana, N. P. 2017. Factors influencing adoption of mobile banking by Jordanian bank customers: Extending UTAUT2 with trust. *International Journal of Information Management*, 37(3), 99–110.

Anjarsari, I., & Ariyanti, M. 2017. Analysis factors affecting Instagram adoption by online shoppers in Bandung City. *International Journal Science and Research (IJSR)*, 6(8), 1378–1382.

APJII. 2018. *Asosiasi Penyelenggara Jasa Internet Indonesia*. [Online].

Arenas-Gaitan, J., Begoña, P.-P., & Ramón-Jerónimo, M. 2015. Elderly and internet banking: An application of UTAUT2. *Journal of Internet Banking and Commerce*, 20(1), 1–23.

Beritasatu. 2018. *Inklusi Keuangan Nasional dan Digital Banking*. [Online]. [Retrieved on September 24, 2018].

BTPN. 2018. *Inovasi Digital Banking di Indonesia*. [Online]. [Retrieved on October 6, 2018].

CNN Indonesia. 2018. *Luncurkan Digibank, DBS Ingin Perluas Pasar Ritel*. [Online]. [Retrieved on October 21, 2018].

Escobar-Rodríguez, T., & Carvajal-Trujillo, E. 2014. Online purchasing tickets for low cost carriers: An application of the unified theory of acceptance and use of technology (UTAUT) model. *Tourism Management*, 43, 70–88.

Farah, M., Hasni, M. J., & Abbas, A. K. 2017. Mobile-banking adoption: Empirical evidence from the banking sector in Pakistan. *International Journal of Bank Marketing*, 36(7), 1386–1413.

Indrawati. 2017. *Perilaku Konsumen Individu*. Bandung: PT Refika Aditama.

Indrawati & Ariwiati. 2015. *Factors Affecting E-commerce Adoption by Micro, Small and Medium-Sized Enterprise in Indonesia*. s.l. International Association for Development of the Information Society (IADIS).

Infobanknews. 2018. *ABC Masuk Dalam Top 50 Digital Online Banks 2017 Dari Financial IT*. [Online]. [Retrieved on October 6, 2018].

Jenius. 2018. *Jenius Raih Penghargaan "The Best Digital Bank in Indonesia 2018."* [Online]. [Retrieved on October 21, 2018].

Qasim, H., & Abu-Shanab, E. 2016. Drivers of mobile payment acceptance: The impact of network externalities. *Information Systems Fronitiers*, *18*(5), 1021–1034.

Sharma, R., & Mishra, R. 2014. A review of evaluation of theories and models of technology adoption. *Indore Management Journal*, *6*(2), 17–29.

Venkatesh, V., Thong, J., & Xu, X. 2012. Consumer acceptance and use of information technology: Extending the Unified Theory of Acceptance and Use of Technology. *MIS Quarterly*, *36*(1), 157–178.

Xu, X. 2014. Understanding users' continued use of online games: An application of UTAUT2 in social network games. *MMEDIA*.

Self-regulated learning for smart learning in a university at Cyberjaya

Lilian Anthonysamy
Faculty of Management, Multimedia University, Cyberjaya, Malaysia

Koo Ah Choo & Hew Soon Hin
Faculty of Creative Multimedia, Multimedia University, Cyberjaya, Malaysia

ABSTRACT: Smart learning is an important part of smart city construction. However, limited studies are found on smart learning. Relying on smart devices is not sufficient. A smart city must develop 21st century skills to promote quality learning. Self-regulation is evidently one of the vital competencies as it helps individuals master their own learning process. The purpose of this paper is to provide a descriptive analysis on the use of self-regulated learning strategies (SRLS) in online learning. A total of 107 undergraduates were purposively selected to examine their use of SRLS in online learning. Results revealed that level of SRLS among students are found to be generally in the middle range. It was also identified that, female students uses more learning strategies than male students. This paper is able to provide a clear picture for future studies on smart learning in a smart city which is a key contemporary learning method.

1 INTRODUCTION

Education plays a very important role in the development and sustainability of smart cities. Smart learning is an important part of the construction of smart city (Hall, Braverman, Taylor and Todosow, 2000). Therefore, young adults in a learning city must acquire strategies, values and attitude to handle changes, challenges and contribute to the development of smart cities (Ljiljana and Adam, 2015) and promote quality learning. Learning and education give individuals empowerment and inclusion to prosper and develop in a smart city from an educational perspective.

Self-regulation learning is evidently one of the most vital competencies for the 21st century (OECD, 2013b). Self-regulated learning strategies helps young adults achieve smart learning. Example of self-regulated learning strategies are time management, peer learning, effort regulation, monitoring and so forth. Since limited studies can be found on smart cities from the educational perspective (Huang, Zhuang, & Yang, 2017), there is a need to assess the extent of SRLS among young adults in a smart city. Little is known about the use of SRLS that are essential to assist young adults in smart learning. This paper seeks to assess the extent of use of SRLS among young adults in Cyberjaya, Malaysia. Cyberjaya is Malaysia's first technology cities. Cyberjaya is the perfect ecosystem conducive for technological creativity that creates and enhances smart living. The present study also seeks to investigate the demographic differences exploring between gender, CGPA and year of study with the use of self-regulated learning strategies.

This paper begins with discussion of self-regulated learning strategies, smart learning and an overview of Cyberjaya. It then discusses the research method used to assess SRLS in blended online learning among university students. Lastly, this paper includes concluding remarks.

2 SELF-REGULATED LEARNING

Self-regulation learning is a significant predictor to learning and performance (Cho, Kim & Choi, 2017). There are four domains of self-regulated learning strategies which are cognitive engagement, metacognitive knowledge, resource management and motivational beliefs. Cognitive engagement refers to mental acquisition of knowledge through learning process (e.g rehearsal, elaboration, etc..). Metacognitive knowledge is an internal guide that enables an individual to be aware of their cognition processes (E,g planning, monitoring and regulation) Resource management comprises of behavioral and environmental components (E.g time management, peer learning, etc..)(Zimmerman and Martinez-Pons, 1986). Motivational beliefs are to help students observe and reach learning goals (E.g self-efficacy beliefs, goal orientation, etc..)(Pintrich, 1999).

2.1 *Smart learning*

Smart learning environment is a high level of digital environment that could be regarded as having three basic characteristics which are easy learning, engaged learning and effective learning as well as "4As" learning environment which are anytime, anywhere, any way and any pace which can also actively provide the necessary learning guidance, hints, supportive tools or learning suggestions for learners (Huang, Yang, & Hu, 2012). The dimension of learning is becoming a part of smart cities developments (Huang et al., 2017). In other words, smart learning should be the driving force to enhance citizens who want to have a better quality of life.

Since blended learning has been considered a smart classroom learnings' learn environment (Tas, 2016), self-regulated learning is needed to better understand and improve studenting experiences. Self-regulated learning which is seen as a constructive process through which learning goals are set, learning engagement occurs through regulating and controlling their cognition, motivation, and behavior in order to accomplish those goals is congruent with the three basic characteristics of smart learning.

3 RESEARCH METHOD

This study employed a purposive sampling method. Participants were 107 undergraduate computer science students from one local private university in Cyberjaya, Selangor, Malaysia. These students have enrolled in at least one blended learning subject. Students were surveyed on their use of self-regulated learning strategies in online blended learning. Random computer science undergraduates from 1st and 2nd year classes were chosen. Paper-based questionnaire were distributed to university students. The data collection was mainly collected during classes in the trimester. The instrument in this study was mainly adapted by Motivated Strategies for Learning Questionnaire (MSLQ) (Pintrich, Smith, Garcia and McKeachie,1991)because it has been mainly used to assess self-regulatory behaviours in online learning environment among undergraduates. It has been tested for reliability and validity in SPSS (v25) and SmartPLS 3.0 by using Cronbach Alpha and composite reliability.

4 FINDINGS AND DISCUSSION

4.1 *Demographic profile of respondents*

A total of 107 computer science undergraduates were involved in this study. 71 were male respondents and 36 were female respondents. 76 participants' age were between 19 and 20 years of age, 28 students were between 21 and 22 years of age and only 3 students were 23 years old and above. Apart from that, there were 76 students of Year 1 and 31 students from Year 2. 32 students were in the 1st class group (3.67-4.00), 33 students were in the second upper class group (3.33-3.66), 28 students were in the second lower class group (2.67-3.33) and 2 students were in the third class group (below 2.00).

4.2 Overall view of students' self-regulated learning levels

Results revealed that level of SRLS among students are found to be generally in the middle range (Cognitive Engagement Mean 3.47, Metacognitive Knowledge Mean 3.13, Resource Management Mean 3.19 and Motivational Belief Mean 3.50). Metacognitive Knowledge is seen to be the least used domain strategy which is consistent with the findings where many university students used ineffective cognitive and metacognitive strategies within the online learning environment (Hashemyolia, Asmuni, Ayub, Daud, and Shah, 2015).

4.3 Students' self-regulated learning levels based on CGPA

In this study, students' with CGPA equals or above 3.33 are assumed to be high achievers, CGPA 2.67 – 3.32 were considered average achievers and low achievers were assumed to have a CGPA less than 2.67. Since there are only two low achievers, only average and high achievers were examined in this study. Results revealed that SRLS mean use SRLS are different across achievers. High achievers (Mean 3.46) tend to use more SRLS as compared to average achievers (Mean 3.34). Literature reported that poor use of SRLS often resulted in online learning difficulty as opposed to those who have self-regulative abilities (Zhu, Au, and Yates, 2016).

The highest scored strategy is "If I can, I want to do better than most of my classmates in my online task"(Mean, 3.75).This strategy belongs to motivational beliefs domain. The possible reason for this is a competitive spirit lies in young adults. The lowest scored strategy is "I will email or talk to my lecturer to clarify concepts or questions I don't understand (Mean, 2.92). This strategy belongs to the resource management domain, under the help seeking strategy. Learners who are self-regulated, are capable of managing the available resource and are able to adapt to learning situation (Credéa & Phillips, 2011). It was observed that high achievers prefer to work on their own (Mean, 3.22) or seek help from their friends (Mean, 3.49).

4.4 Students' self-regulated learning levels based on gender

Based on gender differences, findings showed a higher use of SRLS among female students (n=71, mean 3.5) as compared to male students (n=36, mean 3.36). Even though the use of SRLS is higher in females, male students scored higher in Resource Management Domain and Motivational Beliefs Domain.

The most used strategy among the male students' falls in metacognitive knowledge domain-(regulating) (Mean: 4.15). The least used strategy is under resource management domain (help seeking) (Mean, 2.77). On the other hand, the most used strategy by female students' is resource management domain (peer learning) (Mean: 3.92). This indicates that female students prefer to work in a group (Kuhn and Villeval, 2011). Male students also prefer to work with their friends but not as high as females (Mean: 3.20). The least used strategy by female students is metacognitive knowledge domain (planning) (Mean 3.14).

4.5 Students' self-regulated learning levels based on year of study

Furthermore, this study also investigated whether seniority plays a role in the utilization of SRLS. Findings revealed that Year 2 students (Mean 3.44) were found to utilise SRLS slightly more than Year 1 students (Mean 3.40). Year 1 and year 2 students show some similar pattern on the use of SRL strategies. The highest used strategy falls in the Motivational Beliefs domain whereas the lowest used strategy falls in the resource management domain. This could indicate that Year 1 and Year 2 students still lack in resource management skills such as time and environment management, effort regulation, help seeking and peer learning.

5 CONCLUSION

This paper provides a descriptive analysis on the use of self-regulated learning strategies (SRLS) in online learning among university students. The results indicated that the level of SRLS among university students is only in the middle range although most of the studied sample are average to high achievers. Therefore, these findings should not be neglected and further analysis should be conducted on self-regulated learning strategies in smart learning across various digital learning environments. In addition, although smart learning bases its foundation on smart devices and intelligent technologies, self-regulation is needed to create a more effective digital learning environments that offer convenience to learners and able to keep pace with the changing demands of the digital age in a smart city.

This paper is part of the first author's PhD work at Faculty of Creative Multimedia, Multimedia University, financed via Special Mini Fund Multimedia University, under the grant number MMUI/180227.

REFERENCES

Cho, M. H., Kim, Y., & Choi, D. H. 2017. The effect of self-regulated learning on college students' perceptions of community of inquiry and affective outcomes in online learning. Internet and Higher Education 34: 10–17.

Credé, M., & Phillips, L. A. 2011. A meta-analytic review of the Motivated Strategies for Learning Questionnaire. Learning and Individual Differences 21(4): 337–346.

Cyberview Sdn Bhd. 2016. Smart Living is a Life Fulfilled: Smart City Framework. Retrieved from http://cyberview.com.my/industry/smartcity

Hall, R.E., Braverman, J., Taylor, J., Todosow, H. 2000. The vision of a smart city. 2nd International Life Extension Technology Workshop, Paris.

Hashemyolia, S., Asmuni, A., Ayub, A. F. M., Daud, S. M., & Shah, J. A. 2015. Motivation to Use Self-Regulated Learning Strategies in Learning Management System amongst Science and Social Science Undergraduates. Asian Social Science 11(3).

Huang, R., Zhuang, R., & Yang, J. 2017. Promoting Citizen's Learning Experience in Smart Cities. 15–25.

Huang, R., Yang, J., Hu, Y. 2012. From digital to smart: the evolution and trends of learning environment. Open Education Research 1: 75–84.

Kuhn, P. & M.C. Villeval 2011. European Workshop on Experimental and Behavioral Economics.

Ljiljana, M., and Adam, S. 2015. Building a gamified system for capturing MOOC related data: smart city learning community as its most precious source of intangible cultural heritage. International Conference on Culture and Computing (Culture Computing):175–182.

Mat Saad,M.I., Eng Tek,O. and Baharom.S. 2009. Self-regulated learning: Gender differences in motivation and learning strategies amongst malaysian science students. 1st International Conference on Educational Research.

OECD. 2013b. Skilled for Life? Key findings from the survey of adult skills. Paris: OECD. Retrieved from http://www.oecd.org/site/piaac/SkillsOutlook_2013_ebook.pdf

Pintrich, P. R. 1999. The role of motivation in promoting and sustaining self-regulated learning. International Journal of Educational Research 31 (6): 459–470.

Pintrich, P. R., Smith, D. F., Garcia, T., and McKeachie,W. 1991. A manual for the use of the Motivated Strategies for Learning Questionnaire (MSLQ). Available: http://files.eric.ed.gov/fulltext/ED338122.pdf

Tas. 2016. The contribution of perceived classroom learning environment and motivation to student engagement in science. European Journal of Psychology Education 31:1–21.

Zhu, Y., Au, W., & Yates, G. 2016. University students' self-control and self-regulated learning in a blended course. Internet and Higher Education 30: 54–62.

Zimmerman, B. J., and Martinez-Pons, M. 1986. Development of a Structured Interview for Assessing Student Use of Self-Regulated Learning Strategies. American Educational Research Journal 23(4): 614–628.

Analysis of factors influencing consumer intention toward Indonesia QR mobile payment

H.A. Baskoro & A. Amini
Master of Management, Faculty of Economics and Business, Universitas Indonesia, Jakarta, Indonesia

ABSTRACT: This study aimed to identify the factors influencing consumer behavior toward the QR Code mobile payment system in Indonesia. We proposed a conceptual framework based on the technology acceptance model (TAM), which consists of three factors (perceived usefulness, perceived ease of use, and trust), one social context factor (social influence), and two individual user characteristics (perceived enjoyment and perceived behavioral control). The proposed research framework was empirically tested by data collected from 195 potential QR Code payment service users through an online survey. Data were analyzed using the structural equation modeling (SEM) technique. The result showed a particularly significant positive impact of QR Code payment on perceived usefulness, social influence, perceived enjoyment, and perceived behavioral control. However, trust and perceived ease of use had an insignificant impact on QR Code payment intention. The findings of this study have important theoretical and practical implications, particularly to understand user-centric factors affecting QR Code payment customer intention.

1 INTRODUCTION

1.1 Background

The rapid societal adoption of mobile phones and their role in the development of personal activities has been one of the major technological events in recent decades. A large number of features that help make daily life easier can explain the widespread, growing use of mobile phones. The increasing number of smartphones in the mobile market is a clear example of this trend. According to MDI Ventures and Mandiri Sekuritas, the total number of smartphone users in Indonesia had exceeded the number of bank account users by the end of 2017. Based on this trend, most technologies companies are focusing their efforts on developing service capability, including mobile payments, which are defined as commercial transactions that take place over internet networks and smartphones. The director of financial technology regulation, Otoritas Jasa Keuangan (OJK), claims there were 3 million mobile payment users in Indonesia with a total of 25 trillion transactions by the end of 2018.

Mobile payment providers are developing their payment features; one of them is QR Code payment. This technological development is also affecting people's payment behavior. QR Code is a type of mobile payment system that is gaining popularity in developing countries like Indonesia. As the mobile payment market is increasing and is expected to further increase exponentially in Indonesia, it is imperative to study the concerns of Indonesian consumers for the benefit of QR Code payment providers in Indonesia. This research discusses both theoretical and business contributions. First, this research was based on a theoretical model developed from the technology acceptance model (TAM), including some modifications. Second, this research analyzes factors affecting consumer acceptance of QR Code mobile payment.

2 LITERATURE REVIEW

2.1 Mobile payment

According to MDI Ventures and Mandiri Sekuritas, in 2017, the penetration rate of the smartphone in Indonesia exceeded the number of bank accounts. The abundance of opportunity has attracted numerous enterprises to enter the mobile payment market and release their products. Mobile network operators pioneered mobile payment services in Indonesia about a decade ago; Telkomsel TCash launched in 2007, followed by Indosat Dompetku in 2008 and XL Tunai in 2012. Mobile payment services in Indonesia use server-based electronic money as their underlying currency. Regulated by Bank Indonesia, the central bank of Indonesia, mobile payment services were also designed to improve the unbanked population's access to financial services. Services offered never really extend beyond remittance and payment. Lending through mobile payment services and offering interest on electronic money deposits, for instance, were restricted. The services managed to build strong monthly use cases such as bill payment, telecom top-ups, and virtual transfer. For perspective, the transaction value generated by chip-based electronic money contributed 70% of the total transactions generated by electric money as of 2017, in our estimates. According to MDI Ventures and Mandiri Sekuritas, the mobile payment market in Indonesia was dominated by GoPay and OVO followed by LinkAja, a merged product from TCash, and a few state-owned enterprises.

2.2 TAM model

The technology acceptance model (TAM) is a widely recognized and utilized theoretical model that predicts user acceptance of information technology (Hsu & Lu, 2004). The TAM explains information technology adoption behavior through two primary variables: perceived ease of use (PEoU) and perceived usefulness (PU) (Davis, 1989). Perceived usefulness is defined as the degree to which a person believes that using a particular technology would enhance his/her job performance. Perceived ease of use is defined as the degree to which using the technology will be free of effort (Faqih & Jaradat, 2015). In the TAM, both PU and PEoU influence the individual's attitude toward using new technology. Attitude and PU, in turn, affect the individual's intention to use the technology. Additionally, PEoU enhances PU of information technology (Davis, 1989).

3 METHODOLOGY

3.1 Research approach

The descriptive research used quantitative methods by gathering information from questionnaire research and analytics from other sources.

3.2 Object/subject of the research

The object of this research was to examine the factors of behavioral intention among mobile payment users toward employing QR Code features for payment. The subject of this study was mobile payment in Indonesia.

3.3 Research data collection and sources

The data collected in this research were conducted through broadcasting and delivering online questionnaires. Data were collected from existing mobile payment users already using the QR Code payment feature in Indonesia. Indonesia has plenty of QR Code payment users, and this expedited our data collection process. The study utilized QR Code payment as a specific application feature for mobile payment usage. In order to ensure that participants had sufficient information to form an opinion about their use of QR Code payment technology, we provided participants in our study with descriptions of how QR Code payments work in general, and for what they may be used. Items to measure the focal constructs of the study were drawn from previous research and slightly modified to suit the specific context of this study. These items were first translated into Bahasa by a researcher. When the instrument was developed, it was tested among 30 users with QR Code payment experience. We then used their comments to revise some items in order to improve clarity. Among the online questionnaires distributed, 238 questionnaires were initially collected for input. Later, 43 of the collected questionnaires were dropped due to missing data or invalid responses.

3.4 *Data analysis method*

In analyzing the data, three steps were followed according to Malhotra (2010):

a. Data selection;
b. Data presentation; and
c. Conclusion and verification.

These three steps were carried out using structural equation modeling (SEM). To assess convergent validity, the standardized factor loadings, average variance extracted (AVE), and composite reliabilities (CRs) were examined. Thus, the scale had a good convergent validity. To analyze the causality between the antecedent factors and consumers' intention to use the QR Code payment feature, SEM techniques were used. In this study, the authors used SEM and data obtained from the results of surveys in order to paint a comprehensive picture and get confirmation regarding the conditions of mobile payment users' intention.

4 RESULTS AND DISCUSSION

The results of the conducted analysis showed significant and positive effects of perceived ease of use and trust on perceived usefulness, thereby confirming H3 and H5. H4, that trust has positive effects on behavioral intention to use, was also confirmed. H6 predicting the positive effect of social influence on behavioral intention to use, was supported. Expected perceived enjoyment and perceived behavioral control (PBC) had a positive effect on consumers' intention to use QR Code payment, supporting H7 and H8. Nevertheless, the results showed insufficient evidence in support of H2, suggesting that perceived ease of use played an insignificant role in predicting the intention to use QR Code payment in the presence of the other variables, thereby rejecting H2. The results of the analysis can be seen in Table 1.

The study aimed to analyze the factors that influence users' acceptance of QR Code payment technologies in an integrated manner. The results confirmed the robustness of the TAM, explaining technology acceptance behavior for users within the context of QR Code technologies. Suggesting an extended TAM to which one social context variable (i.e., social influence) and two individual user characteristic variables (i.e., perceived enjoyment and PBC) are added, the study finds that a consumer's perceptions of usefulness, ease of use, behavioral control, enjoyment, and social influence are predictive of his/her intention toward using QR Code payment.

In comparing the path coefficients of antecedents of the behavioral intention toward QR Code payment features, usefulness and enjoyment emerged as powerful, while enjoyment was the most powerful predictor relative to the other belief factors. This supports prior TAM research finding usefulness to be the primary determinant of a person's use of a technology while ease of use, trust, and enjoyment were secondary determinants (Davis, 1989).

This study has both theoretical and practical implications. The main theoretical contribution is it theorizes the factors influencing consumer acceptance of QR Code payment, from a unified perspective, including technology perceptions (perceived usefulness and perceived ease of use), social context variables (social influence), and user characteristics (perceived enjoyment and PBC). It links the constructs of social influence, enjoyment, and PBC to the TAM and successfully extends the TAM to QR Code payment, which differs from the context of other information systems.

Table 1. Structural model table.

Hypothesis	Original Sample	Sample Mean	Standard Derivation	T Statistics	P values
Perceived Usefulness → Behavioral Intention to Use (H1)	0.341	0.343	0.080	4.287	0.000
Perceived Ease of Use → Behavioral Intention to Use (H2)	0.064	0.069	0.075	0.854	0.197
Perceived Ease of Use → Perceived Usefulness (H3)	0.429	0.425	0.092	4.680	0.000
Trust → Behavioral Intention to Use (H4)	0.094	0.097	0.055	1.703	0.045
Trust → Perceived Usefulness (H5)	0.314	0.319	0.076	4.132	0.000
Social Influence → Behavioral Intention to Use (H6)	0.152	0.154	0.049	3.096	0.001
Perceived Enjoyment → Behavioral Intention to Use (H7)	0.361	0.358	0.078	4.635	0.000
Perceived Behavioral Control → Behavioral Intention to Use (H8)	0.121	0.120	0.067	1.804	0.036

This integrated model achieves greater understanding of consumer acceptance of QR Code payment in Indonesia while retaining the parsimony of the model.

5 CONCLUSION

The conclusion from the series of users' behavioral intention toward QR Code payment has benefits for analyzing consumers' behavior while using financial technology. Of the factors mentioned in this research, perceived ease of use has an insignificant relation to behavioral intention. This supports prior TAM research finding usefulness to be the primary determinant of an individual's use of a technology while ease of use, trust, and enjoyment are secondary determinants (Davis, 1989). Therefore, we need to optimize the other factors that might have an effect.

REFERENCES

Ajzen, I. 1985. From intentions to actions: A theory of planned behavior. In *Action Control*. https://doi.org/10.1007/978-3-642-69746-3_2

Davis, F. D. 1989. Perceived usefulness, perceived ease of use, and user acceptance of information technology. *MIS Quarterly*. https://doi.org/10.2307/249008

Gao, L., & Bai, X. 2013. A unified perspective on the factors influencing consumer acceptance of Internet of Things technology. *Asia Pacific Journal of Marketing and Logistics*.

Jun, J., Cho, I., & Park, H. 2018. Factors influencing continued use of mobile easy payment service: An empirical investigation. *Total Quality Management and Business Excellence*, 29(9–10), 1043–1057. https://doi.org/10.1080/14783363.2018.1486550

Kim, C., Mirusmonov, M., & Lee, I. 2010. An empirical examination of factors influencing the intention to use mobile payment. *Computers in Human Behavior*, 26(3), 310–322.

Malhotra, N. 2010. *Marketing Research: An Applied Orientation*. Sixth Edition. Upper Saddle River, NJ: Pearson Education.

Pousttchi, K., & Wiedemann, D. G. 2007. What influences consumers' intention to use mobile payments (pp. 1–16). Los Angeles: Proceedings of the 6th Annual Global Mobility Roundtable.

Venkatesh, V., Morris, M. G., Davis, G. B., & Davis, F. D. 2003. User acceptance of information technology: toward a unified view. *MIS Quarterly, 27*(3), 425–478.

Zhou, T. 2013. An empirical examination of continuance intention of mobile payment services. *Decision Support Systems*. https://doi.org/10.1016/j.dss.2012.10.03

Hsu, C.L. and Lu, H.P. (2004) Why Do People Play On-Line Games? An Extended TAM with Social Influences and Flow Experience. Information & Management.

Faqih, K.M.S. and Jaradat, M.R.M. (2015),"Assessing the moderating effect of gender differences and individualism-collectivism at individual-level on the adoption of mobile commerce technology: TAM3 perspective", *Journal of Retailing and Consumer Services*, Vol. 22.

Patterning consumer behavior to school innovations from two different perspectives

Elfindah Princes & Johan Setiawan
Bina Nusantara University, Jakarta, Indonesia

ABSTRACT: Schools are competing for customers and they must manage their resources, just like private firms (Boyd, 2003). Prior research shows that parental involvement is an essential factor for children's educational success (Erdener & Knoeppel, 2018). Dissatisfaction and wrong policies will lead to failures. Educational success has four critical variables, namely income level, educational background, product knowledge, and customer satisfaction. Mixed-method research was conducted to analyze this issue by patterning consumer behavior from different income levels and educational backgrounds using a purposive sampling method. The research took place in Jakarta due to Jakarta's high social disparity. This research examined two hypotheses on how income level and educational background influence customer satisfaction when consumers receive the same product knowledge. The result showed that extensive product knowledge of school innovations will solve the knowledge gap and prevent dissatisfaction, increasing school performance.

1 INTRODUCTION

1.1 Research background

School restructuring is an international trend (Boyd, 2003). There are huge pressures and incentives to innovate, and this has resulted in many schools adopting reforms for which they did not have the capacity (individually or organizationally) (Fullan, 2005). Unmet capability creates dissatisfaction in educational reform and has become a worldwide phenomenon (Fullan, 2005). By examining the individual and collective settings, we can contend with both the "what" and the "how" of change (Fullan, 2005). Findings suggest that parents have a positive attitude toward parental involvement and are generally aware of the academic and psychological aspects of education. Hence, they have a good relationship with teachers and get involved in their children's education directly and indirectly. Findings also indicate that gender, age, occupation, or level of education generally make no significant difference (Kalaycı & Öz, 2018).

1.2 Research purposes

The purposes of this research were to examine parents' income level and educational background as they relate to parents' level of satisfaction with school performance. It is expected that this study may help the school community to gather general understanding about school innovations and further support school performance.

2 LITERATURE REVIEW

2.1 School innovations

Schools will succeed if superintendents and principals are visionary as educational leaders, and teachers learn new approaches to curriculum improvement (Fullan, 2005). We need powerful, usable strategies for powerful, recognizable change (Fullan, 2005). Mind service has been less used than lip service for the pivotal role of the principal as gatekeeper or facilitator of change (Fullan, 2005).

2.2 Parental involvement

Parental involvement is one of the most significant predictors of students' achievement (Kalaycı & Öz, 2018). Schools need to understand the big picture of educational change by delivering meaning followed by a strategy (Fullan, 2005).

Cultural diversity within communities (Denessen, Bakker, & Gierveld, 2007), lack of awareness of both parents and schools, and overloaded teaching stuff are the three major challenges involving parents (Kabir & Akter, 2014). All of the respondents shared their experience that a personal approach by oral, informal communication is far more fruitful (Denessen et al., 2007), especially for those with limited knowledge (Fullan, 2005).

2.3 School performance and customer satisfaction

Simple systems are more meaningful but less deep. Complex systems generate overload and confusion, but also contain more power and energy. Our task is to realize that finding meaning in complex systems is as difficult as it is rewarding (Fullan, 2005). With regard to communication with parents, it is indicated that formal, written communication does not work (Denessen et al., 2007). To build a relationship with parents, schools use limited strategies, for example, organizing parent conventions and committees (Kabir & Akter, 2014).

3 RESEARCH METHODOLOGY

A mixed-method methodology was used with two hypotheses. Qualitatively, the writer arranged seven interviews with educational professionals. Quantitatively, 800 surveys were distributed with 625 surveys ultimately used. After comparing the results of both methods, hypotheses were confirmed significant or not. For final results, a focus group discussion (FGD) with six people was conducted that resulted in one new variable for future research.

3.1 Research framework

Figure 1. Research framework.

In this study, income level and educational background were taken as the control variables, product knowledge was taken as an intervening variable, and customer satisfaction was taken as a dependent variable.

3.2 Research hypotheses

Hypotheses:
H1. Income level does not directly affect consumer satisfaction with school innovations.
H2. Educational background directly affects consumer satisfaction with school innovations.

4 FINDINGS AND RESULTS

Using a 5-Likert scale and SPSS analysis for data survey taken, the result was as follows:

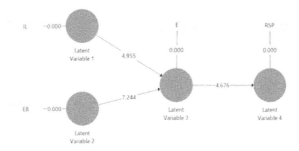

Figure 2. 5-Likert scale.

The result showed a positive correlation between income level, product knowledge, and customer satisfaction, and also a positive correlation between educational background, product knowledge, and customer satisfaction.

The in-depth interview results were as follows:

Table 1. Interview result.

Respondent	Job Position	Summary
Resp. A	Principal	The higher the level of education of parents, the more open-minded the parents are to school innovation.
Resp. B	Teacher	Parents' level of education influences the response to school policies but is not significant because there are other factors.
Resp. C	Teacher	Educational background and income level influence the response to school policies but is not significant depending on the surroundings. The parents will seek other's opinions.
Resp. D	Teacher	If the education is low, understanding and knowledge will be low too, and they will not care about the school policies.
Resp. E	Principal	EQ is more important than IQ, but the most important is SQ.
Resp. F	Teacher	The education level influences the level of acceptance of the school policies.
Resp. G	Teacher	Parents graduating with a bachelor's degree will give more trust to school policies.

The FGD result was as follows:

Table 2. FGD result.

Resp	Education	Income Level	Summary
A	Undergraduate	Medium	Income level influences the most.
B	Undergraduate	Medium	Income level influences parents most because the higher the Ceducation, the better they can understand school policy.
C	Undergraduate	Medium	Income level has the biggest influence.
D	Undergraduate	High	Parents with high income and low education will be difficult to handle; they will be arrogant and underestimate the school. Parents with high income and high education will be able to analyze circumstances and be more collaborative.

(*Continued*)

Table 2. (Continued)

Resp	Education	Income Level	Summary
E	Graduate	Medium	Manner matters the most.
F	Graduate	Medium	Income level. The ones who think they are privileged enough will not respect others. Those who are privileged will be more flexible.
G	Undergraduate	Medium	Educational background is more important than income level. It depends on the individual manner and races.

Significance test of hypotheses:

1. H1 → Not Proven Significant   2. H2 → Proven Significant

5 CONCLUSION

Based on the findings and results, a simple four-quadrant graph can be drawn.

The graph shows the relationship between the income level and educational background of parents to their response to school innovations with the same level of perceived product knowledge. Quadrant 1 shows the low acceptance level of customer satisfaction, while Quadrant 3 shows the high conformity due to lack of power (low-income level and low educational background). Quadrant 2 shows that with higher income level and educational background, the parents will possess high flexibility since they have sufficient product knowledge. Quadrant 4 shows the Critical Thinker of parents with the high educational background but low-income level.

Figure 3. Four-quadrant graph.

After conducting the study, the results showed that our hypothesis was partly not supported. It can be concluded that both income level and educational background affect customer satisfaction; nevertheless, educational background has a more robust effect. It is expected that this result may encourage stakeholders in the education industry to outline more product knowledge for customers, especially those with lower educational background and higher income level, and hence increase their satisfaction.

REFERENCES

Boyd, L. 2003. *Restructuring School.*
Denessen, E., Bakker, J., & Gierveld, M. 2007. Multi-ethnic schools' parental involvement policies and practices. *School Community Journal, 17*(2), 27–44.
Fullan, M. 2005. *The New Meaning of Educational Change.* Third Edition. New York: Teachers College Press.
Johnson, P. N. 2018. Getting it just right! Rigor and college prep for all. pp. 1–8.
Kabir, A. H., & Akter, F. 2014. *International Journal of Whole Schooling, 9*(2), 21–38.
Kalaycı, G., & Öz, H. 2018. Parental involvement in English language education: Understanding parents' perceptions. *International Online Journal of Education and Teaching (IOJET), 5*(4), 832–847. Available at: http://iojet.org/index.php/IOJET/article/view/447/296.
Erdener, M.A., & Knoeppel, R.C. (2018). Parents' perceptions of their involvement in schooling. International Journal of Research in Education and Science (IJRES), 4(1), 1–13. DOI:10.21890/ijres.369197

Analysis of purchase decision-making in different surroundings

Elfindah Princes & Sasmoko
Bina Nusantara University, Jakarta, Indonesia

ABSTRACT: Consumers are always at the top of a company's priority list. Their existence will ensure the company stands firm. However, each consumer has distinct behaviors, and when consumers are put in different surroundings, their behaviors will change too, which is known as social conformity. A mixed-method approach was used to investigate a company's need to make use of social conformity in order to increase purchase decision-making. The higher the purchase decision-making, the greater the company's benefit. The findings showed a direct correlation between social influence and surroundings on consumer behaviors; consumers were categorized into three groups of people: conforming, nonconforming, and situational nonconforming. A company must be capable of creating a reliable plan to increase the purchase rate. Furthermore, when the surrounding is not available, the company must build it. Purchase decision-making is a crucial moment as it proffers a high success rate for the company.

1 INTRODUCTION

1.1 Research background

Today's fast-changing world has had huge impacts on society. A teenager dresses in a K-Pop style because he wants to fit in with the rest of the guys in his social group. A mom leaves a reunion and goes home begging for a new bag from her husband because all her friends are using branded bags. A kid feels the urge to buy a particular cartoon character item because all of his friends are buying the same thing. These are examples of changing consumer behaviors in different surroundings, known as *social conformity*. The term *conformity* indicates an agreement with the majority position, brought on by a desire to *fit in* or be liked (normative) or by a desire to be correct (informational), or simply to conform to a social role (identification) (Bohner, 2014; Levine & Russo, 1981). Conformity can also be known as *majority influence* (or *group pressure*) (Asch, 1955, 1956; Levine, 1999). In some cases, this social influence might involve agreeing with or acting like the majority of people in a specific group, or behaving in a particular way in order to be perceived as normal by the group.

In today's highly competitive markets, firms must analyze the majority trend, perceived as the market trend, to prepare marketing and production plans. Moreover, if possible, firms should create the surrounding where the majority will be part of the surrounding and join the trend. At the end of this process is purchase decision-making by the people joining the surrounding. This, of course, will benefit the company by increasing sales growth and profits.

1.2 Research problems

When a person gets influenced by his surroundings and conforms to fit in, we call this behavior change social conformity. Numerous studies have affirmed that social conformity has changed the face of the culture. People are changing their behaviors to conform to the majority. A problem arises when a company does not take advantage of this phenomenon to benefit. If the company does not recognize this social conformity, it will face market penetration problems because it will lose its market share to the majority power.

1.3 Research purposes and benefits

The purposes of this research were to provide a new strategy for the company to increase its purchase rate and to understand the changing market caused by the majority trend, or social conformity. By using social conformity, the company will see the trend before others and quickly design strategies to benefit from it.

2 LITERATURE REVIEW

An autokinetic experiment conducted by Sherif (1935) states that when two or three people give judgments in the presence of others, the group will establish anorm. Asch (1956) describes social conformity as yielding to group pressures where people lack independence and must give in to the majority influence. Levine and Russo (1981) mask complexities involving movement and congruence conformity, public and private conformity, and motivational determinants of conformity. Social influence in groups is captured in terms of minority and majority influence and of persuasion (Bohner, 2014). Asch's experiments revealed that even though amajority is wrong, the minority still follows (Beran, Drefs, Kaba, al-Baz, & al-Harbi, 2015; Latané & Wolf, 1981; Mori & Arai, 2010). This fact is consistent with the presence of punishment and reward for not following what the majority desires (Bohner, 2014). With so many debates and contradictory results concerning social conformity, the writer assumes the existence of social conformity based on the latest research.

In previous research on innovation, the majority has been viewed as the passive recipient from the minority (Latané & Wolf, 1981). Conformity involves a public behavior change. Attitude changes involve changes in internal perceptions or private beliefs, which are often consistent with one another and affect each other (Montgomery, 2011).

3 RESEARCH METHODOLOGY

A mixed-method approach was utilized to obtain results. The writer distributed 160 survey papers and received 156 filled out. The survey consisted of two questions, one with a 5-point Likert scale and the other with a 9-point Likert scale. The surveys confirmed that most people do purchase by conforming to the majority, as explained in the literature review. Subsequently, nine in-depth interviews of around 15 minutes each were administered to obtain information on how people behave when they crave things. Summarizing the journals from the 1938 until the present, it is known that consumer behavior has changed due to social conformity.

In order to examine the aforementioned phenomenon, this research employed customer behaviors as the independent variable, purchase decision-making as the dependent variable, and social conformity as the intervening variable. The control variable was surrounding or social influence.

4 RESEARCH FINDINGS AND RESULTS

The compilation of data gathered was as follows:

Table 1. Quantitative findings on purchase decision-making.

Choice	A*	B*	C*	D*	E*	Total
Qty.	130	12	2	8	7	159
Percentage	82%	8%	1%	5%	4%	

*A: Because I need them.
*B: Because of the advertisement.
*C: Suggestion.
*D: Because of the outer packaging.
*E: Others.

Table 2. Quantitative findings on how people react to purchase intention.

Choice	A	B	C	D	E	F	G	H	I	Total
Qty.	7	51	8	6	1	5	38	35	5	156
Percentage	4%	33%	5%	4%	1%	3%	24%	22%	3%	

A Do a direct purchase.
B Think calmly and count the budget.
C Close the advertisement as soon as possible.
D Ask for other people's opinion.
E Find a loan if I do not have enough money.
F Wait patiently for a lower price.
G Find a substitution with a lower price.
H Look for more information from the seller.
I Others.

Table 3. Qualitative findings on the factors of purchase behavior.

Do you think your surrounding influences your purchase behavior?	
Resp. A (Female – 41 years)	No. I will not buy things that I do not need. However, if things are good and I heard other people's opinion that it is good, then I will choose the specific brand.
Resp. B (Female – 40 years)	Yes, we always seek other people's advice to find the best brand or the cheapest item to get what we want.
Resp. C (Male – 24 years)	Yes, we need to follow the trend to stay updated. But if we do not have the money, we have to wait until we have enough money.
Resp. D (Female – 30 years)	Sometimes yes and sometimes no. We ask other people first before we buy things or we check Google to check other people's reviews.
Resp. E (Male – 29 years)	Not really; if the product is new and we believe the brand, then we will buy it.
Resp. F (Female – 40 years)	When the trend is on, people will buy everything related to the trend, no matter how much it cost. That is very advantageous.
Resp. G (Male – 32 years)	Yes. I think when we buy things, especially the expensive ones, we will make sure the product works well by asking other people's opinions.
Resp. H (Male – 33 years)	No. I will only buy things that I need. However, if I need the thing, I will read references, see the specifications, and check if the price suits me. After that, I will ask users' opinions to make sure before I buy.
Resp. I (Male – 38 years)	Yes. But not only the surrounding, but also the brand engagement and other social influences. We must make sure first.

5 CONCLUSION

It can be concluded that the surrounding has the most significant influence on a person's decision-making. The result as shown in Table 3 confirmed the prior data in accordance with that showed in Tables 1 and 2. Therefore it is suggested that the entrepreneurial process also pay attention to the surrounding in order to create a majority that will influence other people to conform. Additionally, social conformity will be beneficial to the company when it is prepared to adapt to it. This condition, as mentioned earlier befalls the company when the marketing division can influence the market and create a trend. Third, people from nonconforming groups will have the hardest time and make the most effort to be assured. Some will not even change their minds until the end. However, a beneficial situation might be built with better information (informational conformity) or when the consumer attribute changes.

Three groups of social conformity exist:

1. The conforming group;
2. The nonconforming group (Javarone, 2014; Latané & Wolf, 1981; Montgomery, 2011); and
3. The situational conforming group.

This research has added another group of social conformers. Previous studies show two groups: conformers and nonconformers. The writer adds a group that engages in situational conformity, which is a group of people who are trapped between conformity and nonconformity due to their lack of abilities in making decisions (low income levels, remote locations, rarity of items).

Entrepreneurship should step up to a higher level by paying attention to surroundings as one of the most critical factors of consumers' decision-making. The ability to foresee the majority trend will give the company a competitive advantage in the future market.

REFERENCES

Asch, S. E. 1955. Opinions and social pressure. *Scientific American*, 72416.
Asch, S. E. 1956. *Studies of Independence and Conformity*, 70(9).
Beran, T., Drefs, M., Kaba, A., al-Baz, N., & al-Harbi, N. 2015. Conformity of responses among graduate students in an online environment. *Internet and Higher Education*, 25, 63–69. https://doi.org/10.1016/j.iheduc.2015.01.001
Bohner, G. 2014. Social influence and persuasion: Recent theoretical developments and integrative attempts. *Social Communication* (October), 191–221.
Javarone, M. A. 2014. Social influences in opinion dynamics: The role of conformity. *Physica A: Statistical Mechanics and Its Applications*, 414, 19–30. https://doi.org/10.1016/j.physa.2014.07.018
Latané, B., & Wolf, S. 1981. The social impact of majorities and minorities. *Psychological Review*, 88(5), 438–453. https://doi.org/10.1037/0033-295X.88.5.438
Levine, J. M. , & Russo, E. M. 1981. *Majority and Minority Influence*.
Levine, J. M. 1999. Solomon Asch's legacy for group research. *Personal and Social Psychology Review*, 3 (4), 358–364.
Montgomery, R. L. 2011. Social influence and conformity: A transorientational model. *Social Judgment and Intergroup Relations*, 175–200. https://doi.org/10.1007/978-1-4612-2860-8_8
Mori, K., & Arai, M. 2010. No need to fake it: Reproduction of the Asch experiment. *International Journal of Psychology*, 45(16653054), 390–397. https://doi.org/10.1080/00207591003774485
Sherif, M. 1935. *A Study of Some Social Factors in Perception*. New York.

Is the relationship between entrepreneurial competency and business success contingent upon the business environment?

M.A.S. Ismiraini & L. Yuldinawati
Telkom University, Bandung, West Java, Indonesia

ABSTRACT: Many owners of micro and small to medium-sized enterprises (MSMEs) have problems due to lack of knowledge and competence in business as well as attitudes in running a business in a certain business environment. The purpose of this study was to determine the influence of entrepreneurial competency on business success and business environment as a moderating effect on the relationship between entrepreneurial competency and business success in MSMEs organized by the Cooperatives and MSMEs Office in Bandung. The structural equation modeling (SEM-PLS) procedure was used to test the proposed model. The results showed that entrepreneurial competencies influence business success and business environment does not have a significant moderating influence on the relationship between entrepreneurial competency and business success in MSMEs organized by the Cooperatives and MSMEs Office in Bandung.

1 INTRODUCTION

Business success is certainly influenced by internal and external factors. Entrepreneurial competency is one of the internal factors that have a big impact on achieving business success. Business owners must have knowledge, skill, and attitude in running a business. The theory of entrepreneurial competency describes a link between the behaviors and attributes of the business owner and business success (Ahmad, Ramayah, Wilson, & Kummerow, 2010). Micro and small to medium-sized enterprises (MSMEs) comprise a business sector that plays a very important role in the regional, national, and global economy. Christiana, Pradhanawati, and Hidayat (2014) state that MSMEs in Indonesia are the main players in economic activities. Some MSMEs have proven reliable as a safety valve in times of crisis, through the mechanism of creating work fields. However, not every MSME can achieve success. Seeing the importance MSMEs for the country, the Indonesian government has relied on the Bandung Cooperatives and MSMEs Office to help MSMEs in achieving success. The Office has stated that MSMEs still lack knowledge and attitudes in entrepreneurship. An Office mentoring study found that entrepreneurial competency is one of the factors influencing the success of entrepreneurs, but the Office was still not sure about which competency MSME owners should have. This study has attempted to evaluate the effect of entrepreneurial competencies on business success in MSMEs organized by the Bandung Cooperatives and MSMEs Office.

This study also examined the moderating effect of perceived business environment on the relationship between entrepreneurial competencies and business success. The Office has stated that entrepreneurs cannot face changes and still do not understand their business environment. Some studies show that environment has a significant direct or moderating effect (Entrialgo, Fernandez, & Vazquez, 2001), while others have found weak impact or no impact whatsoever (Baum, Locke, & Smith, 2001; Jogaratnam, 2002).

2 LITERATURE REVIEW

2.1 *Entrepreneurial competencies and business success*

According to Ahmad et al. (2010), entrepreneurial competency can be defined as individual characteristics such as attitudes and habits where entrepreneurs can reach and maintain business success. Entrepreneur competence can be divided into eight competencies: strategic competency, conceptual competency, opportunity competency, relationship competency, learning competency, personal competency, ethical competency, and familial competency.

Success refers to the satisfaction that someone has in the work they have done, where they have freedom to create their own atmosphere or environment. Success is measured from each individual's value, not only market share, cash flow, and revenue (Austhi, 2017).

According to Ahmad et al. (2010), business success can be divided into satisfaction with financial success and satisfaction with nonfinancial success.

Based on various theories put forward by experts and on previous research used in this study, entrepreneurial competence influences entrepreneurial success. Entrepreneurial competency is one of the significant determining factors for success, performance, and growth or failure of a business operation (Kabir, Ibrahim, & Shah, 2017). Ahmad et al. (2010) argue that entrepreneurial competency is a predictor of the success of small to medium-sized enterprises (SMEs) in Malaysia. Sarwoko, Surachman, and Hadiwidjojo (2013) show that entrepreneurial competencies have significant influence on business performance, meaning that the higher the entrepreneurial competencies, the better the business performance. Based on the foregoing argument, it is hypothesized that:

H1. Entrepreneurial competency has a significant influence on business success.

2.2 *Moderating effect of business environment*

According to Ahmad et al. (2010), business environment is an external factor that can influence the mind-set and actions of an entrepreneur. Business environment can be divided into two categories, which are benign versus hostile environment and stable versus dynamic environment.

The effect of business environment on entrepreneurial activities, especially in SMEs, is important, with evidence suggesting that how entrepreneurs run their businesses is affected, to a considerable extent, by the environment in which they operate (Gynawali & Fogel, 1994). Ahmad et al. (2010) argues that entrepreneurial competency is a predictor of business success among SMEs in Malaysia, and its influence is even stronger in stable environmental conditions. Based on the foregoing argument, it is hypothesized that:

H2. Business environment significantly moderates the relationship between entrepreneurial competency and business success.

3 METHODOLOGY

3.1 *Sample*

The sample population consisted of 441 MSMEs organized by the Cooperatives and MSMEs Office in Bandung in 2017, and the final sample of respondents in this study included 82 MSMEs by using the Slovin formula. The sampling technique used in this study was nonprobability sampling with purposive sampling. The definition of MSMEs provided by the Cooperatives and MSMEs Office in Bandung was used to identify appropriate businesses for inclusion in the study.

3.2 *Data collection*

In this study data were collected through questionnaires and a literature review. The survey focused on MSME owners; a total of 82 owners were selected from various MSMEs. The variables selected in this study were entrepreneurial competency, business success, and business environment. The items selected to measure these variables were adopted from Ahmad et al. (2010) with 63 questionnaire items. All of the variables were measured using a 4-point Likert scale with level 1 = strongly disagree and 4 = strongly agree.

3.3 Method of data analysis

Before data analysis was completed, a validity test was performed for each instrument and a reliability test for each variable. The validity and reliability tests were carried out using SPSS 2.4 and all instruments and variables passed. This study used structural equation modeling (PLS 2.0) for data analysis. We considered several things while using SEM-PLS for data analysis techniques in this study: SEM-PLS is practical because it is more efficient in the execution process, can be used to test moderation variables, can use a small sample size, and does not require randomization of samples so that nonprobability approaches can be used. Partial least square (PLS) analysis has two steps, which are the measurement model and the structural model. The measurement model (outer model) was used to test validity and reliability, while the structural model (inner model) was used to test causality (hypothetical testing with prediction model). The first step of SEM-PLS testing was outer model testing, which included convergent validity testing, discriminant validity testing, and reliability testing. The standard for convergent validity testing is that every item must have loading factor > 0.5 and every variable must have an AVE value > 0.5. Discriminant validity testing is measured from its cross-loading value, where each indicator must correlate higher with its own construct compared to other constructs. Reliability testing is obtained from the composite reliability value and the Cronbach's Alpha value, which must be above 0.7 even though 0.6 is still acceptable. The second step was the inner model that showed significance level in hypothetical testing.

4 RESULTS

Out of 82 MSME owners tested in this research, 61% were women and 39% were men. The respondents mostly were between 21 and 30 years old. The majority of the respondents (41.5%) had only a high school education. Many of the respondents (43.9%) were running their businesses in the culinary sector.

The results of the outer model testing, which included convergent validity testing and discriminant validity testing, showed that all indicators and variables were above their standard values, so it can be concluded that every indicator and variable in this research model is valid. Reliability testing gave CR > 0.7 and CA > 0.7, which means all variables are reliable.

Based on the test results shown in Table 1, H_1 has a t-statistic value of 11.123, which is higher than the t-table value (1.96), and H_1 was accepted. It can be concluded that entrepreneurial competency significantly influences business success. This finding supports the previous research from Ahmad et al. (2010). The path coefficient between entrepreneurial competency and business success is positive (0.587) and shows a direct relationship so that the higher the entrepreneurial competency, the higher the business success. Owners of MSMEs should have the ability to prioritize work in alignment with business goals, take reasonable job-related risks, actively look for products or services that provide real benefit to customers, maintain a personal network of work contacts, learn from a variety of means, respond to

Table 1. Hypothesis test result summary.

Hypothesis	Relation	Path	T statistic	T table	R square	Conclusion
H1	EC -> BS	0.587	11,123	1.96	0.345	Accepted
H2	EC -> BS	0.454	4,833			
	BE -> BS	−0.326	2,445	1.96	0.487	Rejected
	EC * BE -> BS	−0.156	0.752			

constructive criticism, take responsibility and be accountable for their own actions, and get support and advice from family and close associates.

Meanwhile on H2 testing, the t-statistic value was 0.766, which is below the t-table value (1.96), so H2 was rejected and it can be concluded that business environment does not significantly moderate the relationship between entrepreneurial competency and business success. This finding does not support the results of previous studies conducted by Ahmad et al. (2010), which showed a moderating influence on the relationship between entrepreneurial competency and business success. Based on the result, business environment has a significant effect if it acts as an independent variable.

5 CONCLUSION

The results of this study illustrate that entrepreneurial competency has a role in achieving business success. Higher entrepreneurial competency of MSMEs will cause higher business success. With these findings, the Cooperative and MSMEs Office in Bandung can figure out which competencies MSME owners should foster to achieve success and how better to help them. Meanwhile, entrepreneurial competency's influence on business success rose from 34.5% to 48.5% after moderation by the business environment variable. But the moderating influence on the relationship between entrepreneurial competency and business success is insignificant.

Even though this research has tested the proposed model empirically, limitations must be considered. First, this research only focused on MSMEs organized by the Cooperatives and MSMEs Office in Bandung, Indonesia. Consequently, to get better generalization, in the future researchers are expected to conduct research with broader scope such as in provinces or nationally scope. Second, entrepreneurial competency's influence on business success is 34.5%, which means that 65.5% of business success can be explained by other factors not included in this research.

REFERENCES

Abdillah, W., & Jogiyanto. 2015. *Partial Least Square (PLS) Alternatif Structural Equation Modeling (SEM) dalam Penelitian Bisnis*. Yogyakarta: ANDI OFFSET.

Ahmad, N. M., Ramayah, T., Wilson, C., & Kummerow, L. 2010. Is entrepreneurial competency and business success relationship contingent upon business environment? *International Journal of Entrepreneurial Behaviour & Research*, 16(3), 182–203.

Austhi, D. 2017. *Motivasi Berwirausaha dan Kesuksesan Berwirausaha pada Wirausahawan Wanita Anne Avantie*. AGORA, 5(1).

Baum, J. R., Locke, E. A., & Smith, K. G. 2001. A multidimensional model of venture growth. *Academy of Management Journal*, 44(2), 292–303.

Christiana, Y., Pradhanawati, A., & Hidayat, W. 2014. Pengaruh Kompetensi Wirausaha, Pembinaan Usaha dan Inovasi Produk Terhadap Perkembangan Usaha (Studi Pada Usaha Kecil dan Menengah Batik di Sentra Pesindon Kota Pekalongan). *Diponegoro Journal of Social and Politic*, 1–10.

Entrialgo, M., Fernandez, E., & Vazquez, C. J. 2001. The effect of organizational context on SMEs' entrepreneurship: Some Spanish evidence. *Small Business Economics*, 16(3), 223–236.

Gynawali, D. & Fogel, D. 1994. Environments for entrepreneurship development: Key dimensions and research implications. *Entrepreneurship Theory and Practice*, 18(4), 43–62.

Hamid, Z. A., Azizan, N. A., & Sorooshian, S. 2015. Predictors for the success and survival of entrepreneurs in the construction industry. *International Journal of Engineering Business Management*, 7(12), 1–11.

Jogaratnam, G. 2002. Entrepreneurial orientation and environmental hostility: An assessment of small independent restaurant businesses. *Journal of Hospitality & Tourism Research*, 26(3), 258–277.

Kabir, M., Ibrahim, H. I., & Shah, K. A. M. 2017. Entrepreneurial competency as determinant for success of female entrepreneurs in Nigeria. *Indonesian Journal of Business and Entrepreneurship*, 3(2), 143–152.

Sarwoko, E., Surachman, A., & Hadiwidjojo, D. 2013. Entrepreneurial characteristics and competency as determinants of business performance in SMEs. *IOSR Journal of Business and Management*, 7(3), 31–38.

A proposed model for measuring cloud accounting adoption among SMEs in Indonesia

E. Widaryanti & Indrawati
Telkom University, Bandung, West Java, Indonesia

ABSTRACT: Many small and medium enterprises (SMEs) in Indonesia have not yet realized the importance of financial records and accounting to identifying potential growth as well as company wellness. Cloud accounting is sure to have a significant impact on the growth of SMEs and can be highly beneficial for SMEs as it offers efficient technology, high security, ease of use, cost efficiency, etc. Nevertheless, the factors that indicate Indonesian SMEs' intention to use cloud accounting are not well understood. Therefore, this study undertook a literature review and interviews with 15 SMEs in Indonesia and suggests modified UTAUT2 as the model. Thus, this study proposed performance expectancy, effort expectancy, social influence, facilitating condition, price value, and perceived security & risk as independent variables influencing Indonesian SMEs' behavioral intention in adopting cloud accounting. Age and scale of business are proposed as moderating variables.

1 INTRODUCTION

The rapid development of technology has provided various alternative solutions for many companies in running and developing their businesses, for example, by building business models that rely on cloud computing. Among the examples of cloud computing application in business are e-commerce and cloud accounting (recording the accounting or transactions online by cloud/internet as a medium). Many micro, small and medium enterprises (MSMEs) in Indonesia are online through e-commerce and those enterprises (including micro enterprises) produce more than 60% of Indonesian gross domestic product (GDP). Since it has a significant role for the nation, MSMEs should level up and grow. These MSMEs are supported by the Indonesian Ministry of Communication and Informatics through a program called "MSMEs Go OnlineCloud accounting is sure to have a significant impact on the growth of SMEs and can be highly beneficial for SMEs as it offers efficient technology, high security, ease of use, cost efficiency (accounting service at a lesser cost), etc. (Rao et al., 2017). Nevertheless, many MSMEs have not yet realized the importance of financial records and accounting to help them identify potential growth as well as monitor company wellness. Several previous studies related to adoption of cloud accounting were not considered because of the limited research scope and the factors that indicate the behavioral intention of Indonesian SMEs in using cloud accounting are not well understood. Thus, this research is aimed at identifying the factors affecting Indonesia SMEs' behavioral intention to use cloud accounting.

2 OBJECTIVES AND METHOD

As mentioned in the introduction, this research is intended to obtain a model for measuring the behavioral intention of SMEs in Indonesia to use cloud accounting. Therefore, the objectives of this research are: 1) To determine whether the proposed model can be used to measure the behavioral intention of SMEs in Indonesia toward cloud accounting based on literature review results; and 2) To determine whether the proposed model can be used to measure the behavioral intention of SMEs in Indonesia toward cloud accounting based on interview results.

To accomplish these objectives, this research used two approaches: a literature review and interviews. This research used a literature review of published articles, mainly from international journals, to attain the most suitable model to predict behavioral intention of SMEs in Indonesia toward cloud accounting. The interviews with SMEs in Indonesia were conducted to discern the perception of SMEs in Indonesia regarding the factors considered in cloud accounting adoption. The semi-structured interview involved 15 randomly chosen Indonesian SMEs.

3 LITERATURE REVIEW RESULTS

The study suggests that the appropriate model is UTAUT2 by Venkatesh et al. (2012), with some adjustment. In the UTAUT2 model, the construct consists of seven independent variables: performance expectancy (PE), effort expectancy (EE), social influence (SI), facilitating conditions (FC), price value (PV), hedonic motivation, and habit; two dependent variables: behavioral intention (BI) and use behavior; and three moderating variables: age, gender, and experience. The following is the explanation of each variable in the modified UTAUT2 model.

PE is described as the degree to which an individual believes that using the system will help him or her to attain advantages in job performance (Venkatesh et al., 2003:447). PE is also the strongest predictor of BI (Venkatesh et al., 2003:447). Venkatesh et al. (2012), Lafraxo et al. (2018), Moryson & Moeser (2016), Mursalin (2012), Martinsa et al. (2014) also noticed that PE is the factor that significantly influence the BI.

EE is the degree of ease associated with the use of the system (Venkatesh et al., 2003:450). The research results from Venkatesh et al. (2003:467), Venkatesh et al. (2012), Lafraxo et al. (2018), Moryson & Moeser (2016), Mursalin (2012), and Martinsa et al. (2014) found that EE has a positive influence on BI to use technology.

SI is defined as the degree to which an individual perceives that important others believe he or she should use the new system (Venkatesh et al., 2003:451). SI is adirect determinant of BI and has a positive effect on BI (Venkatesh et al., 2012; Lafraxo et al. 2018; Moryson & Moeser, 2016; Mursalin, 2012; and Martinsa et al., 2014).

FC are portrayed as the degree to which an individual believes that organizational and technical infrastructure exists to support use of the system (Venkatesh et al., 2003:453). In UTAUT2 model by Venkatesh et al. (2012:162), FC have an effect to BI, which is cited as well as by Mursalin (2012) and Indrawati and Ridwan (2018).

Corresponding to Venkatesh et al. (2012:161), cost and pricing structure may have a significant effect on use of technology by the consumer. PV is positive when the benefits of using a technology are perceived to be greater than the monetary cost, and such PV has a positive impact on intention. PV became the most influential factor in the intention of accepting smart metering moderated by the age variable as well as income (Indrawati & Tohir, 2016). PV moderated by scale of business has a positive influence in the BI of SMEs (Indrawati & Ridwan, 2018).

According to Ajzen (1991, pp.181), intentions are assumed to capture the motivational factors that influence a behavior: they are indications of how hard people are willing to try – of how much effort they are planning to exert – in order to perform the behavior. Venkatesh et al. (2003, pp. 456) expect that behavioral intention will have a significant positive influence on technology usage.

Referring to Moryson & Moeser (2016), perceived security & risks (PSR) significantly impacts the attitude toward use of cloud services. PSR is described as the safety issues that can affect SMEs' perception in regard to using cloud accounting. PSR was the second factor affecting the intention to accept smart metering moderated by age and income (Indrawati & Tohir, 2016). Perceived risk was one of the most important in explaining user's intention (Martinsa et al., 2014).

4 INTERVIEW RESULT

Based on interviews of 15 Indonesian SMEs, all respondents (R1, R2, R3, R4, R5, R6, R7, R8, R9, R10, R11, R12, R13, R14, R15) agreed that PE, EE, SI, FC, PV, and PSR significantly influence BI toward cloud accounting and these are their statements respectively.

1. PE: R1 mentioned that "Using cloud accounting, we hope to see our money in exact. We can check our actual cash, daily business value or business valuation, stock, and balance sheet." R14 explained, "The first benefit of cloud accounting is 24 hours access seven days a week, which means anytime, the person in charge who need to check the finance can access at any time. Second is anywhere access as long as they are connected to the internet, so it doesn't need to be installed on one device but yes when it is connected to the internet it can be checked immediately. Third one is multiple users, meaning every person in charge and has the authority to access finance and financial information are able to check, thus make it easier for stakeholders."
2. EE: R2 testified, "It is easy for me. I only need to enter income and outcome, then it creates the balance sheet. I only put simple inputs. Enter into what it is called a cash flow for expenditure and income." R3 said, "Yes, it is easier to adopt."
3. SI: R7 said, "I just heard about cloud accounting now, there is no socialization, right? I think in Indonesia has not booming yet." R15 said, "Oh yes, indeed we have a recommendation from friends then while in Jakarta we also came to his office in West Jakarta and we knew directly in real terms how they develop the software, so yes because there are recommendations and trust that we see the headquarters directly."
4. FC: R5 explained, "If the company using desktop application which is off line it is possible, but off line off course different to all connected and synchronized with the internet." R15 said, "It is online base thus it can be from headset or anything online since there is application as well."
5. PV: R12 said, "I hope the price if it is monthly is affordable for SMEs." And R15, "Actually it is more customizable, in terms of features are good and affordable price fits with our pockets."
6. PSR: R3 mentioned, "For example, there is a disaster or whatever we still have that data since it is online based, so even if our device is damaged or else, we still have that data." R13 described that "The server is reliable, maintenance only a few times per year." The interview result shows the Behavioral Intention of Indonesian SMEs to use cloud accounting. As stated by R4, "I am very interested in using cloud accounting because it is much easier, for example, if we are going outside and we want to control the store, for example, financial conditions or sales when we don't take notes, it will be useful certainly."

5 RESULT AND DISCUSSION

Venkatesh et al. (2012) explained that whereas the various studies contribute to understanding the utility of UTAUT in different contexts, the need for systematic investigation and theorizing of the salient factors of consumer technology adoption context is necessary. Consequently, the present research adjusted the UTAUT2 model to fit the behavioral intention toward cloud accounting adoption context. To attain the proper model, literature reviews and interviews were conducted in this study.

Cloud accounting is considered as a new technology where there is no certain measurement of SMEs as the users yet. In consequence, researcher excluded use behavior in this study. Along these lines, the variables namely hedonic motivation and habit as well as moderating variables gender and experience were removed from the model since those are considered as not relevant in this research. Nevertheless, contemplated in terms of the previous journals and the interview results, a variable called perceived security & risk along with scale of business as a moderating variable were added into the model.

The variables of the model are performance expectancy, effort expectancy, social influence, facilitating condition, price value, and perceived security & risk along with Behavioral Intention; and the moderating variables are age and scale of business.

6 CONCLUSION

The proposed model consists of six constructs: performance expectancy, effort expectancy, social influence, facilitating conditions, price value, and perceived security & risk; and age and scale of business as the moderating variables influencing behavioral intention.

The next process to be done in the research is composing the measurement tool. A set of questionnaires will be generated and tested through a pilot test. Once the measurement tool is valid and reliable, the main data will be collected according to the proposed model to predict the behavioral intention of SMEs in Indonesia in using the cloud accounting.

REFERENCES

Ajzen, I. 1991. The theory of planned behavior. *Organizational Behavior and Human Decision Process 50* [Online 2019, February 25]. http://www.courses.umass.edu/psyc661/pdf/tpb.obhdp.pdf.

Indrawati & Ridwan, N.F. 2018. *Analyzing Factors Affected Customers' Acceptance Toward High Speed Internet Access in SME Market: A Case Study in XYZ Corporation.* American Scientific Publishers, Vol. 24, 2996–3002.

Indrawati & Tohir, L. M. 2016. Predicting smart metering acceptance by residential consumers: An Indonesian Perspective, *Conference Proceeding of 4th International Conference on Information and Communication Technology (ICoICT) 2016.*

Lafraxo, Y., Hadri, F., Amhal, H., & Rossafi, A. 2018. The effect of trust, perceived risk and security on the adoption of mobile banking in Morocco. *Proceedings of the 20th International Conference on Enterprise Information Systems,* 497–502.

Martinsa, C., Oliveiraa, T., & Popovic, A. 2014. Understanding the Internet banking adoption: A unified theory of acceptance and use of technology and perceived risk application. *International Journal of Information Management,* 34(1),1–13.

Moryson, H. & Moeser, G. 2016. Consumer adoption of cloud computing services in Germany: Investigation of moderating effects by applying an UTAUT model. *International Journal of Marketing Studies,* 8 (1). Canadian Center of Science and Education, ISSN 1918-719X E-ISSN 1918-7203.

Mursalin, M. J. A. 2012. Information system adoption and usage: Validating UTAUT model for Bangladeshi SMEs. *BRAC University Journal,* IX(1&2), 15-24.

Rao, M., Thirmal, J. T. G., & Sivani, M.A. 2017. Impact of cloud accounting: Accounting professional's perspective. *IOSR Journal of Business and Management* (IOSR-JBM). e-ISSN: 2278-487X, p-ISSN: 2319-7668; pp. 53–59.

Venkatesh, V., Morris, M. G., Davis, G. B., & Davis, F.D. 2003. User acceptance of information technology: Toward a unified view. *MIS Quarterly,* 27(3),425–478.

Venkatesh, V., Thong, J. Y. L., & Xu, X. 2012. Consumer acceptance and use of information technology: Extending the unified theory of acceptance and use of technology. *MIS Quarterly,* 36(1),157–178.

Digital literacy deficiencies in digital learning among undergraduates

Lilian Anthonysamy
Faculty of Management, Multimedia University, Cyberjaya, Malaysia

ABSTRACT: Digital technology have changed the way university students approach learning because it has become a necessity and an integral part of their lives. University students are accustomed to using their digital devices for almost anything such as communication, collaboration, accessing multiple source of information for solutions, etc. Although these digital generation are undeniably engaged with technologies and they are very comfortable and confident using technology to accomplish tasks, many educators tend to assume that students today have digital literacy. Surprisingly, current studies reveal that students do not have the required digital literacy skills for digital learning. Furthermore, scholars mentioned that individual's level of digital literacy affects a student's performance positively. This paper presents the deficiencies of digital literacy among university students in a digital learning environment in higher education, implications of these deficiencies and suggests some measures to reduce these deficiencies.

1 INTRODUCTION

Digitisation has transformed the education sector where digital learning is an integral part of education. In a digital world where the technology continues to impact the education sector and students, the need for digital literacy is growing. Digital literacy means the ability and awareness to use emerging technologies to perform academic tasks online while demonstrating proper online attitude in a digital environment (Perera, Gardner, & Peiris, 2016). As society becomes more digital in their everyday task, knowledge, attitude and skills are essential to be digitally literate. Being digitally literate today is not confined to understanding just the hardware and knowing how to use the software. In a digital learning environment, digital literacy represents one of the prerequisites that is necessary for students to navigate their learning process in a digital learning environment. Furthermore, in order to excel in a digital learning environment, students need to equip themselves with digital literacy (Tang & Chaw, 2016; Techataweewan & Prasertsin, 2017).

Scholars mentioned that individual's level of digital literacy affects a student's performance positively (Mohammadyari & Singh, 2015; Scholastica, Nkiruka, Ifeanyichukwu, 2016). Tang and Chaw (2016) also reported that digital literacy is a prerequisite for students to learn effectively in a blended learning environment. When the level of students' digital literacy is high, it can make it easier for students to participate in the learning process, giving learners a more positive feeling about their educational experience. Hence, students' learning performance may be increased.

Although these digital generation are undeniably engaged with technologies and they are very comfortable and confident using technology to accomplish tasks, many educators tend to assume that students today have digital literacy. Surprisingly, current studies show that many undergraduates exhibit poor employment of digital literacy in online learning (Fazli & Norazilah, 2016). More research is needed to invest in digital literacy enhancement for economic growth and competitiveness (European Commission, 2010b; Krish, Liu, Nozibele, Jaya, Li & Chen, 2017; Chelghoum, 2017). Although digital literacy is well known term that has been used interchangeably, digital literacy deficiencies have not been presented. Hence, this paper aims to add to the present growing literature on aspects of digital literacy undergraduates are lacking in a digital learning environment.

This paper is divided into five sections. This section presents a brief introduction of digital literacy and its challenges. The following section of this paper provides an understanding of digital literacy and the digital learning environment. In the third section, a discussion on digital literacy deficiencies, implications and suggestions were presented. Conclusion is included in the last section.

2 DIGITAL LEARNING ENVIRONMENT

Digital learning environment includes a full or partial online environment. Digital learning which encompasses blended learning, mobile learning, online or e-learning, and many others ("Teaching with Digital Technologies", 2017), enables collaboration and access to content that extends beyond the classroom. The learning environment is an educational ecosystem consisting of interactive elements such as teachers, learners, teaching materials, evaluation and technology (Zheng & Ma, 2010). Digital literacy plays an important role in the digital learning environment.

3 DIGITAL LITERACY

Digital literacy means more than just being able to use computers or technologies for a task. An individual need to develop function skills, values, attitude and behavior to become a digitally-literate person. Although digital literacy is used to measure learners' quality in a digital environment, research support the fact that students lack digital literacy such as not engaging in a thoughtful process while learning online (Vissers, Rowe, Islam, & Taeymans, 2017), not being able to evaluate and integrate digital information effectively (Tang & Chaw, 2016; Ng2012; O'Sullivan & Dallas, 2010; Tenku Shariman, Talib & Ibrahim, 2012), not able to critically judge the suitability of large amount of information online (Greene, Yu and Copeland, 2014), not understanding the ethical and social usage of information, interpret reference to a paper or journal, search databases effectively (Shopova, 2014), discern the validity and value of information found online (Tenku Shariman et al., 2012) and not understanding copyright issues when using digital information for sharing purposes (Tenku Shariman et al., 2012).

3.1 *Digital literacy model*

Digital literacy embraces the perspective of cognitive, technical and socio-emotional of learning in an offline or online mode (Ng, 2012). Cognitive aspect includes choosing the technology, searching, assessing and selecting information using critical thinking skills, etc. The technical dimension concerns the skills needed to operate digital technologies for learning. Socio-emotional dimension is associated with behavior of an individual in using digital technologies (Ng, 2012).

4 DISCUSSION

The purpose of this paper was to present the deficiencies of digital literacy among university students in a digital learning environment in higher education, implications of these deficiencies and suggests some measures to overcome these deficiencies.

Table 1 highlights the aspects of digital literacy that students are lacking. Based on Table 1, it can be posited that students have technical or digital skills to manoeuvre through digital technologies but lack in cognitive and socio-emotional skills. This is consistent with literature

Table 1. Digital literacy deficiencies mapping according to Digital Literacy Model (Ng, 2012).

No	Authors(Year)	Digital Literacy Model (Ng,2012)		
		Technical Literacy	Cognitive Literacy	Socio-Emotional Literacy
1	Tan, Melissa and Saw (2010)		√	√
2	O'Sullivan and Dallas (2010)		√	
3	Shariman, Razak and Noor (2012)		√	√
4	Tenku Shariman, Talib and Ibrahim (2012)		√	√
5	Ng (2012)		√	√
6	Shopova (2014)		√	
7	Greene, Yu and Copeland (2014)		√	
8	Tang and Chaw (2016)		√	
9	Prior, Mazanov, Meacheam, Heaslip and Hanson (2016)		√	√
10	Vissers, Rowe, Islam, and Taeyans (2017)	√		

pointing out that many students who enter higher education have no digital literacy needed for digital learning (European Commission, 2013; Tenku Shariman et al., 2012). Students are proactive in using technology for social media or entertainment but not for learning (Prior, Mazanov, Meacheam, Heaslip and Hanson, 2016).

Digital literacy deficiency among students have implications on employment opportunities since more and more employers demand more digital literacy skills (Janks, 2010). Digital literacy can be classified into core and fundamental skills for performing task that are necessary in the workplace (e.g: technical, communication, collaboration, critical thinking and information management,etc..) and conceptual skills which are skills to bring the core skills to its full advantage(e.g: ethical awareness, cultural awareness, flexibility, etc..) (van Laar, van Deursen, van Dijk, & de Haan, 2017).

The following are several suggestions to improve and strengthen digital literacy among undergraduates:

a. Embrace self-regulation: Previous studies have reported that self-regulated learning strategies is a critical component in digital literacy (Greene et al., 2014; Greene, Copeland, Deekens, & Yu, 2018). The digital environment requires students to be self-regulated it is key in promoting digital literacy (Liew, Chang, Kelly & Yalvac, 2010; Greene et al., 2014). In other words, researchers revealed that self-regulated learning strategies are likely to be critical predictors of digital literacy.

b. Develop self-awareness: Students must be aware of the risk and reliability of online sources. They need to know how to determine if an article or website is trustworthy. Students can increase self-awareness by self-exploration through the web or by seeking guidance and advice from others (Fazli, 2016).

c. Exposure in digital literacy activities: Educators can play a role in providing more exposure to students on what it takes to acquire digital literacy because students have limited understanding how technology can be incorporated into learning (Margaryan, Littlejohn & Vojt, 2011; Ng,2012). Assignments or task can be given that comprehensively assess students' digital literacy skills. This may enhance assist in developing students' digital literacy skills with emphasis on cognitive and socio-emotional aspects.

d. Manage Digital Distraction: Distraction takes up many forms (ie. Digital distraction, peer distraction, instructor distraction, etc..) (Frisby et al., 2018).When learners are distracted, their cognitive processing capacity is limited. Hence, managing cognitive load is an important consideration due to its connection to student learning (Frisby et al.,2018) which may help with digital literacy acquisition.

5 CONCLUSION

The level of skill, knowledge and confidence a learner have will affect the quality and use of technology for learning. Three aspects of digital literacy, namely technical, cognitive and social-emotional domains are needed by students in order to participate actively in a digital learning environment. Digital literacy significantly enhances graduate employability because of it empowers graduates to achieve more in a digital economy. Inevitably, the demand for highly skilled people is growing in the current workforce. Future studies can look into providing more empirical evidence on the use of digital literacy competence in a digital learning environment to enhance quality education.

REFERENCES

Chelghoum, A. 2017. Promoting Students' Self-Regulated Learning Through Digital Platforms: New Horizon in Educational Psychology, American Journal of Applied Psychology 6(5):123.

European Commission. 2010b. Europe 2020: A strategy for smart, sustainable and inclusive growth, COM (2010) 2020.

European Commission. 2013. Digital Agenda for Europe: A Europe 2020 Initiative.

Fazli, M. B. & Norazilah, S. 2016. Digital Literacy Awareness among Students.2(October):.57–63.

Frisby, B. N., Sexton, B., Buckner, M., Beck, A.-C., & Kaufmann, R. 2018. Peers and Instructors as Sources of Distraction from a Cognitive Load Perspective. International Journal for the Scholarship of Teaching and Learning 12(2).

Greene, J. A., Yu, S. B., & Copeland, D. Z. 2014. Measuring critical components of digital literacy and their relationships with learning, Computers and Education 76: 55–69.

Greene, J. A., Copeland, D. Z., Deekens, V. M., & Yu, S. B. 2018. Beyond knowledge: Examining digital literacy's role in the acquisition of understanding in science, Computers and Education 117: 141–159.

Janks, H. 2010. Language, power and pedagogy. In N. Hornberger and S. McKay (Eds). Sociolinguistics and Language Education. Clevedon: Multilingual Matters.

Krish C., Liu Q., Nozibele G., Jaya J., Li W., and Chen F. 2017. Bridging the digital divide: measuring digital literacy. Economics Discussion Papers, No 2017–69, Kiel Institute for the World Economy.

Margaryan, A., Littlejohn, A., & Vojt, G. 2011. Are digital natives a myth or reality? University students' use of digital technologies. Computers & Education, 56(2):429–440.

Mohammadyari, S., & Singh, H. 2015. Understanding the effect of e-learning on individual performance: The role of digital literacy. Computers and Education, 82:11–25.

Ng, W. 2012. Can we teach digital natives digital literacy?.Computers and Education, 59(3):1065–1078.

O'Sullivan, M. K., & Dallas, K. B. 2010. A collaborative approach to implementing 21st century skills in a high school senior research class. Education Libraries 33(1): 3–9.

Perera, M. U., Gardner, L. A., & Peiris, A. 2016. Investigating the Interrelationship between Undergraduates' Digital Literacy and Self-Regulated Learning Skills, Proceedings of the Thirty Seventh International Conference on Information Systems: 1–13.

Tang, C.M. & Chaw, L. Y. 2016. Digital Literacy: A Prerequisite for Effective Learning in a Blended Learning Environment? The Electronic Journal of e-Learning 14(1): 54–65.

Tan, Kok Eng, Melissa L.Y. N., and Saw, K. G. 2010. Online activities and writing practices of urban Malaysian adolescents. System 38(4): 548–559.

Teaching with Digital Technologies. 2017, April 10 Retrieved April 20, 2017, from http://www.education.vic.gov.au.

Techataweewan, W., & Prasertsin, U. 2017. Development of digital literacy indicators for Thai undergraduate students using mixed method research. Kasetsart Journal of Social Sciences: 1–7.

Tenku Shariman, T. P. N., Talib, O., & Ibrahim, N. 2012. The Relevancy of Digital Literacy for Malaysian Students for Learning With WEB 2.0 Technology. Proceedings of the European Conference on E-Learning:536–545.

Zheng,T. and Ma,X. 2010. Design of learning environment - Dialogue with Professor Michael F. Hannafin. China. Education Technology 2: 1–6.

Smart city concept based on Nusantara culture

D. Trihanondo & D. Endriawan
Department of Creative Arts, Telkom University, Bandung, Indonesia

ABSTRACT: This paper raises a concept of future urban planning that still maintains the local culture or the culture of the archipelago in Indonesia. This article is the result of research conducted on a traditional village in the northern Bandung area. The method used in this study is a qualitative method that uses a visual cultural transformation theory approach. The results of this study are a new concept regarding smart cities without losing the cultural characteristics of the archipelago. This research is still in its early stages, and requires more discussion and application in more detailed and specific case studies.

1 INTRODUCTION

1.1 Community service program in Girimekar Village

This paper is part of community service activities carried out in Girimekar Village, Bandung regency that took place in 2018–2019. There was a specific section of the village where the activities were carried out, namely in section eight residents. This community service activity aimed to improve the competitiveness of Girimekar Village to improve the welfare of its residents. The main activities carried out were in the form of making murals, planting ornamental plants, and making village websites, as well as making databases and taking photos to support the concept of a smart city or smart village.

All activities that have been carried out involve the residents of Girimekar. This is very important, because the purpose of community service is to empower the community itself, so that after the activity is finished, the community can carry on, with a little monitoring from the organizer.

1.2 Nusantara culture and the smart city concept

The concept of the archipelago culture is used as a basis for developing the Girimekar region. Nusantara culture itself is a concept that can be interpreted broadly. Therefore, the concept of the archipelago culture referred to here is related to where this village is located.

Most of the inhabitants of the Girimekar Village area are Sundanese, a tribe that inhabits the western part of the island of Java in Indonesia. Generally speaking, the island of Java is dominated by two main tribes, namely the Javanese and Sundanese. There is a clear difference between the two tribes. The difference is caused by the environment on the western part of Java which is dominated by highlands and forests, especially in the south of West Java.

Sundanese culture has a variety of proverbs that can be used again today. This is important because now the use of Sundanese is almost displaced by Indonesian, the lingua franca language used throughout Indonesia. Moreover, Sundanese culture can also be described in terms of closeness to nature itself. This seems contradictory when viewed in the present context, where the original nature has almost disappeared especially in the midst of urban development in Bandung.

When associated with the smart city concept, the culture of the archipelago, especially Sundanese culture, is very supportive of the economic development of the village that still pays attention to nature and locale. The smart city concept can be used to support the economy without destroying nature itself, as well as raising unique things that are rarely found in the urban environment of Bandung. The Girimekar community actively produces agricultural and livestock products and have also started to produce processed products from their agriculture, such as bamboo soap, ground coffee, and herbs such as ginger, galangal, etc., that are cultivated by Girimekar residents.

The smart city concept can be developed through a website application specifically developed to support the economy in the Girimekar region. The application must also act as a promotional tool for Girimekar Village.

Another smart city concept being considered is the use of technology to solve problems faced by Girimekar Village, which are mainly the problems of environmental cleanliness, education of young people, and employment. With the smart city concept, which includes a smart village, it is hoped that the village of Girimekar can be self-sufficient. The agricultural power of the village of Girimekar in the form of coffee, bamboo, and spices need to be improved and marketed better so that it can improve the economy of the residents.

2 METHODOLOGY

2.1 Smart city

In several published articles, the definition of the term smart city is very diverse, in accordance with a paper written by Hollands in 2008. Referring to this broad definition, the smart city concept is related to the use of the latest technology in solving problems that exist in cities and surrounding areas.

The nature of this research is participatory action research. This study aims to change the lifestyle and habits of the target community. The community is expected to optimize the use of technology in life in the Girimekar village area. This study used a qualitative method employing a visual cultural transformation theory approach. As part of community service activities, every activity undertaken was recorded and afterwards an evaluation was carried out. In general, there were stages of action planning, activity processes, and evaluation. The results of the research activities carried out were in the form of recommendations for next steps that will be useful in community service activities that may be carried out afterwards.

2.2 Archipelago culture, Sundanese culture, and research outputs

This research is qualitative in nature, which offers views and results that can be applied to further research. The results of other studies also become a reference in relation to this research activity. This research is a preliminary study that is expected to be tested for validity if carried out in many other case studies. The conclusion proposed is expected to be an overview related to the implementation of community service and can improve the quality of community service that will be carried out in the future.

The culture of the archipelago has been widely raised in other studies. The uniqueness of this research is in the form of activities to go directly to the location and see how the culture of the archipelago is seen in a contemporary way. This allows comparison between cultures that have existed and considers how to restore lost old values to create a new syncretic culture that combines the positive values of technology and science with the wisdom and nobility of traditional culture.

Sundanese culture is part of the culture of the archipelago yet has its own peculiarities. Part of Sundanese culture used in this study is the culture of speech proverbs that contain policies related to nature and cultural conservation. This culture is taken from Sundanese proverbial literature by first consulting with the surrounding community. Another thing related to Sundanese culture that will be used is related to visual arts and architecture.

3 DISCUSSIONS

3.1 *Technology in Girimekar smart village*

The use of technology in developing rural areas is a necessity but has many challenges and must be of an appropriate level. The use of more sophisticated technology requires assistance and training to ensure that the technology used can be optimized properly. The technology used by universities is usually the latest product, which requires time to be absorbed by the community. Therefore, it is the responsibility of lecturers and researchers in universities to ensure that the products of higher education can be well utilized by the community.

Activities carried out in the village of Girimekar are expected to be a model for the use of technology that is initiated by universities, and to be utilized by the village community. Communities are introduced to technologies that they have not used before. This is intended to increase community ICT literacy. The community is expected to be able to open their mindset, so that innovations will emerge, ultimately bringing competitiveness to the village community. In addition, the community was guided to make other products, such as processing soap and crafts to encourage innovation. Community service activities initiated by tertiary institutions such as this, indeed, cannot produce results in a short time. An intense meeting between the community and universities will have positive impacts on both parties. On the one hand, the community benefits from access to knowledge, technology, and art that are part of the daily life of universities. On the other hand, universities will get more comprehensive information and data about the surrounding community environment, which is the goal of the higher education institution itself, namely to improve the degree of society. The possibility of obstacles in the process cannot be denied.

That obstacles will arise, especially from the community, is likely. Like a tree, mineral nutrient intake from the soil determines the results that can be obtained. The tree does not take all of the nutrients present in its environment, but only takes nutrients that are beneficial for its growth and development. Information about technology is the same: not all information is felt to be beneficial to the community, and therefore it is very important to emphasize the benefits of the information provided. And it should also be borne in mind that the way of thinking of higher education is sometimes too complex and needs to be reduced to a level that can be consumed by the public, without losing its essence.

Examples of the use of technology in community service activities are actually quite intensive. For example, at each meeting activity held at the mosque, an infocus projector is used to present the programs that will be rolled out to the Girimekar village community. The hope is the visualization provided will allow the community to become more enthusiastic, and become accustomed to new things. Mapping the surrounding environment is also done using drones and digital photography tools. The visual role here is very important, because we come from the field of art and visual literacy is something we want to introduce so that the community can be more visually cultured. This can be something interesting for village tourism activities.

3.2 *Sundanese culture through mural*

The community is not likely to be faced with the process of designing murals directly, but actually the community can gradually participate in the manufacturing process, which also uses a kind of projection technology to make the mural. This actually presents a response that exceeds estimates, where, after the mentoring program, the community was moved to make murals independently, proof that memorable and relevant information on the community can continue to roll out, and in turn produce new products.

Mural activities such as those carried out in Girimekar are not the first time such activities were held. Previously, murals had been carried out in schools and sidewalk walls by a team of writers who had also been published in a scientific article (Trihanondo and Endriawan, 2018). Halsey and Pederick (2010) claim that there is a strategy that requires that murals be meaningful when they visually stop being themselves – that "best" graffiti, bureaucratically, is what functions as a form of eliminating its own meaning. In mural activities like this, to speed up the production process, usually mural designs have been made before in digital form using graphics applications such as Photoshop. The design that has gone through the curation process is projected with the help of a digital projector at night, then the design is traced to the wall. The next day artists color the pattern that has been moved to the wall. From the mural activities we get higher time efficiency, and the quality of the work gets better. Indeed, there are already technologies that can print designs of large size, but digital printing technology still cannot replace the quality of paint and the smoothness of the image produced. Moreover, art made by hand cannot be replaced by a machine. The machine only helps in effectiveness and efficiency, but the automation should not replace humans.

4 CONCLUSIONS

The impact that was felt was the change in social conditions in the form of increasing community appreciation for artistic activities. This was followed by the increasing understanding of citizens towards better artistic values. In the environmental field, residents are increasingly aware of keeping the area clean. Even more creative citizens produce innovative products, as well as increasing community skills in their daily lives. Physical changes are seen in the form of Sundanese-themed murals, structuring of ornamental plants, as well as the cultivation of economically valuable plants such as coffee, bamboo, and spices.

REFERENCES

Endriawan, D., & Trihanondo, D. 2015. Interpretasi Spiritualitas Pada Karya Seni Patung Amrizal Salayan. *Atrat: Visual Art & Design Journa*l, 3(1).

Endriawan, D., Trihanondo, D., & Haryotedjo, T. Seni Rupa Islam dan Tantangannya di Indonesia pada Revolusi Industri 4.0. *In Seminar Nasional Seni dan Desain 2*018 (pp. 6–9). State University of Surabaya.

Halsey, M., & Pederick, B. 2010. The game of fame: Mural, graffiti, erasure. *City*, 14(1–2), pp.82–98.

Hollands, R.G. 2008. Will the real smart city please stand up? Intelligent, progressive or entrepreneurial? *City*, 12(3), pp.303—320.

Nasr, S.H., Sutejo, D., & Tarekat, H. 1993. *Spiritualitas dan seni Islam*. Penerbit Mizan.

Sabana, S. 2002. *Spiritualitas dalam Seni Rupa Kontemporer di Asia Tenggara: Indonesia, Malaysia, Thailand, dan Filipina sebagai Wilayah Kajian*. Disertasi Institut Teknologi Bandung.

Trihanondo, D., & Endriawan, D. 2018, November. The role of higher education in society activation through digital mural in ASEAN cities. *In IOP Conference Series: Materials Science and Engineering*, 434(1), 012284). IOP Publishing.

Trihanondo, D., & Endriawan, D. 2019, February. Cultural and environmental conservation through community service program in Girimekar Village. *In IOP Conference Series: Earth and Environmental Science*, 239(1), 012050). IOP Publishing.

Credit scoring model for SME customer assessment in a telco company

L.A. Baranti & A. Lutfi
University of Indonesia, Jakarta, Indonesia

ABSTRACT: Divisi Business Service (DBS) is one division in PT Telekomunikasi Indonesia (PT Telkom). DBS's scope of work is handling small medium enterprise (SME) customers. In 2018, DBS closed the year with a 79% collection rate. This rate had been declining from prior years. Moreover, DBS had the lowest number, compared to other divisions in Telkom Group,. This number represented a financial loss to Telkom of IDR 580 billion, largely due to a low rate of collection, mostly caused by bad customer assessment at the beginning of the sales process. From those facts, this research is determined to give a better alternative to the customer assessment process by implementing a credit scoring model to be derived by using logistic regression. As the result, there are three financial and three non-financial variables that affect the customer worthiness: quick ratio, DER, EBIT, business reputation, industry growth, and business competition.

1 INTRODUCTION

Divisi Business Service (DBS) is one of the functional divisions in PT Telekomunikasi Indonesia (PT Telkom). DBS plays a role in managing lower middle or corporate customers for micro, small and medium enterprises (MSMEs). In general, DBS manages customers with asset values below Rp 500 billion and annual turnover below Rp 50 billion (PT Telekomunikasi Indonesia, 2016). DBS itself closed 2018 with a collection rate of 79%. The low collection rate has caused a loss for Telkom Group of Rp 580 billion (PT. Telkom Indonesia, 2019). This situation needs attention because the collection rate of DBS in 2018 has decreased significantly from the previous year. In addition, it can also be seen that the collection rate of DBS has decreased from 2017 to 2018.

Moreover, in 2018, DBS is the division with the lowest collection rate compared to other divisions in Telkom. The graph can be seen in Figure 2.

To overcome these problems, DBS has carried out more comprehensive organizational restructuring and customer management. This can be seen by the existence of a Revenue Assurance (Revas) Unit to show DBS is serious about improving collection performance. The Revas Unit is responsible for conducting credit assessments of prospective customers, which results in providing recommendations to management regarding the financial feasibility of prospective customers who will subscribe to Telkom through DBS. Revas Unit is expected to provide "insurance" that the prospective customer concerned will indeed be able to complete the payment to the end. Currently, the Revas Unit is assessing the feasibility of a prospective customer with a series of procedures from studying the prospective customer's internal documents. An assessment mechanism like this will certainly complicate the Revas Unit's process for determining the eligibility of prospective customers. The possibility of a wrong decision becomes quite large because there is still a lot of intuitive interference in decision-making.

Therefore, this study tries to eliminate the qualitative assessments that occur in DBS customer assessments. This study will recommend a more effective customer assessment mechanism with a credit scoring mathematical model developed through a logistic regression approach.

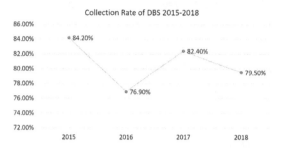

Figure 1. Collection rate of DBS in 2015–2018.

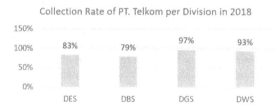

Figure 2. Collection rate of DBS in each division.

2 LITERATURE REVIEW

2.1 Credit scoring system

Credit scoring is a tool commonly used to test credit risk in credit problems in a scientific and quantitative way (Marquez, 2008). Credit scoring is considered quite appropriate in the world of credit because this method can predict which prospective customers are acceptable, and provide recommendations on how much credit is appropriate to the prospective creditor, and it is an appropriate strategy to increase customer collection success (Long, 1973).

2.2 Logistics regression

In logistic regression analysis, mathematical models are assumed to be able to describe how the probability of binary results is related to the linear combination of variables (Michaela Kiernan, 2001).

The standard equation for logistic regression is:

$$ln\left(\frac{p}{1-p}\right) = B_0 + B_1 X_1 + \ldots + B_n X_n \qquad (1)$$

3 METHODOLOGY

This research will begin with determining variables that affect credit assessment at PT Telkom. The variables formed from the previous step must be tested for significance to determine the relationship of each independent variable to the dependent variable. If it is known that there are variables that do not significantly affect the dependent variable, these variables may not be included in the next test. Furthermore, the initial model will be built from the selected variables. If in this initial test, a good model cannot be formed (indicated by the significance value of each independent variable of more than 0.05), then it is necessary to conduct a multivariate test.

Then, the next step is to build a model with variables that have passed the multivariate test and then test the model formed with various statistical tests. These tests are carried out to see the performance of the model in terms of the accuracy of the results and its ability to represent real conditions. This study will take 200 data to be sampled. This sample will be taken from the last four years, namely 2015–2018. The distribution of samples will mimic population data patterns.

4 RESULTS AND DISCUSSIONS

The initial variables used to start the study are the financial variables: current ratio, quick ratio, debt equity ratio (DER), earnings before interest and tax/total assets, return on assets, return on equity, and revenue growth; and the non-financial variables: business reputation, prospective customer behavior, industrial development, and industry competition level.

A multivariate test is a test that is applied to understand data structures or check whether the variables that are related to each other. The test is done by selecting the independent variables by removing the independent variables in turn to determine the variables with the best significance value.

As the result, only significant variables are found, namely quick ratio, DER, EBIT/TA, business reputation, industrial development, industrial competition level. Thus, the mathematical model will be as follows.

Ln P/1-P = -48.938 + 19,132 QUICK_RAT -24,934 DER + 47,159 EBIT + 4,628 REPUTATION + 3,285 DEVELOPMENT + 2,480 COMPETITION

In the quick ratio variable, statistical data shows that the higher the value of the quick ratio owned by a prospective customer, the higher the potential customer is to be considered "feasible". This is consistent with what Ross, Westerfield, and Jaffe stated in their book *Corporate Finance*, namely that quick ratio is one of the liquidity measures that is useful to see how liquid a company is. Liquidity measure can describe the ability of companies to pay short-term debt without getting additional capital (Ross, Westerfield, & Jaffe, 2013).

In the DER variable, statistical data indicate that the DER value has a negative impact on customer eligibility. Customers who have a high DER value tend to be judged unfeasible on this model. These results are consistent with what Ross, Westerfield, and Jaffen stated: the Debt to Equity Ratio is one of the debt ratios commonly used to measure long-term solvency. Long-term solvency measures are measurement methods to determine a company's ability to settle long-term debt or its financial leverage (Ross, Westerfield, & Jaffe, 2013).

In the EBIT/TA variable, statistical data indicate that the variable has a positive influence on customer eligibility. The higher the value of EBIT/TA, the greater the tendency for customers to be declared "feasible" by this credit scoring model. Altman has also formulated this in the Altman Z-Score. EBIT/TA is one of the variables in the Altman Z-Score. This variable is used to represent operating efficiency. Operating efficiency is important in the Altman Z-Score to predict the sustainability of the company (Bodie, Kane, & Marcus, 2014).

On the business reputation variable, statistical data shows that the business reputation variable has a positive influence on customer eligibility. This is consistent with what Khemakhem and Boujelbene have stated in their study, namely that the age and size of the company, the legality, and the relationship of the company to the bank related to credit can affect the probability of default (Khemakhem & Boujelbene, 2017).

On industrial development variables, statistical data show that the relationship between industry development variables and customer eligibility is positive. This is also consistent with Annamaria and Trenca's statement that the type of industry also influences the probability of default. The type of business that is developing or stable and business processes that favor political and social conditions tend to have better sustainability so they can settle their debts (Annamaria & Trenca, 2009).

At the variable level of industry competition, statistical data show that the tight industry competition variable has a positive relationship with customer eligibility. This is in line with research conducted by Mahmoudzadeh and Syefi, showing that market competion has a positive relationship with a company's financial performance (Mahmoudzadeh & Seyfi, 2017). Other research conducted by Sarkar and Sensarma states that competition has a negative effect on the default risk of a company (Sarkar & Sensarma, 2015).

Then the credit scoring model is tested to see its accuracy. The results of these tests can be summarized as follows.

Table 1. Recap of model feasibility test results.

Test	Result
ROC Curve	The accuracy of the model is 70.2%
Omnibus Test	With a 95% confidence level, there is at least one independent variable in the model that significantly influences the dependent variable
Nagelkerke R Square	91.2% of the dependent variable has been explained by the independent variable
Hosmer and Lemeshow Test	With a 95% confidence level, the logistic regression model formed is acceptable because there is no significant difference between the model and its observational value
Classification Plot	The accuracy of this research model is 75%

5 CONCLUSION

As the result, only significant variables are found, namely quick ratio, DER, EBIT/TA, business reputation, industrial development, and industrial competition level. Thus, the mathematical model will be as follows.

$$\text{Ln } P/1-P = -48.938 + 19{,}132 \text{ QUICK\_RAT} - 24{,}934 \text{ DER} + 47{,}159 \text{ EBIT} + 4{,}628 \text{ REPUTATION} + 3{,}285 \text{ DEVELOPMENT} + 2{,}480 \text{ COMPETITION}$$

Moreover, from the five tests that have been carried out, it can be stated this model quite accurately describes the real situation, shows that the independent variable is sufficiently precise in shaping the dependent variable, and can represent reality.

REFERENCES

Annamaria, D.-B., & Trenca, I. 2009. *Using Credit Scoring Method for Probability Companies Default Estimation at Industry Level*. Cluj-Napoca: Babes-Bolyai University.

Araújo, C. U. (2011). Application of credit scoring models in the analysis of insolvency of a Brazilian microcredit institution. *Journal of Accounting and Auditing*, 7(8),799–812.

Bodie, Z., Kane, A., & Marcus, A. 2014. *Investments* 10th Edition. New York: McGraw Hill.

Khemakhem, S., & Boujelbene, Y. 2017. Predicting credit risk on the basis of financial and non-financial variables and data mining. *Review of Accounting and Finance*, 316–340.

Lewis, E. M. 1992. *An Introduction to Credit Scoring*. California: Fair Isaac and Co Inc.

Long, M. 1973. *Credit Scoring Development for Optimal Credit Extension and Management Control*. College of Industrial Management. Atlanta Georgia: Purdue University.

Mahmoudzadeh, M., & Seyfi, A. 2017. The effect of product market competition on the relationship between capital structure and financial performance of companies. *International Journal of Economics and Financial Issues*, 523–526.

Marquez, J. 2008. An Introduction to Credit Scoring for Small and Medium Size Enterprises.

Michaela Kiernan, H. C. 2001. Do logistic regression and signal detection identify different subgroups at risk? Implications for the design of tailored interventions. *Psychological Method*, 6(1),35–48.

PT Telekomunikasi Indonesia. 2016, May 5. *DBS Review*. (Divisi Business Service, Performer) Telkom DBS, Jakarta Barat, Jakarta, Indonesia.

PT. Telkom Indonesia. 2019. *Historikal Collection CFUE 2018*. Jakarta: PT. Telkom Indonesia.

Ross, S., Westerfield, R., & Jaffe, J. 2013. *Corporate Finance* 10th Edition. New York: McGraw Hill.

Go Beyond training and its impact on the performance of employees in PT Telkomsel

Akrom Dharmiko & Nidya Dudija
Faculty of Economics and Business, Telkom University Bandung, Indonesia

ABSTRACT: In 2017, Telkomsel implemented Go Beyond training for its employees. Training feedback shows that there were employees who refused to recommend that training be conducted again. Telkomsel's revenue growth has declined since 2016. The purpose of this research is to review Go Beyond training and its impact on the performance of employees in Telkomsel West Java. This study used a questionnaire from 110 employees. Data analysis using SmartPLS version 3.2.8. It was found that content of training, training methods, and training facilities had a positive and significant influence on employee performance. Training duration had a negative and not significant influence on employee performance, and instructor attitude and skill had a negative and significant influence on employee performance. The conclusion obtained in this study is that companies need to maintain training content, training methods, and training facilities. Then, they need to make improvements for training duration, and instructor attitudes and skills.

1 INTRODUCTION

Data from the Indonesian Ministry of Communication and Informatics shows a high number of active internet users in Indonesia. In 2017, there were 143.26 million active internet users in Indonesia (Press Release #53/HM/KOMINFO/02/2018). In 2016, there were 132.7 million active internet users in Indonesia. This shows an increase of around 10.56 million internet users in 2017. Telkomsel's internet data usage increased by 126.2% and revenue from Telkomsel's digital business increased by 28.7% in 2017.

Along with this, due to the increasing use of internet data, Telkomsel is transforming its business into a digital company. In 2017, Telkomsel implemented Go Beyond competency training for its employees. Go Beyond competency consists of eight factors: global mindset, own customer, business savvy & ecosystem understanding, execution focus, youthful thinking, out of the box, networking, and data driven. The process of implementing this competency was carried out by providing Go Beyond competency training for all employees in the West Java region. Training feedback shows that there were employees who refused to recommend that training be conducted again. In addition, Telkomsel's revenue growth in the West Java region has declined since 2016. In 2015, revenue growth was 15%, in 2016 it was 14%, in 2017 it was 7.5%, and in 2018 it was –4.2%. This shows the phenomenon of decreased company performance and employee refusal to endorse this training.

2 THEORETICAL FRAMEWORK

According to Notoatmodjo (2015, p. 124) *performance* is work that can be displayed or the appearance of an employee's work. Thus, the performance of an employee can be measured from the results of work, or the results of activities in a certain period of time. To measure employee performance, several performance criteria can be used. Indicators for measuring employee performance, according to Mathis and Jackson (in Yusup 2017) are: quantity, quality, presence, and employee cooperation.

According to Mondy (2008, p. 210), training is an activity designed to provide the knowledge and skills needed by learners to be able to do their work at this time. The dimensions of an effective training program given by the company to its employees, according to Byars & Rue (in Wardhana, 2014, p. 102) can be measured through training content, training methods, instructor attitudes and skills, training duration, and training facilities.

3 RESEARCH QUESTION

The following research questions will be addressed in this study.
1. How good is Go Beyond competency training for employees at PT Telkomsel West Java?
2. How good is the performance of employees at PT Telkomsel West Java?
3. How high is the influence of Go Beyond competency training on the performance of employees at PT Telkomsel West Java?

4 HYPOTHESIS

In accordance with the review and existing theoretical basis, we can develop a model of the following hypothesis:

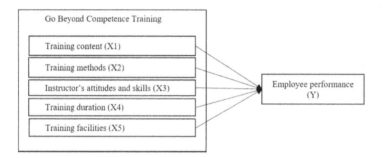

Figure 1. Hypothesis model.

5 RESEARCH VARIABLES

There are five independent variables used in this study: training content, training methods, instructor's attitudes and skills, training duration, and training facilities. The one dependent variable is employee performance. This research model was adopted from theories taken from Byars and Rue (in Wardhana, 2014, p. 102) and Mathis and Jackson (in Yusup, 2017).

6 POPULATION AND SAMPLE

The population in this study was Telkomsel employees who had attended Go Beyond training in the West Java region, and that was 151 people. The sample in this study was obtained from the Isaac and Michael formula (Sugiyono, 2018), such that the total number of respondents in this study was 110 respondents. This study uses a non-probability sampling technique to determine the relationship between research variables. More specifically, this study uses quota sampling according to the considerations of researchers, so that samples can answer research problems.

7 DATA ANALYSIS

The analytical method used was the variance-based structural equation model (VB-SEM) method, which is intended to make predictions from the construct or independent and dependent variables in a model (Hair et al., 2010, in Indrawati, 2015). Statistical analysis using partial least square (PLS) included in VB-SEM. The tool used to analyze these variables was SmartPLS 3.2.8 software.

8 RESEARCH FINDING

Descriptive analysis results showed that training content (85.82%), training methods (84.14%), and instructor's attitudes and skills (86.18%) were in the very good category. Training duration (80.36%), training facilities (83.73%), and employee performance (81.73%) were in the good category.

Hypothesis testing was done by comparing t-count values and t-table values. If the value of t-count> t-table, the significance level ($\alpha = 5\%$), then the estimated value of the path coefficient is significant (accepted). The t-table value is 1.96 (Hair et al., in Abdillah & Hartono 2015).

Based on Table 1, the hypothesis test shows that training content has a positive and significant influence on the performance of employees at PT Telkomsel West Java. Training method has a positive and significant influence on employee performance. This result is supported by previous research by Triasmoko et al. (2014) which states that training materials significantly affect employee performance variables, or by increasing training materials, the employee's performance will increase significantly. Triasmoko et al. also stated that training methods significantly affect employee performance variables. This means that improving good training methods will also improve the performance of these employees.

Instructor attitude and skills have a negative and significant influence on the performance of employees. The reality in the field that supports the negative attitudes and skills of training instructors is due to differences in instructor backgrounds compared to trainee backgrounds. Instructors who provide training are often HR consultants, while the training participants had backgrounds in various divisions. Training participants data obtained from HCM Telkomsel West Java shows this difference: participants have backgrounds in the ICT division (36.3%), sales (31.8%), finance (9.3%), GA (6%), project (4.6%), marketing (4%), Y&C (3.3%), AM (2.6%), CS (1.3%), and HR (0.7%). This difference shows that instructors' abilities in the areas of ICT, sales, and others are also needed.

Training duration has no significant influence on employee performance. The reality in the field that supports this result is due to several factors, as follows:

a. Time management is an indicator of training duration. From the company's evaluation data on the Digital Transformation Go Beyond Workshop for facilitators in 2017, it is known that the facilitator's score for time management is the lowest compared to other categories. Four of the five facilitators have the lowest time management value.

Table 1. Statistical test results - path coefficients and T statistics.

		Original Sample (O)	T Statistics (|O/STDEV|)	Note
Training Content ->	Employee Performance	0.363	2.910	Significant
Training Method ->	Employee Performance	0.544	5.536	Significant
Instructors attitude & skills ->	Employee Performance	−0.196	2.029	Significant
Training Duration ->	Employee Performance	−0.041	0.485	Not Significant
Training Facilities ->	Employee Performance	0.229	2.019	Significant

b. The Go Beyond competency training was carried out over three days for all employees in all divisions. Previous research conducted by Faradita (2013) used the duration of training for three days for only the administrative section; this means more training duration is needed.

Training facilities have a positive and significant influence on employee performance, while employee ability also has a positive and significant influence on performance. Higher employee ability will improve employee performance.

Table 2. R square.

	R Square
Employee Performance	0.728

The value of R^2 for employee performance is 0.728, which means that the training content, training method, instructor attitudes and skills, training duration, and training facilities are able to explain employee performance by 72.8%. While the remaining 27.2% is affected by other factors.

9 CONCLUSION AND RECOMMENDATION

Based on the results of descriptive analysis, the study shows that training content, training methods, and the attitudes and skills of the instructor are in the very good category. Training duration, training facilities, and employee performance are in the good category.

Based on statistical analysis, there are positive and significant relationships for training content, training methods, and training facilities. Then, there is a negative and significant relationship for instructor's attitudes and skills, and also a negative and insignificant relationship for training duration. Regarding instructor's attitudes and skills, the instructor must have the right abilities for each division of employees. Regarding training duration, improvements to time management are needed and additional training is needed that is more specific to each division. Training methods can be added with other methods such as on-the-job training, while training content can be added with more detailed tactical training, and training facilities can be better added from the inside.

REFERENCES

Abdillah, W., & Hartono, J. 2015. *Partial Least Square (PLS) Alternative Structural Equation Modeling (SEM)*. Dalam Penelitian Bisnis, Andi, Yogyakarta.
Indrawati. 2015. *Metode Penelitian Manajemen dan Bisnis*. Konvergensi Teknologi Komunikasi dan Informasi, Refika Aditama, Bandung.
Faradita, A. L. 2013. Progam Training (Pelatihan) Terhadap Kualitas Karyawan, *eJurnal Administrasi Bisnis Universitas Mulawarman*, 1(1),1–7.
Hakim, Lukman Nul. 2014. Pengaruh Pelatihan dan Motivasi Terhadap Kinerja Operator Produksi di PT Alam Lestari Unggul. Tesis Program Pascasarjana Magister Manajemen, Universitas Mercu Buana.
Kementrian Komunikasi dan Informatika Republik Indonesia 2018, Pers Release No. 53/HM/KOMINFO/02/2018), Jakarta, viewed 12 July 2018.
Mathis, R. L., & Jackson, J. H. 2006. *Human Resource Management*, 10th Ed, Salemba Empat, Jakarta.
Mondy, R. W. 2008. *Manajemen Sumber Daya Manusia*, Erlangga, Jakarta.
Notoatmodjo, S. 2015. *Pengembangan Sumber Daya Manusia, PT*. Rineka Cipta, Jakarta.
Sugiyono. 2018. *Metode Penelitian Kuantitatif*, Alfabeta, Bandung.
Telkomsel. 2018. Annual Reports, Jakarta, viewed 28 June 2019; https://www.telkomsel.com/about-us/investor-relations.

Triasmoko, D., Mukzam, M. D., & Nurtjahjono, G. E. 2014. Pengaruh Pelatihan Kerja Terhadap Kinerja Karyawan (Penelitian pada Karyawan PT Pos Indonesia Cabang Kota Kediri)., *Jurnal Administrasi Bisnis Fakultas Ilmu Administrasi Universitas Brawijaya*, 2(1).

Wardhana, A. 2014. *Manajemen Sumber Daya Manusia*, PT Karyamanaunggal Lithomas, Bandung.

Yusup, M. 2017. Pengaruh Budaya Organisasi The Telkomsel Way Terhadap Kinerja Karyawan PT. Telkomsel Di Area 2 Jabotabek Jabar. Tesis Program Pascasarjana Magister Manajemen, Universitas Telkom.

Exploring jewelry design for adult women by developing the pineapple skin

A.S.M. Atamtajani & S.A. Putri
Telkom University, Bandung, Indonesia

ABSTRACT: Pineapple is a tropical fruit commonly found and consumed in Indonesia. The pineapple skin is a waste product from this process and usually mixed with banana peel to make organic fertilizer. While the unused waste can pollute the environment, the pineapple skin has potential to be used as the main material in jewelery-making due to its unique texture. The possibility to use pineapple skin in jewelry products is highly likely, not just for reducing the waste that might pollute the environment but also to develop the potential of the region that produces pineapples. In this experiment using pineapple skin as the main material for a jewelry product, the target market is adult women aged 19–35 years. This research process uses qualitative methods, data collecting through literature studies, interviews, and questionnaires. The data is analyzed with the material experiment method. This research aims to show the repurposing of an organic waste as an alternative solution in managing waste.

1 INTRODUCTION

Pineapple skin has a texture that resembles arranged scales. This unique texture is a distinctive feature of the pineapple skin and can be incorporated into jewelry products. Pineapple skin is rarely processed into artistic creations. There are several obstacles experienced during the waste treatment process. Pineapple skin is organic waste that contains a lot of water, meaning it rots easily. Many experimental processes have been carried out to get the desired results, one of which, fumigation, has been successfully applied to pineapple skin.

This experiment was inspired by D. S. Moeljanto's (1992) book titled *Prahara Budaya*. This book discusses fumigation to preserve fish so it does not easily rot. This fumigation was successfully applied to pineapple skin so the pineapple skin becomes dry and not rotten. The shortcomings of this experiment are that the dry pineapple skin becomes brittle and easily broken so the authors add resin as the final preservation stage which strengthens the processed pineapple skin, gives a glossy appearance, accentuates the texture, and also makes pineapple skin more durable.

2 METHOD

The pineapple skin processing method was adopted from the fish-smoking method. The basic principle of this smoking method is to reduce water content and kill bacteria contained in pineapple skin so it will not rot. The smoking method started with a salting process. Salting is part of the process of preserving, where salt reduces the water content and inhibits the growth of bacteria. There are two ways to do this: dry salting and brine salting. Pineapple skin that has been salted and drained is placed on a roasting shelf surrounded by hot smoke. Indirect heating causes evaporation of the remaining liquid of any pineapple pieces that might still attached to the pineapple skin. This has a preservative effect because bacteria are more active in damp environments. Therefore, the drying process has an important role and the quality of pineapple skin resilience depends on the amount of water evaporated.

Pineapple skin can be treated with either a warm smoking or a cold smoking method. In cold smoking, the temperature is low and the evaporation is slow. To increase the durability of the pineapple skin, the time spent on fumigation of the pineapple skin must be extended.

In warm smoking, there is less distance between the source of fire (smoke) and pineapple skin so the temperature is higher. High temperatures can stop enzyme activity and change the color of the pineapple skin. Warm smoking dried the pineapple skin faster than the cold smoking. The comparison between these two smoking methods can be seen below.

Table 1. Comparison between cold & warm smoking methods.

No	Smoking Method	Temperature	Processing time duration
1	Cold Smoking	40–50°C	1–2 week
2	Warm Smoking	70–100°C	2–4 hour

(case: https://www.academia.edu/5428885/Pengasapan, 2000)

The smoking was followed by pressing to flattening the smoked pineapple skin to achieve consistent thickness and make it easier to process. This pressing process also smoothes the wrinkles in the pineapple skin from the smoking process. The thickness of the pineapple skin depends on how long the pressing process is carried out and how many times the pressing is repeated.

After being pressed to the desired thickness, the smoked pineapple skin is cleaned using a small, soft brush to remove dust and dirt that covers the natural base color of the pineapple skin. The whole process ended with pouring resin onto the pineapple skin to achieve a glossy look and harden the pineapple skin.

3 EXPLORATION

The exploration scheme was made as a reference sequence for the exploration of pineapple skin waste in this research. The following is the sequence scheme.

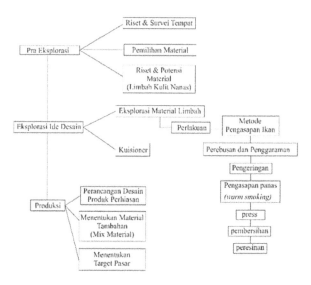

Figure 1. Exploration scheme.

4 DESIGN DEVELOPMENT

As the first step in making silver jewelry, craftsmen trace the design by attaching it to the plate and then cutting the plate using a jigsaw. They then trim and smooth the edges of the plate using sandpaper. The purpose of making a plaque using a plate is to make a silver frame.

The following image is a wireframe that has been made and then affixed to paper using paper glue to help the filigree filling. The silver craftsmen have different tasks, usually adapted to their respective abilities. For example, there are craftsmen who are in charge of making the outer framework, craftsmen making the inner framework, some who are making filigree entries, and those who are in charge of finishing parts or polishing silver.

Figure 2. Making the wireframe & filling in the frame.

After the frame has been completed, the craftsmen braze between one part and another. The process uses silver powder and borax water as adhesive. Then the silver is ignited. When the framework is ready, a craftsman fills in the silver frame.

Terms used in making silver consist of *waleran*, which means a deep frame, *odoh* which means frame, *ketep*, which means complementary decoration with a round shape. The type of silver filling consists of *krawangan*, zigzag, *unther*, unther round, unther cock, fan, *unthuk*, *eses*.

How to fill the filigree also depends on the ability of each craftsman. There are those who fill the filigree by adding paper glue so it does not easily fall apart or break down, and some who fill the filigree without using paper glue. Actually, filling the filigree by using paper glue is not recommended, because in the desoldering stage, the craftsmen sometimes have difficulty because the glued silver sticks to the paper instead of the framework, so after the paper glue is removed, the pieces don't stick completely and are damaged.

Figure 3. Product result.

5 CONCLUSION

This experiment could be very influential for the environment and for society, aiming to reduce the waste of pineapple skin that can adversely affect the environment and also the local community if not handled wisely. This research is to show the reduction of organic waste can be a solution to a problem.

From the experiments that have been done, it is highly recommended to further develop pineapple skin waste reuse, due to the potential of the material to be used as excellent jewelry. It is quite unique as no one else has developed this product into jewelry or jewelry accessories or used it in the form of other art crafts.

Introducing and providing education on how to preserve pineapple skin waste to the community can add insight in the field of arts and jewelry. Besides that, the goal of developing pineapple skin waste into jewelry could be a very good achievement.

REFERENCES

Creswell, J. W. 2014. *Research Design*. Los Angeles, London, Newdelhi, Singapore, Washington DC: SAGE Publications.
Adawyah, R. 2007. *Pengolahan dan Pengawetan Ikan*. Bumi Aksara. Jakarta.
Ahde-Deal, P. 2013. *Women And Jewelry*. Finlandia: Aalto Arts Books Helsinki.
Atamtajani, A. S. M. 2018. Filigree Jewelry Product Differentiation (Case Study Filigree Kota Gede Yogyakarta). Bandung Creative Movement (BCM) Journal 4(2).
Sufyan, A. 2013. Tinjauan Proses Pembuatan Perhiasan dari Desain ke Produksi (Studi Rancangan Aplikasi Logo STISI Telkom pada Liontin). *Jurnal Seni Rupa & Desain*, 5 (Mei-Agustus).
Sufyan, A. "The Design Of Kelom Kasep (Differentiation Strategy In Exploring The Form Design Of Kelom Geulis as Hallmark Of Tasikmalaya)." *Balong International Journal of Design* 1(1).

Identifying factors that influence interest using credit scoring by implementing customer journey mapping

L. Nadeak & D. Tricahyono
Telkom University, Bandung, Indonesia

ABSTRACT: This study aims to identify the factors that influence the experience of PT.TS credit scoring (TSCS) customers by implementing customer journey mapping (CJM). This is an exploratory study in order to understand the phenomenon of customer experience. The research was conducted by in-depth interviews, observation, and a literature study about the topic. Three financial institutions were subjects of this study. These subjects were the first three customers of TSCS in 2018 to use the credit scoring service. Customer journey mapping is the tool being used by this study to map the touchpoints based on an adoption process. The implication of the study is discussed at the end.

1 INTRODUCTION

PT.TS was established on May 26, 1995. Since then, it consistently served Indonesia by presenting telecommunications access to the Indonesian people. PT.TS develops digital services by utilizing telecommunication data processing and in-depth analysis to assist business decision-makers. One of the digital data services of PT.TS is credit scoring (CS) under unit TSCS. Until 2018, TSCS had 10 customers of CS and plans to have 90 customers by the end of 2019. CS customers buy the service to protect them from non-performing loans (NPL). The problem that occurs in prospective customer credit risk assessment carried out by banks is 80% of prospective customers do not have a credit history and 20% of the market is for highly competitive credit with a low margin for customers (PT.TS, 2018).

The customer process to buy is a point of contact that can have a direct and an indirect effect on purchasing decisions (Lemon & Verhoef, 2016). Mapping customer trips is done to identify the main interactions between the customer and the company; raise questions about customer feelings, motivation, and considerations at the touchpoint of the customer's journey; and to find the customer's biggest motivation to buy (Richardson, 2010). Customer journey mapping (CJM) is a tool used to help companies better understand each step taken by customers, thinking of all stages of customer experience and the number of touchpoints between companies and customers (Hong, 2016). The stages, perceptions, and experience of the customer is a journey wherein the customer considers and decides to use or not use CS from TSCS. This study aims to identify the factors that influence the customer journey based on the experience of CS's customers.

2 LITERATURE

2.1 Credit scoring

Credit scoring is an assessment of prospective borrowers or debtors regarding proposed loan applications that are classified into credit risk categories (Mario, 2017). Credit scoring must meet several credit evaluations as a way to reduce the bad credit burden by using credit risk analysis based on 5C (Abbadi & Karsh, 2013). Peprah (2017) explained the 5C consists of capacity, character, capital, collateral, and condition.

2.2 Customer journey

Customer journey (CJ) maps the customer experience from the first communication to building a continuous relationship with the company. It is a recurring and dynamic process from pre-purchase to post-purchase that combines past experience and external factors (Meredith, 2016; Lemon & Verhoef, 2016). CJ is the sequence of events that the customer goes through to gather information and buy and interact with products or services, where the interaction exists at each contact point (touchpoint) between the company and the customer (Norton and Pine, 2013). Touchpoint is also an important moment during a customer's journey to touch deals and companies (Rawson et al., 2013). According to Richardson (2010), touchpoints are based on product, interactions, messages, and settings. According to Lemon and Verhoef (2010), touchpoint can be based on brand, partners, customer, and social/external. Meanwhile, according to Halvorsrud, Kvale, and Folstad (2016), touchpoints are based on the attributes of initiator, time, and channel.

2.3 Customer journey mapping

Customer journey mapping (CJM) is a tool for evaluating customer experience (CE) based on customer perceptions, with visual or graphical images of emotions or feelings and interactions that occur between customers and companies, products, services, and brands in a certain period of time through existing channels (Isaacson, 2012). Lemon and Verhoef (2016) developed a framework that describes CJM based on adoption process, which are pre-purchase, purchase, and post-purchase. Pre-purchase is the stage before the act of buying encompassing the need for recognition, search, prior experience, and consideration. Purchase is the stages during the purchase with choices, orders, and payments and an effective shopping experience that occurs at this stage. Post-purchase is the stage after the purchase with use, consumption, involvement after purchase, and demand for services and products to be the main touchpoints. Experience (consumption, repurchase, product return, and/or service selection) becomes an important aspect in carrying out analysis after purchase for the emergence of the loyalty loop.

3 RESEARCH METHODOLOGY

3.1 Framework

Figure 1 shows the framework of the present study based on Lemon and Verhoef's (2016) work. It consists of three main processes on the customer's decision journey using PT.TS

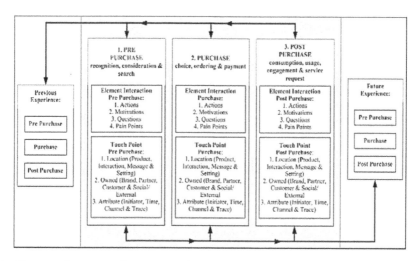

Figure 1. Framework customer journey mapping.

credit scoring, namely pre-purchase, purchase, and post-purchase. Pre-purchase focuses on recognition, consideration, and search. Purchase focuses on choices, ordering, and payment. Post-purchase focuses on consumption, use, service usage, and demand agreements. Each main process has an interaction element and a touchpoint. The operational variables can be seen in APPENDIX 1.

3.2 *Population and sample*

Determination of samples in qualitative research is not based on statistical calculations, but the sample chosen serves to obtain maximum information (Yin, 2012). Currently, there are only 10 customers of TSCS. Thus, the present study chooses three early customers representing banking, a credit company, and a financing company as the informants. APPENDIX 2 shows the informants of the present study. The in-depth interview was conducted to the informants.

4 RESULT AND IMPLICATION

The results of the present study can be seen in APPENDIX 3, which shows the CJM of credit scoring customers at TSCS by identifying pre-purchase, purchase, and post-purchase stages in credit scoring customers' journey. If we focus on the pain points, the present study reveals that in pre-purchase there are three issues: expensive price, incomplete data, and undefined standards for results. In the purchase stage, there are two issues: technical disruption, and slow getting results. Finally, in post-purchase stage, there are two pain points: expensive price and low quality in data and implementation.

This present study shows that some issues are discovered when we implement CJM to identify the key factors that influence the customer's interest on CS product. From APPENDIX 3, we may identify actions, motivations, questions, pain factors, and touchpoint elements. TSCS can identify what key elements should be scrutinized and improved in every stage. For example, several issues have been found for CS customers, namely, price, quality of data, and undefined standard for results.

REFERENCES

Abbadi SM, Abu Karsh SM. Methods of evaluating credit risk used by commercial banks in Palestine. *International Research Journal of Finance and Economics*. 2013; 111: 146–159.
Hong, T. 2016. *Customer Experience as a Competitive Differentiator in Subscription Services: Thinking Beyond the Paywall*. Helsinki: Metropolia University of Applied Sciences.
Isaacson, W. 2012. *Steve Jobs*. New York: Simon & Schuster.
Lemon, K.N., & Verhoef, P.C. 2016. Understanding customer experience throughout the customer journey. *Journal of Marketing*, 50.
Malhotra, N.K. 2013. *Marketing Research: An Applied Orientation*. New Jersey: Pearson Education.
Mario, D. 2017. Evalusi Performa Kredit Menggunakan Data Mining untuk Menilai Permohonan Kredit Fasilitas Layanan Pembiayaan Perumahan: Studi Kasus Pada PT.TSBank XYZ. Jakarta: Tesis Fakultas Ilmu Komputer Universitas Indonesia.
Meredith, B. 2016. Marketing and the customer experience. *NZBusiness + Management Magazine*, 40.
Norton, D.W., & Pine II, B.J. 2013. Using the customer journey to road test and refine 87 Customer Journey: A new approach for retailers the business model. *Strategy & Leadership, Vol 41*(2), 12–17.
Peprah. 2017. Ranking the 5C's of credit analysis: Evidence from Ghana banking industry. *International Journal of Innovative Research and Advanced Studies (IJIRAS)*, 4(9), 78–80.
Priyanto, I.W. 2018. Analisa Strategi Pemasaran Video Market Place (VMP) Untuk Mencapai Target Customer-Based Studi Kasus: PT.TS. Bandung: Tesis Fakultas Ekonomi dan Bisnis - Universitas Telkom.
PT.TS. 2018. *Internal Report*.

Rawson, A., Duncan, E., & Jones, C. 2013. The truth about customer experience. *Harvard Business Review*, *91*(11), 26.

Richardson, A. 2010. Touchpoints bring the customer experience to life. *Harvard Business Review*, *91*(12), 30.

Salsberry, M. 2017. *Customer Journey Mapping Framework 1*.

Yin, R.K. 2012. *Applications of Case Study Research*, Edisi 3. Washington DC: SAGE Publications.

APPENDIX 1. Operational variables.

No	Variable Indicators	Source	
1	Element interaction	Actions, motivation, questions, pain points	Salsberry (2017)
2	Touchpoint location	Products, interaction, message, setting	Richardson (2010)
3	Touchpoint owned	Brand, partner, customer, social/external	Lemon & Verhoef (2010)
4	Touchpoint attribute	Initiator, time, channel, trace	Halvorsrud et al. (2016)

APPENDIX 2. Informants.

No	Type of Company	Position	Reason
1.	Crediting	Risk coordinator	Responsible for determining the company's risk position, calculating risk criteria and targets, underwriting accountability
2.	Financing	Head of credit Development	Responsible for determining the development and management of credit agreements and underwriting
3.	Banking	Dept head Credit analyst	Responsible for managing the analysis and underwriting credit approval and development

APPENDIX 3. Results of CJM CS at TSPS.

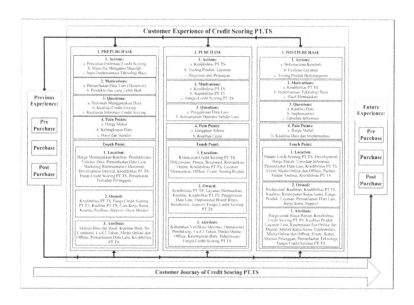

Assessing Indonesia's textile company listed in IDX market: The valuation case

P.W. Satriawana & R. Hendrawan
Telkom University, Bandung, Indonesia

ABSTRACT: This study is aimed to estimate the fair value of textile and garment companies share. The companies being studied are listed in Indonesia Stock Exchange, using data from 2013-2017 as the basis for the 2018-2022 projection. The methodology for company valuation was the Dividend Discounted Model and Relative Valuation. RV was applied using the combination of Price Earnings Ratio and Price Book Value. What is the fair value of SRIL-PBRX-TRIS-RICY shares by using the DDM method and comparing with the RV method in pessimistic, moderate and optimistic scenarios? The results showed that using the DDM method, the fair value of RICY-TRIS-PBRX-SRIL in all scenarios were overvalued. Except for RICY, in the optimistic scenario was undervalued. The RV PER-PBV method shows that the value of RICY-TRIS-PBRX-SRIL is still within the IDX market range Q1-2018. The recommendation for investors is to buy shares at undervalued conditions and sell them in overvalued conditions.

1 INTRODUCTION

The textile industry is one of the five priority sectors for the implementation of the industrial system 4.0 by the Indonesian Ministry of Industry. The industrial system 4.0 is an industry that combines automation technology with cyber technology. This is in line with digital developments in the industrial world and a trend in data automation and exchange in manufacturing technology. The domestic textile and garment industry experienced significant growth in 2017. Therefore, the textile industry has potential to grow and contribute significantly to national economic growth.

Dividends and capital gains are the two wealth-building tools of the stock market. Laurens (2018) state that dividend per share (DPS) is the sum of declared dividends issued by a public listed company's share outstanding. DPS is vital because the number one goal of a company is to return value to its shareholders.

The main problem in this study is how to calculate of the intrinsic value of the stock price, so that investors can determine the right investment decision. According to Djaja (2018), Discounted Cash Flow is a method calculating the present value of a future cash flow to determine company value today. The third variations of DCF are Dividend Discounted Model (DDM), Free Cash Flow to Equity (FCFE) and Free Cash Flow to Firm (FCFF), Neaxie & Hendrawan (2017). This study using the DDM method and comparing with the Relative Valuation method because the textile companies to be studied are companies that routinely distribute dividends during the study period. The DDM method will be very limited, if the company to be studied does not distribute dividends. Using these methods, three scenarios were developed, namely; pessimistic, moderate and optimistic.

The purpose of this study is to answer the research questions that have been formulated as follows: to identify the intrinsic value of RICY, TRIS, PBRX and SRIL shares in three scenarios, namely; pessimistic, moderate and optimistic; to provide recommendations for investors about the four shares; to provide benefits both theoretically and practically for interested party

2 LITERATURE REVIEW

The values and conditions of a company are strongly influenced by macro conditions, among others: the political, economic, social conditions of the country where the company carries out business activities and the industrial conditions of the company. The company's value is an investor's perception towards the level of company's success in managing the existence resources frequently associated with its stock price. High stock prices make the company's value become high and increase market confidence in the company's cash flow performance and prospects in the future.

Damodaran (2006) states that in general there are three approaches to valuing an asset, namely: Discounted Cash Flow Valuation, Relative Valuation, Contingent Claim Valuation. There are three variations of DCF, namely; Dividend Discounted Model (DDM), Free Cash Flow to Equity (FCFE) and Free Cash Flow to Firm (FCFF). The approach chosen in this study is Discounted Cash Flow Evaluation with Dividend Discounted Model (DDM) approach and Relative Valuation through the Price to Earning Ratio (PER) and Price Book Value (PBV) approaches.

2.1 Dividend Discounted Model

The dividend discount model is a valuing model based on assumption that a stock value represents the discounted sum of all its future dividend payments (Irons, 2014). The intrinsic value of a stock is a present value of future dividends discounted by required rate of return. According to Gacus & Hinio (2018), the following formula can be used to calculate the intrinsic value of shares with the assumption that there is no development of revenue in the company and dividends are still divided:

$$Po = \frac{D}{k - g} \quad (1)$$

where Po = intrinsic value; D = estimation dividend; k = required rate of return; g = expected dividend growth rate

2.2 Relative Valuation

Relative Valuation is one of the most commonly used valuation methods by comparing similar companies or with the industries in which the company is located (Damodaran, 2006). Here are some approaches that can be used to valuate the Relative Valuation method, namely; Price to Earning Ratio (PER), Price to Book Value Ratio (PBV) and Multiple EBITDA.

Another alternative in valuation to calculate the intrinsic value of a stock or fundamental value is to use the profit value of the company (earnings). The formula for determining the intrinsic value of stock through PER is as follows (Copeland, 2000):

$$Po = Estimation\ EPS\ x\ PER \quad (2)$$

where Po = intrinsic value; EPS = earning per share; PER = price earning ratio

PBV is an alternative approach to determine the value of a stock with the Relative Valuation method. The formula for PBV, namely:

$$Po = PBV\ x\ BV \quad (3)$$

where Po = intrinsic value; PBV = price book value; BV = book value of equity.

We have some references from the previous research, such as: Neaxie and Hendrawan (2017) used the Discounted Cash Flow method with the Free Cash Flow to Firm and Relative Valuation approaches to calculate the fair price of telecommunication company shares listed on the Indonesia Stock Exchange (IDX). The results of this study indicate that the DCF method with the FCFF approach is optimistic, resulting in fair value of TLKM in an undervalued condition, fair value of ISAT under overvalued conditions and fair value of EXCL companies in undervalued conditions.

3 METHODOLOGY

3.1 Types of research

The type of research used is verification research with quantitative methods that aim to explain existing phenomena by using numbers, namely valuation to obtain the intrinsic value of shares of companies engaged in the textile and garment sub-sector business using the DCF method with Dividend Discounted Model approach and Relative Valuation with PER and PBV approaches.

3.2 Population and data collection

The population taken is all shares of the textile and garment sub-sector companies on the Indonesia Stock Exchange (IDX), while the data sample uses a purpose sampling technique which is the four textile and garment sub-sector companies that routinely distribute dividends and still have active transactions until 2018, namely SRIL, PBRX, TRIS and RICY.

The data source in this study uses secondary data, namely five year historical data in the form of annual reports and financial reports which are object research sourced from IDX.com, the world of investment and official website object research.

4 RESULT AND DISCUSSION

4.1 Intrinsic value with Dividend Discounted Model method

According to Zemba and Hendrawan (2018), the optimistic scenario describes if the company's growth is greater than expected. The scenario is called moderate if the company's growth is normal. While the company's growth is smaller than forecast, its called a pessimistic scenario.

If the market share price is lower than its intrinsic value, the value of the stock is an undervalued condition, and vice versa if the market share price is higher than its intrinsic value, it can be said that the stock is an overvalued condition.

Table 1. Intrinsic value with DDM method.

Company	Market Value	Optimistic		Moderate		Pessimistic	
		Intrinsic	Status	Intrinsic	Status	Intrinsic	Status
RICY	Rp. 150	Rp. 349	Undervalued	Rp. 149	Overvalued	Rp. 70	Overvalued
TRIS	Rp. 308	Rp. 149	Overvalued	Rp. 104	Overvalued	Rp. 71	Overvalued
PBRX	Rp. 515	Rp. 513	Overvalued	Rp. 420	Overvalued	Rp. 148	Overvalued
SRIL	Rp. 380	Rp. 257	Overvalued	Rp. 175	Overvalued	Rp. 120	Overvalued

Based on Table 1, RICY has undervalued condition at an optimistic scenario, and its overvalued condition at other scenarios, while TRIS, PBRX and SRIL have overvalued conditions at all scenarios.

4.2 Relative Valuation method with PER and PBV approach

Based on the results of calculation, processing, and analysis of overall stock valuation data using the Relative Valuation with PER and PBV approaches, the intrinsic value is obtained for the textile and garment companies RICY, TRIS, PBRX and SRIL using a pessimistic, moderate and optimistic scenario, which are presented in Table 2 below:

Table 2. PER and PBV on Relative Valuation method.

Company	Optimistic		Moderate		Pessimistic	
	PER	PBV	PER	PBV	PER	PBV
RICY	10.38	2.33	5.06	0.99	2.71	0.47
TRIS	75.73	0.49	58.01	0.34	41.42	0.23
PBRX	14.74	0.99	14.08	0.82	6.55	0.29
SRIL	2.67	0.68	2.16	0.46	1.64	0.32

Besides that, it shows that with the calculation results using the Relative Valuation (RV) with PER and PBV approaches which is still in line with the range market, the assumption that is built on the valuation processing analysis based on the first method, The DDM is fulfilled, because in the PER calculation and this PBV, one of the most important component, is the value of earnings data from one of the final results of the DCF method calculation.

5 CONCLUSION

Based on the results of calculation of all stock valuation data using the DDM and RV with PER and PBV approaches, the summary of all valuation methods are presented in Table 3 below:

Table 3. The Summary of all valuation methods.

Company	Scenario	Intrinsic DDM	Market DDM	Intrinsic PER	Industry PER	Intrinsic PBV	Industry PBV
RICY	Pessimistic	70	150	2.71	13.36	0.47	1.1
	Moderate	149	150	5.06	13.36	0.99	1.1
	Optimistic	349	150	10.38	13.36	2.33	1.1
TRIS	Pessimistic	71	308	41.42	13.36	0.23	1.1
	Moderate	104	308	58.01	13.36	0.34	1.1
	Optimistic	149	308	75.73	13.36	0.49	1.1
PBRX	Pessimistic	148	515	6.55	13.36	0.29	1.1
	Moderate	420	515	14.08	13.36	0.82	1.1
	Optimistic	513	515	14.74	13.36	0.99	1.1
SRIL	Pessimistic	120	380	1.64	13.36	0.32	1.1
	Moderate	175	380	2.16	13.36	0.46	1.1
	Optimistic	257	380	2.67	13.36	0.68	1.1

- The results of this research indicate that the fair value of shares using the DCF with DDM approach in all scenarios was overvalued condition. Except for RICY at optimistic scenario was at the undervalued condition.
- The Relative Valuation method with the PER and PBV approaches shows that with all calculation scenarios namely pessimistic, moderate and optimistic, the textile and garment companies RICY, TRIS, PBRX and SRIL have PER and PBV values that are still in the market ratio PER and PBV according to the data Indonesia Stock Exchange (IDX) in Quarter-1 2018.
- As an investment decision, based on the Discounted Cash Flow method with Dividend Discounted Model approach and compared by the Relative Valuation method with PER and PBV approaches, it is recommended to buy RICY and sell TRIS, PBRX and SRIL.

REFERENCES

Copeland, T., Koller, T., Murrin, J., 2000. *Valuation Measuring and Managing The Value of Companies on Third Edition*, USA: John Wiley and Sons Inc.
Damodaran, A., 2006. *Damodaran on Valuation on Second Edition*, USA: John Wiley & Sons Inc.
Djaja, I., 2018. *All About Corporate Valuation on Second Edition*, Jakarta: PT. Elex Media Komputindo.
Gacus, R.B., Hinio, J.E. 2018. The Reliability of Constant Growth DDM in Valuation of Phillippine Common Stock. *International Journal of Economics & Management Sciences: Vol.7 Issue 1*: 1–9.
Irons, R. 2014. Enhancing the Dividend Discounted Model to Account for Accelerated Share Price Growth. *Journal of Accounting and Finance: Vol.14(4)*: 153–159.
Laurens, S. 2018. Influence Analysis of DPS, EPS and PBV Toward Stock Price and Return, *Journal The Winner: Vol.19, No.1*: 21–29.
Neaxie, L.V., Hendrawan, R. 2017. Stock Valuations in Telecommunication Firms: Evidence from Indonesia Stock Exchange, *Journal of Economic & Management Perspectives 11 (3):* 455.
Tandelilin, E., 2010. *Portofolio dan Investasi Teori dan Aplikasi*, Yogyakarta: Kanisius.
Zemba, S., Hendrawan, R. 2018. Does Rapidly Growing Revenues Always Produce An Excellent Company's Value? DCF & P/E Valuation Assessment on Hospital Industry. *Journal e-Proceeding of Management: Vol.5, No.2*: 2045.

Influence of perceived quality of mobile payment application towards loyalty

T.N. Sandyopasana & M. Gunawan Alif
Master of Management, Faculty of Economics and Business, Universitas Indonesia, Jakarta, Indonesia

ABSTRACT: This study analyses the influence of perceptions of the quality of the mobile payment application on customer loyalty. The cognitive-affective-behaviour model was used in this study where cognitive stages included perceptions of quality divided into delivery quality and outcome quality. Customer trust, satisfaction and switching barriers are included in the affective stages, and loyalty including as behavioural intention. All data used in this study were collected from 418 respondents who were users of mobile payments in Indonesia. Then, the model is estimated using SEM and showing that the delivery quality and outcome quality formed through customer interaction with mobile payment services positively affect trust and customer satisfaction, which in the end will affect loyalty. The results also show that satisfaction can mediate the effect of trust on loyalty where the influence is greater than the direct influence between trust and loyalty. Therefore all hypotheses formed in this study were accepted.

1 INTRODUCTION

1.1 Background

Nowadays Indonesia is facing digital disruption which is directly or indirectly changing the behavior of its people in their daily lives. The digital disruption also forces business practitioners to be able to adapt and follow digital developments so that their business can catch to the changing consumers' behavior. The digital disruption is shown by the number of emerging digital businesses or startups in Indonesia such as fintech (e.g GOPAY) and e-commerce that try to answer the needs of consumers today. The convenience provided by digital businesses has a significant impact on people's consumption behavior in Indonesia because it can address pre-existing problems. According to research data conducted by MDI Ventures, until 2017, the number of GO-PAY and TCASH users has reached around 10 million users and continues to grow until now.

However, even though mobile payment users have experienced a rapid increase, the loyalty of the users is still questionable, and even can be considered as low. This argument is supported by the number of mobile payment users who have more than one mobile payment brand in their smartphones. The tendency of the users to use mobile payment is currently more dominated by the number of discounts or cashback provided by mobile payment service providers. Currently mobile payment service providers are competing to increase the number of users using price war strategy by providing discounts and cashback for consumers who consequently require large amounts of funding to provide subsidies for these discounts.

Through this research, business practitioners could recognize what factors that could influence the users' loyalty. So that the business could improve their user growth by using strategies other than discounts or cashback which is arguably not effective and efficient. Better strategy later could be derived from improving several factors found by this research.

2 LITERATURE REVIEW

2.1 C-A-B model

This research model refers to the C-A-B (cognition-affect-behavior) model which is mentioned in previous studies that customer awareness of the quality of e-services directs their attitude which can affect the loyalty of the service providers used (Kao & Lin, 2016). The cognition stage is the initial stage based on an evaluation of several aspects of the object. The second stage is "affect" which involves emotions and preferences obtained from evaluating the use/consumption of good/pleasant/beneficial. Then the third stage or behavior is explained that this stage is related to intention/desire/will (Lopez-Miguens & Vazquez, 2017). Thus, in e-services, especially those based on mobile, customer assessment of service quality (cognition) formed through its interaction with the mobile payment application will have a positive impact on customer trust and satisfaction that results in loyalty (behavior).

3 METHODOLOGY

3.1 Research approach

This research using quantitative method by gathering data using online questionnaire.

3.2 Object/Subject of the research

The object of this research is mobile payment application in Indonesia. While the subject of this study is mobile payment application user in Indonesia.

3.3 Research data collection & sources

A total of 418 respondent data used in this study were obtained by convenience sampling technique through an online survey with a total of 44 items/questions and with a 5 Likert scale format.

3.4 Data analysis method

Before the questionnaire was distributed, a pilot test was conducted on 30 respondents to determine the validity and reliability of each item using SPSS software. The data processing technique is done by the SEM method using AMOS software.

4 RESULTS AND DISCUSSION

The results of data analysis in this study explained that at the measurement model stage, the proposed research model met the criteria ranging from model fit/goodness of fit to the reliability and validity of the indicators of each construct.

Table 1. Goodness of fit results.

Goodness of Fit	Nilai Kriteria	Uji
GFI	.800	Marginal Fit
NFI	.858	Marginal Fit
CFI	.903	Good Fit
PNFI	.789	Good Fit
RMR	.041	Good Fit
RMSEA	.065	Good Fit

4.1 Results on the effect of delivery quality on trust

The test results on this hypothesis explain the importance of the quality of mobile payment applications, especially when consumers use the application which is assessed through several things such as ease of use, completeness of information, speed of loading page and application design that will affect consumer confidence in mobile applications. payment. That is because consumers will evaluate some/initial experience in using a service when in this delivery process before they evaluate the overall quality of service (Collier & Bienstock, 2006).

4.2 Results of the effect of outcome quality on trust

In the previous study, there was a significant positive effect between quality on trust (Lopez-Miguens & Vazquez, 2017). This is supported by this study where the results of the processed data show results that are also positive and significant. These results indicate that the trust of mobile payment users in this study can increase with the increasing quality of the final results it receives. In other words, the trust of mobile payment users can increase when their expectations of the application can be fulfilled.

4.3 Results the effect of outcome quality on satisfaction

Test results in this study indicate that there is a significant positive effect between outcome quality on satisfaction in the context of mobile payment. These results are supported by previous studies that explain that outcome quality has an influence on satisfaction because at this stage consumers will evaluate and assess whether the consumer is satisfied with the services used not only when using the process but also by assessing the final results or the outcome received (Collier & Bienstock, 2006). So this means that the evaluation of the outcome of using the mobile payment service completes the previous evaluation at the stage when using the service.

4.4 Results of trust influence on satisfaction

In a previous study in the context of online banking it was mentioned that when a consumer trusts the banking website of the financial institution where he is registered to make a transaction, the consumer assumes that the financial institution will meet the expectations of these consumers and strive to mutually benefit and not act opportunistically (Andaleeb, 1996). The situation will also lead consumers to build greater satisfaction with the services provided (Lopez-Miguens & Vazquez, 2017) and this argument is also proven in the context of mobile payment. So that the more consumers believe and trust in a service, the greater the satisfaction felt by these consumers when the customer's expectations can be met.

4.5 Results of trust influence on loyalty

The test results in this study indicate that in the context of mobile payments, trust can increase customer loyalty directly. This result is supported by the results of previous studies which prove the direct positive influence between trust on loyalty (Kao & Lin, 2016). When users feel confident about a brand of mobile payment applications, those users will tend to be loyal to the brand of mobile payment applications that they use and will make repeated use in the future.

4.6 Results of the effect of satisfaction on loyalty

In a previous study it was also mentioned that satisfaction influences loyalty and is a major indicator in influencing loyalty (Gummerus, Liljander, Pura, & Van Riel, 2004). This argument is also strengthened from the results of hypothesis testing in this study where in the context of mobile payment, satisfaction has a significant positive effect on loyalty. Therefore, in this study, customer satisfaction can be a mediator that can affect loyalty.

4.7 Results of the effect of barrier switching on loyalty

Switching barriers in the online context are lower in value than in the offline context (Liang & Chen, 2009). This is in accordance with this study where consumers can easily switch brands because to be able to register using a mobile payment application does not require a lot of formalities and most of them can be used free of charge. In previous studies the effect of this relationship produced a significant positive effect (Lopez-Miguens & Vazquez, 2017). Likewise in this study where switching barriers are important factors that can affect consumer loyalty.

Figure 1. Research model results.

5 CONCLUSION

This research proves that the quality of the mobile payment application is an important factor in increasing the trust and satisfaction of its customers and subsequently generates loyalty to the application. The results show that the delivery quality of mobile payment applications has a significant influence on increasing consumer confidence. Therefore, it is important for mobile payment application service providers to improve the quality of their applications in terms of variations in payment types, information quality, ease of use to technical matters such as the speed of opening an application (loading page). Significant positive effect between outcome quality on trust and satisfaction is evidenced in this study. This means that it is important for service providers to be able to meet and maintain the expectations of consumers in using their services in order increase customers' loyalty. This is important because outcome quality will affect the overall evaluation process of the quality of the service.

REFERENCES

Andaleeb, S. S. (1996). An experimental investigation of satisfaction and commitment in marketing channels: The role of trust and dependence. *Journal of Retailing*, 72(1), 77–93. https://doi.org/10.1016/S0022-4359(96)90006-8

Collier, J. E., & Bienstock, C. C. (2006). Measuring service quality in E-retailing. *Journal of Service Research*, 8(3), 260–275. https://doi.org/10.1177/1094670505278867.

Gummerus, J., Liljander, V., Pura, M., & Van Riel, A. (2004). Customer loyalty to content-based Web sites: The case of an online health-care service. *Journal of Services Marketing*, 18(3), 175–186. https://doi.org/10.1108/08876040410536486

Jones, M. A., Mothersbaugh, D. L., & Beatty, S. E. (2000). Switching barriers and repurchase intentions in services. *Journal of Retailing*, 76(2), 259–274. https://doi.org/10.1016/S0022-4359(00)00024-5

Kao, T. W., & Lin, W. T. (2016). The relationship between perceived e-service quality and brand equity: A simultaneous equations system approach. *Computers in Human Behavior*, *57*, 208–218. https://doi.org/10.1016/j.chb.2015.12.006

Liang, C. J., & Chen, H. J. (2009). A study of the impacts of website quality on customer relationship performance. *Total Quality Management and Business Excellence*, *20*(9), 971–988. https://doi.org/10.1080/14783360903181784

Lopez-Miguens, M. J., & Vazquez, E. G. (2017). An integral model of e-loyalty from the consumer's perspective n Gonzalez Vazquez. *Computers in Human Behavior*, *72*, 397–411.

Implementation of marketing innovation at PT. Pegadaian in the revolutionary 4.0

M.A. Sugiat & K. Sudiana
Telkom University, Bandung, Indonesia

ABSTRACT: PT. Pegadaian sees the need to innovate to get a new market segment: millennial customers. PT. Pegadaian transforms digital technology by launching new products based on digital finance and changing the company's image. This research aims to analyze the marketing innovation strategies of PT. Pegadaian concerning the transformation of its digital technology. The data were taken from surveys, interviews, and secondary data online. The method used is SWOT analysis. The results show the strength of the company's long 118-year experience, implying high trust from customers, most of whom are aged over 45. The opportunities of this company are changes in behavior towards gold and pawning, emergence of new opportunities and innovative business models to serve customers, increasing uses of digital business, and changes in regulations. The threats are weaknesses in technology adaptation of the customers, other private mortgage companies, advancement of banking as competitors, and the presence of financial technology.

1 INTRODUCTION

Millennials are the generation that was born and grew up in an age of technological developments. Technological developments have an impact on changes in human behavior in transactions, communication, and culture. In the last ten years, the development of digital banking has shown a rapid increase. The emergence of digital-based banking products known as revolution 4.0, such as mobile banking, digital money, and cashless transaction through QR code, became a part of millennial generations. Every company tries to adapt to this digital development, including PT. Pegadaian, which has been established in Indonesia since 1901 to serve the middle and lower classes, and is now experiencing a shift in segmentation and innovation to get the market and survive in the current competition. The millennial generation is the target and has great potential to be taken as a customer, where the transformation of digital technology at PT. Pegadaian can fulfill millennial needs in transactions.

2 PT PEGADAIAN AND DIGITAL TRANSFORMATION

From its founding in 1901 until now, PT. Pegadaian has succeeded in becoming an integrated business solution in the community, especially as a fiduciary-based micro-market leader and to help the middle to lower social groups. PT. Pegadaian has a vision, namely providing the fastest, easiest, safest financing and always assisting middle- and lower-class businesses to encourage economic growth through convenience and comfort in service. Thereby, PT. Pegadaian helps the government in improving the welfare of the lower middle class. The success of PT. pawnshop can be seen and proven through performance indicators such as more than 118 years of experience, reaching 9 million customers, and stable growth performance making this company ready to face the challenges ahead. However, the digital era has made business conditions and dynamics change in the revolutionary era 4.0, and PT. Pegadaian is carrying out a digital transformation to answer the challenges of the times. Now, the speed of technological change is beating the speed of change in individual humans, business organizations, and government. PT. Pegadaian is trying to adapt to this change by developing services in the field of digital finance. This digital transformation has an impact on the shifting of the customer target, both in terms of age and type of work.

According to the data, PT. Pegadaian's main customers are over 45 years old, but to follow this digital transformation, the company must expand its market to the millennials who are very close to technological developments. This is the main reason PT. Pegadaian must be digitally transformed due to changes in the pawnshop business landscape as a result of technological changes, changes in people's lifestyles, and changes in global business. Therefore a paradigm shift from PT. Pegadaian, as follows: 1) changes in behavior towards gold and mortgage, 2) the emergence of new opportunities, 3) increased use of digital cards, 4) the appearance of innovative business models to serve customers, 5) changes to regulation.

3 MILLENNIALS

Millennials are also known as Generation Y or Gen Y or Generation Immortal. In general, this generation was born between the 1980s and 1990s. This generation has the characteristics of increasing use and familiarity with communication media and digital technology resulting an a generation more impressed by individuals, ignoring political issues, and focusing on materialistic values. However, from the positive side, this generation has good self-confidence, can express their feelings, and is liberal, optimistic, and accepting of ideas and ways of life. They even have freedom in determining work schedules. Judging from Indonesia's current economic growth, the millennial generation has a significant and strategic role in building the Indonesian economy through its creative industry. Opportunities in the economy and productivity of the millennial generation build interrelated relations, especially regarding digital finance. The daily activities of the millennial generation are highly dependent on the use of digitalization both for communication and online transactions, and this is an attractive market for PT. Pegadaian.

4 METHOD

The method used in this study is a SWOT analysis that begins with data collection from PT. Pegadaian. We also use secondary data and literature to strengthen the results of the analysis. Through SWOT, we will analyze the transformation of digital technology as part of the marketing innovation of PT. Pegadaian to get millennial customers. Part of the company's strategic plan, considering the latest data, is that in 2018, their customer base will increase to 11.5 million people, and this year that process is assisted by customers from the millennial age group. After analysis, breakthroughs made were assessing in four types of strategies: strategies for strengths-opportunities (SO), weaknesses-opportunities (WO), strengths-threats (ST), and weaknesses threats (WT). From this analysis, the extent of the implementation of marketing innovations at PT. Pegadaian to get new customers by approaching digital transformation was determined.

5 RESULTS AND DISCUSSION

A. Strengths
   As a company has being established since 1901, PT. Pegadaian has a good reputation, including trust in it. This company has as many as 11.5 million customers in 2018. Besides, this company has a stable growth performance and, from 2015–2017, it is increasing.
B. Weaknesses
   The results of the weakness analysis are the presence of customers aged over 45 years, who are loyal customers and have a quantity that very much supports PT. Pegadaian. However,

there are obstacles including lack of adoption of technological advances, including digital finance. Other weaknesses are generally the result of disruptive effects, almost on average, short-lived. Changing the consumer target to millennials requires an appropriate marketing innovation strategy, so as not to eliminate old customers, but to add new customers who are young, dynamic, and productive.

C. Opportunities

Opportunities at PT. Pegadaian, among others, includes the development of information technology to open up new opportunities in the marketing of a pawn company. New businesses are disruptive, so they open up new opportunities in terms of opportunities to get customer segmentation. There is a change in behavior towards gold and mortgage, where the survey results show 89% of consumers consider gold pawning only as an emergency loan. There are opportunities for new growth, increased domestic credit for the private sector by banks (a 33% increase in World Bank data sources), and increased retail, micro, and SME loans. The latest development, a fact in the field, is that there is an increase in the use of digital customers, so it is necessary to penetrate pawn company products based on mobile applications. PT. Pegadaian tried to breakthrough via pawn company digital services (PDS) and Prima Pawn as a form of digital technology transformation in pawnshops. On the other hand, the emergence of innovative business models to serve customers from other banks, triggered pawnshops to likewise provide innovative and comfortable services for customers. Finally, there has been a change in regulation so that the OJK has made it easier to set up a pawn company.

D. Threats

In addition to analyzing strengths, opportunities, and weaknesses, this study analyses the threats to PT. Pegadaian, in the form of being unable to keep up with technological developments or the inability of customers to keep up with rapid technological developments resulting in customers getting out and uncomfortable with new services. In addition, there are private plots that are faster and more attractive, and then there are threats from banking and fintech.

6 CONCLUSION

Based on the results of the SWOT analysis conducted, it can be concluded that PT. Pegadaian's strength in expanding the market through technology transformation is having 118 years of experience, so there is high trust from old customers. However, there is a weakness, considering that the age of existing customers is over 45 years, so changes in segmentation to millennials needs to be considered as users who can adapt, without eliminating old customers. The opportunities of this company are changes in behavior towards gold and pawn, the emergence of new opportunities, the increasing use of digital business, the emergence of innovative business models to serve customers, and changes in regulations. Then, for the threat, there are weaknesses in technology adoption by customers, the existence of private mortgage, the advancement of banking as competitors, and the existence of fintech.

ACKNOWLEDGMENT

The authors would like to thank both fellow lecturers at Faculty of Economics and Business Telkom University, and also our colleagues from Faculty of Economics and Business Padjadjaran University for giving their suggestions and input for the content of this article.

REFERENCES

Amuzu, C.S. 2017. Engaging the workforce: Baby boomers, Generation Xers and Millennials (doctor of philosophy dissertation). ProQuest Dissertation Publishing. UMI No. 10287782).

Aon Hewitt. 2015. *2015 Trends in Global Employee Engagement: Making Engagement Happen.* http://www.aon.com/human-capital-consulting/thought-leadership/talent/2015-global-employee-engagement.jsp.

Aon Hewitt. 2016. *Managing Millennials: Changing Perspectives for a Changing Workforce.* http://images.respond.aonhewitt.com/Web/AonHewitt/%7Bc10490e0-08ab-4d32-bd44-c61909f45206%7D_2016-Managing-Millenials.pdf.

Aon Hewitt. 2017. *2016 Trends in Global Employee Engagement: Making Engagement Happen.* http://www.modernsurvey.com/wp-content/uploads/2017/04/2017-Trends-in-Global-Employee-Engagement.pdf.

Badan Ekonomi Kreatif. 2017. *Data Statistik dan hasil Survei Ekonomi Kreatif.* Jakarta, DKI. http://www.bekraf.go.id/berita/page/17/infografi-data-statistik-dan-hasil-survei-khusus-ekonomi-kreatif.

Badan Pusat Statistik 2016. Survei Angkatan Kerja Nasional. https://www.bps.go.id/linkTabelStatis/view/id/1904.

Brigham, T. J. 2015. An introduction to gamification: Adding game elements for engagement. *Medical Reference Services Quarterly, 34*(4), 471–480.

Bryman, A. 2012. *Social Research Methods* (4th ed.). New York: Oxford University Press Inc.

Clark, T. R. 2012. *The Employee Engagement Mindset.* USA: McGraw-Hill.

Deloitte. 2016. *The 2016 Deloitte Millennial Survey.* https://www2.deloitte.co/id/ed/pages/about-deloitte/articles/millennialsusvey.html.

Duchscher, J. E., & Cowin, L. 2004. Multigenerational nurses in the workplace. *Journal of Nursing Administration, 34*(11), 493–501. https://journals.lww.com.

Fountain, D. M. 2016. Relationship among work engagement, drivers of engagement, and bullying acts in registered nurses working in hospital settings (doctor of philosophy dissertation). ProQuest Dissertation Publishing. (UMI No. 10597139).

Howe, N., & Strauss, W. 2000. *Millennials Rising.* New York: Vintage Books.

Lemmeshow, S., Hosmer, D. W., Klar, J., Lwanga, S. K. & WHO. 1990. *Adequacy of Sample Size in Health Studies.* USA: World Health Organization.

Orosco, J. S. U. 2014. Examination of gamification: Understanding performance as it relates to motivation and engagement. ProQuest Dissertations Publishing. (UMI No. 3669296).

Pratama, A. H. 2017. Laporan Kondisi *Start Up* Indonesia Q2 2017. Techinasia. Diakses dari https://id.techinasia.com/laporan-kondisi-startup-indonesia-q2-2017.

Presiden Republik Indonesia. 2009. Instruksi Presiden tentang Pengembangan Ekonomi Kreatif (Inpres Nomor 6 Tahun 2009. Jakarta, DKI: Sekretaris Kabinet Bidang Hukum. Diakses dari http://www.kemenpar.go.id/userfiles/file/7193_2610 -Inpres6Tahun2009.pdf.

Preston, C. C., & Colman, A. M. 2000. Optimal number of response categories in rating scales: Reliability, validity, discriminating power, and respondent preferences. *Acta Psychologica, 104*, 1–15.

Silverman, S. N., & Silverman, L. L. 1994. Using total quality tools for marketing research: A qualitative approach for collecting, organizing, and analyzing verbal response data. *Advanced Research Techniques Forum.*

Simamora, B. 2005. *Analisis Multivariat Pemasaran.* Jakarta (ID): PT Gramedia Pustaka Utama.

Simbolon, D. A. A. 2016. Analisis *Quality of Work Life* pada Generasi X dan Y Alumni Fakultas Ekonomi dan Manajemen IPB (Skripsi). Bogor (ID): Institut Pertanian Bogor.

Suliyanto. 2005. Analisis Data dalam Aplikasi Pemasaran. Bogor (ID): Ghalia Indonesia.

Tanduwaldikar, A. 2013. Gamifying business to drive employee engagement and performance. *Cognizant.* https://www.cognizant.com/InsightsWhitepapers/Gamifying-Business-to-Drive-Employee-Engagement-and-Performance.pdf.

Analysis of customer intention in adopting automated parcel station services

P.M.T. Sitorus & M.O. Alexandra
Telkom University, Bandung, Indonesia

ABSTRACT: The rapid development of e-commerce has a significant influence on the logistics industry and supply chain management. Therefore, some service providers have developed a technological innovation, namely automated parcel station (APS) services. However, currently, direct home delivery is still the main choice in several countries. The purpose of this study was to discover some factors that could influence customer intention in adopting APS services with attitude as an intervening variable, and to find out how much attitude could influence customer intention. This study used quantitative method and its respondents were e-commerce users who had never used APS services. As a conclusion, based on these results, the factors that influence customer intention towards the beginning of APS adoption through attitude as an intervening variable, were relative advantage and trialability. In addition, attitude also had a key role that could influence customer intention in adopting APS services.

1 INTRODUCTION

The rapid development of e-commerce in developed and developing countries has a significant influence on the logistics industry and supply chain management (Yu et al., 2016) because it has an impact on the high volume of daily orders in small sizes. Therefore, the number of package delivery vehicles in customers' residential areas and homes also increases (Cho et al., 2008; Weltevreden, 2008; Deutsch and Golany, 2017). Some service providers have developed a technological innovation, namely automated parcel station (APS) services (Wang et al., 2017). In the next few years, the popularity of parcel lockers will increase and the need for parcel lockers will not be avoided because it has a lot of benefits. However, currently, direct home delivery is still the main choice in several countries (Yuen et al., 2018). Therefore, an analysis of customer intention in adopting APS service is one of the solutions to help determine customer's characteristics and factors that can influence customer intention in adopting new technology to improve competency and satisfaction of logistics service providers.

This study refers to the theory of reasoned action (TRA) model and diffusion of innovations (DOI). Data was obtained quantitatively through a questionnaire filled out by the people of Bandung city who frequently do online shopping through e-commerce, but have not used APS services.

This study consists of three parts. It began with the literature review showing the construct explanation to form a conceptual model. It was then followed by the calculation of data obtained from e-commerce users through a questionnaire using structural equation modeling (SEM). Finally, analysis and interpretation were conducted followed by conclusions and suggestions.

2 LITERATURE REVIEW

Theory of Reasoned Action (TRA) (Ajzen & Fishbein, 2005)

TRA theory states that specific behavioral determinants are largely guided by a reasoned action approach, assuming that people's behaviors follow naturally from beliefs, attitudes, and intentions (Ajzen & Fishbein, 2005).

Diffusion of Innovations (DOI) (Rogers, 2003)

DOI theory (Rogers, 2003) is used to expand the variable beliefs in TRA where the variables used include compatibility, relative advantages, complexity, observability, and trialability. An intervening variable in this study, namely attitude, was used in the following hypotheses:

Hypothesis 1: Belief (explained by compatibility, relative advantages, complexity, observability, and trialability) influenced intention towards the beginning of APS adoption through attitude as an intervening variable.

H1a : Compatibility positively and significantly influenced intention towards the beginning of APS adoption through attitude as an intervening variable.

H1b : Relative advantage positively and significantly influenced intention towards the beginning of APS adoption through attitude as an intervening variable.

H1c : Complexity negatively and significantly influenced intention towards the beginning of APS adoption through attitude as an intervening variable.

H1d : Observability positively and significantly influenced intention towards the beginning of APS adoption through attitude as an intervening variable.

H1e : Trialability positively and significantly influenced intention towards the beginning of APS adoption through attitude as an intervening variable.

Hypothesis 2: Attitude directly influenced intention towards the beginning of APS adoption. Therefore, the following is the framework used in this study.

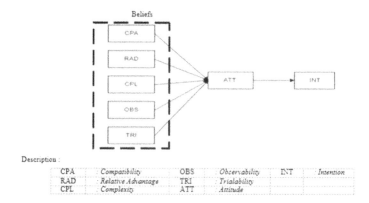

Figure 1. Research framework.
Source: Wang et al (2017)

3 METHODOLOGY

This study used the quantitative method. Samples were determined purposively and data were obtained from 100 e-commerce service user respondents in Bandung city using a questionnaire.

4 RESULTS AND DISCUSSION

The results of the study indicate that the people of Bandung city who like to do online shopping through e-commerce have the desire to use APS services. The following is the result of the hypothesis test that has been carried out.

Based on Table 1, it can be seen that variables that can influence customer intentions in adopting APS services are relative advantage (Zhu et al,, 2006; Lawson-body et al., 2014; Jerayaj et al., 2016; Wang et al., 2017; and Yuen et al., 2018), trialability, and attitude (Weitjers et al., 2007; Wang et al., 2017). Relative advantage is defined as the extent to which an innovation is considered better than existing ideas and trialability is defined as the extent to which an innovation can be tested in a limited way (Rogers, 2003). Most of the respondents were students and employees who frequently spent time outside their homes. Therefore, the use of APS services was considered to be more profitable than direct delivery home and could minimize a high risk of failed delivery. If the respondents used APS service, they could determine when they would take the package in accordance with the time convenient for them. In addition, the location of APS in various public locations made it easier for respondents to find this service if they wanted to try using it.

Attitude also has a significant influence on intention. It is undeniable that, besides the need for knowledge in operating this technology, many benefits were provided if people used APS as an alternative choice in their package delivery. Most respondents stated that using APS service would be a satisfying way, a pleasant experience, a good idea, and a popular trend. In addition, respondents also felt that they had a positive bias towards using a parcel locker. The process of receiving a unique package with APS service was considered to provide a pleasant experience. Compatibility, complexity, and observability had no significant influence on customer intention in adopting the service. Based on the results of the study, it could occur because the majority of respondents who only did online shopping once a month did not need to use this service. Other than that, respondents were also more likely to want to be practical by directly receiving packages at home. The majority of the respondents were students and employees who very close to technology, so the APS service was not so difficult to learn. Besides that, activities in shopping or obtaining packages were considered as private activities by some people, which did not need to be seen by other people, especially unknown people, so compatibility, complexity, and observability did not have a significant influence on increasing customer intention to adopt APS services.

5 CONCLUSION

In conclusion, based on the conducted study, the factors that influenced customer intention towards beginning the adoption of APS with attitude as an intervening variable, based on a combination of the theory of TRA and DOI, were relative advantages and trialability. Three other factors, namely compatibility, complexity, and observability, showed insignificant results. Furthermore, attitude had a significant influence on customer intention towards the beginning of APS adoption.

Table 1. Result of hypothesis test.

Hypothesis	Variable	Original Sample (O)	T-Statistics	T-Table	Description
H1a	CPA -> Attitude->Intention	0.097	1.325	1.65	Rejected
H1b	RAD -> Attitude ->Intention	0.227	3.767	1.65	Accepted
H1c	CPL -> Attitude ->Intention	0.033	0.855	1.65	Rejected
H1d	OBS -> Attitude ->Intention	0.061	0.953	1.65	Rejected
H1e	TRI -> Attitude ->Intention	0.382	5.027	1.65	Accepted
H2	Attitude -> Intention	0.816	28.166	1.65	Accepted

Theoretically, this study enriched the literature by introducing a DOI framework combined with TRA theory. As a recommendation for further study, it would be helpful to discover the factors causing some people of Bandung City who have already known about APS services to not use this service. In practical aspects, this study is useful for APS service providers, freight forwarding service providers, and businessman. APS service providers must further improve socialization regarding the various benefits provided and how to use this service. APS service providers can also provide attractive promotion considering the majority of respondents as potential adopters are students and employees aged 20 to 29 who are chronically looking for promotion from various services. Then, freight forwarding service providers better cooperate with APS in delivering goods. Besides being able to reduce failed delivery, APS can also reduce operating costs. It is also recommended that businessmen encourage the use of APS as one of the channels for shipping goods because APS can minimize customer waiting time in obtaining ordered goods, so as to increase customer satisfaction.

REFERENCES

Ajzen & Fishbein. 2015. The influence of attitudes on behavior. *The Handbook of Attitudes*, 173–221. [online]. https://researchgate.net/publication/26400974_The_Influence_of_Attitudes_on_Behavior [13 Jan 2019].

Cho, J. J. K., Ozment, J., & Sink, H. 2008. Logistics capability, logistics outsourcing and firm performance in an e-commerce market. *International Journal of Physical Distribution & Logistics Management*, 38(5), 336–359. [online]. https://doi.org/10.1108/09600030810882825 [16 Nov 2018].

Deutsch, Y., and Golany, B. 2017. A parcel locker network as a solution to the logistics last mile problem. *International Journal of Production Research*, 56(1–2), 251–261. [online]. https://doi.org/10.1080/00207543.2017.1395490 [3 Jan 2019].

Jeyaraj, A., Rottman, J. W., & Lacity, M. C. 2006. A review of the predictors, linkages, and biases in IT innovation adoption research. *Journal of Information Technology*, 21, 1–23. [online]. https://link.springer.com/article/10.1057/palgrave.jit.2000056 [05 Jan 2019].

Lawson-body, A., Willoughby, L., Illia, A., & Lee, S. 2014. Innovation characteristics influencing veterans' adoption of e-government services. *Journal of Computer Information Systems*, 54(3), 34–44. [online]. http://dx.doi.org/10.1080/08874417.2014.11645702.

Rogers, E. M. 2003. *Diffusion of Innovations*, 5th edition. Free Press. United States.

Wang, X., Yuen, K. F., Wong, Y. D., & Teo, C. C. 2017. An innovation diffusion perspective of e-consumers' initial adoption of self-collection service via Automated Parcel Station. *International Journal of Logistics Management*, 29(1), 237–260. [online]. https://doi.org/10.1108/IJLM-12-2016-0302 [25 Nov 2018].

Weijters, B., Rangarajan, D., Falk, T., & Schillewaert, N. 2007. Determinants and outcomes of customers' use of self-service technology in a retail setting. *Journal of Service Research*, 10(1, August), 3–21. [online]. https://doi.org/10.1177/1094670507302990 [1 March 2019].

Weltevreden, J. W. J. 2008. B2c e-commerce logistics: The rise of collection-and-delivery points in The Netherlands. *International Journal of Retail & Distribution Management*, 36(8),638–660. [online]. http://dx.doi.org/10.1108/09590550810883487 [20 Nov 2018].

Yu, Y., Wang, X., Zhong, R. Y., &Huang, G. Q. 2016. E-commerce Logistics in Supply Chain Management: Practice Perspective. *Procedia CIRP*, 52, 179–185. [online]. https://doi.org/10.1016/j.procir.2016.08.002 [20 Nov 2018].

Yuen, K. F., Wang, X., Wendy Ng, L. T., & Wong, Y. D. 2018. An investigation of customers' intention to use self-collection services for last-mile delivery. *Transport Policy 66*, 1–8. [online]. https://doi.org/10.1016/j.tranpol.2018.03.0 01 [23 Dec 2018].

Zhu, K., Dong, S., Xu, S. X., & Krame, K. L. 2006. Innovation diffusion in global contexts: Determinants of post-adoption digital transformation on European companies. *European Journal of Information Systems*, 15, 601–616. [online]. https://link.springer.com/article/10.1057/palgrave.ejis.3000650 [5 March 2019].

Predicting VR adoption on e-commerce platforms using TAM and porter five forces

C.S. Mon
UCSI University, Kuala Lumpur, Malaysia

ABSTRACT: Virtual Reality (VR) is an emerging technology which has been widely used by many sectors such as education and entertainment. Online shopping or e-commerce platforms are one of the potential sectors to adopt virtual reality to convince more shoppers to purchase online. However, there is no research or framework which investigate how e-commerce can adopt virtual reality based on competitive analyses in related area. This research develops a conceptual framework to predict virtual reality adoption on e-commerce platforms using Technology Acceptance Model (TAM) and Porter Five Forces. This paper is the beginning stage of the research and initial review on literature has been conducted. Research problems and objectives have been identified as well as conceptual framework and hypotheses have been developed. Research methodology has been constructed and conclusion which include contribution and limitations are presented in this research.

1 INTRODUCTION

E-commerce or electronic commerce platforms are modern ways to carry out shopping from anywhere and anytime. Alibaba, Lazada, Shopee and others are popular e-commerce platforms which are widely used by consumers all over the world. According to (Dave Chaffey, 2017), consumers prefer to shop online because of convenience and availability of price comparison on the platform and especially due to the different groups of generation such as baby boomers to Gen X and Millennials. Besides, there are many other reasons such the option to shop anytime, able to save time, varieties of choices, avoid crowds and able to shop from all over the world, etc. As from the sellers point of view, sales from e-commerce platforms has been increasing since 2014. In 2017, it was about US$2.3 trillion of sales from e-commerce platforms and it was expected to be US$4.88 trillion in 2021 (Statista, 2019). Besides, due to the arrival of e-commerce, offline retailers suffer major downside in the sales and in order to survive in the market, there is no other choice than sell the products online. There are a variety of ways to attract the customers' purchase intention and e-commerce platforms have used multiple technologies such as personalized advertisement, freedom of choice in delivery methods and locations, etc. Virtual reality is one of the new technologies which e-commerce platforms or online retailer can consider to attract the customers. One of the well-known furniture companies, IKEA, developed a VR App to provide virtual reality kitchen experience to its customers. This way, the potential customers will be able to experience how the furniture will be like in the actual environment and also attract them to buy. This paper aims to provide a conceptual framework to predict the adoption of virtual reality technology in e-commerce platforms using TAM and Porter Five Forces.

2 BACKGROUND

2.1 Virtual reality and E-Commerce platform

Technology innovations have been advanced, and development of virtual reality has been widely used to shape the future of e-retailing (Martínez-Navarro et al., 2019). The research has been done to study the effectiveness of virtual reality devices and formats in virtual store environment or v-commerce platforms. According to their researches, the different types of VR devices used have an effect on the purchase intension but not on the presence.

2.2 Porter five force

This framework, introduced by Michael (Michael E Porter, 1980), is a tool to analyze competition in industrial organization. The framework determines the intensity of the competition as well as attractiveness of an industry regarding profitability. The framework depicts five variables, namely bargaining power of supplier, bargaining power of buyer, threat of new entry, threat of substitute products or services as well as rivalry among competitors.

2.3 Technology acceptance model

Technology Acceptance Model or TAM was introduced by (Davis, Bagozzi and Warshaw, 1989), one of the most frequently used models which draws the attention of the information system community. TAM describes how the users can accept when the new technology has been introduced to them and also suggests the factors which determine their decision on the use of new technology. There are two main variables in TAM; (a) perceived usefulness and (b) perceived ease of use. These factors are affecting the intention of new technology usage and decision on actual usage. There were several researches conducted using TAM such as Technology Acceptance Model in the e-commerce segment (Fedorko, Bacik and Gavurova, 2018), Study of online shopping adoption among youth by using TAM (Blagoeva and Mijoska, 2017)'. The research was conducted in Macedonia. Besides a conceptual framework of TAM on E-commerce extension (Fayad and Paper, 2015).

3 RESEARCH PROBLEMS AND OBJECTIVES

Based on the literature which has been reviewed, there were no significant research conducted on the combination of Technology Acceptance Model (TAM) and Porter Five Forces (Michael E Porter, 1980) to predict if there will be an adoption of virtual reality in e-commerce platforms. Hence, the problems statement of the research has been constructed as follow: (a) there is an unclear research on whether or not virtual reality should be adopted in e-commerce platforms.; (b) there is no research framework on how e-commerce platforms intend to adopt virtual reality;. and (c) there is unknown research where this combined new conceptual framework can predict virtual reality adoption in e-commerce platforms. The objectives of the research are to address the research problems mentioned above.

4 RESEARCH CONCEPTUAL FRAMEWORK AND HYPOTHESIS

The conceptual framework which will be used in this research is illustrated in following Figure 1. The illustration is based on combination of TAM and Porter Five Forces.

The research hypotheses for this research have been developed as follows:

H1: Bargaining power of buyer can positively predict perceived benefits of virtual reality.
H2: Bargaining power of buyer can positively predict perceived ease of adoption of virtual reality.
H3: Bargaining power of supplier can positively predict perceived benefits of virtual reality.

Figure 1. Research conceptual framework.

H4: Bargaining power of supplier can positively predict perceived ease of adoption of virtual reality.
H5: Threat of new entry can positively predict perceived benefits of virtual reality.
H6: Threat of new entry can positively predict perceived ease of adoption of virtual reality.
H7: Rivalry among competitors can positively predict perceived benefits of virtual reality.
H8: Rivalry among competitors can positively predict perceived ease of adoption of virtual reality.
H9: Threats of substitute products or services can positively predict perceived benefits of virtual reality.
H10: Threats of substitute products or services can positively predict perceived ease of adoption of virtual reality.
H11: Perceived ease of adoption of virtual reality can positively predict perceived benefits of virtual reality.
H12: Perceived benefits of virtual reality can positively predict intention to adopt virtual reality in E-commerce platforms.
H13: Perceived ease of adoption of virtual reality can positively predict intention to adopt virtual reality in E-commerce platforms.

5 RESEARCH METHODOLOGY

5.1 Sampling procedure and questionnaire design

Quantitative research methodology will be used in this study and online survey will be distributed to e-commerce platforms administrators as well as retailers who sell their goods/services online. Emails will be sent to the target audience with the hyperlink for the survey questionnaire. The estimated time for the data collection will be around 4 weeks with total of 200 usable samples for quantitative data analysis. A total of 50 questions will be included in the online survey which consists of 10 demographics questions and 40 questions based on the variables defined in conceptual research framework.

5.2 Data analysis techniques

There will be several data analysis tasks which will be carried out in this research. Tests such as normality, descriptive statistics analysis, and reliability and validity test will be carried out using a statistical software called **SPSS** (Statistical Package for Social Sciences). Besides, another software called SmartPLS will be used to evaluate the relationships among variables, which is defined in the research framework and to conduct multiple regression analysis, path analysis, factor analysis and total effects analysis.

6 CONCLUSION

This paper presented the initial phase of the research which identified the research gap in predicting the adoption of virtual reality in e-commerce platforms. Virtual reality is the popular technology which has been used in many sectors. However, its adoption among e-commerce platforms is rare. This paper presented the research framework by using a combination of Porter Five Forces and Technology Acceptance Model in predicting the scenario. In conclusion, this paper will attempt to address the important gap among e-commerce sectors and open the research opportunities for later stage.

REFERENCES

Blagoeva, K. T. and Mijoska, M. 2017. *"Applying TAM to Study Online Shopping Adoption Among Youth in the Republic of Macedonia," pp. 543–554.* Available at: http://www.hippocampus.si/ISBN/978-961-7023-71-8/167.pdf.

Dave Chaffey. 2017. *The reasons why consumers shop online instead of in stores.* Available at: https://www.smartinsights.com/ecommerce/ecommerce-strategy/the-reasons-why-consumers-shop-online-instead-of-in-stores/ (Accessed: June 17, 2019).

Davis, F. D., Bagozzi, R. P. and Warshaw, P. R. 1989. *"User Acceptance of Computer Technology: A Comparison of Two Theoretical Models," Management Science, 35(8), pp. 982–1003.* doi: 10.1287/mnsc.35.8.982.

Fayad, R. and Paper, D. 2015. *"The Technology Acceptance Model E-Commerce Extension: A Conceptual Framework," Procedia Economics and Finance. Elsevier B.V., 26(961), pp. 1000–1006.* doi: 10.1016/s2212-5671(15)00922-3.

Fedorko, I., Bacik, R. and Gavurova, B. 2018. *"Technology acceptance model in e-commerce segment," Management and Marketing, 13(4), pp. 1242–1256.* doi: 10.2478/mmcks-2018-0034.

Martínez-Navarro, J. et al. 2019. "The influence of virtual reality in e-commerce," Journal of Business Research, 100, pp. 475–482.

Michael E Porter. 1980. *Competitive Strategy.* New York.

Statista. 2019. *Retail e-commerce sales worldwide from 2014 to 2021 (in billion U.S. dollars).* Available at: https://www.statista.com/statistics/379046/worldwide-retail-e-commerce-sales/ (Accessed: June 17, 2019).

Transportation mathematical model teaching using inquiry-based learning

Rio Aurachman & Nopendri
Universitas Telkom, Bandung, Indonesia

ABSTRACT: Some prominent research has shown that students are unable to solve real problems because of lack of understanding of physical world concepts. This research tries to contribute in designing inquiry-based learning on logistic engineering, industrial engineering, and computer programming learning activities, especially for transportation problems. It was found that some improvement to inquiry-based learning could be accomplished by assignments learning method, by accommodating complicated concepts, and by explaining the connection of the current course with the previous course.

1 INTRODUCTION

Topics used in this research are transportation and distribution management systems. This course focuses on discussing the engineering of transporting goods/products from producers to consumers. The courses cover the concept of transportation, designing mathematical models, and cases and their application in the industrial world. This course aims to provide knowledge and skills for students to plan and manage effective and efficient transportation and distribution activities. Some of the implementation of transportation concepts are DRP (distribution requirement planning) (Muttaqin et al., 2017), vehicle routing problem (Desiana et al., 2016), queueing theory (Aurachman & Ridwan, 2016), and so on.

Facing industry revolution 4.0, concerns should be raised so that education institutions can equip students with suitable competencies to prevent unemployment (Aurachman, 2018; Aurachman, 2019). In order to produce competent engineers, in addition to rigorous learning process, rigorous preparation and selection for lecturers (Wankat & Oreovicz, 2015) is also required to ensure that the learning outcome target will be achieved. Various methods can be applied in the learning process of engineering courses. In general, engineering courses will be taught by practicing design process so that students will improve their hands-on capability. Engineering students should complete exercises on the design process. However, to do the design process, they need to comprehend the concept beforehand. Sometimes the student cannot handle new and hard basic concepts. One method that may suitable for this case is inquiry-based learning. In order to solve this problem we need to conduct some research, for example DBER (discipline-based education research). DBER applies scientific findings in the teaching field to the engineering field specifically (Singer & Smith, 2013)

The traditional engineering learning process applies the deductive principle, which begins with providing the theory and ends with the application of the theory; the latest teaching method uses inductive principles, where case studies and direct observations are used in understanding the theory (Prince & Felder, 2006). The application of case studies learning methods, enhances various high-level thinking skills and also increases motivation to learn (Yadav, et al., 2010). Interactive learning methods enhance students' performance by encouraging collaboration and introducing scientific contexts. (Balasubramanian et al., 2006). In many fields, the deductive method is used to help students understand the concept of engineering, one of which is in learning information systems and databases (Batory & Azanza, 2017). The application of these methods is also supported by using innovative tools (Eddy et al., 2015).

Interactive learning could support suitable workshop, such as Makerspaces, which is equipped by 3D Scan and 3D printing, whose students value positively towards the development of creativity (Saorín et al., 2017). Interactive method using games could help student to understand the concept and improve their attitude (Bodnar, et al., 2016). Some improvement in class and schedule may needed to ensure student engagement. For example, in pharmacy learning, series teaching increase students' interest in the experiment activity more than parallel class method (Zhou et al., 2018)

One method that is philosophically close to those deductive methods is inquiry-based learning. Inquiry-based teaching aims to increase student involvement in learning by helping students to develop hands-on, mind-on skills needed for the 21st century. This approach accommodates complex learning work. Inquiry-based teaching prioritizes the knowledge and experience that students bring to class and encourages active problem solving, communication, and joint development of new idea. Karplus and Atkin (1962), developed a learning style called "guided inquiry." This learning style focuses on developing ideas and observing students as the basis for designing the learning process. Guided inquiry covers the stages of exploration and discovery. Atkin and Karplus devised a method to support students in drawing on their own experiences and findings a way to develop learning conclusions and understand the phenomena that occur. One of the models of inquiry-based learning is Model 5E. The 5E model is developed by the biological sciences curriculum study (BSCS) and covers five stages of the learning cycle, each begun with E. These stages are engagement, exploration, explanation, elaboration, and evaluation (Allen & Rodriguez, 2018). Some developments are also carried out on inquiry-based learning using supportive gadgets and technology that to create a class just like a game adventure in a real world (Hwang et al., 2013).

There are several problems in engineering education. Some prominent research shows that students are unable to solve a real problem because of lack of understanding about physical world concept (Prince & Vigeant, 2006). There are several studies that discuss inquiry-based learning, especially in engineering education. One work indicates that universities should provide skills that are necessary in knowledge-building processes using inquiry-based learning (Bernold, 2007). It can be done by comparing the concept that is taught in the class with real-world problems (Madhuri et al., 2012). Experiment and inquiry-based learning can be run through an online system using virtual lab (Kollöffel & Jong, 2013).

This research tries to contribute in applying inquiry-based learning on logistic engineering learning, industrial engineering, and computer programming activities, especially for transportation problems. It is hoped that through this research, a recommendation will be obtained to improve the effectiveness of inquiry-based learning.

2 METHOD

The research method is an experimental research. We did the experiment in one transportation course in a Bachelor of Industrial Engineering Major. The class contained 36 students, studying eight topics in eight weeks. For the first four topics were taught using an inquiry-based learning design. The other four didn't use inquiry-based learning. Then we compared the grades of students in terms of the topic and the method to understand the effect of inquiry-based learning.

Based on the experience using inquiry-based learning for the transportation class, there were several steps to be prepared. The stages of implementation were divided into administrative preparations, preparation of learning content, preparation of learning tools, implementation of learning, evaluation of learning, and the use of learning evaluation results to improve subsequent learning content.

3 FINDING AND DISCUSSION

One of the topics used to implement inquiry-based learning methods was an introduction to the concept of transportation systems. This was an introductory-level topic and was delivered at the beginning of the semester. This topic was the starting gate for students to understand other topics.

In its implementation, the learning process was carried out according to the plan stated in table 5E. Some adjustments were made but in general the learning process followed the planned scenario. Each student carried out the roles and processes that had been written in framework 5E. In the learning process, students were actively involved in each agenda. The role of the lecturer was not based on providing material. Lecturers had to responsively review the work processes carried out by the students. Lecturers also needed to be empathetic to student comments and patient with student failures and mistakes.

Evaluation was done by online quiz. Before the online quiz was held, students took part in the lecture session. The lecture session used the inquiry-based learning method or not. The first and second lecturer of the four lectures involved in this study used the inquiry-based learning method. In addition to lectures, students were supported with lecture videos and lecture notes that could be accessed online. Students were expected to be able to learn content independently.

Online quizzes were designed so that students could carry out unlimited attempts. We recorded the score of the first attempt and the best score attempt. The setting used was that the student would get direct information about whether the attempt had a good score or a bad score, so that after each attempt, the student could evaluate themselves and then try again to get a better score. After many attempts, the system would take the best score among all attempts, so we could compare the first attempt and understand the process of student improvement.

4 CONCLUSION

By implementing this method in class we found some important conclusions that could be used as a guide to implement inquiry-based learning in another class and subject.

1. Inquiry-based learning uses examples that are very simple and close to the everyday lives of students. This is good for building understanding. However it lacks in terms of preparing students to face actual and more complicated problems.
2. it is necessary to develop methods to explain complex and detailed concepts. Inquiry-based learning discusses simple concepts and daily connections, but, in elaborate and detailed conceptual explanations, simple concepts and everyday links are not enough. Then the development of teaching methods is needed so that inquiry-based teaching can also be used in the introduction of complex and detailed concepts.
3. In inquiry-based learning, the first stage of learning is an explanation of the relation of topics to everyday life.
4. It is also necessary to give assignments so that students better understand the objectives of the previous meeting learning and their relevance to the next meeting.

REFERENCES

Allen, K. & Rodriguez, S. 2018. Classroom Strategies for Inquiry-Based Learning. [online] https://www.edx.org/course/classroom-strategies-for-inquiry-based-learning.

Atkin, J., & Karplus, R. 1962. Discovery or invention? *The Science Teacher*, 29(5), 45–61.

Aurachman, R. 2018. Perancangan Influence Diagram Perhitungan Dampak Dari Revolusi Industri 4.0 Terhadap Pengangguran Kerja. *Jurnal Teknologi dan Manajemen Industri*, 4(2), 7–12.

Aurachman, R. 2019. Model Matematika Dampak Industri 4.0 terhadap Ketenagakerjaan Menggunakan Pendekatan Sistem. *Jurnal Optimasi Sistem Industri*, 14–24.

Aurachman, R., & Ridwan, A. Y.,2016. Perancangan Model Optimasi Alokasi Jumlah Server untuk Meminimalkan Total Antrean pada Sistem Antrean Dua Arah pada Gerbang Tol. *JRSI (Jurnal Rekayasa Sistem dan Industri)*, 3(2), 25–30.

Balasubramanian, N., Wilson, B. G., & Cios, K. J. 2006. Innovative methods of teaching science and engineering in secondary schools.

Batory, D. & Azanza, M. 2017. Teaching model-driven engineering from a relational database perspective. *Software & Systems Modeling*, *16*(2), 443–467.

Bernold, L. E. 2007. Preparedness of engineering freshman to inquiry-based learning. *Journal of Professional Issues in Engineering Education and Practice*, *133*(2), 99–106.

Bodnar, C. A., Anastasio, D., Enszer, J. A., & Burkey, D. D., 2016. Engineers at play: Games as teaching tools for undergraduate engineering students. *Journal of Engineering Education*, *105*(1), 147–200.

Desiana, A., Ridwan, A. Y., & Aurachman, R. 2016. Penyelesaian Vehicle Routing Problem (vrp) Untuk Minimasi Total Biaya Transportasi Pada Pt Xyz Dengan Metode Algoritma Genetika. Bandung, Telkom University.

Eddy, S. L., Converse, M., & We, M. P. 2015. PORTAAL: a classroom observation tool assessing evidence-based teaching practices for active learning in large science, technology, engineering, and mathematics classes. *CBE–Life Sciences Education*, *14*(2).

Hwang, G. J., Wu, P. H., Zhuang, Y. Y., & Huang, Y. 2013. Effects of the inquiry-based mobile learning model on the cognitive load and learning achievement of students. *Interactive Learning Environments*, *21*(4), 338–354.

Kollöffel, B,. & Jong, T. D. 2013. Conceptual understanding of electrical circuits in secondary vocational engineering education: Combining traditional instruction with inquiry learning in a virtual lab. *Journal of Engineering Education*, *102*(3), 375–393.

Madhuri, G. V., Kantamreddi, V. S., & Prakash Goteti, L. N. 2012. Promoting higher order thinking skills using inquiry-based learning. *European Journal of Engineering Education*, *37*(2), 117–123.

Muttaqin, B. M., Martini, S., & Aurachman, R. 2017. Perancangan Dan Penjadwalan Aktivitas Distribusi Household Product Menggunakan Metode Distribusi Requirement Planning (DRP) Di PT. XYZ Untuk Menyelaraskan Pengiriman Produk Ke Ritel. *JRSI (Jurnal Rekayasa Sistem dan Industri)*, 56–61.

Prince, M. J., & Felder, R. M. 2006. Inductive teaching and learning methods: Definitions, comparisons, and research bases. *Journal of Engineering Education*, *95*(2), 123–138.

Prince, M., & Vigeant, M. 2006. Using inquiry-based activities to promote understanding of critical engineering concepts. Chicago, s.n.

Saorín, J. L., et al. 2017. Makerspace teaching-learning environment to enhance creative competence in engineering students. *Thinking Skills and Creativity*, *23*, 188–198.

Singer, S., & Smith, K. A. 2013. Discipline-based education research: Understanding and improving learning in undergraduate science and engineering. *Journal of Engineering Education*, *102*(4), 468–471.

Wankat, P. C., & Oreovicz, F. S. 2015. Teaching engineering. s.l.: Purdue University Press.

Yadav, A., Shaver, G. M., & Meckl, P. 2010. Lessons learned: Implementing the case teaching method in a mechanical engineering course. *Journal of Engineering Education*, *99*(1), 55–69.

Zhou, J., Wang, K., Zhang, X., & Wang, C. 2018. The comparison between series and parallel: Integrated experimental teaching model for pharmaceutical engineering students based on criteria for accrediting engineering programs in China. *Journal of Cleaner Production*, *172*, 4421–4434.

Associative learning method for teaching operation research

Rio Aurachman
Universitas Telkom, Bandung, Indonesia

ABSTRACT: Operation Research 2 is a course that teaches quantitative methods such as game theory, dynamic programming, and the Markov Chain. This method is needed in the complex managerial decision-making process. For most university students, this topic is a very new concept. This paper tries to propose associative learning methods in teaching operation research. We use associative tools such as films, games, and other creative media.

1 INTRODUCTION

One problem that occurs in lectures is that students tend to need a longer time to understand and be able to work on a mathematical problem. We experienced this when teaching an Operation Research 2 class. One chapter, took more than one week to get students' understanding, yet, the time available for discussion of the chapter was only one week because there were other discussions that needed to be done in the following weeks.

On the other hand, students still have not had the chance to train in the analytical skills of the problems studied with the mathematical approach. With the available time, they are able to work on the calculation problem, but there is not enough time to be able to analyze the case studied. As a result, learning outcomes are not achieved.

In addition to information systems in the form of open courseware and online learning systems, a teaching method that facilitates participants' understanding of the subjects is also needed. The method need to be not complicated and should be easily associated with students' experience. There are several studies that discuss the association process in teaching engineering. The learning results of the current experiment are accomplished through setting up an association with actual phenomenon in daily life (Madhuri et al., 2012). Associative learning is the method through which an individual or living organism learns a relationship between many stimuli or events (Plotnik & Kouyomdijan, 2012). Giving lessons using associative learning is expected to also strengthen association with the concept and facilitate the ability of students to complete new problem types. Significant study demonstrates that learners often enter the course with tightly held misconceptions about the physical world that traditional education does not efficiently address. As a consequence, learners are often able to fix issues that are clearly taught but are unable to apply the ideas of the course to solve actual issues that are not specifically taught in the class. Failure to understand the preconditions also leaves learners unprepared for further research (Prince & Vigeant, 2006). The association can be built using everyday problems. Traditional training in engineering is deductive, starting with theories and advancing to the theories' implementations. Inductive is an alternative learning method. Specific observations, case studies, or issues are used to introduce topics (Prince & Felder, 2006). Case studies methods have been discovered to enhance critical thinking skills and problem-solving abilities, higher order thinking abilities, conceptual modification, and also motivation for learning (Yada et al., 2010). The association can also be built using the forms of games. Research on game execution in undergraduate engineering schools has shown that, despite various types of evaluation, there is a trend that game-based engagement activities increase not only student learning but also attitudes (Bodnar et al., 2016). Research on associative learning was also developed according to the development of the cognitive sciences. There is research that studies regularity extraction species-wide: mechanisms of associated learning expressed by humans and animals (Rey et al., 2019).

Meanwhile there is no specific research that examines how to apply learning methods in operation research. Therefore, this research attempts to contribute by proposing operation research learning methods.

2 METHOD

The participants in this study were 25 bachelor students of an international class in the industrial engineering study program of Telkom University. The respondent were randomly chosen by the process of class enrollment. They are fourth-semester students in their second year. The industrial engineering study program is designed so that alumni are able to achieve certain competencies. As referring to ABET (Anon., 2016), the competences of the bachelor degree of industrial engineering are as follows:

"The curriculum must prepare graduates to design, develop, implement, and improve integrated systems that include people, materials, information, equipment and energy. The curriculum must include in-depth instruction to accomplish the integration of systems using appropriate analytical, computational, and experimental practices."

One of the required and standardized courses is the Operation Research course, taught in order, namely Operation Research 1 and Operation Research 2. Operation Research 1 equips students to be able to use mathematical models in decision-making processes in integrated systems. In Operation Research 1, several deterministic optimization models were studied while Operation Research 2 teaches the ability to make decisions with the help of stochastic mathematical models and several other methods.

Some sub-topics in Operation Research 2 are stochastic dynamic topics. This is in line with the basic topics being developed now in the field of machine learning, which is about Reinforcement Learning. Some of the chapters discussed include: game theory, deterministic dynamic programming, dynamic stochastic programming, Markov Chain, queuing theory. The learning process of Operation Research 2 can be the foundation and inspiration in determining the reinforcement learning teaching method.

The participants in this study have taken basic lectures, calculus 1 and calculus 2 in the first year, and have followed Operation Research 1 and probability theory in the previous semester. This becomes the basis and prior knowledge for students to take part in Operation Research 2 and reinforcement learning topics. Of course, this is different from the participants who just newly interacted with the topics of Operation Research 2.

To understand the effect of the method, the analysis will be conducted using descriptive statistic analysis, but in this research we focus on creating a method to implement associative learning. To get the best method, we searched in all possible game, simulation, and another tools. All the tools were compared by measuring the cost and expected benefit. Then, the best tools were implemented.

3 RESULT

Different associative approaches are carried out between one chapter to another. Game theory is also a new concept learned by students. The way of thinking and approach in game theory was not found during high school or in the first year of bachelor degree in industrial engineering program.

In order for students to be able to understand the concepts and works of game theory, the game process is associated with XY game. XY game is a game of four people competing against each other. In each round, each student has to draw either an X or Y card. The number of X and Y cards drawn and cards submitted by individuals affect the score received by each member. Each group member will compete and be independent. At the end of the game, participants will understand the way the game works and how it relates to everyday life:

it turns out that if each member thinks individualistically (or self-servingly), an optimal outcome will not be achieved. In XY game, if team members work together and agree for mutual benefit, they will get more points. This is in accordance with the philosophy and how the game's theory works.

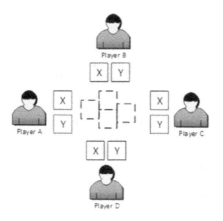

Figure 1. Illustration of XY game.

Table 1. Rule of XY Game.

I	4 Group choose Card-X	Each Group Lost	- $ 1,000
II	3 Group choose Card-X	Each Group Lost	- $ 1,000
	1 Group choose Card-Y	Group Win	+ $ 3,000
III	2 Group choose Card-X	Each Group Win	+ $ 2,000
	2 Group choose Card-Y	Group Lost	- $ 2,000
IV	1 Group choose Card-X	Group Win	+ $ 3,000
	3 Group choose Card-Y	Each Group Lost	- $ 1,000
V	4 Group choose Card-Y	Each Group Win	+ $ 1,000

In the deterministic dynamic programming course, the association used is a mine game. In this game, participants are asked to pass an area containing mines. The mines locations are unknown. Each member must try to navigate a file plot by plot. When the participant steps on a plot containing mines, the participant has to go back to the beginning, and the other participants continue to try to catch the mine. Participants are not allowed to communicate. Each participant may learn the safe steps by observing his team's experience. Gradually the participants gather information on the mines' whereabouts and will gradually reach the destination. This game is used to introduce the concept of dynamic programming to participants. Through this game, it is expected that students will understand that in solving a problem, one decision-making stage and the others are interrelated. Choosing the incorrect step at the beginning will result in a nonoptimal final solution. This is in accordance with the principle of dynamic programming where decision-makers must behave in each phase and each phase is interrelated. Another case study can be used to learn this dynamic programming topic (Aurachman, 2018), (Aurachman, 2019).

Stochastic dynamic programming is also used to complement the associations achieved in deterministic dynamic programming. These two concepts are delivered in sequence. The association used is through watching the movie *Edge of Tomorrow*. The film tells the story of a person who was able to repeat one day continuously, making a different decision in each repeated day. When he died, he would wake up the day before and continue in the cycle so that he could look for the best decisions and learn from the experience that happened on the same day. Decision-making acts are seen in real-life situations. Life is probabilistic. This film is suitable for illustrating an association with stochastic dynamic programming. In taking dynamic decisions, decision-makers are guided to see all alternative decisions and see the implications of the decision, especially the implications for future decision-making. Advanced implementation of stochastic dynamic programming is queueing theory. Some papers report case studies to understand the concept of this topic (Aurachman & Ridwan, 2016).

4 CONCLUSION

The success of this research (in the form of percentage) was obtained through quantitative methods, including determining the effectiveness of the application of methods used both in academic and non-academic aspects. This study also identified a learning system that allows students to repeated learning until they have the expected competencies. In addition, the learning outcome assessment system allowed students to repeat self-evaluations to measure their abilities and improve themselves until they acquired the expected competencies.

This study utilized some content in the form of games, films, and simulations as means of increasing students' understanding using the principle of association. It is expected that the content can be reused in other lectures and can even be used by other subjects. The level of success measured is in accordance with the increase in value that occurs.

REFERENCES

Anon., 2016. Accreditation Board for Engineering and Technology. [online] https://www.abet.org/accreditation/accreditation-criteria/criteria-for-accrediting-engineering-programs-2016-2017/#3

Aurachman, R. 2018. Perancangan Influence Diagram Perhitungan Dampak Dari Revolusi Industri 4.0 Terhadap Pengangguran Kerja. *Jurnal Teknologi dan Manajemen Industri*, *4*(2), 7–12.

Aurachman, R. 2019. Model Matematika Dampak Industri 4.0 terhadap Ketenagakerjaan Menggunakan Pendekatan Sistem. *Jurnal Optimasi Sistem Industri*, 14–24.

Aurachman, R., & Ridwan, A. Y. 2016. Perancangan Model Optimasi Alokasi Jumlah Server untuk Meminimalkan Total Antrean pada Sistem Antrean Dua Arah pada Gerbang Tol. *JRSI (Jurnal Rekayasa Sistem dan Industri)*, *3*(2), 25–30.

Bodnar, C. A., Anastasio, D., Enszer, J. A., & Burkey, D. D. 2016. Engineers at play: Games as teaching tools for undergraduate engineering students. *Journal of Engineering Education*, *105*(1), 147–200.

Madhuri, G. V., Kantamreddi, V. S., & Prakash Goteti, L. N. 2012. Promoting higher order thinking skills using inquiry-based learning. *European Journal of Engineering Education*, *37*(2), 117–123.

Plotnik, R., & Kouyomdjian, H. 2012. *Discovery Series: Introduction to Psychology*. Belmont: CA: Wadsworth Cengage Learning. p. 208. ISBN 978-1-111-34702-4.

Prince, M. J., & Felder, R. M. 2006. Inductive teaching and learning methods: Definitions, comparisons, and research bases. *Journal of Engineering Education*, *95*(2), 123–138.

Prince, M., & Vigeant, M. 2006. *Using Inquiry-Based Activities to Promote Understanding of Critical Engineering Concepts*. Chicago, s.n.

Rey, A. et al. 2019. Regularity extraction across species: Associative learning mechanisms shared by human and non-human primates. *Topics in Cognitive Science*.

Yadav, A., Shaver, G. M., & Meckl, P. 2010. Lessons learned: Implementing the case teaching method in a mechanical engineering course. *Journal of Engineering Education*, *99*(1), 55–69.

Optimization model of the number of one-way servers to minimize cost and constraint

Rio Aurachman
Universitas Telkom, Bandung, Indonesia

ABSTRACT: In determining the optimal number of servers on a highway, a method is needed to automatically provide information when the queue is predicted to be too long. The approach taken is optimization mathematical modeling by modifying the queue formulation that has been developed. The mathematical modeling process uses methodology proposed by Daellenbach, starting from system modeling and followed by mathematical modeling along with verification and validation. We find mathematical models that can optimize the allocation server, adapting to the flow of vehicles. The focus of the findings is on the M/M/C queue system (the number of arrivals per unit memoryless time with Poisson distribution, memoryless service time with binomial distribution, and more than one server).

1 INTRODUCTION

The queuing system and mathematical model for calculating the queue are developed using the principle of arrival flow originating from one direction which is entered into a system where the queue system becomes the barrier gate. Some queuing systems not only become the barrier gate for entry but also the limiting gate for exit.

Some queuing systems allow existing servers to act simultaneously as input and output. When the server acts as input, the output function does not work: when the server acts as an output function, the input function does not work. Such dynamics may occur especially in the queuing system whose utility is seasonal. Sometimes the number of input arrivals is larger and at times the arrival of the output current is greater. This flexibility will be beneficial because it is possible to obtain adaptive combinations to reflect environmental conditions. Permanent configuration of input and output will cause a service paradox. It is possible that there will be a time when an input server is busy but the output server is empty and off. This inequality and paradox can be prevented by using flexible allocations. This case was researched in a previous publication (Aurachman & Ridwan, 2016).

For now, this article will propose the mathematical model to solve the optimization problem of the one-way tollgate system. This time, the mathematical model will recognize the function to minimize cost and is constrained by customer preference, average waiting time.

The mathematical model for calculating the length of the queue has evolved over time. However, an initiative to provide a solution to the calculation of the problems mentioned above needs to be done. There is a need for a policy foundation for decision-makers to determine the right allocation and optimizing queue performance for both sides simultaneously, not individually.

Several studies have been conducted to suggest better management of tollgate queues. Simulation has become one of the methods for analyzing and improving system performance. The previous simulation results show that the number of tollgates operating on a highway and the length of processing at gates greatly influences smooth vehicle flow (Sediono et al., 2004).

Research that recommends opening tollgates adjusted to the rate of arrival of vehicles as proposed by Sediono et al. (2005) has also been conducted. The adoption of the open–close strategy for tollgates is quite effective in anticipating the changing pace of the arrival of vehicles. Moreover, the dynamics of the open–close strategy and the rate of vehicle arrival that constantly changes can be recorded properly through the normal traffic density curve (Sediono et al., 2005). This dynamism can also be optimized by adjusting to vehicle arrival on the highway by modifying the number of toll plazas operating to correspond with the density shift of vehicle arrivals. (Satya & R., 2011). In addition, cost factors can be taken into consideration in determining the number of tollbooths opened. Optimization of the costs of tollbooths operating can be done by adjusting the number of tollbooths operating to the arrival of vehicles (Satya & R., 2011). Traffic behavior and characteristics change over time. It is possible to obtain the equivalent value of passenger cars on the toll road, where the example of the study is carried out on the Jakarta-Cikampek toll road (Prima & Hikmat Iskandar, 2014).

Simulation and analysis can be supported by using software as proposed by Sediono et al. (2005). Modeling of queue conditions at tollgates has been successfully implemented into simple software. On the basis of easy-to-understand principles, by making appropriate data structures and flowcharts, this software is quite flexible. From the discussion above it can be concluded that the desired queue length can be adjusted by controlling in-and-out flow of vehicles (Sediono et al., 2005).

Hardware is used to automatically detect vehicle arrival and then processed into data and information that is useful for decision-makers. This tool uses a system of metal-detection sensors, planted on toll roads which work by detecting a vehicle's metal undercarriage. Information on density and congestion can also be accessed using short message services via cellphone. The system applied to this tool can continuously and automatically update the process of monitoring and reporting congestion occuring on the toll road segment (Sugandhi, 2010).

The goal of toll plaza regulation is to increase customer satisfaction. There are previous studies that proposed a method of evaluating customer satisfaction using several assessment attributes referred to as the five main dimensions of service quality, namely tangibles, reliability, responsiveness, assurance, and empathy. The survey data obtained were then analyzed using the importance performance analysis (IPA) method. Based on the IPA, some service attributes that needed to be addressed in order to improve performance were as follows:

a. Smooth flow of traffic.
b. Short and fast payment queue at the toll plaza.
c. Clean and neat toll road.
d. Safe and comfortable bends, inclines, and derivatives of roads.
e. Flat and not slippery road surface.
f. Proper street lighting to make sure the driver comfortable at night.
g. Availability of median safety fences and toll road guardrails. (Pancawati & Kartika, 2013)

Another technology that can be used to improve toll road performance is the e-toll concept.

Widyani (Widyani, 2015) examined the effect of e-toll implementation. The following is one of the results of the study. Perceived usefulness positively impacts the attitude of Bandung city, West Java Province, toll road users when implementing e-toll and perceiving behavioral control and monitoring; perceived usability, subjective standards, and attitude influence the intention to use E-Toll cards.

Readiness of Bandung city toll road users' when the e-toll card is implemented at the tollgate was studied using the integration model of technology acceptance model (TAM) & planned behavior theory (TPB) (Widyani, 2015)

The study of the queue other than the toll road system is also carried out on other queuing systems. One of them is research conducted by Kurniawan (Wahyu et al., 2015) for the Solo Grand Mall exit system. The effect of vehicle queues and vehicle volume across the Solo Grand Mall exit system was reviewed in terms of vehicle arrival rates and service levels. This case was settled by using queuing theory with the first in/first out (FIFO) queue discipline. Based on the Kolmogorov-Smirnov test, the calculation of statistical analysis at the entry stage was based on the Poisson distribution of the Chi Square test and the statistical analysis of the service level.

2 METHOD

The process of developing a mathematical model begins with reviewing a real problem that occurs in a system (Murty et al., 1990), then taking a system approach to the problem so that a holistic viewpoint and helicopter view are obtained. The viewpoint of this system causes the resulting solution to be not only locally but also globally optimal. In the toll road queuing system, the embedded system approach is in the form of reviewing not only the incoming queue but also the substation at the same time. This system approach considers that the relevant system reviewed is one substation at the end of the toll road but does not review the system as a whole.

After a system approach, the objectives or problems to be resolved and the system characterization are determined. Some diagrams can be used to describe this, one of which is the influence diagram (McNickle et al., 2012).

The next step is to design mathematical models based on the influence diagram. After the mathematical model is designed, the next stage is analysis, validation, and verification. The validation method used is checking each relationship in the influence diagram on the mathematical model. Verification can be done by ensuring that the units of the equation, both the left and right segments, are the same.

At the validation stage a conclusion can be reached whether the model needs to be repaired or whether the model is valid and verified. After the iterative process occurs, the last step is the formation of the correct mathematical model so that it can be used in solving the problems.

3 FINDING AND DISCUSSION

The tollgate has a number of substations for the exit lane, exit here meaning the vehicle is leaving the highway. There are also a number of substations that are used as inflow gates to the highway. Each substation has its own unique function. Almost all substations now have automatic features, namely GTO (Gerbang Tol Otomatis). Often exiting vehicles experience chronic congestion while vehicles enter smoothly. This is a phenomenon of an imbalance between the entry and exit substation. Balance can create a loss in waiting time which does not accumulate in a handful of vehicles but can be shared evenly with other vehicles so that the burden does not feel heavy. However, in some conditions, there are gates that only contain one type of lane, whether it's an entrance gate or an outer gate; for example, Padalarang tollgate, Baros tollgate, and a number of other tollgates.

Next is the process of developing a mathematical model based on the conceptual model, using influence diagram format. Each arrow symbol from the influence diagram shows the relationship between variables and/or parameters. The linkage is expressed in the form of mathematical equations.

The objective function is the total operating cost. For this purpose the minimization procedure is implemented.

$$\text{Min } Z = c \times cost \tag{1}$$

Subject to:
To explain the condition of the system that the number of substations is constrained by the area. And total average waiting time should not exceed preference of customer

$$c \leq cmax \tag{2}$$

$$ENq \leq ENqmax \tag{3}$$

The relationship that occurs between the number of vehicles waiting in line and the number of substations is that the lower the number of substations, the greated number of vehicles waiting in line.

$$\frac{\lambda^c}{c!\mu^c} \frac{1}{\sum_{n=0}^{c} \frac{-\lambda^n}{n!\mu^n} + \frac{1}{c!} \left[\frac{\lambda}{\mu}\right]^c \left[\frac{1}{-\frac{\lambda}{c\mu}}\right]} \frac{\lambda}{c\mu} \frac{1}{\left[-\frac{\lambda}{c\mu}\right]^2} = EN_q \qquad (4)$$

The following equation to explain that the number of operating substations can not be less than one substation.

$$c \geq 1 \qquad (5)$$

$$c, \text{ integer} \qquad (6)$$

4 CONCLUSION

A mathematical model has been designed using a system approach. The model can be a solution for the scheduling and allocation of a toll plaza at each end of the toll road. The model can minimize the existing cost (Z). The minimization is expected to increase consumer satisfaction by constraining average waiting time (EN_q in equation 3). The model is also expected to support decision-making to increase utilization of tollgates. This increase in utility is expected to increase the efficiency of the system.

In the development of further research, the researcher can test the model using real data so that the increase in performance of the system after using this mathematical model can be proven to support decision-making. Another development that can be done is by entering cost factors as a consideration of developing mathematical models.

REFERENCES

Aurachman, R., & Ridwan, A.Y. 2016. Perancangan Model Optimasi Alokasi Jumlah Server untuk Meminimalkan Total Antrean pada Sistem Antrean Dua Arah pada Gerbang Tol. *JRSI (Jurnal Rekayasa Sistem dan Industri)*, *3*(2), 25–30.
H.D., McNickle, D., & Dye, S. 2012. *Management Science: Decision-Making through Systems Thinking*. s.l.: Palgrave Macmillan.
K., Wahyu, H., Sumarsono, A., & Mahmudah, A.M. 2015. Evaluasi Panjang Antrian Kendaraan Pada Pelayanan Pintu Keluar Parkir Dengan Atau Tanpa Perubahan Pintu Keluar Parkir Di Solo Grand Mall. *Matriks Teknik Sipil 3*(1).
Murty, D.N.P., Page, N.W., & Rodin, E.Y. 1990. *Mathematical Modelling: A Tool for Problem Solving in Engineering, Physical, Biological, and Social Sciences*. Oxford: Pergamon Press.
Pancawati, E., & Kartika, A.G. 2013. Analisis Layanan Jalan Tol Berdasarkan Kebutuhan Pengguna (Studi Kasus Ruas Jalan Tol Surabaya–Gresik). s.l., s.n.
Prima, G.R., & Hikmat Iskandar, T.R.J. 2014. Kajian Nilai Ekivalensi Mobil Penumpang Berdasarkan Data Waktu Antara Pada Ruas Jalan Tol. *Jurnal Jalan-Jembatan*, *31*(2), 74–82.
Satya, D., & R., R. 2011. Penentuan Loket yang Optimal pada Gerbang Selatan Tol Pondok Gede Barat dengan menggunakan Teori Antrian untuk Meminimasi Biaya. *Jurnal Teknik Industri*, *1*(3), 224–230.
Sediono, W., & Handoko, D. 2004. Pemodelan dan Simulasi Antrian Kendaraan di Gerbang Tol. s.l., s.n., pp. 11–14.
Sediono, W., & Handoko, D. 2005. Pendekatan Simulasi dalam Mengoptimalkan Pengoperasian Gerbang Tol. s.l., s.n., pp. 91–94.
Sugandhi, S.A. 2010. Perancangan Model Alat Pemantau Dan Petunjuk Informasi Kemacetan Pada Ruas Jalan Tol Dalam Kota Jakarta. SKRIPSI S-1 TEKNIK ELEKTRO.
Widyani, D. 2015. Analisis Adopsi Aplikasi RFID Pada Produk E-Toll di Jalan Tol Kota Bandung. Open Library Telkom University.

Employee behavior towards big data analytics: A research framework

W. Ahmed, S.M. Hizam, H. Akter & I. Sentosa
Universiti Kuala Lumpur Business School (UBIS), Kuala Lumpur, Malaysia

ABSTRACT: Big data analytics (BDA) is heralded to generate vast-scale business prospects. Employee acceptance of BDA system is of utmost significance, as digital transition in organizations can be crippled by employee behavior. Therefore, an in-depth analysis of theory-based research is vital to comprehend the major instigations towards digital tools integration from the employee viewpoint. This study aims to develop the conceptual framework of BDA adoption factors by taking the technology acceptance model (TAM) as the point of departure. External constructs of digital optimization such as digital dexterity, technology–task fit, big data dimensions and self-efficacy are incorporated with TAM variables, namely perceived usefulness, perceived ease of use and behavioral intention. Hypotheses are developed to outline the relationships among the variables. The proposed conceptual framework of the extended TAM model will specify and assess employee behavior in acceptance of big data analytics. The research endeavors to oversee talent in the digital workplace.

1 INTRODUCTION

In this era of disruptive innovation, organizations are striving to achieve the maximum level of digital optimization in business activities. Big data analytics (BDA) is one modern-day business analytics technique to perform various analyses on huge volumes of data sets, such as weather data for accurate forecasting, patient data for timely diagnostics, customer transaction data for buying preferences, airline user data for traveling patterns, and employee activities data for job engagement, etc. BDA applies the complex machine-learning algorithm to datasets to perform predictive analysis for pertinent decision-making to overcome future challenges and mitigate costs. It empowers the transportation sector to be more efficient and smart, as road congestion, traffic level, safety measures, etc., can be predicted with data analytics by analyzing the data of daily traffic, driver patterns of roads usability, commuter usage frequency of public transport, number and nature of traffic violations, accidents on highway, etc. It envisages real-time estimation to assist in route planning, congestion management, and suggesting secure areas to travel, etc. Organizations step into the big data analytics to acquire the industry advantages by predicting customer preferences, enhancing operational efficiency, and achieving employee enablement (Bonnet et al., 2015).

As enterprises work at digital optimization and corporate revolution endeavors, the disparity of IT and business crumbles. Each section and segment of the organization, then, can be equipped with digital tools to perform the various activities. Durodolu (2016) explained that the adoption of digital tools advances the skill capability of employees. Big data analytics tools realize employee productivity uplift with new skills development (Ren et al., 2017). By implementing innovative systems, do organizations succeed in realizing suitable digital talent at the workplace? The answer to this expression is provided by Gartner's (2018) report that 83% of firms are found unable to reach their required level of digital transformation and missed their objectives in spite of system implementation. The main reason discerned was none other than human behavior and preferences towards the use of technology. Employees are a core part of innovative initiatives, and their intuition and mindset affect organizational policies. Individuals' technological ability and ambition inspires them to perform the task in the digital environment.

This behavioral manifestation is coined as digital dexterity at work. It has proven an active catalyst to make sure of system adoption and integration with advantages. However, there is a clear distinction between the deployment of a digital tool and its adoption (Bonnet et al., 2015). Adoption is backed by a belief system (i.e., behavioral intention), to accept or reject the innovation influenced by several factors. The study will seek to uncover the relevant constructs of behavior shaping for smooth digital transition and will answer the following questions:

1. What are employee's main motivations in terms of behavior towards the adoption of BDA?
2. Do digital workplace factors such as digital dexterity, big data dimension, task–technology fit, and self-efficacy influence BDA acceptance behavior of employees?

There are various theoretical expressions and frameworks regarding digital tools (i.e., big data analytics adoption), but it still requires more vigorous and systematic work to explain and depict the willingness of employees to accept big data (Al-Rahmi et al., 2019). The literature on big data in the transportation industry is quite sparse. The technology acceptance model (TAM) is proved to be a suitable predictor of user behavior towards technology acceptance (Sentosa & Mat, 2012) such as employee behavior towards big data (Okcu et al., 2019). However, previous studies couldn't portray the individual contexts in digital implementation, such as digital dexterity that guarantees the employee enablement in digital skills management (Bonnet et al., 2015). The objective of this study is to conceptually propose a technology adoption framework that could depict the significance of the digital workplace setting and direct the key factors in the adoption of big data analytics. The study is envisaged through a theoretical lens of TAM. This underpinning theory is broadly validated for employee behavioral assessment and corroborated through external factors of the adopting technology scenario. Therefore, numerous variables of digital tool implementation are analyzed and included in extended TAM model; for instance, big data dimensions, digital dexterity, task–technology fit and self-efficacy. This research will explore an internal belief system of adoption that builds on personal understanding of technology, technical ability, and ambition to perform the task with proper use of technology features.

2 LITERATURE REVIEW

2.1 *TAM model*

TAM model proposed to assess the usability of the information system in the organization. TAM has proved to be a profound approval measure for adoption analysis with its basic two constructs that extol the level of advantage gain in the productivity (i.e., perceived usefulness) and level of convenience achieved by accepting technology (i.e., perceived ease of use). Convenience in system use also enhances the usefulness of the technology. Big data analytics embrace is measured by adding the environmental context factors with TAM (Verma et al., 2018). It can best predict data analytics used by combining it with other theories and concepts. Recent studies have proven the impact of perceived usefulness and ease of use on behavioral intention (Al-Rahmi et al., 2019). Thus, we hypothesized that

H1: Perceived usefulness will positively influence the behavioral intention towards BDA.
H2: Perceived ease of use will positively influence the behavioral intention towards BDA.
H3: Perceived ease of use will positively influence the perceived usefulness of BDA.

2.2 *Digital dexterity*

Digital dexterity is a novel term describing the ability and yearning agility of the workforce to adopt and actively be involved in the technology transition process for advantageous organizational objectives. It is a combination of attitude, mental approaches, and behaviors to enable employees to perform effectively. Individuals with a higher level of digital dexterity are open to innovation, flexible to the environment, focused and skillful in computational process, and display shrewdness and practical knowledge of data to make the decision. Elevated digital dexterity in employees' performance appears to predict more effective adoption and integration with an innovative system to make better results by keeping the organizational objective in mind (Gartner, 2018). Digital dexterity helps the organization to improve organizational efficiency and ensure a high digital talent level by accepting and making use of digital skills like big data analytics (Soule et al., 2016). Therefore, we hypothesize that:

H4: Digital dexterity will positively influence the behavioral intention towards BDA adoption.

2.3 Task–technology fit (TTF)

The efficient level of acceptance of innovation depends on rectification of proper 'fit' between the implemented technology and employees' assigned task. This task–technology fit enables the effective adoption of any decision-making system. Such a system's value can dwindle to the least level if employee task level is unable to match the technology provided. TTF has demonstrated its contribution in various recent technology adoption studies including BDA (Shahbaz et al., 2019). Therefore we hypothesize that:

H5: TTF will positively influence the behavioral intention towards BDA.

2.4 Big data dimensions

Big data analytics functions through 3V's model. Initially deducing the efficient value from the enormous volume of data sets of a wide variety with higher velocity control and analysis, this model is known as the dimensions of big data. Volume explains large data sets pertaining the certain valuable information. Variety describes the amalgamation of various kinds of data obtained through several informants. Velocity is analyzing the speed of data to conclude information. These dimensions have proven the association towards perceived usefulness of the TAM model in updated studies (Okcu et al., 2019). Therefore, we hypothesize that:

H6: Big data dimensions will positively influence the perceived usefulness of BDA.

2.5 Self-efficacy

The term self-efficacy refers to individuals internal belief that they are capable of performing certain activities in order to accomplish their tasks. The concept is mainly connected with the adoption of new technology at the workplace and employees' acceptance of their ability to complete their task successfully. The relationship between perceived ease of use and self-efficacy is much tested and proven in various innovative systems. In big data analytics, self-efficacy is also an effective factor towards perceived ease of use (Okcu et al., 2019). Thus, we hypothesize that:

H7: Self-efficacy will positively influence the perceived ease of use towards BDA.

3 CONCEPTUAL FRAMEWORK

The proposed framework (Figure 1) illustrates the relationship of TAM with external constructs of employee behavior towards the adoption of big data analytics (BDA) through seven hypotheses.

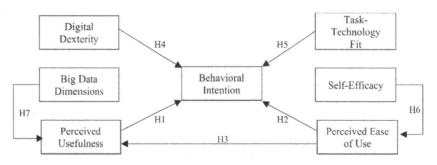

Figure 1. Conceptual framework.

4 CONCLUSION

This study develops the conceptual framework for describing the constructs that influence employee behavior for big data analytics acceptance. The study adds that the digital capability of the organization does not depend on mere deployment of resources and training programs but rather that gaining insight into employee intention in terms of digital tool acceptance entails unconventional factors. By highlighting digital dexterity, task–technology fit, big data dimensions and self-efficacy, this theoretical analysis will contribute to technology adoption literature from the point of view of employee enablement in the digital environment, an element that has not been focused on in previous research. The proposed conceptual framework is intended to support businesses in successful adoption of digital technology in the transportation sector. The empirical study in this regard can better validate the significance of relationships with behavioral intention and explanatory power of the model. It will also postulate the foundation for further research into acceptance of digital tools from the viewpoint of employee enablement. The study supports the policymakers, academics, governments and society in understanding and reaping benefits from smart mobility. The research has presented the novel factor (i.e., digital dexterity), to better understand talent in the digital workplace. Future studies can deduce better ways of managing digital dexterity at the workplace by understanding the role of managers and organizational policies.

REFERENCES

Al-Rahmi, W.M., Yahaya, N., Aldraiweesh, A.A., Alturki, U., Alamri, M., Bin Saud, M.S., Kamin, Y. Bin, Aljeraiwi, A.A., & Alhamed, O.A. 2019. Big data adoption and knowledge management sharing: An empirical investigation on their adoption and sustainability as a purpose of education. *IEEE Access*, 7, 47245–47258.

Bonnet, D., Puram, A.D., Buvat, J., KVJ, S., & Khadikar, A. 2015. Organizing for digital: Why digital dexterity matters organizational design is key to reaping the rewards from technology adoption. Capgemini Consulting, 1–17.

Durodolu, O.O. 2016. Technology acceptance model as a predictor of using information system to acquire information literacy skills. *Library Philosophy and Practice (e-journal)*, (November).

Gartner. 2018. *Executive Guidance: Digital Dexterity at Work*.

Okcu, S., Koksalmis, G.H., Basak, E., & Calisir, F. 2019. Factors Affecting Intention to use Big Data tools: An Extended Technology Acceptance Model. Springer, 401–416.

Ren, S.J.-F., Fosso Wamba, S., Akter, S., Dubey, R., & Childe, S. J. 2017. Modelling quality dynamics on business value and firm performance in big data analytics environment. *International Journal of Production Research*, 55(17), 1–16.

Sentosa, I., & Mat, N.K.N. 2012. Examining a theory of planned behavior (TPB) and Technology acceptance model (TAM) in internet purchasing using structural equation modeling. *Journal of Arts, Science & Commerce*, 2(2), 62–77.

Shahbaz, M., Gao, C., Zhai, L.L., Shahzad, F., & Hu, Y. 2019. Investigating the adoption of big data analytics in healthcare: The moderating role of resistance to change. *Journal of Big Data*, 6 (1).

Soule, D., Puram, A., Westerman, G., & Bonnet, D. 2016. *Becoming a Digital Organization: The Journey to Digital Dexterity*. MIT Center for Digital Business. No. 301.

Verma, S., Bhattacharyya, S.S., & Kumar, S. 2018. An extension of the technology acceptance model in the big data analytics system implementation environment. *Information Processing & Management*, 54(5), 791–806.

Indicators for measuring green energy: An Indonesian perspective

Indrawati, C. Januarizka & D. Tricahyono
Faculty of Economics and Business Telkom University, Bandung, Indonesia

S. Muthaiyah
Multimedia University, Selangor, Malaysia

ABSTRACT: Energy has a very important role in people's lives because it is an important parameter for development and economic growth. In cities of Indonesia, energy is becoming a very important issue, given the high levels of energy consumption and use of fossil fuels, and the impact on the environment of using fossilfuel–based energy, such as pollution. The concept of green energy is expected to be a solution to these problems, as is the creation of green environments and sustainability in energy use. The green energy concept is considered able to handle issues faced by cities around energy and the environment. This research uses an exploratory qualitative method. This research found 6 variables and 20 indicators to measure green energy. Further research is recommended to measure the green energy index using the found variables and indicators.

1 INTRODUCTION

The level of urbanization in the world rises high. More than half the world's people now live in cities and the number will increase to more than two-thirds by 2050 according to United Nations estimates. According to the data of Asia Development Bank (2015), Indonesia ranks third, with an urbanization rate of 54.46%. According to Agung, Hartono, and Awirya (2017), urbanization is positively related to energy consumption. An increased level of urbanization will increase energy consumption, especially from developing countries in Asia, one of which is Indonesia. Based on International Energy Agency (2017), Indonesia is the largest energy consumer in Southeast Asia, equal to 36%. Based on data from the Ministry of Energy and Mineral Resources of Indonesia, Indonesia's greatest energy needs were for electricity (Badan Pengkajian dan Penerapan Teknologi 2017).

West Java, one of the provinces in Indonesia, has the largest urbanization rate growth prediction from 2010 to 2035 at amounted to 89.3% (Bappenas, 2013). Based on statistics, Bandung city has the highest population density in West Java and became the third most populous city in Indonesia after Jakarta and Surabaya (Badan Pusat Statistik Kota Bandung, 2017). Based on data from Badan Pusat Statistik Kota Bandung (2017) total energy use in the city until 2016 amounted to 4,180 MWh, and energy use will continue to increase every year.

As reported by Putra (2014), Indonesia's energy reserve is estimated to be decreasing and potenitally exhausted by 2025. Indonesia is targeting new and renewable energy use to begin to reduce dependence on fossil energy sources by 23% by 2025.

2 RESEARCH OBJECTIVE AND QUESTIONS

Green City is the concept of sustainable and environmentally friendly urban development. One aspect of the green city is green energy. The concept of green energy is expected to become a solution to the problem of environmental energy, the creation of green environment, and to realize efficiency in energy use. The use of green energy is one of solutions for government and other stakeholders for energy needs. Unfortunately, based on literature conducted in this research, to date there is still no standard of variables and indicators that can be used to measure the application of green energy in a green city of Indonesia, that will fit with Indonesia characteristics, especially in Bandung. The research objective of this study is to determine the variables and indicators that can be used to measure green energy in Bandung, Indonesia.

3 RESEARCH METHODOLOGY

The type of research here is qualitative research study with exploratory methods. This research method involves the analysis of data in the form of descriptions and the data is qualitative and not directly quantifiable (Indrawati, 2015)

In selecting interviewees, this study used the purposive sampling approach. Interviewees were selected based on several criteria (Indrawati, 2015), including knowledge and awareness of the green city concept, especially green energy. Interviewees should at least have the academic background equivalent to a master's degree or equivalent experience relevant to the focus area of green energy. This study divided interviewees into four groups based on the concept of a quadruple helix. There are government, business experts, academics/experts/researchers, and civil society.

The collected data are processed through four analysis stages according to the model of Miles and Huberman (1984). The first step starts from collecting verbatim interviews to process and preparing the data to be analyzed. After that, the date is reduced, giving a clearer picture. In this process there are stages of coding, categorizing the collected data based on variables or indicators found in the literature. The second step is presenting the data in the form of brief descriptions, graphs and tables, then leading to the conclusion as the last step.

4 GREEN ENERGY VARIABLES AND INDICATORS

The first step of this study was doing a literature review. The literatures of study are taken from Datta et al. (2011), Demirtas (2013), Economist Intelligence Unit (2012), El Ghorab& Shalaby (2015), Evans et al. (2008), Held et al. (2010), Kumar et al. (2010), Liu (2014), Liu et al. (2013), Martin & Felgueiras (2016), Midilli et al, (2007), Nguyen & Ha-Duong (2009), Oncel (2016), and Vera & Langlois (2007). From this literature, this study found 6 variables and 19 indicators, as in Table 1 below.

Table 1. Green energy variables and indicators from the literature review.

Variables	Concept
Economic stability	Economic measure of the use and production patterns of energy, as well as the quality of energy services, affect progress in economic development in a country. (Vera & Langlois, 2007, Held et al., 2010, Liu et al., 2013, Demirtas, 2013, Liu, 2014, El Ghorab1 & Shalaby*, 2015, Martin & Felgueiras, 2016) Indicators: Energy Pricing, Production and energy supply, Taxation and subsidies, Operation and maintenance costs, Private sector development
Social impact and stability	Social impacts are important to identify and quantify acceptance of risks and consequences, and will allow better understanding of some technologies. (Evans et al, 2008, Kumar et al, 2010, Liu et al, 2013, Demirtas 2013, Oncel, 2016) Indicators: Higher living standards, aware of the social and ethics issues, public awareness for green energy
Social dimension	Social dimension measures the impact that energy-available services may have on social well-being. Availability of energy services has implications in terms of poverty, employment opportunities, education, and demographic transition. (Midilli et al., 2007, Vera & Langlois, 2007, Kumar et al., 2010, Demirtas, 2013)

(Continued)

Table 1. (Continued)

Variables	Concept
Environment dimension	Indicators: Equity and social inclusion, risk of poverty, demographic transition, society's environmentally benign support for green energy programs and activities Environmental impacts of energy activities vary greatly depending on how energy is produced and used, structure of the energy system, and related energy regulations and pricing. (Midilli et al., 2007, Vera & Langlois, 2007, Nguyen & Ha-Duong, 2009, Kumar et al., 2010, Datta et al., 2010, Liu et al., 2013, Liu, 2014, Demirtas, 2013, Martin & Felgueiras, 2016).
Technology use	Indicators: Clean environment, lower emissione, energy consumption, clean energy policy Green energy–based technology is expected to play a key role in global stability and sustainable energy for the future. The most important factor that will determine the role of energy and green technology is likely to be energy demand. (Midilli et al., 2007, Oncel, 2016, Evans et al., 2008, Datta et al., 2010, Liu et al., 2013, Demirtas, 2013, Martin & Felgueiras, 2016, Oncel, 2016).
Industrial aspect	Indicators: Clean technology innovation, well adapted with current technology, availability and technology limitation Measuring indicators that have an impact on the industrial sector in the use of renewable energy (Vera & Langlois, 2007, Midilli et al., 2007, Liu et al., 2013, Martin & Felgueiras, 2016) Indicators: Energy use in industry, efficient application

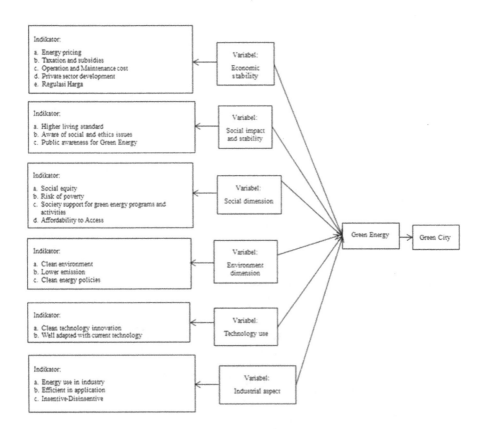

Figure 1. Proposed model for variables and indicators of green energy.

The second step is the interview and focus group discussion (FGD). Based on the results from the interview and FGD process with 16 resources person, who came from four categories as described above, this study confirmed the variables and indicators obtained from literature review and suggests adding new indicators confirmed by the majority of interviewees. In addition to the existing indicators, the following new indicators should be added: regulation price, affordability to access, and incentive-disincentive. Thus, the proposed model that should be used for green energy measurement consists of 6 variables and 20 indicators, as presented in Figure 1.

5 CONCLUSION

Based on the literature and interviews, as well as focus group discussions with 16 resources person, this study concludes as follows:

1. Development of a model for measuring the green energy dimensions of green city of Indonesia, especially Bandung city has been presented in Figure 1
2. The proposed model for measuring green energy in this study consists of 6 variables and 20 indicators.

REFERENCES

Agung, P.S., P., Hartono, D., & Awirya, A.A. 2017. Pengaruh Urbanisasi terhadap Konsumsi Energi dan Emisi CO2: Analisis Provinsi di Indonesia. *Jurnal Ekonomi Kuantitatif Terapan.*, p. 10.
Asian Development Bank. 2015. *Green City Development Tool Kit.* Mandaluyong City: Asian Development Bank.
Badan Pengkajian dan Penerapan Teknologi. 2017. *Indonesia Energy Outlook 2017.* Jakarta: Pusat Teknologi Sumber Daya Energi dan Industri Kimia (PTSEIK).
Badan Pusat Statistik Kota Bandung. 2017. *Kota Bandung Dalam Angka 2017.* Bandung: s.n.
Bappenas. 2013. *Proyeksi Penduduk Indonesia 2010-2035.* Jakarta: BPS - Statistics Indonesia.
Datta, A., Ray, A., Bhattacharya, G., & Saha, H. 2011. Green energy sources (GES) selection based on multi-criteria decision analysis (MCDA). *International Journal of Energy Sector Management.*
Demirtas, O. 2013. Evaluating the Best Renewable Energy Technology. *International Journal of Energy Economics and Policy*, *3*, 12.
Economist Intelligence Unit. 2012. *European Green City Index.* Munich: Siemens AG.
El Ghorab, H.K., & Shalaby, H.A. 2015. Eco and Green cities as new approaches for planning and developing cities in Egypt. *Alexandria Engineering Journal*, p. 9.
Evans, A., Strezov, V., & Evans, T.J. 2008. Assessment of sustainability indicators for renewable energy technologies. *Renewable and Sustainable Energy Reviews*, p. 7.
Held, A., et al., 2010. *RE-Shaping: Shaping an effective and efficient European*, Karlsruhe: Intelligent Energy Europe.
Indrawati. (2015). Metode Penelitian Manajemen dan Bisnis, Konvergensi Teknologi Komunikasi dan Bisnis. Bandung: PT Refika Aditama.
International Energy Agency. 2017. *Energy Efficiency 2017. Special Report: Energy Efficiency In Indonesia.* OECD/IEA.
Kumar, A., et al. 2010. Renewable energy in India: Current status and future potentials. *Renewable and Sustainable Energy Reviews*, p. 9.
Liu, G. 2014. Development of a general sustainability indicator for renewable energy systems: A review. *Renewable and Sustainable Energy Reviews*, 21 January. p. 11.
Liu, G. et al. 2013. General sustainability indicator of renewable energy system based on grey relational analysis. *International Journal Of Energy Research*, 12 March. p. 9.
Martins, F., & Felgueiras, C. 2016. Indicators used in the energy sector. *Journal of Clean Energy Technologies*, *4*, 4.
Midilli, A., Dincer, I., & Rosen, M. 2007. The role and future benefits of green energy. *International Journal of Green Energy*, p. 24.
Miles, M., & Huberman, M. 1984. *Qualitative Data Analysis.* London: Sage Publication.
Nguyen, N., & Ha-Duong, M. 2009, February 11. Economic potential of renewable energy in Vietnam's power sector. *Energy Policy*, 13.

Oncel, S.S. 2016. Green energy engineering: Opening a green way for the future. *Journal of Cleaner Production*, p. 13.

Putra, Y.M. 2014. *Cadangan Energi Fosil Indonesia Diperkirakan Habis 2025*. [Online] http://www.republika.co.id/berita/ekonomi/makro/14/06/03/n6lis0-cadangan-energi-fosil-indonesia-diperkirakan-habis-2025.

Vera, I., & Langlois, L. 2007. Energy indicators for sustainable development. *ENERGY: Third Dubrovnik Conference on Sustainable Development of Energy, Water and Environment Systems*, 32, p. 8.

Indicators for measuring green waste: An Indonesian perspective

Indrawati, F. Andriawan & D. Tricahyono
Faculty of Economics and Business, Telkom University, Bandung, Indonesia

S. Muthaiyah
Multimedia University, Malaysia

ABSTRACT: The influence of urbanization is increased population density in a city, which causes several issues including garbage problems. The waste problem must be resolved immediately by starting waste management with the concept of *Reduce, Reuse, Recycle*, which is commonly applied in a city that uses the green city concept. One of the green city elements is *green waste*, outlining waste reduction targets, and strategies to improve the quality and coverage of cleaning services. The purpose of this research is to find variables and indicators to measure the green waste in Bandung. The method used is an exploratory qualitative. This research proposes a model for measuring green waste with 5 variables and 22 indicators. Further research is recommended to test and measure the green waste index of Bandung by using the model.

1 INTRODUCTION

The level of world urbanization has increased significantly, according to United Nations (UN) estimates. Global urbanization impacts the environment in many ways, one of which is waste management. Based on data from World Bank Open Data (2018), Indonesia holds the third highest urbanization rate in Southeast Asia with 54.46%. Based on data from Bappenas (2013), the level of urbanization in Indonesia will increase and reach 66.6% in 2035, with the highest rate coming from Java (Jakarta, West Java, Yogyakarta, and Banten) by more than 80% and the largest urbanization coming from West Java, the most populated of the provinces, from 2010–2035. A major impact of urbanization is population density in the city: with uncontrolled density as well as inadequate development of infrastructure, waste problems are inevitable. This will affect the environment of the city, including sanitation, drainage systems, and health, as well as tourism. Bandung is a metropolitan city that contributes as much as 1,494 tons of waste per day to be transported to landfill and around 264.09 tons per day of garbage that are not managed in the temporary disposal sites and scattered in the corners of the city or the rivers (Jabarprov 2017).

2 RESEARCH OBJECTIVE AND QUESTIONS

Green city is a concept of environmentally friendly urban development and Bandung is currently implementing a sustainable green city concept. One of the dimensions of a green city is green waste, which regulates waste management, and therefore is an important program for reducing and handling garbage disposal. The basic principle of green waste is to produce zero waste through the reduce, reuse, and recycle (3Rs) concept. Unfortunately, according to the literature review conducted in this research, currently there are no applicable standard-related indicators to measure the application of green waste in the green city in Indonesia, at least none that will fit the characteristics of Indonesia, specifically in Bandung. The research objective of this study is to find the variables and indicators to measure green waste in Bandung, Indonesia.

3 RESEARCH METHODOLOGY

This study is conducted through a qualitative method, a research method that involves the analysis of data in the form of descriptions and where the data is not directly quantifiable so qualitative data are collected by assigning a code or a category (Indrawati 2015).

Based on the nature of the investigation, this study applied a case-study research approach. This research process took a cross-sectional survey that collected data one-by-one in an interview format (Creswell 2014). The resource people in this study are the government, the researcher, the business people, and the citizen. The resource people selection technique was purposive sampling. Purposive sampling is the technique of taking a data source with certain characteristics.

The data collected was analyzed through the steps described by Creswell (2014). The first step was to prepare data for data analysis, which involved the transcription of the interview process, selecting the relevant data or selecting the data into different types, depending on the source of information. The second step was reading the data to establish a general well-rounded sense of the obtained information and reflect on its meaning as a whole. After that, we performed the coding process for categorizing data. This step involved describing the interview transcript. Further, we took the narrative approach in delivering the results of the analysis.

4 GREEN WASTE VARIABLES AND INDICATORS

The first step of this study was a literature review. Based on the literature review related to the concept of green waste, we obtained five variables described in terms of several indicators for each, as in the table below.

Table 1. Green waste variable and indicator literature review.

Variables	Concept
Environment Control	Environmental protection is the primary driver when the focus is on waste treatment and disposal, but environmental protection in collection and in reuse and recycling is considered as well. (Wilson et al., 2013; Wilson et al., 2014; Guerrero, 2013; Wilson et al., 2012; Matete & Trois, 2008). Indicator: Source separation of "dry recyclables," waste captured by the solid waste management and recycling system, degree of controlled of treatment and disposal, waste collection coverage rate, waste recycling rate, quality of recycled organic materials.
Financial Sustainability	Financial sustainability refers to a quantitative range of the data points available related to the solid waste budget and the effectiveness and affordability of cost recovery mechanisms. (Wilson et al., 2013; Wilson et al., 2014; Guerrero, 2013; Wilson et al., 2012; Matete & Trois, 2008; Furqon, 2013; Soezer, 2016; Zurbrügg, 2012). Indicator: Access to capital for investment, local cost recovery – from businesses and institutions, move from a linear economy of waste management to the circular economy, local cost recovery – from households.
Operational Management	Operational systems include material handling and treatment processes by the which the waste generated from different sources is collected, transported, processed, and disposed of regularly. (Ashok,2008; Furqon,2013). Indicator: Degree of clean technology used, suitable infrastructure, develop and keep improving a waste information system, number of disposal sites complying with standards defined operation.
Public Participation and Awareness	Public awareness campaigns for waste management plus the points provided by the respondent on the five-point variables: reduction campaigns in schools and recycling awareness campaigns. (Ashok, 2008; Furqon, 2013; Wilson et al., 2012; Wilson et al., 2013; Wilson et al., 2014; Guerrero, 2013).

(*Continued*)

Table 1. (*Continued*)

Variables	Concept
Institutional and Policy Framework	Indicator: Effectiveness in achieving behavior change, stakeholder participation, integration of community and informal recycling sector, public education & awareness Measure of the institutional strength and coherence of a city's solid-waste management functions, with the individual criteria, including organizational structure, institutional processes to the make the public aware of ISSWM through active participation in the system. (Wilson et al., 2012; Wilson et al., 2013; Wilson et al., 2014; Furqon, 2013; Khan et al., 2016; Memon, 2010) Indicator: Institutional arrangements, adequacy of national framework, endorsed national waste strategies.

The second step was holding the interviews. This study used a sentiment analysis process to interpret the interview script. The sentiment analysis was carried out by three people for each source in order to further label and visualize the opinions of each speaker.

The author set the applied threshold value at above 60% tend to agree or tend to disagree below 2% according to the discussions with the green city team of the research group. Indicators included quality of recycled organic materials, integration of community, and adequacy of national framework. According to the literature, indicators that have a tendency to agree with a value above 60% will be applied as an indicator to measure green waste in Bandung. From the results of the in-depth interview there are four newly obtained indicators at the tendency above 60%. The four new indicators are the reduction in waste disposal life per unit, the cost of recycling operation, the degree of communication and collaboration for the recycling program, and the degree of performance monitoring.

According to the results of the literature review, in-depth interviews, and sentiment analysis, the variables and indicators of green waste were generated, as shown in Figure 1.

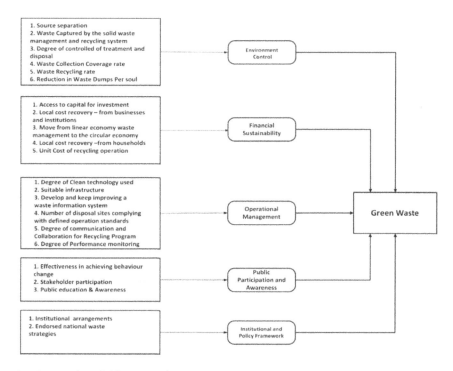

Figure 1. Proposed model for measuring green waste.

5 CONCLUSION

Based on the literature review, in-depth interviews, and sentiment analysis, the variables and indicators of green waste were found, and consisted of 5 variables and 22 indicators as shown in Figure 1.

REFERENCES

Ashok. 2008. Sustainable solid waste management: An integrated approach for Asian countries. *Waste Management*, 29 (2009) 1438–1448.

Bappenas. 2013. 2010–2035, *Proyeksi Penduduk Indonesia*, Jakarta: BPS.

Costas, A. Velis. 2015. Circular economy and global. *Waste Management & Research*, 389–391.

Creswell, J. W. 2014. *Research Design; Qualitative, Quantitative, and Mixed Methods Approaches*, 4th Edition. Los Angles: Sage.

Furqan, M. 2013. Supporting indicators for the successful solid waste management based on community at Rawajati, South Jakarta. *JURNAL WILAYAH DAN LINGKUNGAN*, 3 December, 1, 245–250.

Guerrero, L. A. 2013. Solid waste management challenges for cities in developing countries. *Waste Management*, 220–232.

Indrawati. 2015. Metode Penelitian Manajemen dan Bisnis Konvergensi Teknologi Komunikasi dan Informasi. Bandung: Refika Aditama.

Jabarprov. 2017. Bandung produksi sampah capai 1500 ton per hari. [Online] http://www.jabarprov.go.id/index.php/news/21468/2017/02/21/Bandung-Produksi-Sampah-Capai-1500-Ton-Perhari.

Khan, I. U., Waseer, W. A., Ullah, S., & Khan, S. A. 2016. "Wasteaware" indicators: An assesment of the current solid waste management system in Lahore, Pakistan. 3(3), 85–94. Retrieved from malikinamullahkhan@gmail.com.

Matete & Trois, C. 2008. Towards zero waste in emerging countries. *Journal of Waste Management*, 1480–1492.

Memon, M. A. 2010. Integrated solid waste management based on the 3R approach. *J Mater Cycles Waste Manag*, 30–40.

Soezer, A. 2016. *A Circular Economy Solid Waste management Approch For Urban Area in Kenya*. Kenya: United Nations Development Programme.

United Nations Environment Programme. 2015. *Global Waste Management Outlook*. Austria: International Solid Waste Association.

Wilson, D.C. et al. 2012. Comparative analysis of solid waste management in 20 cities. *Waste Management & Research*, 235–254.

Wilson, D.C. et al. 2013. Benchmark indicators for integrated & sustainable waste management (ISWM). Waste Management, (7 December), 1–16.

Wilson, David C. et al. 2014. "Wasteaware" benchmark indicators for integrated sustainable waste. Waste Management, pp. 1–14.

World Bank Open Data. 2018 [Online] https://data.worldbank.org.

Zurbrügg, C. 2012. Determinants of sustainability in solid waste management; The Gianyar. *Waste Management*, 2126–2133.

Essence and implementation of ERP in culinary industry: Critical success factors

R. Suryani, R.W. Witjaksono & M. Lubis
Telkom University, Bandung, Indonesia

ABSTRACT: Many companies apply information and communication technology to improve their work efficiency, so that it becomes one of the solutions that will later be able to increase the level of competition in the company (Sihotang and Sagala, 2015). Software that is widely used is Enterprise Resource Planning (ERP). Currently adopters of ERP systems that experience a lot of success are middle-class companies engaged in the culinary industry. The business process of a company that is mutually integrated can make it easier for companies to run their business, thus making ERP the main business strategy (Purwidiantoro and Widiati). This study aims to determine the main factors for the success of ERP implementation amid the many failures experienced by the company using diffusion of innovation theory by Rogers with qualitative methods. The object of this research is a company engaged in the culinary industry using observation and interview methods.

1 INTRODUCTION

ERP in Indonesian is a company resource plan, an information system in a company consisting of several modules to integrate and automate business processes related to the aspects of operations, production or distribution directly from a company (Hardjono, 2017). ERP has become an information system that has been widely adopted by companies since the 1990s to improve the company's competitive advantage through the efficiency of business processes (Govindaraju, 2017). ERP implementation requires several modules to be implemented and integrated into business processes, including modules in human resource management, supply chain management, financial and accounting, purchasing, and many more. By implementing ERP, the company can integrate data in every part of the existing business process. Although the ERP system offers benefits for the adopting company, not all companies directly benefit. Many companies are competing to implement ERP and have not been able to achieve the expected efficiency that has caused losses. Some causes of failure of ERP implementation include companies that do not have the right change management policies, vendors that do not understand the company's business, do not have training programs for employees, and many more (Widiyanti, 2013). The object used in this study is a company engaged in the culinary industry including Wendy's, Burger King, and KFC. The three companies are companies that have implemented ERP so that researchers can identify the factors that influence the success of ERP implementation and its essence in the culinary industry. ERP implementation in companies is expected to support operational and marketing activities. The business process of a company that is mutually integrated can make it easier for companies to run their business, thus making ERP the main business strategy (Purwidiantoro and Widiati). This study aims to examine more deeply the main factors for the success of ERP implementation in the culinary industry by using the diffusion theory of innovation invented by Rogers. Rogers said that diffusion is a process in which an innovation is communicated through various channels and a certain period of time in a social system (Rogers and Everett, 1983).

2 LITERATURE REVIEW

2.1 Enterprise Resource Planning

Enterprise Resource Planning (ERP) is an information system in a company that consists of several modules in order to integrate and automate business processes related to the aspects of operations, production or distribution of a company directly (Hardjono, 2017).

2.2 Culinary industry

The word industry according to Law Number 5 of 1984 is economic activities that process raw materials, raw materials, intermediate goods, or finished goods into goods with a higher value for their use, including industrial design and engineering activities. Whereas culinary is an absorption element of English, namely culinary which means it is related to cooking, so the culinary industry can be defined as an economic activity that processes food ingredients to be of higher value.

2.3 ERP in culinary industry

The implementation of ERP in companies engaged in the culinary industry is very useful to meet customer satisfaction with the services provided. By implementing ERP, companies can minimize the time needed to serve customers because ERP is realtime. In addition, the use of ERP software can also monitor employee performance and provide appropriate evaluations for each employee in accordance with work performance.

3 DIFFUSION OF INNOVATION THEORY

Rogers revealed that diffusion is a process in which an innovation is communicated through certain channels from time to time among members of the social system. While innovation is an idea, practice, or object that is considered new by individuals or other adoption units (Rogers and Everett, 1983). So diffusion of innovation is a process in which an idea, practice, or object is communicated through certain channels from time to time among members of the social system. There are five attributes that affect adoption rates, namely:

1. Attributes of innovation,
2. Type of innovation decision
3. Communication channels
4. Nature of Social System
5. Promotion efforts of agent change rates

4 PROBLEM CLASSIFICATION

Every company has problems that arise after implementing the ERP system that can be classified as follows:

1. Wendy's: Internet connection (technical), virus (technical), employee presence (managerial)
2. Burger King: Internet connection (technical), employee presence (managerial), infrastructure (technical), update data by center (managerial)
3. KFC: Internet connection (technical), cash register (technical), understanding of system operator (behavior), employee presence (technical), system upgrade (administration)

5 DEPENDABILITY ANALYSIS

Table 1. Dependability analysis.

Rogers'	Means	Thread	Consideration	Weakness	Opportunity
Relative advantages	Processing data becomes easier, time efficeny	Data is very important, when there is an error when entering data by the employee there will be a problem during the audit later	The success of a system is inline with the benefits and convenience obtained, if the user feels more comfortable using the new system, it can be said that the system is successful	If there is an error when entering data, the employee must contact the authorities to edit the wrong data, this is a bit inconvenient for the employee to do his job	The system must be in accordance with the needs of the company by considering the shortcomings of the system used previously, so that the company can increase profits and simplify the work of each employee
Compatibility	The system used based on previous references, providing a new system usage manual	If the employee cannot use the system correctly errors will often occur when inputting data	System development must be in accordance with the needs of the company by considering the shortcomings of the previous system	The system must continue to be developed that is tailored to every change in the company's operations	Sustainability is of the system with the operations of the company's work can be more efficient, and can increase sales, and the guide book can help employees get to know the basics of using the system
Complexity	Establish system operatos	The designated system operator does not have fulls power over the system, so it requires other parties involved to make changes to the data	Limiting the authority of the system operator can maintain the confidentiality of company data	If en error occurs when the employee inputs data, the data can only be changed by third parties who have the authority to change the data	Giving full authority to employees appointed as system operators can facilitate their work
Trialability	Companies can control each branch by monitoring	Lack of direct interaction between central employees	By monitoring each branch, it can prevent unilateral	Often cause misunderstandings because indirect monitoring is carried out in the field	The system that has been implemented can prevent

(Continued)

Table 1. (Continued)

Rogers'	Means	Thread	Consideration	Weakness	Opportunity
		and branch employees	conversion of data by employees		data manipulation
Observability	After the system implementation employees can go home from work on time because of the ease of processing data	There is a difference in data during the audit if the authorities have not made the wrong data changes entered by the employee	If an error occurs when entering data, it should be immediately replaced by being monitoring by the boss	When the system is having problems or the internet is dead, the data processing is done manually	The system can be upgraded so that processing data can be done without internet
Communication on channel	Introduction of systems by vendors to employees or IT teams	Vendors do not immediately come when there are problems with the system that it can hinder the work of employees	Vendors who are in each area will help employees there are problems with the system	The system can be remotely repaired but it doesn't always work	If the party responsible when the system has an error in each region, there so will not be many obstacles related to store operations

6 CONCLUSION

Based on the research result, it can be concluded that companies in the culinary industry apply an ERP system due to problems that arise before implementing an ERP system that is an error when processing data. In addition the company implements an ERP system so that each department can be integrated with each other and the company can control and monitor the operations of each branch through the implemented ERP system. The success factors of ERP implementation is influenced by several factors, namely the suitability of the system with the company's business processes that can have a positive impact on the company in the form of time efficiency, ease in processing data, work becomes mutually integrated, and control of each branch that can be done from the head office. Another factor that influences the success of ERP implementation is the external factor, namely the existence of training for employees before using the system that is equipped with the manuals that have been provided, in addition to the direct involvement of vendors in case of problems with the system also affects the success of system implementation. The essence of ERP implementation can be measured by how comfortable the user operates the system, of course the comfort is obtained from the ease of the system and the compatibility of the system with the company's business processes so that the company can achieve the targets set by relying on the system.

REFERENCES

Sihotang, H. T. and Sagala, J. R. (2015). *Penerapan Tata Kelola Teknologi Informasi Dan Komunikasi Pada Domain Align, Plan and Organize (APO) Dan Monitor, Evaluate and Assess (MEA) Dengan Menggunakan Framework Cobit 5 Studi Kasus: STMIK Pelita Nusantara Medan*, J. Mantik Penusa Desember, vol. 18, no. 2, pp. 2088–3943.

Hardjono, C. (2017*). Perancangan Dan Implementasi ERP (Enterprise Resource Planning) Modul Sales And Werehouse Management pada CV. Brada, vol. 4, no. 3, pp. 4983–4993.*

Govindaraju, R. (2017). *Analisis Faktor Pendorong Realisasi Manfaat Implementasi ERP Di Perusahaan Indonesia, vol. XII, no. 204, pp. 7–14.*

Widiyanti, S. (2013). *Kesuksesan dan Kegagalan Implementasi Enterprise Resource Planning (ERP) Pada Perusahaan.*

Purwidiantoro, M. H. and Widiati, I. S. *Penerapan Enterprise Architecture Planning Untuk Meningkatkan Strategi Sistem Informasi Pada Perusahaan Makanan, pp. 232–238.*

Rogers, Everett M. (1983). *Diffusion of Innovations.* Collier Macmillan Publishers: New York.

Interpersonal skills in project management

A. Said, H. Prabowo, M. Hamsal & B. Simatupang
Bina Nusantara University, Jakarta, Indonesia

ABSTRACT: Expeditious change is an accelerating problem for projects in most industries. In fact, a project leader becomes the key person to project success. The aim of this paper is to explore how the interpersonal skills of a project manager influence project performance. This quantitative study was conducted to test the relationship between communications and project performance. This paper also examined the effect of leadership transformation on project performance, by presenting three different characteristics of project success (time, budget, and customer satisfaction). An online questionnaire was used as a survey tool to obtain data from project managers/PMO/PD and team leaders of various projects in Indonesia. The dataset was subsequently analyzed using SmartPLS-SEM. This paper proposes a conceptual integrated research model of interpersonal skills of a project manager to promote transformational leadership which can positively impact on project success.

1 BACKGROUND

Having new ideas and describing a strategy are very important for an organization (Landscape et al., 2018). According to De Smet, Lurie, and St George (2018), with more complex growth in the world at this time, many emerging things become uncertain. There are two main concepts when discussing the project, namely, project success and project management success (Radujković & Sjekavica, 2017).

The project manager has the greatest power to bring project success by his/her own style and manner of leadership (Gruden & Stare, 2018). For the project manager, interpersonal skills are very necessary. Strohmeier (1992) suggests four interpersonal skill problems, namely, motivation, conflict, communication, and team work. Thus, to avoid project failure, the project manager must have good interpersonal skills in carrying out activities and getting people involved during the project.

2 LITERATURE REVIEW

2.1 Project management method

Project management is defined not only as a theory, but also as a guideline to the basic standards in executing a project (Kostalova et al., 2015). Though guidelines have been standardized by some organizations, the overall goal is to focus organizational resources on the completion of certain projects (D. C. Smith, 2014).

With a growing number of organizations adopting the project management approach, the demand for project managers will grow because of the need for them and for standards for the development and assessment of project management competencies (Crawford, 2005).

2.2 Project success

Although project failure in application development is very low, it continues to be a problem for organizations (Engelbrecht et al., 2017). According to Aga, Noorderhaven, and Vallejo (2016), traditionally, project management has been associated with construction and engineering, where the criteria for project success are objective, well received, and measurable. Joslin and Müller (2016) argue that historical understanding related to project success has developed from the concept of simple triple constraints, into something that includes many additional criteria of success such as quality, stakeholder satisfaction, and knowledge management.

According to Joslin and Müller (2015), in addition to the triple constraints, there are five additional dimensions of project success criteria, namely, project efficiency, organizational benefits, project impacts, stakeholder satisfaction, and future potential. In contrast,to Berssaneti and Carvalho (2015), there have been many criticisms in recent years regarding the success of a project, not only from time, cost, and quality constraints. The success of the project relies mainly on the expertise of the project manager to get good outputs or outcomes (Sanchez et al., 2017).

2.3 Communication skills

Effective communication is a broader skill and involves a lot of knowledge that is not unique to the project context (Edum-Fotwe & McCaffer, 2000). Implementation or form of this communication function can include preparing project direction, attending meetings, general project management, marketing and sales, public relations, record management such as minutes, memos/letters/newsletters, reports, specifications, and contract (Siregar et al,, 2018). Good communication will lead to customer satisfaction while poor communication will result in customer dissatisfaction (Huijgens et al., 2017).

2.4 Transformational leadership

Van Kelle et al. (2015) divided leadership style in project management into the following:

1. Transformational leadership. This style refers to an adaptive leadership style that has such characteristics as motivating, inspiring, expressing a vision, and emotional involvement of followers, while focusing on long-term commitment and involvement.
2. Transactional leadership. This style refers to social environments where expectations and rewards are clearly stated, and the existence of short-term focus.

Furthermore, Aga et al. (2016) concluded that, in the previous studies, transformational leadership had general agreement that there were four dimensions shaping transformational leadership: ideal influence, intellectual stimulation, inspiration motivation, and individual consideration.

3 RESEARCH PROBLEM

For someone in the position of project manager, technical knowledge is a lesser priority, while interpersonal skills are really essential and even become critical (Strohmeier, 1992).

4 METHODOLOGY AND RESEARCH MODEL

This study uses 3 latent variables with 50 indicators and employs a quantitative approach using key stakeholder analysis units which include project directors, PMOs, and project managers who have experience as project leaders.

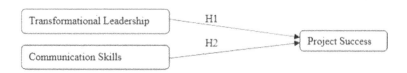

Figure 1. Research model.

5 RESULTS AND DISCUSSION

Of the 80 questionnaires distributed, 17 respondents did not fill out the questionnaires and did not qualify as research units: 53 respondents filled in the questionnaires completely and qualify for analysis in this study.

In the results of smartPLS in looking at the loading factors of each indicator against variable latency, loading factors of less than 0.700 were excluded from the model while those greater than 0.700 remained included in the model.

The results of testing the relationship between variables based on the existing model, T-test for transformational leadership was 2.245 > 1.96, so it can be assumed that transformational leadership has a positive and significant relationship with project success, while for communication skills, T-Test was revealed 1.474 <1.96, so it is concluded that communication skill is not significant for project success.

This study found a significant relationship between transformational leadership and project success, as nowadays many team members tend to work independently and are allowed to execute a project in their own. Hence, at this time, there has been a shift from transactional to transformational leadership.

There are two possible limitations of this study. First, this study does not have industry specifications to be examined, and second, this study does not specify which method approach is to be selected.

REFERENCES

Aga, D.A., Noorderhaven, N., & Vallejo, B. 2016. Transformational leadership and project success: The mediating role of team-building. *International Journal of Project Management, 34*(5), 806–818.

Berssaneti, F.T., & Carvalho, M.M. 2015. Identification of variables that impact project success in Brazilian companies. *International Journal of Project Management, 33*(3), 638–649.

Crawford, L. 2005. Senior management perceptions of project management competence. *International Journal of Project Management, 23*(1), 7–16.

De Smet, A., Lurie, M., & St George, A. 2018. *Leading agile transformation: The new capabilities leaders need to build 21st-century organizations,* (October), 27.

Edum-Fotwe, F., & McCaffer, R. 2000. Developing project management competency: Perspectives from the construction industry. *International Journal of Project Management, 18*(2), 111–124.

Engelbrecht, J., Johnston, K.A., & Hooper, V. 2017. The influence of business managers' IT competence on IT project success. *International Journal of Project Management, 35*(6), 994–1005.

Gruden, N., & Stare, A. 2018. The influence of behavioral competencies on project performance. *Project Management Journal, 49*(3), 98–109.

Huijgens, H., van Deursen, A., & van Solingen, R. 2017. The effects of perceived value and stakeholder satisfaction on software project impact. *Information and Software Technology, 89*, 19–36.

Joslin, R., & Müller, R. 2015. Relationships between a project management methodology and project suc- cess in different project governance contexts. *International Journal of Project Management, 33*(6).

Joslin, R., & Müller, R. 2016. The relationship between project governance and project success. *International Journal of Project Management, 34*(4), 613–626.

Kostalova, J., Tetrevova, L., & Svedik, J. 2015. Support of project management methods by project man- agement information system. *Procedia: Social and Behavioral Sciences, 210*, 96–104.

Landscape, V.D., Cost, H., & Performance, L. 2018. *PULSE OF THE PROFESSION ® | 2018: "Success in Disruptive Times | Expanding the Value Delivery Landscape to Address the High Cost of Low Performance,"* 35.

Radujković, M., & Sjekavica, M. 2017. Project management success factors. *Procedia Engineering, 196* (June), 607–615.

Ribeiro, A., & Domingues, L. 2018. Acceptance of an agile methodology in the public sector. *Procedia Computer Science, 138*, 621–629.

Siregar, M., Ichsan, M., Riantini, L.S., & Indonesia, U. 2018. Project Management Office (PMO) Practices in Moderating the Project Communication: An Empirical Study in Oil and Gas Industry in Indonesia, *119*(15), 2997–3004.

Smith, D.C., Bruyns, M.B., & Evans, S. 2014. A project manager's optimism and stress management and IT project success. *International Journal of Managing Projects in Business, 4*(1), 10–27.

Stare, A. 2014. Agile project management in product development projects. *Procedia - Social and Behavioral Sciences, 119*, 295–304.

Strohmeier, S. 1992. Development of interpersonal skills for senior project managers. *International Journal of Project Management, 10*(1), 45–48.

Van Kelle, E., Visser, J., Plaat, A., & Van Der Wijst, P. 2015. An empirical study into social success factors for agile software development. *Proceedings: 8th International Workshop on Cooperative and Human Aspects of Software Engineering*, CHASE 2015, 77–80.

Employees' post-adoption behavior towards collaboration technology

K. Rapiz & Arviansyah
Faculty of Economics & Business, Universitas Indonesia, Depok, Indonesia

ABSTRACT: To encourage broad-feature use of a collaboration technology at work, the drivers of employee adoption behavior once the system is successfully implemented needs to be investigated. We integrate the unified theory of acceptance and use of technology (UTAUT) with job crafting theory and examine the role of job crafting behavior moderating employees' acceptance. The model tested using PLS on survey data collected from 115 employees working in an Indonesian airline company. The analysis confirmed positive influence of performance expectancy on intention; and intention alongside facilitating conditions toward use behavior. Having significance but negative influence, our finding did not confirm proposed moderation effect of job crafting on social influence as the predictor of employee intention.

1 INTRODUCTION

Post-adoption of technology refers to interaction with a system the individual is acquainted with (Tams et al., 2018). Two salient behaviors with a familiar system have been noted: deep structure usage and trying to innovate with IT (Tams et al., 2018).

Collaboration technology is the technologies that facilitate communications, social interactions, and resource sharing via electronic means (Brown et al., 2010). Firms invest in such technologies to leverage the intellectual resources of their employees (Maruping & Magni, 2015). Although the potential benefits of collaboration tools have been documented in existing literature, those can be realized only if their intended users actually take advantage of them and incorporate the full functionality of system features into their work routines (Tan & Kim, 2015).

As one of the most established and mature areas of IS research, technology adoption research has yielded many competing models. Venkatesh et al. (2003) had put forth a consolidated model of UTAUT after examining commonalities across eight prominent models. UTAUT posits that performance expectancy, effort expectancy, and social influence were directly influencing behavioral intention; and that behavioral intention alongside facilitating conditions are directly influencing actual use behavior. In addition, those relationships of influencecan be moderated by gender, age, experience, and voluntariness of use.

We review the theory of behavior of employees at work, by which individuals initiate changes in the level of job demands and job resources to make their job more meaningful, engaging, and satisfying (Demerouti, 2014); this has also been previously identified by the term "job crafting" (Wrzesniewski & Dutton, 2001). Therefore, we opt to answer the following research questions. First, what salient factors influence employees as current users of implemented collaboration technology at work to learn, use, and extend the full range of the system's built-in features? Second, how would job crafting be linked to the post-adoptive use of technology by employees at their work routines?

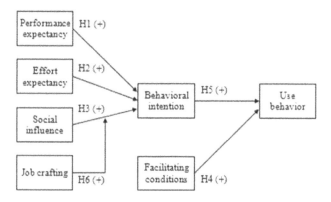

Figure 1. The proposed research model.

2 RESEARCH MODEL & HYPOTHESES

This research is grounded on the UTAUT (Venkatesh et al., 2003) to explain the post-adoptive use of collaboration technology by employees.

Our proposed model includes all relationships on UTAUT main constructs: performance expectancy (PE), effort expectancy (EE), social influence (SI), facilitating conditions (FC), behavioral intention (BI), and use behavior (UB). We omit the original moderators of UTAUT; instead, we introduce a new moderator linked with job crafting (JC) as summarized in Figure 1.

3 METHODS

We use convenience sampling to collect data from employees of an airline headquartered at Tangerang, Indonesia. They use Office 365 software products provided by the company as one tool to perform collaborative work tasks.

All participants were invited to an online survey through a messaging application. A total of 187 visitors to the survey page was recorded. Eventually, only 115 (61.5%) respondents completed the survey questionnaire with valid answers.

Measurement items are adapted from the literature: PE, EE, SI, FC, and BI are adopted from Venkatesh et al. (2003); UB from Brown et al. (2010); and job crafting scales (JC) on dimension of increasing social job resources from Tims et al. (2012). Items of PE, EE, SI, FC, BI, and JC are measured using seven-point Likert scales, ranging from (1) strongly disagree to (7) strongly agree. For UB, three items, measuring intensity, frequency, and preference, were also assessed using seven-point Likert scales while one item, measuring duration, used groups of nominal scales.

4 RESULTS

We employ partial least square (PLS) since it is appropriate for analyzing an exploratory model with small to medium sample sizes and for examining moderation effects (Chin & Newstedm 1999, Henseler & Chin 2010, Hair et al. 2011). SmartPLS v3.2.8 (Ringle et al. 2015) is the software used for analysis.

4.1 Measurement model

All constructs were modeled as reflective measurement and assessed for composite reliability, indicator reliability, convergence validity (AVE), and discriminant validity. The final results confirmed the model's accepted composite reliability with all constructs exhibiting values between 0.7 and 0.9 (Hair et al., 2011).

All measurement items meet indicator reliability with loading factor above 0.4 as the minimum acceptable value. Item JC3 was then dropped to improve the construct convergence to pass the minimum 0.50 AVE value (Hair et al., 2011).The Henseler et al. (2015) heterotrait–monotrait ratio confirmed discriminant validity with all correlations of two constructs resulting below 0.90 (Ringle et al., 2015).

Table 1. Summary of measurement model analysis.

Variable	Items	Loading	Cronbach's alpha	Composite reliability	AVE	Discriminant validity
BI	BI1	0.958	0.955	0.955	0.877	Yes
	BI2	0.930				
	BI3	0.921				
EE	EE1	0.780	0.910	0.910	0.718	Yes
	EE2	0.959				
	EE3	0.797				
	EE4	0.841				
FC	FC1	0.760	0.823	0.824	0.545	Yes
	FC2	0.802				
	FC3	0.551				
	FC4	0.809				
PE	PE1	0.868	0.872	0.879	0.650	Yes
	PE2	0.879				
	PE3	0.862				
	PE4	0.576				
SI	SI1	0.789	0.834	0.834	0.557	Yes
	SI2	0.683				
	SI3	0.778				
	SI4	0.730				
JC	JC1	0.606	0.812	0.816	0.529	Yes
	JC2	0.725				
	JC4	0.843				
	JC5	0.717				
UB	UB1	0.924	0.864	0.871	0.640	Yes
	UB2	0.959				
	UB3	0.722				
	UB4	0.517				

4.2 Structural model and hypotheses testing

Analysis was carried out separately on the main constructs and moderation (Martins et al., 2014). We also follow Fassott et al. (2016) in moderation analysis using unstandardized estimates. The summary for statistical significance on path coefficients are shown in Figure 2.

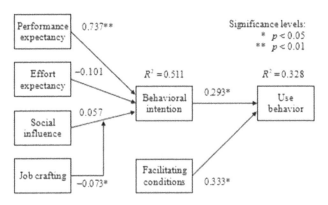

Figure 2. Estimation results on the structural model.

5 DISCUSSION & CONCLUSION

Performance expectancy was found significant in explaining intention, as well as intention alongside facilitating conditions in predicting use behavior. Other direct effects were insignificant. Moderation effect of job crafting was found significant but with an opposite sign, thus, not supporting our hypothesis. Overall, the results only supported hypotheses H1, H4, and H5.

Our analysis reveals several noteworthy findings. First, it suggests that our respondents had concerns with the outcome of using technology in supporting their jobs, as theorized by Venkatesh et al. (2003). Likewise, in line with prior studies, neither associated effort nor influence from others became concerns to our respondents in the post-adoption stage (Brown et al., 2010, Baptista & Oliveira, 2015).

The role of social influence in driving intention to use the technology tends to diminish as employees craft their job. We conclude that this is an unexpected impact of individuals job crafting, by which Tims et al. (2012) has suggested that they were aimed at improving their person–job fit and work motivation.

Our results are somewhat specific to the context and should be interpreted as explaining the post-adoptive intention. Moreover, the use of employees in an Indonesian airline company on Office 365 software products as a technology for collaborative job tasks may limit generalizability.

REFERENCES

Baptista, G., & Oliveira, T. 2015. Understanding mobile banking: The unified theory of acceptance and use of technology combined with cultural moderators. *Computers in Human Behavior*, 50, 418–430.

Brown, S.A., Dennis, A. R., & Venkatesh, V. 2010. Predicting collaboration technology use: Integrating technology adoption and collaboration research. *Journal of Management Information Systems*, 27(2), 9–54.

Chin, W.W., & Newsted, P. R. 1999. Structural equation modeling analysis with small samples using partial least squares. *Statistical Strategies for Small Sample Research*, 1(1), 307–341.

Demerouti, E. 2014. Design your own job through job crafting. *European Psychologist*, 19(4), 237–247.

Fassott, G., Henseler, J., & Coelho, P. S. 2016. Testing moderating effects in PLS path models with composite variables. *Industrial Management & Data Systems*, 116(9), 1887–1900.

Hair, J.F., Ringle, C.M., & Sarstedt, M. 2011. PLS-SEM: Indeed a silver bullet. *Journal of Marketing Theory and Practice*, 19(2), 139–152.

Henseler, J., & Chin, W.W. 2010. A comparison of approaches for the analysis of interaction effects between latent variables using partial least squares path modeling. *Structural Equation Modeling: A Multidisciplinary Journal*, 17(1), 82–109.

Henseler, J., Ringle, C.M., & Sarstedt, M. 2015. A new criterion for assessing discriminant validity in variance-based structural equation modeling. *Journal of the Academy of Marketing Science*, 43(1), 115–135.

Martins, C., Oliveira, T., & Popovič, A. 2014. Understanding the Internet banking adoption: A unified theory of acceptance and use of technology and perceived risk application. *International Journal of Information Management, 34*(1), 1–13.

Maruping, L.M., & Magni, M. 2015. Motivating employees to explore collaboration technology in team contexts. *MIS Quarterly, 39*(1), 1–16.

Ringle, C.M., Wende, S., & Becker, J.M. 2015. SmartPLS 3. www.smartpls.com.

Tams, S., Thatcher, J.B., & Craig, K. 2018. How and why trust matters in post-adoptive usage: The mediating roles of internal and external self-efficacy. *The Journal of Strategic Information Systems, 27*(2), 170–190.

Tan, X., & Kim, Y. 2015. User acceptance of SaaS-based collaboration tools: A case of Google Docs. *Journal of Enterprise Information Management, 28*(3), 423–442.

Tims, M., Bakker, A.B., & Derks, D. 2012. Development and validation of the job crafting scale. *Journal of Vocational Behavior, 80*(1), 173–186.

Venkatesh, V., Morris, M.G., Davis, G.B., & Davis, F.D. 2003. User acceptance of information technology: Toward a unified view. *MIS Quarterly, 27*(3), 425–478.

Wrzesniewski, A., & Dutton, J.E. 2001. Crafting a job: Revisioning employees as active crafters of their work. *The Academy of Management Review, 26*(2), 179–201.

Designing RecyclerApp: A mobile application to manage recyclable waste

A. Jalil, A.A.A. Hussin & R. Sham
UCSI University, Kuala Lumpur, Malaysia

ABSTRACT: Digital social innovation is an emerging field that uses digital tools for social change. This innovation could be developed to help society to dispose and collect recyclable waste as well. From our survey on waste disposal in Malaysia, however, we have found that there is no application available for the society to properly dispose and collect the household recyclable waste. This study aims to design a solution in order to provide them with a mobile application that could be used to track content levels of recycling bins in their locations. The application is developed based on rapid application development methodology. At the end of the study, we propose and discuss a model that could be applied by other societies to achieve the same goal. We conclude that an affordable digital social solution could be implemented by leveraging current open-source tools, low-cost hardware, crowdsourcing and cloud technologies.

1 INTRODUCTION

1.1 Recyclable waste

There are three different types of recyclable wastes generated in Malaysia: paper, plastic, and bottle. However, Malaysian's waste recycling rate is way below the average level due to several factors, including the inconvenience of recycling facilities and services (Teo, 2016). In Kuala Lumpur for example, the current recycling rate is at 5 percent of the waste generated. Although the government has taken many recycling campaigns and initiatives to improve the recyclable waste management problems, still more needed to be done for a sustainable solution (Jereme et al., 2015).

In order to encourage a higher recycling rate, Malaysian society should play a role in the process of disposal and collection of recyclable waste. They should know where to dispose of recyclable waste, as well as how to collect the waste to generate side income. Hence, there is need of a digital social innovation to cater to this kind of problem.

2 LITERATURE REVIEW

2.1 Digital social innovation

Digital social innovation is still an emerging field, with little knowledge of who the innovators of the solution are and how they use digital technology and tools to have a postive influence on society. The growth of this field is supported by knowledge co-creation platforms, social networking, open source technology, wireless sensor networks, and open data infrastructure.

Currently, there are many research projects that implement digital social innovation to address some issues in education, health (Mason et al., 2015), and the environment. However, there is no innovation that could be a sustainable solution to disposal and collection of household recyclable waste in Malaysia (Jereme et al., 2015, Teo, 2016). Therefore, this article presents a research project that aims to design RecyclerApp, a digital solution for society to encourage them to dispose of and collect their recyclable waste properly.

2.2 Mobile application

Nowadays, almost everybody owns a smartphone, a mobile computing device that is highly functional and available at a variety of price points. Each smartphone has pre-installed applications and there are other applications that users download after purchase based on their needs. Users use the mobile applications for many reasons, including for learning (Jalil et al., 2015, Jalil et al., 2016), entertainment, business, social communication, and map navigation (Islam et al., 2010).

With the current mobile technology, most of the smartphones have WLAN/WiFi that allow users to connect to the Internet or local network with high data transmission rates. This network technology is assumed to provide better connectivity for mobile users in future. Therefore, a mobile device that can host customized application is potentially a good digital tool for tracking content levels of recycling bins.

3 METHODOLOGY

3.1 Rapid application development (RAD)

This research project adopted rapid application development methodology, which breaks down the process into four phases: requirement planning, user design, rapid construction, and cutover. The methodology is widely adopted in current information system development projects (Beynon-Davies et al., 1999). The task conducted in each phase is described as follows:

A. Requirement planning

At this phase, the current situation was researched by reading journals and website articles to gather information related to the project, including reviewing hardware and software minimum specifications. Next, a discussion with team members was held in order to determine the proper locations for the bins to be placed in UCSI University.

B. User design

At this phase, the model of the system was produced (Satyawati et al., 2017). The recycling bin design was sketched by a team member to include a unit to protect the Arduino sensor from rain and heat. For the mobile application, an outline of the system design was developed via the drag-and-drop features in Android Studio. Before system development began, UML diagrams were drawn out to serve as a guide for efficient development. Later, the system design was finalized so that approval construction could be obtained from all team members.

C. Rapid construction

In this phase, each of the recycling bins was equipped with an Arduino sensor to detect either the bin is empty or full. At the same time, the mobile application was developed and then tested on the Android Studio emulator and Android smartphones. The initial version of the application was then refined several times and tested by team members before finalizing a complete version of it (Caliwag et al., 2018).

D. Cutover

At this implementation phase, RecylerApp.apk files were distributed to the team members to be installed on their smartphones. The team members were also trained on how to use the application. Meanwhile, all the three recycling bins were placed on their respective locations and all the sensors were switched on as well.

Figure 1. The program is written in Arduino IDE.   Figure 2. RecyclerApp screen interface.

3.2 *RecyclerApp prototype experiment*

RAD methodology was chosen because this research project included development of a mobile application prototype that needed to be tested in an experiment (Ahmad et al., 2014). In the experiment, three recycling bins were placed in three difference locations in the university. Each recycling bin was equipped with an Arduino sensor that could detect that the bin was either empty or full. As shown in Figure 1, an Arduino program was written to send the status data to a real time database in cloud, which was Google's Firebase. Then, the data were read and displayed in the RecyclerApp prototype in real time when users clicked on each marker to check the status of the bin. As shown in Figure 2, the prototype displayed the status data from one of the three markers indicating the locations of the three recycling bins respectively. During the experiment, the data from the sensor were monitored through Arduino IDE in order to make sure that it is sent to the cloud.

4 RESULTS AND DISCUSSION

When the experiment was being conducted, the results were recorded from each sensor. Table 1 shows the results of the experiment.

As stated in the Table 1, both data from the first and second recycling bins could be sent by sensors to the database in cloud. The data could also be fetched by RecyclerApp and displayed on its screen. However, the third recycling bin failed as its sensor could not send the status data to the cloud to be read by RecyclerApp due to the weak network signal. Figure 3 shows a snapshot of one of the results which were monitored for each bin.

From this project, we propose a model of social innovation by adopting RecyclerApp to provide a better way of monitoring recyclable waste disposal and collection. Figure 4 illustrates the model, which contains the necessary components for implementation. Other societies could follow this model if they need to adopt a mobile application for the same issue.

Table 1. The success of sending and displaying data for each bin.

Bin	Status	Data Sent	Data Read
Bin 1	Empty	Yes	Yes
Bin 2	Full	Yes	Yes
Bin 3	Empty	No	N/A

Figure 3. FULL status was sent to cloud database.   Figure 4. Proposed model.

5 CONCLUSION

A mobile application is an affordable digital social innovation that could be implemented as a solution to the issue of recyclable waste disposal and collection in Malaysia. The solution demonstrates how to leverage current open-source tools, low-cost hardware, crowdsourcing, and cloud technologies in an effort to solve a social need, as suggested in previous research (Taylor et al., 2011).

From the experiment, we have successfully presented the process of sending data from sensors to a mobile application. We also have learned that the success of this implementation depends on the availability of reliable network coverage in order to send to and read data from the cloud (Desouza & Smith, 2014). Therefore, good network coverage is an important factor for the success of social innovation. For future works, we suggest that a better design of recycling bin should be provided so that the sensor could be protected from temperature, theft, and vandalism. The sensor also must be smart to distinguish the waste so that only the appropriate waste type is allowed to be put in the bin.

REFERENCES

Ahmad, W.F.W., Muddin, H.N.B.I., & Shafie, A. 2014. Number skills mobile application for down syndrome children. In *2014 International Conference on Computer and Information Sciences (ICCOINS)*. IEEE, pp. 1–6.
Beynon-Davies et al. 1999. Rapid application development (RAD): An empirical review. *European Journals of Information Systems*, 8(3), 211–223.
Caliwag, J.A. et al. 2018. A mobile expert system utilizing fuzzy logic for venereal and sexually transmitted diseases. *Journal of Advances in Information Technology*, 6(3), 57–61.
Desouza, K. C., & Smith, K. L. 2014. Big data for social innovation. *Standford Social Innovation Review* 9, 39–43.
Islam, M.R. et al. 2010. Mobile application and its global impact. *International Journal of Engineering & Technology*, 10(6), 72–78.
Jalil, A., Beer, M., & Crowther, P. 2015. Mobile learning in a seminar or workshop: A case study for evaluating MOBIlearn2 basic components and their application. *Bulletin of the IEEE Technical Committee on Learning Technology*, 17(4), 30.
Jalil, A., Beer, M., & Crowther, P. 2016. Improving design and functionalities of MOBIlearn2 application: A case study of mobile learning in metalwork collection of Millennium Gallery. In *IEEE Eighth International Conference on Technology for Education (T4E)*. IEEE, pp.19–25.
Jereme, A.I. et al. 2015. Assessing problems and prospects of solid waste management in Malaysia. *Journal of Social Sciences and Humanities*, 10(2), 70–87.
Mason, C. et al. 2015. Social innovation for the promotion of health equity. *Health Promotion International*, 30(2), 116–125.
Satyawati, E., Lyna, & Cahjono, M.P. 2017. Development of accounting information system with Rapid Application Development (RAD) method for micro, small and medium scale enterprises. *Review of Integrative Business and Economics Research*, 6(1), 166–175.
Taylor, D.G., Voelker, T.A., & Pentina, I. 2011. Mobile application adoption by young adults: A social network perspective. *International Journal of Mobile Marketing*, 6(2), 60–70.
Teo, C.B.C 2016. Recycling behavior of Malaysian urban households and upcycling prospects. *Journal of International Business, Economics and Entrepreneurship*, 1(1), 9–15.

Stock market prediction using multivariate neural network backpropagation

Tendra Kristian & Farida Titik Kristanti
Telkom University, Bandung, Indonesia

ABSTRACT: Since stock market prices are unpredictable, there are no consistent patterns in the data to create any near-perfect model of stock prices. Technical indicators play important roles in building a strategy, hence this research employed technical indicators, namely Accumulation Distribution Oscillator (ADOSC), Commodity Channel Index (CCI), Larry William R% (WILLR), Momentum (MOM), Rate of Change (ROC), Relative Strength Index (RSI), Simple Moving Average (SMA), Moving Average (MA), Weighted Moving Average (WMA), as the variables input in Artificial Neural Network (ANN) Backpropagation with Multivariate Regression. The models were evaluated using three statistical performance evaluation criteria. The daily data on Indonesia Stock Exchange were chosen, especially those in 10 years of trading days, to predict daily closing price. Experimental results showed that the ANN Backpropagation with Multivariate Regression obtained a promising performance in the closing price prediction on the training and validation data compared with other models.

1 INTRODUCTION

IDX High Dividend 20 (IDXHIDIV20) is one of the new indices issued by the Indonesia Stock Exchange (IDX), which can be used as a reference for investors. IDXHIDIV20 was chosen as the object of this research due to its suitability for novice investors. If the company always distributes dividends, the share price can always be active, and investors gain dividends every year. If shares with large dividends are demanded, then the choice of shares is on the IDXHIDIV20. Based on the phenomena, IDXHIDIV20 using ANN Backpropagation is required to help investor to obtain appropriate information to determine the process of purchasing shares, and to maximize returns and minimize risk. It also can be used to support investment decisions.

2 LITERATURE REVIEW

The stock market price is formed through the mechanism of demand and supply in the capital market (Sartono 2001: 70). According to Tiphimmala (2014), there are four groups of techniques to predict stock price movements, namely fundamental analysis, technical analysis, time series analysis, and machine learning. Widiatmojo (2004) suggests that short-term stock price movements cannot be ascertained precisely. Determining stock prices can be done through technical and fundamental analysis. Jogiyanto (2008: 167) states that fundamental analysis is to calculate the intrinsic value of shares using company financial data.

A number of research have examined stock price predictions using ANN, such as Atsalakis & Valavanis 2009, Cao & Parry 2009, Chang et al. 2009, Chavarnakul & Enke 2008, Enke & Thawornwong 2005, Hassan et al. 2007, Kim 2006; Tsang et al. 2007, Vellido et al. 1999, Yudong & Lenan 2009, Zhang et al. 1998, Zhu et al. 2008. ANN is a data-driven and non-parametric model. It does not require strong modeling because it can map non-linear functions without prior assumptions and is proven effective in making predictions (Cao et al.

2005, Dai et al. 2012, Enke & Thawornwong, 2005). ANN Backpropagation is the most popular neural network training algorithm used for financial forecasting (Atsalakis & Valavanis 2009, Cao & Parry 2009, Chang et al. 2009, Le & Chen 2002, Lee & Ciu 2002, McNelis 2004, Vellido et al. 1999, Yudong & Lenan 2009, Zhang et al. 1998, Lu Chi-Jie 2010). Performance Evaluation of ANN Backpropagation employs the MSE method (Mingyue 2014).

3 DATA, VARIABLES AND METHODOLOGY

3.1 Data sample

This research used stock market data from the IDXHIDIV20 in the period of August 2009 to January 2018. The samples were selected using a purposive sampling technique based on the criteria of big five capitalization market. The data were taken from the IDX Research and Development Division report statistic, second quarter of 2018.

3.2 Variables

There were nine technical indicators, such as ADOSC, CCI, WILLR, MOM, ROC, RSI, SMA, MA, and WMA applied as the input variable used in this research. The indicators were built using Technical Analysis Library for Python from open, high, low, close, and volume.

3.3 Artificial neural network backpropagation architecture

Stock Market Prediction using ANN Backpropagation with Multivariate Regression in this research employed the following architecture (Figure 1).

Based on architecture in the Figure 1, the data were split into Training Data 70%, and Test and Validation Data 30%, scaling, and normalization were applied to improve performance and make better conditioned data for convergence. Pearson Correlation was used to choose variable input. This correlation performed a direct correlation measurement and was a standard correlation measurement because it gave an accurate correlation between two variables (Rika et al. 2010: 99). Neural Network was built using Tensor Flow, Keras, and Python as the Open Source Machine Learning Platform.

3.4 Model building

This research applied four model selections, each models used five sub models. Therefore, there were total numbers of 20 trial simulations. The complexity and speed of the network in

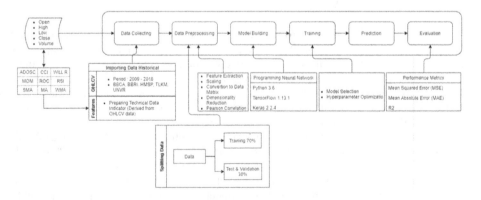

Figure 1. Artificial neural network backpropagation architecture.

the computational process, both the training process and prediction process were conducted to determine the number of hidden layer networks. For those with insufficient hidden layers, it was possible that the network did not have sufficient capabilities to make a fit model. However, if there were too many hidden layers, the model was likely to be overfit and the computation process would run slowly (Nanayakkara et al. 2014).

4 FINDINGS AND DISCUSSIONS

4.1 *Hyperparameter optimization*

The selection of network structures on the appropriate ANN Backpropagation was important and difficult. Selection of inputs which were consistent with the results of the Pearson Correlation determined the accuracy of a model. Choosing the appropriate Hyperparameter Optimization could determine a better accuracy of a model.

4.2 *Result simulation models*

This research employed calculating the level of accuracy method, namely MSE, MAE, and R2. The experiment was conducted using twenty prediction models for one study sample research company, TELKOM. Based on Table 1, the third model with four hidden layers, with batch normalization and RMSProp Optimizer generated the better result with 61.75% accuracy. It was the best model for stock price predictions in companies included in IDXHI-DIV20 for the period of August 2009 to January 2018.

Table 1. Result simulation models.

No	Experiment	Epoch	Batch Normalization	Optimizer	TLKM Accuracy	MSE	MAE	R2
1*	1	100	No	SGD=0.01	52.19%	0.2523	0.4957	-0.011
	2	200	Yes	SGD=0.001	51.37%	0.2556	0.4945	-0.0589
	3	300	Yes	SGD=0.0001	57.10%	0.245	0.4804	-0.0072
	4	100	Yes	Adam	56.83%	0.2484	0.4892	-0.0081
	5	100	Yes	RMSprop	54.37%	0.2502	0.4868	-0.0053
2*	1	100	No	SGD=0.01	57.65%	0.2433	0.4902	0.0019
	2	200	Yes	SGD=0.001	50.55%	0.2617	0.496	-0.0581
	3	300	Yes	SGD=0.0001	46.99%	0.2854	0.5075	-0.1582
	4	100	Yes	Adam	56.83%	0.2449	0.49	0.0107
	5	100	Yes	RMSprop	51.37%	0.2525	0.4964	-0.0236
3*	1	100	No	SGD=0.01	55.46%	0.2461	0.4922	-0.0099
	2	200	Yes	SGD=0.001	54.64%	0.2566	0.482	-0.0569
	3	300	Yes	SGD=0.0001	51.91%	0.2618	0.4918	-0.0519
	4	100	Yes	Adam	54.92%	0.2654	0.4765	-0.072
	5	100	Yes	RMSprop	61.75%	0.2328	0.4693	0.0334
4*	1	100	No	SGD=0.01	59.29%	0.2438	0.4919	-0.0102
	2	200	Yes	SGD=0.001	57.92%	0.2439	0.4902	-0.0009
	3	300	Yes	SGD=0.0001	59.84%	0.2429	0.4906	-0.0107
	4	100	Yes	Adam	57.10%	0.2455	0.4933	-0.0023
	5	100	Yes	RMSprop	59.29%	0.2438	0.4919	-0.0102

* Model:
1: 32-32-1
2: 256-128-64-32-1
3: 1000-1000-1000-1000-1
4: 1000-1000-1000-1000-1

5 CONCLUSIONS AND RECOMMENDATIONS

The research results revealed that hyperparameter optimization assisted to improve performance model accuracy. Third model with four hidden layer, with batch normalization and RMSProp Optimizer contributed the better result with 61.75% accuracy. It is necessary for further research to analyze the stock price prediction model related to the neural network model using other types of models, such as Recurrence Neural Network (RNN) and Long Short-Term Memory (LSTM), so that it could improve the performance and accuracy of stock price predictions.

REFERENCES

Atsalakis, G. S., & Valavalanis, K. P. 2009. "Surveying stock market forecasting techniques -Part II: Soft computing methods". Expert System with Applications, Vol 36 (3), page 5932–5941.
Cao, R. & Parry, M. E. 2009. Neural network earnings per share forecasting models: A comparison of backward propagation and the genetic algorithm. Decision Support Systems. 47. 32–41. 10.1016/j.dss.2008.12.011.
Chang, P. C., Liu, C. H., Lin, J. L., Fan, C. Y., & Ng, C. S. P. 2009. A neural network with a case based dynamic window for stock trading prediction. Expert Systems with Applications, 36(3), 6889–6898.
Chavarnakul, T., & Enke, D. 2008. Intelligent technical analysis-based equivolume charting for stock trading using neural networks. Expert Systems with Applications, 34(2), 1004–1017.
Enke, D., & Thawornwong, S. 2005. The use of data mining and neural networks for forecasting stock market returns. Expert Systems with Applications, 29(4), 927–940.
Kim, K. J. 2006. Artificial neural networks with evolutionary instance selection for financial forecasting. Expert Systems with Applications, 30(3), 519–526.
Hassan, M. R., Nath, B., & Kirley, M. 2007. A fusion model of HMM, ANN and GA for stock market forecasting. Expert Systems with Applications, 33(1), 171–180.
Jogiyanto, H. 2008. Teori Portofolio dan Analisis Investasi. Yogyakarta: BPFE.
Lee, T. S., & Chiu, C. C. 2002. Neural network forecasting of an opening cash price index. International Journal of Systems Science, 33, 229–237.
Lu, C. 2010. Integrating independent component analysis-based denoising scheme with neural network for stock price prediction. Expert Systems with Applications 37 (2010) 7056–7064.
McNelis, P. D. 2004. Neural networks in finance: Gaining predictive edge in the market. New York: Academic Press.
Nanayakkara, S., Chandrasekara, V., & Jayasundara, D. D. M. 2014. Forecasting Exchange Rates using Time Series and Neural Network Approaches. European International Journal of Science and Technology, Vol 3 No 2. page 6–7.
Rika, S., Maharani, W., & Kurniati. A. P. 2010. Analisis Perbandingan Metode Pearson dan Spearman Correlation Pada Recommender System. Konferensi Nasional Sistem dan Informatika, Bali.
Sartono, A. 2001. Manajemen Keuangan Teori dan Aplikasi. Yogyakarta: BPEF.
Tiphimmala, S. 2014. Forecasting Stock Price Index Using Artificial Neural Networks in the Indonesian Stock Exchange. Thesis, UAJY.
Tsang, P. M., Kwok, P., Choy, S. O., Kwan, R., Ng, S. C., Mak, J., Tsang, J., Koong, K., & Wong, T. 2007. Design and implementation of NN5 for Hong Kong stock price forecasting. Engineering Applications of Artificial Intelligence, 20(4), 453–461.
Vellido, A., Lisboa, P. J. G., & Vaughan, J. 1999. Neural networks in business: A survey of applications (1992–1998). Expert Systems with Applications, 17, 51–70.
Widiatmodjo. 2004. Jurus Jitu Go Public: Bagaimana Meningkatkan Kekayaan Pemegang Saham Dan Perusahaan Tanpa Kehilangan Kontrol. Jakarta: PT Elex Media Komputindo. ISBN-10: 979-20-5813-3 (pbk).
Yudong, Z., & Lenan, W. 2009. Stock market prediction of S&P 500 via combination of improved BCO approach and BP neural network. Expert Systems with Applications, 36(5), 8849–8854.
Zhang, G., Patuwo, B. E., & Hu, M. Y. 1998. Forecasting with artificial neural networks: The state of the art. International Journal of Forecasting, 14, 35–62.
Zhu, X., Wang, H., Xu, L., & Li, H. 200). Predicting stock index increments by neural networks: The role of trading volume under different horizons. Expert Systems with Applications, 34(4), 3043–3054.

IDR/USD forecasting: Classical time series and artificial neural network method

M.F.Q. Alam & B. Rikumahu
School of Economic and Business, Telkom University, Bandung, Indonesia

ABSTRACT: This research focuses on the performance comparison of classical time-series and the artificial neural network method to forecast rupiah and US dollar exchange rates based on 2008–2017 history data. General characteristic of the data is fluctuating and has heteroskedasticity pattern as typical of financial time-series data. Several ARIMA (p,q) model is verified and ARIMA-GARCH(0,0) gave the best result for a classical time-series analysis. Meanwhile, after testing several network combinations, then backpropagation ANN with two hidden layers, cost of training (0.0001) and epoch (100) were selected for the neural network model. The result from mean square error performance measurement shows if BP-ANN has a better performance compared to classical ARIMA-GARCH in short, medium, and long projections. Although both models demonstrate performance decreasing along with the projection time duration, the ANN has several advanced methods (i.e., LSTM model) to improve long-duration projection performance.

1 INTRODUCTION

While forecasting by fundamental (financial) assessment has, in many cases, failed to make a decent projection for short-term periods (Shamah, 2008), fund managers are starting to see a technical method to help them to create a good forecasting model. The technical process of forecasting currently achieved by classical time-series and artificial neural network (ANN), is now gaining momentum from the rapid improvement of modern technology such as artificial intelligence and big data.

The exchange rate is time-series data that typically has a fluctuating pattern. Historical data of Indonesian rupiah to US dollar exchange rates from 2008 to 2017, shows the rupiah experienced depreciation and had several fluctuation periods. To make typical history data useful as training data for forecasting will require special attention. Hence, one model might be more appropriate compared to another as a forecasting model depends on its training data pattern.

Furthermore, based on financial reports of the Indonesian government from 2008 to 2017, deviation occurred between the government's macro assumption and the realization value of Indonesia rupiah to US dollar exchange rate, and the difference is as much as 6.22%, on average. This fact confirmed that exchange rate forecasting is a challenging task to perform, and this research can contribute by proposing the appropriate and fit forecasting method of the financial sector, especially in the exchange rate.

2 LITERATURE REVIEW

Classic time-series modeling has two integrated models: autoregressive (AR) and moving average (MA), written as ARIMA (p, d, q) (Box & Jenkins, 1970). To anticipate heteroskedasticity symptom in the ARIMA model, the generalized autoregressive conditional heteroskedasticity – GARCH is proposed (Bollerslev, 1986). Otherwise, the artificial neural network (ANN) proposes a neuron network connectivity as projection model that can be implemented to non-linear function of time-series data (Palit & Popovic, 2005).

Research has been done on exchange rate forecasting methods using time-series and ANN. Korol (2014) used JPY/USD, GBP/USD, and CHF/USD exchange rates as a dataset to measure the performance of ANN and GARCH with the result that mean absolute error (MAE) of ANN is higher than GARCH. Regarding ANN parameter determination, Panda and Narasimhan's (2007) research regarding ANN method and GARCH method comparison toward rupee India/USD exchange rates, produced results if ANN projection had better performance compared to GARCH model. In term of periods of projection, Zhang (2001) research toward ARIMA, ANN, and hybrid (ARIMA+ANN) comparison study to forecast pound sterling and USD time-series data concluded the hybrid model was more accurate compared to ARIMA and for a long-term period; the hybrid model has higher accuracy if compared to both ANN and ARIMA models.

3 METHODOLOGY

3.1 Classical time-series analysis

The procedure of classical time-series analysis is defined as below:

1. Find a general characteristic of the sample (training) data.
2. Perform a stationarity test for training data and differencing data (data return if lag-1).
3. Perform an autocorrelation test for training data and differencing data (data return if lag-1)
4. ARIMA model (Box & Jenkins, 1970) selection and testing.

$$wt = \varphi 1wt-1 + \varphi 2wt-2 + \cdots + \varphi pwt-p + at - \theta 1at-1 - \theta 2at-2 - \cdots - \theta qat-q \qquad (1)$$

5. Do the heteroskedasticity test for selected ARIMA model.
6. If step 5 results in heteroskedasticity, then perform a GARCH model (Engle & Mc Fadden, 1994) selection and testing.

$$\sigma_t^2 = \omega + \sum \alpha_i \varepsilon_{t-j}^2 + \sum_{i=0}^{q} \beta_j \sigma_{t-1}^2 \qquad (2)$$

3.2 Artificial neural network model

There are several steps to get the best model of ANN:

1. Defining a network layer using backpropagation neural network.
2. Data normalization, so it can be matched with the activation function value.
3. Weighting and bias allocation, initially by random value, then adjusted during training.
4. Choosing the activation function.
5. Choosing the optimizer method.
6. Modifying learning rate, cost of training, and number of training iteratively to find the best forecasting performance using MSE.

4 FINDINGS AND DISCUSSIONS

4.1 Classical time-series analysis

The data training presented in Table 1 indicates some general characteristics as follows:

1. From the range value of minimum and maximum, it can be assumed there was high volatility exchange rate of IDR/USD during that period of time.
2. From the skewness value of +0.52, it can be assumed that if data spreading is not symmetrical and with the value > 0, there will be some unit data with values higher than average of overall unit data.
3. From the kurtosis value of –1.24, it can be assumed the distribution value of data training has a fat-tail (heteroskedastic) characteristic.

In time series analysis, it is a must that a time series follows a certain rule. The most important rule that a time series must follow is stationarity which means that the statistical properties of a process generating a time series do not change over time. It does not mean that the series does not change over time, just that the *way* it changes does not itself change over time. The formal test for stationarity of a time series is Augmented Dickey Fuller (ADF) test. If according to the test that according the ADF test a time series is not stationary, then it has to undergo a differencing process. Since we used the exchange rate (price of a currency in terms of the currency of another country), the differencing process itself means that we take the difference between the current price (y(t)) and the price one period earlier (y(t-1)) and then test the result (Δy) for stationarity using the ADF test one more time. The process of differencing and testing is repeated until the test indicate that the time series has become stationary. In the case of this research, we only do one differencing since, the time series has become stationary after only one differencing (called first differencing). In finance, there is a measurement that is similar with first differencing that is called return (percentage change of prices). So, because our data became stationary after first differencing, then the return data will also be stationary and so we can use the return data in our analysis.

Furthermore, to prove the assumption of positive autocorrelation, it is necessary to test the data by the statistical test of autocorrelation called Durbin-Watson test. Because d(0.000042) < dL(1.925480), then it can be verified if the data is positively autocorrelated. Furthermore, when the Durbin-Watson test is performed on the data return, then the result is d(2,064366) > dU(1.927470). Another rule for the analysis of time series is that for the time series to be considered good for analysis, the time series should not have autocorrelation (means that the price of one particular day does not have any correlation with other prices in the time series). This is important since if the time series does not have autocorrelation, it can help in uncovering hidden patterns in data, helps in selecting the correct forecasting methods (help identify seasonality and trend in the time series data). Additionally, analyzing the autocorrelation function (ACF) - and partial autocorrelation function (PACF) in conjunction - is necessary for selecting the appropriate ARIMA model for the time series prediction.

The data return (one lag) that is proposed as a model will be tested by a correlogram test to find MA order by auto-correlation function (ACF) and AR order by partial auto-correlation function so that (p,d,q) of the model can be decided. From the correlogram test, the ARIMA (1,1,1) became the model of forecasting, thus the forecasting model of data return for IDR/USD exchange rate proposed based on (1) is as (3):

$$return(t) = (0.235247 * return(t-1)) + 1.383801 + \varepsilon(t) - (0.27363 * \varepsilon(t-1)) \quad (3)$$

The next step process is a heteroskedasticity test on return data and it has a positive result by the Park test of heteroskedasticity: alpha (0.05) > p-value (0.000128), thus GARCH should be proposed to get a better forecasting model. From previous correlogram tests, it is definitely finding the anticipated AR and MA for GARCH(p,q) is AR=1 and MA=1, so the expected GARCH model is GARCH (1,1). Thus, GARCH (1,1) forecasting model that fits to (2) is:

$$\sigma_t^2 = 0.0030378 + 0.1885 * \varepsilon_{t-1}^2 + 0.8115 * \sigma_{t-1}^2 \quad (4)$$

Hence, time-series forecasting model of data return for IDR/USD exchange rate is:

$$return_t = (235247 * return_{t-1}) + (1.383801 - (0.27363 * \varepsilon_{t-1}))$$
$$\pm \sqrt{(0.0030378 + 0.1885 * \varepsilon_{t-1}^2 + 0.8115 * \sigma_{t-1}^2)} \quad (5)$$

4.2 Backpropagation neural network model

The overall data (2,448 unit data) is separated into training data (2,207 unit data) and testing data (241 unit data). By those proportions of each subset, we expect to get the best model and projection result where training data is more than unit data (Kuhn & Johnson, 2013). Furthermore, testing data will be segregated into three types of testing data: short-term (21 unit data), medium-term (122 unit data) and long term (241 unit data). The model of backpropagation NN chosen in this research is as follow:

- Number of hidden layers: 2 hidden layers
- Data normalization scheme: minimum-maximum scheme
- Activation function: rectified linear unit (RelU)
- Optimizer method: gradient descent optimizer
- Cost of training: 0.0001
- Number of iterations (Epoch): 100

4.3 Performance comparison

The result of performance measurement of classical time-series and backpropagation neural network on this research is as in Table 1. Backpropagation artificial neural network can perform non-linear projection based on a model trained on the history data, while ARIMA (1,1,1)-GARCH model forecasted more on the linear pattern. Model comparison also indicates if BP-ANN performance beats time-series for all time periods: 99.59% on the short-term, 99.53% on the medium-term, and 99.44% on the long-term projection. Considering those results, BP-ANN will be recommended as a model to predict the IDR/USD exchange rate rather than the time-series model.

5 CONCLUSIONS AND RECOMMENDATIONS

During 2008–2017, the exchange rate of rupiah to USD showed different levels of fluctuation: this characteristic shows heteroskedasticity symptom, thus the ARIMA-GARCH model is used as classical time-series analysis. Furthermore, the ANN proposed is backpropagation ANN with two hidden layers, and after several tests, cost of training (0.0001) and epoch (100)

Table 1. Performance of time-series model and BP-ANN.

Model	Short Term		Medium Term		Long Term	
	MSE	Accuracy	MSE	Accuracy	MSE	Accuracy
IMA(1,1)	12,267	99.21%	52,325	98.39%	84,238	97.94%
ARI(1,1)	12,255	99.21%	51,983	98.39%	83,336	97.95%
ARIMA(1,1,1)	11,919	99.22%	49,958	98.43%	78,903	98.01%
ARIMA(1,1,2)	12,680	99.19%	54,750	98.35%	89,582	97.88%
ARIMA (2,1,1)	12,865	99.19%	41,230	98.55%	54,183	98.35%
ARIMA(1,1,1), GARCH	11,528	99.23%	49,172	98.44%	77,904	98.02%
BP-ANN	5,188	99.59%	5,541	99.53%	8,257	99.44%

were selected as a model. From the results of this research, the backpropagation neural network can beat classical time-series performance, thus the artificial neural network is considered to be a great tool to forecast the exchange rate.

For future research, performance measurement tools to analyze data projection accuracy and performance stability should be added, while the duration of the data sample was also considered properly chosen to capture the behavior of the pattern. One suggestion regarding the neural network model is to use different models such as recurrence neural network and long short-term memory to accommodate long-term projection (Hochreiter & Schmidhuber, 1997).

REFERENCES

Bollerslev, T., Engle, R.F., & Nelson, D.B. 1994. ARCH models. *Handbook of Econometrics, 4*, 2959–3038. https://doi.org/10.1016/S1573-4412(05)80018-2.

Hochreiter, S., & Schmidhuber, J. 1997. Long short-term memory. *Neural Computation*, 9(8), 1735–1780. http://dx.doi.org/10.1162/neco.1997.9.8.1735.

Korol, T. 2014. A fuzzy logic model for forecasting exchange rates. *Journal Elsevier Knowledge-Based Systems 67*(2014) 49–60.

Kuhn, M., & Johnson, K. 2013. *Applied Predictive Modeling*. Springer Science+Business Media, LLC. 12.

Palit, A. K., & Popovic, D. 2005. Computational intelligence in time series forecasting theory and engineering applications. Springer Science+BusinessMedia, LLC. 127.

Panda, C., & Narasimhan, V. 2007. Forecasting exchange rate better with an artificial neural network. *Journal of Policy Modeling, 29*.

Zhang, G. P. 2001. Time series forecasting using a hybrid ARIMA and neural network model. *Journal Elsevier Neurocomputing 50*(2003) 159–175.

Development of Stock Opname application with integration to SAP Business One using Scrum

Leonardo, M. Lubis & W. Puspitasari
Telkom University, Bandung, Indonesia

ABSTRACT: SAP Business One is one of the most-used enterprise resource planning systems for small and medium enterprises because of cost, compared to other SAP packages. But SAP B1 does not handle some feature like Stock Opname. Stock Opname is a process to check whether stock in the system is the same in reality. Usually companies do Stock Opname once or more per year, depending on their needs. Therefore, an add-on to manage Stock Opname is needed. This add-on will be developed in website and Android mobile applications. The development uses Scrum as a framework to manage time and quality. To support mobile application, there is barcode scanner and a solution disconnected scenario feature to make application still usable without internet. This system will allow workers to maximize their productivity while doing Stock Opname and also integrate with SAP B1 so the company will have an integrated system.

1 INTRODUCTION

In this era of modern technology, a lot of organizations are racing to have system that can help their business. ERP (Enterprise Resource Planning) is a business management system that consists of a series of comprehensive software modules designed to integrate and manage all business functions in an organization. SAP Business One is one of many ERP solutions used for small and medium-sized enterprises (SMEs). But SAP B1 is not the best choice when organizations have specific business processes like Stock Opname. There are a few solutions for this problem, but the best option is to build an add-on application for SAP B1. This add-on is an Android application that can handle Stock Opname process. Android is chosen as the platform to carry this application because it can handle all the feature needed, such as a camera for barcode scanner and is much cheaper than other platforms like iOS. There is also a feature to make this application still runnable without internet connection called solution disconnected scenario. This feature will save data to a temporary database when there is no internet connection, and when connection is established, the user can upload the data to the main database. Stock Opname Android application will be developed using Scrum as a framework. Scrum can accelerate the delivery process for this application and is adaptive to changes that exist during the development process. With this application as an add-on to SAP Business One, organizations will still have an integrated system that can handle Stock Opname processes.

2 LITERATURE REVIEW

2.1 *SAP Business One*

Enterprise resource planning (ERP) is an information system used by companies whose role is to connect and optimize business processes in factories, logistics, distribution, accounting, finance, and human resources. ERP is a superset of SAP Business One. Unlike many other small business solutions on the market today, SAP B1 is a single application, which eliminates the need for separate installations and complex integration of several moduleS (Wan & Hou,

2012). SAP B1 application integrates all core business functions across all companies including financials, sales, inventory, and operations that ERP solutions focus on.

2.2 Stock Opname

Stock Opname is a calculation and adjustment of the stock of goods and assets owned by the store or company in a warehouse or storefront with stock data contained in the company's system database (Wibisono et al., 2013). Stock Opname on goods is done to check the difference between the amount of goods in the warehouse and the amount recorded in the database, while the Stock Opname on assets is done to control assets owned by the company, such as furniture, computers, equipment, and company equipment. During the Stock Opname process, entry and exit of goods cannot be carried out. Stock Opname activities in the company can take a long time depending on the amount of goods and assets owned by the company, so that the company can only do Stock Opname when it is closed or not in crowded conditions (non-peak season).

2.3 Android

Google released Android, which is an open-source mobile operating system with a Linux-based platform. Android consists of operating systems, middleware, user interfaces, and application software (Holla, 2012). Some of the advantages of Android are open source, support by a lot of hardware manufacturers, high scalability, and a wide range of prices from low to high-end smartphone.

2.4 Scrum

Scrum is a framework where people can overcome complex problems adaptively, productively, and creatively to provide high-quality products (Adi, 2015). There are few steps to develop an application using this framework as pictured by Figure 1 below.

3 APPLICATION DESIGN

3.1 Use cases diagram

Use cases diagram is created to identify the important features in this application.

3.2 Class Diagram

Class diagram is created to represent a model to this application such as application structure, class, attribute, operation, and relationship between objects.

4 APPLICATION RESULT

The researchers used Scrum as the framework to develop this application. During development, the researchers referred to Scrum guidelines such as product backlog, sprint backlog,

Figure 1. Scrum process diagram.

sprint, and sprint review. Using Scrum, the development took about three weeks, for three sprints. The first sprint is for the application base user interface and second and third sprints are for all the features in this application. Here are some screenshots of the application:

Figure 2. Stock Opname Android application result.

5 APPLICATION TESTING

Application testing was done to make sure the product and all the features have been completed. During testing, it is common to find some bug or crash within the application. Scenario testing was used to test this application. There are several paths from scanning an item to saving data. Figure 3 below shows one of the testing scenarios when staff want to scan item.

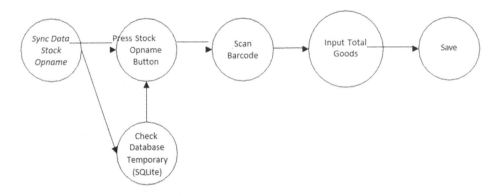

Figure 3. Scenario testing path.

Table 1. Scenario testing result.

Scenario Name	Sync Stock Opname	Database SQLite	Stock Opname	Scan Barcode	Input Total Item	Save
Successfully save using Sync Stock Opname	Yes	-	Yes	Yes	Yes	Yes
Successfully save using database SQLite	No	Yes	Yes	Yes	Yes	Yes
Failed to sync and no data in temporary database	No	No	-	-	-	-
Successfully sync but no assignment for staff	Yes	-	No	-	-	-
Incorrectly scan Barcode	Yes	-	Yes	No	-	-
Incorrectly input or not all total item entered	Yes	-	Yes	Yes	No	-

6 CONCLUSION

With development of Stock Opname Android application as an add-on to SAP Business One, organizations will have an integrated system to handle Stock Opname processes with SAP Business One as their main ERP system. This application will help organizations to get real-time data from Stock Opname processes to inform top-level management about the current amount of inventory. Based on scenario testing, the application can scan items using barcode, use the app without internet connection, and save scan result to the SAP Business One database.

REFERENCES

Holla, S. 2012. Android-based mobile application development and its security. *Continuum*, *26*(5), 741–752. doi: 10.1080/10304312.2012.706462.

Wan, J., & Hou, J. 2012. Research on SAP Business One implementation risk factors with interpretive structural model. *Journal of Software Engineering and Applications*, *553022*(March), 147–155. doi: 10.4236/jsea.2012.53022.

Wibisono, M. C., Noertjahyana, A., & Handojo, A. 2013. Pembuatan Aplikasi Pencatatan Stock Dengan Menggunakan Barcode Pada Android. *Jurnal Infra*, *1*(2), 1–4.

How to reduce the risk of fintech application through cooperative organizations?

Sugiyanto
Master of Management, Institut Manajemen Koperasi Indonesia (Ikopin), Jatinangor, Indonesia

ABSTRACT: The impact of information technology innovation in the field of financial services has become a trend in Indonesia. One of the financial technology business models is P2PL. As a relatively new financial service business, it is, indeed, not free from problems, such as legality issues and business risks that are predicted to increase as results of various violations. The purpose of this research is to descriptively study the P2PL business phenomenon, which is supported by references and regulations on fintech and cooperative business, so that various risks and ways to reduce them can be identified. The results of this research indicate that if P2PL business is conducted by a cooperative, then by changing the model of conventional financial service business into using information technology, the business entity becomes legal; service users are the members of cooperatives who can function as lenders and other members as borrowers, and the cooperative is owned by the members.

1 BACKGROUND

Financial technology (fintech) is one of the fastest growing financial service businesses in Indonesia. The Financial Stability Board (FSB) classifies fintech businesses into: (1) peer-to-peer lending (P2PL) (KoinWorks), (2) market aggregator (DuitPintar.Com), (3) investment and risk management, and (4) payment, clearing, and settlement (e-wallet, GO-PAY). The fintech applications are mostly illegal and were closed down in the last two years; however, 113 fintech applications are legal. In addition to legal issues, other problems encountered by financial businesses are community illiteracy; reaping benefits from loans; data hacking; and business risks dealing with the default, fraud, misuse of client data, and high interest rates. The risk of fintech business as described above is predicted to increase in intensity.

Financial Service Authority Regulation No. 77/POJK.01/2016 regarding Information Technology-Based Money Lending and Borrowing Services explains that fintech can legally be managed through a legal entity such as limited liability companies or cooperatives. Most fintech applications are formally in the form of limited liability companies and not many of them are in the form of cooperatives. In fact, in the form of cooperative legal entity, the problems and risks of fintech applications can be minimized. Loan providers and borrowers are members as users of the cooperative service, and they are also owners (dual identity of member principle) (Dulfer, 1994).

2 LITERATURE REVIEW

The growth of fintech in Indonesia is supported by the government and banks, but for conventional financial service players such as savings and loan cooperatives. Fintech, as a company, not only uses information technology as a differentiator but is also directed to provide more efficient services, especially related to the process and utilization of the nonbank financial services market. An economic industry composed of companies that use technology to make financial systems is more efficient (McAuley, 2015).

Fintech is a service sector that uses mobile-centered information technology to enhance the efficiency of the financial system (Kim et al., 2016). According to Bank Indonesia (2017) and Kominfo (2017), fintech is a fusion phenomenon that occurs between technology and financial features that change business models and reduce barriers to utilizing financial services. Wilson (2017) states that fintech is a company which primarily uses technology to generate income through the provision of financial services.

Fintech in the P2PL platform is two sides of financial service market players that are not different from the conventional banking system with special challenges (Klafft, 2008). Lenders and borrowers are the main target group of all these financial services. So some research focuses on stakeholders and the determinants of the success of the evangelism process (Freedman & Jin, 2008; Iyer et al., 2009). Its business activities begin, on the one hand, with loan providers looking for opportunities to invest their funds to obtain the highest possible profit at a certain level of risk and, on the other hand, with borrowers looking for a source of liquidity with a different risk of non-payment. Lenders or borrowers sometimes engage in groups and form communities to strengthen their interests (Greiner & Wang, 2009). The challenge is to overcome the principal–agent problem (Jensen & Meckling, 1976), Eisenhardt, K. M (1989) states that agency theory uses the assumption of three human traits: humans are generally selfish, humans have limited thinking power about future perceptions (limited rationality) and humans always avoid risk (risk averse). Agency problems occur initially with the existence of asymmetric information which is a basic problem in P2PL online.

The problems of legality and various business risks can be reduced by utilizing a cooperative organization that has a special identity as an organization characteristic. Cooperative organizations should be understood as business entities that must be distinguished from other forms of companies (ICA, 2001). Dulfer, 1994; Hanel, 1995; and ICA, 2001, state that as an organization, a cooperative is characterized by an economic and social institution. A cooperative has two households, namely, a cooperative company and members' household: members function not only as users but also as owners (double identity of a member). In cooperatives that apply P2PL, both providers and borrowers of funds are members who function as users of cooperative services. As an owner, the member capitalizes and finances the operations of the cooperative, and controls and makes decisions on all cooperative policies.

3 METHODS, DATA, AND ANALYSIS

The method used in this study was descriptive qualitative, data were collected from the results of reviews on various regulations and literature on P2PL applications and cooperatives, supplemented by surveys on P2PL businessmen and cooperatives. Data was processed and analyzed with a qualitative descriptive approach.

4 RESULT AND DISCUSSION

According to the results of literature reviews and surveys on P2PL business and cooperatives, various information can be obtained:

(1) Anatomy of P2PL business, classified as illegal and legal. In the past two years, the Financial Services Authority has closed down as many as 947 illegal P2PL units. The owners were either Indonesians or foreigners. Foreign owners usually appoint Indonesians as managers accompanied by credit analysts and debt collectors. Loan terms and conditions are determined by the lender. Legal P2PL businesses consist of lenders, borrowers, and P2PL companies. The lender is the owner of capital and lends to other parties who need funds (borrowers). P2PL serves to bring together lenders and borrowers, with fees up to 3% of loans. P2PL does not manage loan money, only regulates payment traffic. The lender distributes the loan through a temporary account and receives it back through this account.

(2) Loan conditions and loan terms set by illegal P2PL include interest of up to 60% per month, passport, personal photo, ID card, and a period of disbursement of funds of 15–30 minutes. Legal

P2PL loan terms include interest on loans up to 24% per month, passport, personal photo, ID card, and terms of disbursement of funds of 15–30 minutes. Bank loan terms are 1% commercial interest per month, borrower's ID card, family card, marriage certificate, bank account, photocopy of proof of salary, 3 to 7–day disbursement period; and cooperative loan terms are that the borrowers need only to apply and the interest is around 1.5 % per month and time for disbursement between 3 and 30 days.

(3) P2PL business problems and risks and legal issues include many illegal P2PL organizers in Indonesia, indicating that this business is widely used by irresponsible parties. There is a problem of borrowing funds and P2PL is often used by various parties to simplify and accelerate the loan process. Owners capital use this business to seek maximum profits, and those who need funds often ignore the heavy requirements set by the capital owner. Data hacking, the P2PL application system, is often misused to access personal data from the borrower's cellular phone, from the contact list, photo gallery, short message history, to family history. The misuse of data includes the telephone numbers of colleagues. It also includes the family of customers who can be used in consumer intimidation. The offer of loans that are easy and quick encourage prospective borrowers to apply for loans, which in turn traps the borrower who have difficulties in paying back the debt. Customers have problems to get funds to repay loans, interest, and late fees. This condition usually continues with intimidation, humiliation of customers in front of friends and colleagues, to sexual harassment. Risk of failure to pay refers to the default risk, which is very high especially for illegal P2PL, with high penalties and penalties that will trap customers. Fraud Risk will frequently occur because the parties do not know each other, the agreements have the tendency to be detrimental to the customer and there is no guarantee. The misuse of client data is carried out by P2PL organizers to obtain other benefits. Customer personal data can be monetized when offered to other investors, as revenue niches for the illegal fintech business. High interest rates and penalties burden customers.

(4) P2PL applications in cooperatives have lenders and borrowers as members of cooperatives; in this case members are users. As an owner of the cooperative who applies P2PL, or contributes capital, controllers and decision-makers work through member meetings to apply cooperative business policies. The members who have excess funds look for other members who need funds to finance their business through cooperatives.

As users, the function of members is as lenders to other members who need funds. On the other hand, members also function as owners of cooperative P2PLs, and they must contribute capital as well as control and make decisions through member meetings.

(5) The risk of P2PL can be reduced through cooperative organizations by consistently upholding the cooperative self-identity in the form of values and principles, and as members participating as users as well as owners. Cooperatives are governed by democratic principles: for example, "one member, one vote"; the direction of decision-making is principally "bottom up"; and governance is ideally by members who make up the general assembly and elect the board and supervisory group of directors. The members exonerate these boards and decide on the use of surpluses accrued, and they alone can change the constitution of their joint venture.

The cooperative formula is particularly suited as a vehicle for achieving diverse community objectives. It may correct market failures and enable the efficient organization of markets by enabling those who are in a weaker market position to combine their purchasing and selling power, uniting small enterprising activities into bigger marketable and more efficient units whilst allowing them to retain their autonomy. The cooperative model gives market power to lay people or small enterprises where homogenous services or products are needed, by enabling those who have little capital to influence economic decision-making, enabling citizens to affect or determine the services they need, integrating large sections of the population to economic activity, and generating trust and creating and maintaining social capital due to democratic governance and economic participation.

A cooperative organization is a legal entity. Through member education as one of the cooperative principles, cooperatives can improve member literacy regarding information technology-based financial services. Both lenders and borrowers are members of the cooperative: they make joint decisions related to requirements, interest rates, and disbursement of loans democratically through member meetings, so that they do not harm each other. Hacking data can be minimized, because data is stored by cooperatives under the control of members; there will be no intimidation to borrowers because the members value togetherness and helping one another. The borrower is also the owner of a cooperative. The borrower has no difficulty repaying the loan because of low interest and no penalties; relatively low interest, in accordance with the agreement between members, is expected to reduce the risk of default. Lenders and borrowers are members of cooperatives, so it is expected that they will get to know each other thus reducing the risk of fraud. All members must implement the values of honesty and responsibility (cooperative values). Misuse of client data does not occur because P2PL operators are cooperatives owned and controlled by members both as lenders and as borrowers. Interest rates and penalties can be determined by members, either as lenders and or borrowers. In addition, members must participate as users and owners. In cooperative organizations, there are cooperative companies and household members.

5 CONCLUSIONS AND SUGGESTIONS

This study indicates that P2PL as a fintech business has numerous problems and risks, even a tendency to harm the wider community. This condition can be minimized by applying P2PL through cooperative organizations that have different identities and characteristics from other corporate legal entities. The application of identity will consistently reduce the various risks and problems of the P2PL application.

The suggestions that can be proposed include the need for education and dissemination of information to the community on an ongoing basis; announcing a list of illegal P2PL perpetrators through the OJK website; requesting the blocking of illegal P2PL websites; reporting to authorities; P2PL business direction through cooperative legal entities; and cooperatives that already have loan and saving services applying the fintech business with the P2PL pattern.

REFERENCES

Dulfer, E.1994. Managerial economics of cooperatives. *International Handbook of Cooperative Organization* p. 587–592.
Eisenhardt, K. M. 1989. Agency theory: An assessment and review. *Academy of Management Review*, 14(1), 57–74.
Freedman, S., & Jin, G. Z. v. 2008. Dynamic larning and selection: The early years of prosper. *working paper*. College Park, MD. http://www.prosper.com/downloads/research/Dynamic-Learning-Selection062008.pdf.
Greiner, M. E., & Wang, H. 2009. The role of social capital in people-to-people lending marketplaces. *Thirtieth International Conference on Information Systems* (p.18). Phoenix: Association for Information Systems.
Hanel, A. 1985. Basic aspects of cooperative organization. *Policies for Their Promotion in Developing Countries*, Fakultas Ekonomi-Unpad.
ICA. 2001. *Jatidiri Koperasi* (Prinsip-prinsip Koperasi untuk Abad ke-21), LSP2I, Jakarta.
Iyer, R., Khwaja, A. I., Luttmer, E. F. P., & Shue, K. 2009. *Screening in New Credit Markets: Can Individual Lenders Infer Borrower Creditworthiness in P2PL Management*. Cambridge, MA.
Jensen, M. C., & Meckling, W. H. 1976. Theory of the firm: Managerial behavior, agency costs and ownership structure. *Journal of Financial Economics*, 3(4), 305–360.
Kim, Y. – Park, Y. J. – Choi, J. 2016. The adoption of mobile payment services for "fintech". *International Journal of Applied Engineering Research*, 11(2), p. 1058–1061.
Klafft, M. 2008. Peer to Peer Lending: Auctioning microcredits over the Internet. Proceedings of the 2008 *Int'l Conference on Information Systems, Technology and Management* (pp. 1–8). Dubai: IMT. http://papers.ssrn.com/sol3/papers.cfm?abstract_id=1352383
Kominfo. www.kominfo.go.id, accessed on July 18, 2017.
McAuley, D. 2015. What is FinTech? *Wharton FinTech*, 22.10.2015.
Prawirasasra, Kannya Purnamahatty. 2018. Financial technology in Indonesia: Disruptive or collaborative? *Reports on Economics and Finance*, 4(2), 83–90. HIKARI Ltd, www.m-hikari.com https://doi.org/10.12988/ref.2018.818.
Surat Edaran Bank Indonesia. No.6/18/DPNP.
Wilson, J. D. 2017. *Creating Strategic Value through Financial Technology*, 1st edition. Wiley Finance. Canada. https://doi.org/10.1002/9781119318682.

Factors affecting adoption of mobile banking in Bank Mandiri customers

I. Fitri & T. Widodo
Telkom University, Bandung, Indonesia

ABSTRACT: A variety of technologies have been developed in the banking industry, for example, mobile banking. But the rate of adoption for mobile banking does not reach the expected level, especially in developing countries. The purpose of this study was to measure the model based on UTAUT2 theory, both its direct and indirect effects. The variables that were measured were price value, hedonic motivation, social influence, trust, performance expectancy,, effort expectancy, and adoption. This research used a quantitative method. A total of 241 questionnaires with 34 total indicators from Mandiri mobile banking users were used to test the model. This research showed that performance expectancy, effort expectancy, social influence, trust, and price value had insignificant impacts on behavioral intention. Trust showed a significant impact on performance expectancy. Behavioral intention and facilitating conditions had significant impacts on adoption of mobile banking.

1 INTRODUCTION

As the most innovative and newest technology, mobile banking is a m-commerce application produced by financial institutions or banks that enable users to make financial transactions remotely from a smartphone. This service is used to make a payment, check a balance, or make a transaction or money transfer. So, an innovative technology can reduce costs and can improve service quality to customers (Malaquias & Hwang, 2019). But the rate of adoption for mobile banking does not reach the expected level, especially in developing countries, and customers are less interested in this service (Laukkanen, 2007). KPMG International's 2009 survey showed that of 4,000 cellular service customers in 19 countries around the world, the users of mobile banking are 19% smartphone users (Hanafizadeh et al., 2014). Thus, it could be said that the biggest challenge for a technological success is to convince customers to use its service as an alternative (Alalwan et al., 2016).

2 THEORETICAL RESEARCH FRAMEWORK AND HYPOTHESIS DEVELOPMENT

2.1 Performance expectancy

Previous research showed that customer intention for using mobile banking is significantly predicted by performance expectancy (Zhou et al., 2010).

H_1: Performance expectancy positively impacts behavioral intentions.

2.2 Effort expectancy

Individuals' intention to accept a new system is not only determined by whether the systems are valued positively but also by whether the system is easy to use. Mobile banking requires certain knowledge and skill level. Effort expectancy has an important role in determining customer intention for using the technolgy (Alalwan et al., 2016b).

H$_2$: Effort expectancy positively impacts behavioural intention.

H$_3$: Effort expectancy positively impacts to performance expectancy.

2.3 Social influence

Social influence can be conceptualized as the impact of social environment on customer intention to adopt mobile banking, for example, reference groups, family, leaders, friends, and colleagues (Zhou et al., 2010).

H$_4$: Social influence positively impacts behavioral intention.

2.4 Facilitating conditions

The customer can be more motivated to use a mobile banking if it has service support level, enough resources dan compatible technology (Yu, 2012).

H5: Facilitating conditions positively impacts adoption of mobile banking

2.5 Hedonic motivation

In the previous research, it was shown that there was a relationship between hedonic motivation and intention to use technology with intrinsic impact (such as excitement, happiness, entertainment, and enjoyment) and with extrinsic impact (such as efficiency, usability, and performance expectancy) (Sancaka and Subagio, 2014).

H6: Hedonic motivation positively impacts behavioral intention.

2.6 Price value

The customer's tendency to adopt mobile banking is determined by financial obstacles. The relationship between service value and price value has been tested from the research about online banking channels (Luo et al., 2010).

H7: Price value positively impacts behavioral intention

2.7 Trust

The research about trust shows that trust has a direct effect on intention to use mobile banking and indirectly facilitates the role of performance expectancy (Luo et al., 2010).

H8: Trust positively impacts behavioral intention.

H9: Trust positively impacts performance expectancy

2.8 Behavioral intention

The previous research showed that the adoption of mobile banking can be predicted by the customer's willingness. This relationship has been proven by research of online banking (Jaruwachirathanakul & Fink, 2005). This research showed that there were some indirect effects between trust, hedonic motivation, price value, effort expectancy, social influence, and performance expectancy to adoption, with behavioral intention as mediator variable. Thus, this research was done to measure some indirect effects to strengthen the relationships between variables.

H10: Behavioral intention positively impacts adoption of mobile banking.

H11: Trust indirectly impacts adoption of mobile banking.

H12: Hedonic motivation indirectly impacts to adoption of mobile banking.

H13: Price value indirectly impacts adoption of mobile banking.

H14: Effort expectancy indirectly impacts adoption of mobile banking.

H15: Social influence indirectly impacts adoption of mobile banking.
H16: Performance expectancy indirectly impacts to adoption of mobile banking.

3 RESEARCH METHODOLOGY

This a quantitative research project using a questionnaire as data collection tool. This research had a conclusive purpose and causal inquiry type (Indrawati, 2015). A total of 241 questionnaires were distributed to Mandiri mobile banking users. A six-point Likert scale was used for this research and Lisrel 8.80 was used to evaluate the model. This research had valid variables because it had already qualified in a construct reliability test. A structural model could be a good fit if at least five indices from Goodness of Fit measurement qualified for the model. In this research, there were nine indices from Goodness of Fit qualified as good fit for the model.

4 RESEARCH RESULT

To test the hypothesis, it was measured by the path coefficient. Coefficients are significant if t- values ≥1.65 or ≤ 1.65. With this measurement, the hypothesis would be accepted. One variable could be affected by the other variables directly and indirectly (Widodo, 2015). The following are the results of the hypothesis:

Table 1. Hypothesis result.

Hypothesis	Coefficients Regression	T-value	Result
H1: PE-BI	0.73	1.03	Rejected
H2: EE-BI	0.28	1.5	Rejected
H3: EE-PE	-0.09	-0.62	Rejected
H4: SI-BI	-0.16	-1.09	Rejected
H5: FC-SS	0.46	3.27	Accepted
H6: HM-BI	0.34	2.97	Accepted
H7: PV-BI	-0.19	-0.56	Rejected
H8: TR-BI	0.034	0.037	Rejected
H9: TR-PE	0.99	3.47	Accepted
H10: BI-SS	0.58	3.87	Accepted
H11: TR-BI-SS	0.44	2.03	Accepted
H12: HM-BI-SS	0.20	2.41	Accepted
H13: PV-BI-SS	-0.15	-0.58	Rejected
H14: EE-BI-SS	0.12	1.46	Rejected
H15: SI-BI-SS	-0.09	-1.03	Rejected
H16: PE-BI-SS	0.42	1.03	Rejected

5 SUMMARY AND RECOMMENDATION

Based on the results, behavioral intention had a significant impact on adoption of mobile banking. Mandiri Bank should optimize customer intention to use mobile banking through promotion, advertisement, and socialization. The impact of hedonic motivation on adoption could be an indication that it indirectly makes the customer feel that mobile banking is interesting and exciting. Trust had a significant impact on performance expectancy. From this result, Mandiri Bank could increase customer trust by giving benefits when they use mobile banking. The respondents paid attention to facilities offered by Mandiri Bank, such as fast access: this could be something that needs attention to improve service quality.

REFERENCES

Alalwan, A. A., Dwivedi, Y. K., & Rana, N. P. 2017. Factors influencing adoption of mobile banking by Jordanian bank customers: Extending UTAUT2 with trust. *Int. J. Inf. Manag. 37*, 99–110. https://doi.org/10.1016/j.ijinfomgt.2017.01.002.

Alalwan, A. A., Dwivedi, Y. K., Rana, N. P., & Algharabat, R. 2018. Examining factors influencing Jordanian customers' intentions and adoption of internet banking: Extending UTAUT2 with risk. *J. Retail. Consum. Serv. 40*, 125–138. https://doi.org/10.1016/j.jretconser.2017.08.026.

Alalwan, A. A., Dwivedi, Y. K., Rana, N. P. P., & Williams, M. D., 2016a. Consumer adoption of mobile banking in Jordan: Examining the role of usefulness, ease of use, perceived risk and self-efficacy. *J. Enterp. Inf. Manag. 29*, 118–139. https://doi.org/10.1108/JEIM-04-2015-0035.

Alalwan, A. A., Dwivedi, Y. K., & Williams, M. D. 2016b. Customers' intention and adoption of telebanking in Jordan. *Inf. Syst. Manag. 33*, 154–178. https://doi.org/10.1080/10580530.2016.1155950.

Hanafizadeh, P., Behboudi, M., Abedini Koshksaray, A., & Jalilvand Shirkhani Tabar, M. 2014. Mobile- banking adoption by Iranian bank clients. *Telemat. Inform. 31*, 62–78. https://doi.org/10.1016/j.tele.2012.11.001.

Indrawati. 2015. *Metode Penelitian Manajemen dan Bisnis Konvergensi Teknologi Komunikasi dan Informasi*. PT Refika Aditama, Bandung.

Jaruwachirathanakul, B., & Fink, D. 2005. Internet banking adoption strategies for a developing country: the case of Thailand. *Internet Res. 15*, 295–311. https://doi.org/10.1108/10662240510602708.

Laukkanen, T. 2007. Customer preferred channel attributes in multichannel electronic banking. *Int. J. Retail Distrib. Manag. 35*, 393–412. https://doi.org/10.1108/09590550710743744.

Luo, X., Li, H., Zhang, J., & Shim, J. P. 2010. Examining multi-dimensional trust and multi-faceted risk in initial acceptance of emerging technologies: An empirical study of mobile banking services. *Decis. Support Syst. 49*, 222–234. https://doi.org/10.1016/j.dss.2010.02.008.

Malaquias, R. F., & Hwang, Y. 2019. Mobile banking use: A comparative study with Brazilian and U.S. participants. *Int. J. Inf. Manag. 44*, 132–140. https://doi.org/10.1016/j.ijinfomgt.2018.10.004.

Sancaka, M., & Subagio, H. 2014. Analisa Faktor Yang Mempengaruhi Penerimaan Dan Penggunaan. Kompas Epaper Oleh Konsumen Harian Kompas Di Jawa Timur Dengan Menggunakan Kerangka Unified Theory Of Acceptance And Use Of Technology (Utaut). *J. Manaj. Pemasar. Petra 2*, 1–7.

Widodo, T. 2015. The effect of transformative IT capability on sustainable competitive advantage, in: 2015 3rd International Conference on Information and Communication Technology (ICoICT). *Presented at the 2015 3rd International Conference on Information and Communication Technology (ICoICT)*, IEEE, Nusa Dua, Bali, Indonesia, pp. 352–357. https://doi.org/10.1109/ICoICT.2015.7231450.

Yu, C.-S. 2012. Factors affecting individuals to adopt mobile banking: Empirical evidence from the UTAUT model. *J. Electron. Commer. Res. 13*, 104–121.

Zhou, T., Lu, Y., & Wang, B. 2010. Integrating TTF and UTAUT to explain mobile banking user adoption. *Comput. Hum. Behav. 26*, 760–767. https://doi.org/10.1016/j.chb.2010.01.013.

EVA approach in measuring distribution company performance listed in IDX

Prima E. Sembiring & Junino Jahja
University of Indonesia, Jakarta, Indonesia

ABSTRACT: This research aims to evaluate the financial performance of PT. TigaraksaSatria, Tbk (TGKA) and PT. Enseval Putra Megatrading (EPMT) as distribution companies using an economic value added (EVA) approach during the period of 2014 to 2017. This research uses a financial report that was obtained from IDX website, company websites, and interviews. The research proves that the financial performance of the two distribution companies for period 2014–2017 show a positive results (EVA>0). In 2014, TGKA's EVA was Rp162.674.706.763 and in 2015, EVA increased to Rp191.395.182.514. In 2016, TGKA created EVA Rp175.608.568.953 and increased to Rp180.518.201.890 in 2017. In 2014, EPMT created EVA Rp266.090.979.215, EVA in 2015 Rp287.795.995.923, and for 2016 EVA of EMPT increased to Rp310.978.500.588. As for 2017, EVA of EPMT reduced to Rp227.805.590.090. The result of the research shows that TGKA and EPMT as distribution companies create value for investors.

1 INTRODUCTION

Every company is racing to optimize profit for a financial performance that gives return to investors. Financial performance is an indicator for the investor to decide whether they will invest or decline to invest in a company. Various fundamental metrics are used by investors in assessing financial performance such as PER (price to earnings ratio), PBV (price to book value), ROE (return on equity) and also ROI (return on investment). The fundamental measurements method is found to have some limitations: it only calculates the rate of return on capital that has been spent and does not consider the cost of capital itself so it will be difficult to know whether value is created for an investment for the investor. Companies that focus on improving financial performance only refer to improvements in company profits, but what is done by the company has not been able to create value for the investor. Stern Stewart & Co., in 1993, published a new method in measuring or evaluating financial performance in a company by introducing a concept called economic value added (EVA). EVA is a technique for analyzing, evaluating, and making additional measurements of economic value that can be determined from the difference between net operating profit after tax and cost of capital (COC) or capital costs. COC is calculated based on debt and equity and weighted average costs of capital (WACC) and the amount of capital used. The use of methods that also calculate the cost of capital is expected to produce more tangible company values

2 THEORETICAL FOUNDATION

Financial performance, according to Helfert (1996), is a complete picture of the condition of the company in a certain period reflecting all operational activities in the company and how the company is utilizing all the resources it has. These results will later be taken into consideration by investors to determine the form of investment to be chosen. EVA can be used to measure the level of difference, in terms of finance, between a company's capital and its rate of return. This is similar to the measurement of profits that exist in conventional accounting measurement methods but there are differences that are considered important. Many people argue that EVA is only a measurement of performance, but in fact, it is more than that.

Reddy et al. (2011) conducted a study by comparing EVA with traditional performance measurements to conclude that EVA provides a definite picture of the amount that shareholders will get at the end of the year, and the report suggests that EVA is the most appropriate measure-to-measure shareholder value. Madhavi's (2015) research on the consumer goods industry in India found that EVA is a better indicator than traditional measurements of increased market value and should be used as a consideration in shareholder value creation or as a measurement of company performance.

3 OBJECTIVE

From the explanation above, the objectives of this study are to evaluate and analyze the financial performance of distribution companies using the EVA metric approach and to compare EVA, EVA margin, and EVA momentum trends in distribution companies from 2014 to 2017 due to a slowdown in economic growth in Indonesia.

4 METHODOLOGY

The study will use the EVA approach to evaluate the financial performance of distribution companies listed in BEI for the period 2014–2017. The study use primary data from company interviews and secondary data collected from published sources. To calculate EVA, EVA momentum and EVA margin, we use the following equation:

Economic Value Added (EVA)

$$EVA = NOPAT - (WACC \times InvestedCapital) \quad (1)$$

Where NOPAT = net operating profit after tax; and WACC = weighted average cost of capital

Weighted Average Cost of Capital (WACC)

$$EVA = (W_d \times CoD) + (W_e \times CoE) \quad (2)$$

Where W_d = debt structure; CoD = cost of debt; W_e = equity structure; and CoE = cost of equity

EVA Momentum

$$EVAMomentum_{t1} = \Delta EVA_{t1} \div Sales_{t_1 - t_0} \quad (3)$$

EVA Margin

$$EVAMargin_t = EVA_t \div Sales_t \quad (4)$$

5 RESULT

Financial performance of the distribution companies used as the object in this research, namely PT. TigaraksaSatria, Tbk and PT. Enseval Putra Megatrading, Tbk, is able to produce positive EVA or EVA > 0. This indicates that during the period of 2014 to 2017, the companies were able to produce positive performance. Positive EVA will provide added value to the investors. However, the authors see that economic growth in Indonesia will play an important role in determining the performance of companies, especially distribution companies. Inflation values

Table 1. EVA, EVA margin and EVA momentum results.

Year	TGKA				EPMT			
	2017	2016	2015	2014	2017	2016	2015	2014
EVA	180.518	175.609	191.395	162.675	227.806	310.979	287.796	266.091
EVA Margin	1.80%	1.83%	2.01%	1.72%	1.16%	1.64%	1.65%	1.56%
EVA Momentum	0.05%	-0.17%	0.30%	-0.44%	0.13%	0.13%		

Sources: Author

that are too high will cause a decrease in purchasing power associated with price increases. The company will increase the sales price to cover the costs incurred.

Based on Table 1, EVA of TGKA fluctuates between 2014 and 2017. In 2014, TGKA created EVA Rp162.675 billion; in 2015, it created EVA Rp191.395 billion. The value of EVA TGKA decreased in 2016 to Rp175.609 billion before increased in 2017 to Rp180.518 billion. EPMT in 2014 created EVA Rp266.091 billion and increased to Rp287.769 billion in 2015. In 2016 EVA increased to Rp310.979 billion and decreased to Rp227.806 in 2017 due to increased operating expenses.

EVA is a tool to evaluate a company's financial performance without the need for comparison with other company. EVA will calculate the owner's capital expectation expressed by the weighted measure of capital structure guided by the book value not the market value. Companies can also use EVA to calculate employee incentives.

6 CONCLUSION

Financial performance of distribution companies used as the object in this research, namely PT. TigaraksaSatria, Tbk and PT. Enseval Putra Megatrading, Tbk, is able to produce positive EVA values or EVA> 0. This indicates that during the period of 2014 to 2017, the company was able to produce positive performance. Positive EVA will increase value to the investors. The trend of EVA and EVA metrics of the two companies are worth almost the same, because the two companies are in the same industry. EVA margin describes the company's ability to produce EVA from sales value. EVA momentum describes the level of change in the company's EVA compared to sales in the previous period. Investors can make comparisons based on EVA margin trends and EVA momentum from both companies.

REFERENCES

Adiningsih, S. I. & Sumarni. 2005. Hubungan economic value added (EVA) and market value added (MVA) Pada Perusahaan Publik Yang Terdaftar di Bursa Efek Jakarta. *TelaahBisnis, 6*(1).
Bodie, Z., Kane, A., & Markus, A. J. 2009. *Investments* 9th edition. New York: McGraw-Hill.
Blume. (1975). Betas and their regression tendencies. *The Journal of Finance 30*(3).
Darmadji, T. & Fakhrudin. 2001. *Pasal Modal di Indonesia*. Jakarta: SalembaEmpat.
Damodaran, A. 1996. *Investment Valuation: Tools & Technique for Determining the Value of Any Asset* 2nd ed. New York: John Willey.
Damodaran, A. 2007. *Return on Capital (ROC), Return on Invested Capital (ROIC) and Return on Equity (ROE): Measurement and Implication*. Stern School of Business.
Damodaran, A. 2002. *Investments* 9th Edition. New York: McGraw-Hill.
Helfert, E. A. 1996. *TeknikAnalisisKeuangan*.Erlangga: Jakarta.
Higgins, R. C. 2007. *Analysis for Financial Management*. New York: Irwin McGraw Hill.
Krisdjoko, R. 2004. AnalisisPenentuan Nilai Intrinsik PT Indofood SuksesMakmurTbk Dengan PendekatanKonsep Economic Value Added. Universitas Indonesia.

Lehn, K., & Mahkija. 1996. EVA &MVA: as Performance Measures and Signals For Strategic Change. June. *Fortune*.

Madhavi & Prasad. 2015. An imperical study on economic value-added and market value added of selected Indian FMCG combanies. *The IUP Journal of Accounting Research & Audit Practices, XIV*(3).

Margaretha, Farah. 2011. *Manajemen Keuanganuntuk Manajer Non Keuangan*. Jakarta: Erlangga.

Nurauliawati. 2010. AnanalisaKinerja Perusahaan Industri Telekomunikasi Dengan Metode Economic Value Added (Periode 2006–2009). Universitas Indonesia.

Putri. 2010. AnalisisPengaruh Return On Equity, Economic Value Added, Dan Momentum EVA Terhadap Return Saham Perusahaan Indeks LQ45 Periode 2005–2009. Universitas Indonesia.

Reddy, R., Rajesh, M., Reddy, N. 2011. Valuation through EVA and traditional measures: An empirical study. *International Journal of Trade, Economics and Finance*, 2.

Rupert, B. 2005. EVA as a management incentive. *Management Accounting*, 12.

Saibaba & Chary. 2015. Evaluation of financial performance using economic value added metrics in pharmaceutical and IT companies: A comparative study. *IPE Journal of Management*, 5(2).

Stern. 1993. EVA, Share Option That Maximize Value. Corporate Finance.

Stewart, G.B. 1991. *The Quest for Value: A Guide for Senior Managers*. Amerika Serikat: HerperCollins Publisher Inc.

Stewart, B. 2009. EVA momentum: The one ratio that tells the whole story. *Journal of Applied Corporate Finance*. Morgan Stanley Publication, *21*(2).

Analysis of Sharia banking health with the risk based bank rating and bankometer

Muhammad Ichsan Hidayat & Farida Titik Kristanti
Telkom University, Bandung, Indonesia

ABSTRACT: This study aims to determine the health of Sharia banking (Islamic Commercial Bank) in Indonesia. There are 2 methods used in this research, namely the Risk Based bank Rating (RBBR) and Bankometer methods. This study will further compare the influence of financial ratio factors between Islamic Commercial Banks listed on the Jakarta Stock Exchange (IDX) and those not listed. Ratios measured and compared include factors in RBBR, namely: Non Performing Financing (NPF), Financing to Deposite Ratio (FDR), Good Corporate Governance (GCG), Net Interest Margin (NIM), Capital Adequate Ratio (CAR), Return on Assets (ROA), and Operating Expenses Operating Income (BOPO). While the Bankometer only has 1 formula, the S-Score Bankometer.

1 INTRODUCTION

Sharia Commercial Bank has been known in Indonesia since 1991. The purpose of establishing the Sharia Public Bank is to provide an alternative banking system for the people of Indonesia. With a very rapid development, at the end of 2017 already operated 12 Sharia Commercial Banks (BUS), 12 Sharia Business Units (UUS) and 1886 Sharia offices. In the 2013-2017 period, Islamic bank assets in Indonesia increased 75% with an average growth of 17% per year (OJK, 2018).

Judging from the banking performance indicators, performance data of Sharia Commercial Banks in Indonesia from 2007-2017 show indicators up and down. This is indicated by the value of Return on Assets (ROA), Non Performing Financing (NPF), Financing Deposit Ratio (FDR) and Operational Expenses and Operating Income (BOPO) that fluctuates.

To predict financial difficulties, various methods are known, based on Bank Indonesia regulation no 13/1/PBI/2011 known as the Risk Based Bank Rating (RBBR) method, while the International Monetary Fund (IMF) recommends the Bankometer method. Both of these methods will be applied to predict financial difficulties at Islamic Banks in Indonesia.

The main problem in this study is how to conduct a health analysis of Islamic Commercial Banks in Indonesia based on the reports they issued on their website from 2013 to 2017. The object group of this study was divided into 2 groups, namely banks listed on the Jakarta Stock Exchange (IDX) and banks not listed on IDX.

The purpose of this study is to explore the difference in the level of soundness of the bank with the RBBR method and bankometer model between Syariah Banks that are listing and non-listing on IDX. This is important because according to Kristanti, Effendi, Herwany, and Febrian (2016) bankruptcy causes social impacts.

2 LITERATURE REVIEW

Definition of bank health according to Bank Indonesia in accordance with RI Law No. 7 of 1992 concerning Banking Article 29 is a Bank said to be healthy if the bank meets the Bank's health requirements by taking into account the aspects of Capital, Asset Quality, Management Quality, Rentability Quality, Liquidity, Solvency, and other aspects related to bank business. From the explanation above it can be seen that the health of a bank is to know the condition of a bank at a time when it is needed for certain interests. Assessment to determine the condition of a bank, usually using a variety of measuring tools. One of the main measurement tools used to determine the condition of a bank is known as risk profile, governance, earning and capital or RGEC or also known as Risk Based Bank Rating (RBBR)

According to Bhattacharya (2012: 445), distress is an acute financial difficulty or crisis. A company experiencing difficulties/not healthy is a situation like a company when unable to meet debt. In other words, when the total value of a company's assets is insufficient to meet total external obligations, the company can be said to be a "stressed company".

In the condition of financial distress the company will experience a decline in functions, ranging from systems, components, processes and resources that will be interconnected, so that when something goes wrong, it will have an impact on the overall level of the company (Sofat and Hiro, 2012: 387).

In a banking or economic context, the Financial Ratio is measured by comparing total debt and total assets. Literature that focuses on financial difficulties provides specific evidence of the relationship between financial leverage and corporate financial difficulties (Kristanti and Effendi, 2017: 4)

Budiman, Kristanti and Wardana (2017) using RBRR to evaluate the healthy of Sharia Bankin in Indonesia. Laila and Widihadnanto (2017) use Altman's Z-Score to predict financial difficulty. Other studies include Arulvel and Balaputhiran (2013) in Sri Lanka, Hanif, Tariq, Tahir and Momeneen (2012) in Pakistan, Fayed (2013) in Egypt. All of these studies show that the Bankometer (S-Score) model is appropriate for the prediction of financial banking.

Budiman, Herwany and Kristanti (2017) uses Bankometer model to measure the health level of Islamic banks in Indonesia. In the 2012-2016 study, the level of health of Islamic banks in Indonesia was found to be in a healthy condition. However, several financial performance indicators such as Return On Assets (ROA), Operating Expenses and Operating Income (BOPO) and Loan to Deposite Ratio (LDR) are in an unhealthy condition.

3 METHODOLOGY

This study uses secondary data from financial statements collected from the websites of each Islamic Bank in Indonesia for the period between 2013-2017. Descriptive analysis is used to describe the measurement results. 11 Islamic banks were selected for the study sample using purposive random sampling with the following criteria: they must have published complete data for the period between 2013-2017. The RBBR and Bankometer models are used together to explore the financial condition of the bank. The final measurements obtained from the RBBR and S-Score are then compared with the criteria used in this model.

To test the hypothesis in this study, an independent t-test was used. This test compares the difference between the two values which gives the average standard error and the standard error of the sample. The first step involves testing whether there are differences between the two variances in the population. Then, the t-test is run to determine the mean difference significantly.

Measured with RBBR contains of 7 factors:

1. Non Performing Financing (NPF)

$$\text{NPF Ratio} = \frac{\text{NPF}}{\text{Total Financing}} \times 100\%$$

2. Financing to Deposite Ratio (FDR)

$$\text{FDR Ratio} = \frac{\text{Total Financing}}{\text{Total Deposite}} \times 100\%$$

3. Good Corporate Governance: Is a self-assessment that includes Governance structure, governance processes and governance outcomes.
4. Return on Asset

$$\text{ROA Ratio} = \frac{Net\ Income}{Average\ Total\ Asset} x 100\%$$

5. Net Interest Margin

$$\text{NIM Ratio} = \frac{Net\ Interest}{Average\ Invested\ Asset} x 100$$

6. Operating Expenses and Operating Income (BOPO)

$$\text{BOPO Ratio} = \frac{Operating\ Expense}{Operating\ Income} x 100\%$$

7. Capital Adequacy ratio

$$\text{CAR Ratio} = \frac{Total\ Capital}{Risk\ Weighted\ Asset} x 100\%$$

Measurement with Bankometer using formula:

$$\text{S-Score} = 1,5(\text{CA}) + 1,2(\text{EA}) + 3,5(\text{CAR}) + 0,6(\text{NPL}) + 0,3(\text{CIR}) + 0,4(\text{LA}) \tag{1}$$

Where:
  CA   = Capital Asset Ratio
  NPL = Nonperforming Loan
  EA   = Equity to Asset Ratio
  CI   = Cost to Income Ratio
  CAR = Capital Adequacy Ratio
  LA   = Loan to Asset Ration

4 RESULT

This test involves variables in RBBR namely NPF, FDR, GCG, ROA, NIM, BOPO and CAR

Table 1. Summary.

SUMMARY			Alpha	0,05				
	Count	Mean	Std Dev	Std Err	Mean Diff	t-sta	P	Result
NPF	50	3,8	2,2	0,3	0,0	0,1	0,93	Non Sig
FDR	50	86,3	28,7	4,1	0,8	0,2	0,85	Non Sig
GCG	50	1,9	0,5	0,1	0,1	1,7	0,09	Non Sig
ROA	50	0,2	3,0	0,4	1,4	3,4	0,00	Sig
NIM	50	7,2	6,7	1,0	3,6	3,8	0,00	Sig
BOPO	50	98,8	20,6	2,9	(14,5)	(5,0)	0,00	Sig
CAR	50	21,4	13,5	1,9	2,1	1,1	0,29	Non Sig

The calculation result of Non Performing Financing (NPF) is 3.8 which means healthy. There is no significant difference in the value of Non Performing Financing (NPF) between Islamic Banks listed on IDX and non listed on IDX. The Financing to Deposite Ratio (FDR) is 86.3 with the title quite healthy. There is no significant difference in the value of Financing to Deposite Ratio (FDR) between Islamic Banks listed on IDX and non listed on IDX. The result of Good Corporate Governance (GCG) is 1.9 with a pretty good predicate. There is no significant difference in the value of Good Corporate Governance (GCG) between Islamic Banks listed on IDX and non listed on IDX. The Return on Asset (ROA) calculation is 0.05 with an unhealthy predicate. There is a significant difference in the value of Return on Assets (ROA) between Islamic Banks listed on IDX and non listed on IDX. The Net Interest Margin (NIM) measurement is 6.7 with a very healthy predicate. There is a significant difference in the value of Net Interest Margin (NIM) between Islamic Banks listed on IDX and non listed on IDX. The measurement results of Operating Expenses Operating Income (BOPO) is 100.17 with an unhealthy predicate. There is a significant difference in the value of Operational Operating Income (BOPO) between Islamic Banks listed on IDX and non listed on IDX. The Capital Adequacy Ratio (CAR) measurement is 22.7 with a very healthy predicate. There is a significant difference in the value of the Capital Adequacy Ratio (CAR) between Islamic Banks listed on IDX and non listed on IDX

Table 2. Measurement for Bankometer.

	SUMMARY			Alpha	0,05			
	Count	Mean	Std Dev	Std Err	Mean Diff	t-statistic	P	Result
S-Score	50	142,08	19,65	2,779207269	34,98	12,59	0,00	Sig

The measurement result of the S-Score Bankometer is 138.9 with the title of not experiencing financial distress. From the test results there are significant differences between banks listed on IDX and non listed IDX.

5 CONCLUSION

By using the Risk Based bank Rating method, it is found that there are no significant differences between the listed banks on IDX and non-Listed parameters, namely: NPF, FDR and GCG. While other RBBR parameters show significant differences, namely ROA, NIM, BOPO and CAR. While using a Bankometer shows a significant difference between listed and non-listed banks.

REFERENCES

Arulvel, K., Balaputhiran.. (2011). Market Efficiency or Not: A Study of Emerging Market of Colombo Stock Exchange (CSE) in Sri Lanka. *International Conference on Beyond the Horizon*.
Bhattacharya, A. K. (2016). *Financial Accounting for Business Managers*, 5th Edition. PHI Learning Pvt. Ltd.
Bank Indonesia. (2001). Surat Edaran Bank Indonesia Nomor 3/30/DPNP tanggal 14 Desember 2001. Lampiran 14. Bank Indonesia.
Budiman, T. Herwany, A. & Kristanti, F.T. An Evaluation of FFinancial Stress for Islamic Banks in Indonesia using a Bankometer Model. *Journal of Finance and Bankig Review*. 2 (3).14–20.
Budiman, T. Kristanti, F.T. & Wardhana. (2017). Islamic Bank Listed in Financial Market: Risk, Governance, Earning and Capital. *Journal of Islamic Economics*. 9(1). 1–12.
Kristanti, & Effendi, (2017). A Survival of Indonesia Distressed Company Using Cox Hazard Model, Int. Journal of Economics and Management 11(S1).pp.157–169.

Kristanti, F.T., Effendi, N., Herwany, A. & Febrian, E. Does Corporate Governance Affect the Distress of Indonesian Company? A Survival Analysis Using Cox Hazard Model with Time-Dependent Covariates. *Advanced Science Letters*. 22. 4326-4329.

Laila, & Widhihadnanto. (2017). Financial Distress Prediction Using Bankometer Model on Islamic and Conventional Banks: Evidence from Indonesia. *Int. Journal of Economic and Management*. 11(SI), 169–181.

Sofat, R., & Preeti Hiro. (2012). *Strategic Financial Management*. PHI Learning Pvt, Ltd.

Implementation of a decision supporting system against satisfaction of Tower Provider

Sarman & Achmad Manshur Ali Suyanto
Telkom University, Bandung, Indonesia

ABSTRACT: This study examines the results of the Implementation of Telkomsel Trouble Ticket (TOTI) as a Decision Supporting System Information System in Telkomsel Jabotabek on Tower Provider satisfaction with the Delone & McLean model approach. Evaluation of TOTI success is measured through user usage and satisfaction. TOTI is implemented to measure and increase the performance of Tower Provider in collaboration with Telkomsel, which in the end there is a sanction application if the performance produced by Tower Provider is not in accordance with the agreement. However, although some time has been implemented, the expected performance did not increase. This study will prove the effect of information quality, system quality, and service quality on user usage and satisfaction variables. This research will also examine the effect of usage and user satisfaction on the net benefits obtained. In the future, TOTI Information System still needs further development, especially to improve the system quality.

1 INTRODUCTION

At this time, the implementation of the information system in a company is a must, in order to improve the company's performance. However, in implementing an information system, it requires an investment in the form of costs and resources that are not small. Therefore, companies try to evaluate the implementation of the information system that they use. So far, the use of cost-benefit analysis cannot be done perfectly because not all benefits can be quantified. Benefits such as more timely decisions, increased employee expertise and employee satisfaction with information systems can escape financial analysis calculations (Laudon and Laudon, 2011: 266). One model of information system measurement that is continuously being developed by researchers to meet the business needs of the information system success measurement tools, is the DeLone and McLean models. According to Jogiyanto (2008: 2), the DeLone and McLean model is a parismonic model that is simple and valid.

In this study, the Implementation of TOTI as an Information System in the Jabotabek regional Telkomsel to Tower Provider will be measured using the adapted DeLone and McLean (2003) models, namely by issuing a variable of user interest (intention to use). This variable was not included in this study because the TOTI information system has become a tool or tool that must be used by Tower providers in the Jabotabek Region in interacting operationally up to the tower rental billing process. In the DeLone and McLean (2003) model, the success of the information system is proxied by user satisfaction which will then have an impact in the form of net benefits both to individuals and to organizations. According to DeLone and McLean (2003), information system success is influenced by information quality, system quality, and service quality.

1.1 *Research questions*

1. Does the quality of information produced by IS TOTI affect the use and user satisfaction?
2. Does the quality of the IS TOTI system affect user usage and satisfaction?

3. Does the Quality of IS TOTI Services affect the use and ser satisfaction?
4. Does the use and user satisfaction affect the net benefits?

2 THEORITICAL BACKGROUND

According to James O'Brien (2010: 4), information systems can be a regular combination of people, hardware, software, communication networks, and data resources that collect, change, and disseminate information within an organization. Humans depend on information systems to communicate with physical equipment (hardware), information processing instructions or procedures (software), communication networks (networks), and data (data resources). As explained by Ward and Prepard (2002: 3) that the concept of information systems that exist in every organization has emerged before the development of information technology. Information technology is related to technology, especially hardware, software, and telecommunications networks. Information systems are defined as people or organizations that use technology to collect, process, store, use and disseminate information.

According to Jogyanto (2003), the Decision Supporting System (DSS) is an information system to help middle-level managers to make half-structured decisions to be more effective by using analytical models and available data. In addition (Alter, 2002), the Decision Supporting System (DSS) is an interactive information system that provides information, modeling and manipulation of data that is used to assist decision making in semitructured and unstructured situations where no one knows exactly how decisions should be made. In the classification of DSS information systems, including the Management Information System (MIS) which is tasked with providing support for decision making on the operations management of the company James O'Brien (2010). TOTI Is an integrated analyzer and dashboard used by Telkomsel Regional Jabotabek ICT for optimization of production equipment contained in the Tower Provider. Seeing this function, TOTI is able to provide information on the current condition of Tower Provider's performance on production equipment, process information, as well as being able to provide input for a decision in the field of semi-structured and unstructured data. TOTI was created in 2016 and is continuously being refined to help the Jabotabek Regional ICT Telkom team in their daily work. Now TOTI is one of Telkomsel's official portal tools. TOTI can be accessed through the address http://www.toti.telkomsel.co.id. In addition, DeLone and McLean also combine two dimensions, namely individual influence and organizational influence into net benefits. The DeLone and McLean models are simple but valid models that are included as parismonic models (Jogiyanto, 2007: 2). The framework is based on the information system success model DeLone and McLean (2003), namely the successful implementation of information systems is proxied by the satisfaction of the use of the relevant information system (user satisfaction) and use (use). In this study, researchers modified the Delone and McLean model (2003) by issuing the intention to use variable, because TOTI is a mandatory DSS in Telkomsel Regional ICT Jabotabek, which means that the TOTI application must be used to monitor performance against Tower providers and influence between use and user satisfaction, because this application is mandatory, the use variable and the user satisfaction variable will not affect each other.

3 HYPOTESIS

Hypothesis 1
Ho: Information quality has no effect on usage
H1: Information quality affects usage
Hypothesis 2
Ho: The quality of the system has no effect on usage
H1: The quality of the system has an effect on usage
Hypothesis 3
Ho: Service quality has no effect on usage
H1: Quality of service has an effect on usage

Hypothesis 4
Ho: Information quality has no effect on user satisfaction
H1: Information quality affects user satisfaction
Hypothesis 5
Ho: The quality of the system has no effect on user satisfaction
H1: The quality of the system influences user satisfaction
Hypothesis 6
Ho: Quality of service does not affect user satisfaction
H1: Service quality affects user satisfaction
Hypothesis 7
Ho: Use has no effect on net benefits
H1: Use affects net benefits
Hypothesis 8
Ho: User satisfaction does not affect net benefits
H1: User satisfaction influences net benefits

4 METHODOLOGY

The research conducted included qualitative, verification and causal research to examine the effect of system quality, information quality and service quality on the use and satisfaction of TOTS DSS system users and its effect on net benefits, namely the performance of individuals and organizations. In this study, researchers did not intervene the existing data, both primary data and secondary data. The variables used in testing the success of the TOTS DSS system are system quality, information quality, service quality, use, user satisfaction and net benefit. The population in this study consisted of 100 employees from 50 Tower Provider companies who use TOTI or who have access (authorization) to the TOTS DSS during the study period. This study uses all members of the existing population, so the sample is not enforced here. In this study, a descriptive analysis was performed, and hypothesis testing was performed using a structural equation model (Structural Equation Model/SEM).

5 DISCUSSION

From the results of the calculation of the respondent's answer percentage, it can be seen that almost all question indicators score above 80%, which states that the indicators of the six variables are very good.

The highest rating is in the System Quality Variable with Item SQ2 which states that the TOTI System can be run (accessed) through other devices, other than the computer/laptop used at this time. This indicates that the ease of accessing TOTI is very important for users.

Whereas the lowest valuation is in the Net Benefit Variable with item NB 8 which states that the TOTI System helps save company costs. This shows that the savings obtained by the Tower Provider company after the TOTI implementation is not very significant, especially if the Tower Provider company is subject to sanctions for fines for not meeting the agreed SLA targets. Information Quality is very directly influential on User Satisfaction compared to Service Quality and System Quality, while for Use, the most influential is Service Quality, while the System Quality has a negative effect, then, User Satisfaction is more influential compared to Use on Net Benefit. the variable that has a total effect on the User Satisfaction variable is still Information Quality, the same as the use variable, the variable that most influences is the Service Quality variable, then it can be seen that the Net Benefit variable, although the variable that most influences is still User Satisfaction, but the variable Information Quality also has a big influence on the expected variable.

Based on the result of previous studies, according to Mc.Gill(2013) stated that the quality system significantly influences user satisfaction and actual use, information quality significantly influences user satisfaction. Testing of this hypothesis is done by comparing the p value on the AMPS SPSS results, using two-way testing (twotailed) and the significance of $p \leq 0.05$. with the results of the comparison of the value of p as follows Hypotesis 1, H2, H3. H4, H6, H7, H8, shows that H0 is rejected, and only on Hypothesis 5 H_0 is accepted.

6 CONCLUSION AND SUGGESTION

Based on data analysis and hypothesis testing, the answers to each question raised in the study are as follows:

1. The quality of information generated by the TOTI Information System influences User Use and User Satisfaction. This proves that TOTI users are very dependent on the quality of information produced by TOTI in the process of using it, which impacts on user dissatisfaction if the quality produced by TOTI is not in line with expectations.
2. IS TOTI System Quality does not affect the Use (Use), but affects the User Satisfaction. This proves the user does not place importance on TOTI System Quality, because TOTI must be used in daily work. For good or bad system quality, TOTI must still be used. However, if the System Quality produced by TOTI is good, it will greatly affect user satisfaction.
3. IS TOTI Service Quality affects the Use (User) and User Satisfaction. This proves that users place great importance on the Quality of Service distributed by TOTI. With the good quality of service, it will guarantee continuity in the use of TOTI and also increase satisfaction to TOTI users.
4. Usage and User Satisfaction affect the net benefits (Net Benefits). This proves that TOTI Information System is very useful for Tower Provider companies as TOTI users, both in terms of performance improvement and by emphasizing operational cost expenditures. With TOTI, the company can monitor directly the field conditions and the performance of its employees.

The conclusion of this study provides many benefits for the Tower Provider company in the Jabodetabek Regional Telkomsel environment, so it is necessary to further develop the TOTI information system. For this development, suggestions that can be given to companies are as follows:

1. The quality of the TOTI System is very influential on the use and satisfaction of customers, it is recommended to further improve the quality of information on the dimensions of security and ease of interface with TOTI users.
2. The quality of information generated by TOTI has not been proven to affect usage, but it still has an effect on user satisfaction. It is recommended to improve the accuracy of the information produced by TOTI continuously. Improving the accuracy of the data on TOTI can be done by increasing the speed of the system in communicating users, so that interaction will be faster and can shorten the time in solving problems that occur.
3. To improve TOTI Service Quality, an SLA must be made which includes recovery time guarantees, contact center guarantees, and connectivity guarantees so that service satisfaction from TOTI users can be measured.
4. In increasing the net benefits generated by TOTI, it is recommended that regular joint discussions be held between Telkomsel and Tower Provider, so that the needs of both parties in optimizing the benefits of TOTI, both for each employee and company coverage can be met.

REFERENCES

Alter, Steven. 2002. Information System, Foundation of E-busines. Prentice Hall, London.
DeLone, W.H., and McLean E.R. 2003. 'The DeLone and McLean Model of Information System Success: A Ten-Year Update'. Journal of Management Information System Vol.19, No. 4, 9–30.
Jogiyanto.2007. Model Kesuksesan Sistem Teknologi Informasi. Andi Offset(1).Yogyakarta.

O'Brien, J.A. dan Marakas, G. M.2009. Management Information Systems (9thEdition). McGraw-Hill. New York.
Wiyono, A.S, Hartono, Jogiyanto dan Ancok, Djamaludin. 2008. "Aspek Psikologis pada Implementasi Sistem Teknologi Informasi". E-Indonesia Initiative 2008 (eII2008). Konferensi dan Temu Nasional Teknologi Informasi dan Komunikasi untuk Indonesia. Jakarta. 64, pp. 12–23.
http://www.telkomsel.co.id & http://Toti.telkomsel.co.id

Cosmopolitanism in the fashion industry in Bali as an impact of the tourism sector

A. Arumsari, A. Sachari & A.R. Kusmara
Bandung Institute of Technology, Bandung, West Java, Indonesia

ABSTRACT: Bali is the most well-known destination in Indonesia, either for local or international tourists. Bali's attraction comes from the combination of its beautiful natural enevironment and the very authentic traditional culture implemented in every aspect of the life of the society up to the present time. These conditions make tourism the most dominant sector in Bali that triggers other sectors found, including fashion. The influence of the tourism sector on fashion arises to meet the need of keeping up with the ever-changing lifestyle in Bali into something more cosmopolitan, brought by tourists and migrants from different places in the world. This study uses sociological research methods to explore some comprehensive data through a literature study of different related literature sources, direct field observation to various Bali fashion centers, as well as in-depth interviews of different respondents in Bali.

1 INTRODUCTION

Bali is a provinces in Indonesia. Besides the main island of Bali, the province also has other small islands, for a total area of 5,634.40 ha with a beach length of 529 km. Mountains and hills cover most of Bali Province. The chain of mountains in the center of the island geographically divides this area into two different areas. Northern Bali is adorned with narrow lowland of the foothills and mountains while Southern Bali is comprised of wide and sloping lowland. The geographical condition, mostly covered by mountain in the north and beach in the south, turns numerous areas in the island into attractions for both local and international tourists.

In addition to its beauty and nature, Bali also possesses other attractions, including Balinese culture and traditions upheld by the people. It is, for example, shown by the background and the general description of the people that in their daily life, their religious values are Hindu. Hindu Dharma in Balinese covers different aspects of theological conception and religious values, socio-religious structures, and the religious heritage including material culture, both tangible and also intangible (Paeni, 2009). Various aspects of Balinese belief is represented by ritual ceremo- nies and numerous cultural activities in the form of either tangible heritage such as *pura* (temple), *keris* (kris, traditional Indonesian weapon), *patung* (statue), and *arca* (statue artifact), as well as traditional fabrics, or intangible heritage such as ritual ceremonies, traditional dances, and traditional performances that tourists see as an authentic and attractive elements.

As cited from Raharjo et al. (1998, p. 131), tourism in Bali is divided into cultural and natural tourisms. However, the main attraction is cultural tourism, from which other sectors are derived.

Based on the data accessed from the official site of the provincial government of Bali at http://www.baliprov.go.id/v1/, the history of tourism in Bali started with Dutch colonialism in Bali in the 20th century marked by the falling of the Klungkung Kingdom in the Puputan Klungkung War in 1908. Under Dutch colonialism, Bali was opened for foreigners. Suryawan (2010, p. 42) suggests that the tourism of Bali started in 1924 with the opening of a weekly cruise route of Singapore – Batavia (Jakarta) – Semarang – Surabaya – Buleleng (Singaraja Port), and Makassar.

Tourists, researchers, and cultural observers from various countries came to Bali to sightsee the beauty of the nature as well as to study the unique art and the culture of Bali. It was the migrants that spread the uniqueness of Bali to the world. Tourists, who initially visited Bali for sightseeing, fell in love with the nature and the culture and eventually stayed and settled down there. The migrants staying in Bali are mostly foreign citizens from different countries, particularly from America, Europe, and Australia.

In 1930, the Bali Hotel was built in the center of Denpasar. The establishment was aimed at supporting the tourism that began to develop at that time. Besides being a place for tourists to stay, the hotel was also used as a place to promote the tourism of Bali to other countries. After the war, as a part of Indonesia, the Balinese also struggled with taking power from the Dutch and Japan.

Tourism in Bali underwent reconstruction after the Independence of Indonesia (i.e., in the decade of the 1950s), and started to regrow in 1963 with the establishment of the so-called Grand Bali Beach Hotel at Sanur Beach. Along with the development of tourism in that era, other areas in Bali also started some development. One of the examples is Nusa Dua, which turned into a one-stop international tourism center. Legian and Kuta developed places for homestays and guest houses in the middle of local housing. Furthermore, Ubud and Gianyar naturally grew into tourist housing with a village ambiance.

Ida Bagus Mantra, the initiator of the cultural tourism ideology, says that the main sources of culture in Bali have normative and operational functions. As a normative function, the culture is expected to possess the capacity and potential to provide some identity, basic reference, and a control pattern to achieve the balance and security of the culture. In terms of operational function, however, the culture is expected to be the main attraction for tourism growth. This emphasizes the importance of culture for the development of tourism, especially for Bali. What is expected to happen is tourism for culture, not the other way around. Furthermore, the culture is not merely for pleasure purposes, but also used as a medium for mutual understanding and respect (Mantra, 1996, p. 35).

Table 1 shows illustrates the development of the tourism sector in Bali provided in the number of either local or international tourists visiting Bali in four recent years.

The data show that the number of tourists, either local or international, tends to rise every year. This resulted in the growing tourism industry in Bali year by year. In addition, the data of international tourists show that 10 countries dominating the annual number are Australia, China, Malaysia, Japan, Singapore, South Korea, France, England, USA, and Germany.

2 MATERIAL AND METHOD

The study uses sociological methodology to explore some comprehensive data through a literature study of different related literature sources, direct field observation of numerous fashion centers in Bali, as well as in-depth interviews of different respondents in fashion social places. Therefore, it is expected that through this study there will be some description of how dynamics of social change in an area influence development in other sectors, which this study is specifically focused on.

The background of the culture of Bali and the beginning of its development in the tourism sector, was obtained from the literature study. The discussion in this study is also expected to

Table 1. Number of foreign visitor to Bali by year, 2013–2016 by Bali Government Tourism Office.

Year	Number of Foreign Visitor to Bali	Growth
2013	3,278,598	13.36%
2014	3,766,638	14.88%
2015	4,001,835	6.24%
2016	4,927,937	23.14%

lead to an analysis on the course of development in the future to enrich knowledge in fashion design specifically, as well as in culture in general.

3 RESULT AND DISCUSSION

Over the course of the development of tourism in Bali, many migrants who initially visited merely to sightsee finally settled there. Most of them ran a business in hospitality, culinary arts, fashion, craft, interior design, and other sectors that support tourism. Their businesses then kept growing into the main motor of the economy sector in Bali. In this context, 'industry' is defined as all sectors related to human activities in the economy which are categorized as productive and commercial. The classification of the industry may then be based on the criteria of raw materials, work forces, market, capital and the type of technology (Hadijah, 2014).

Mapping of the role of migrants, particularly foreign citizens of the fashion industry in Bali, is as explained in detail by Tjokorda Istri Ratna Cora Sudharsana: the global fashion and clothing companies grew rapidly; they were local companies but they were owned by foreigners, predominantly from Italy, Australia, America, Germany, and the Netherlands. Global fashion and clothing in cosmopolitan Kuta became degraded through four stages in the cycle of globalized economic capitalism, as far as the products of global fashion and clothing were concerned. They became degraded at the level of the buyer (stage 1), at the level of the buyer representative (stage 2), at the level of the agent (level 3), and at the level of the agent representative (level 4) (Sudharsana, 2016)

The involvement of migrants in industries and markets, along with their lifestyles, bring cosmopolitan influences to the lifestyle in Bali in general. Epistemologically, the term of cosmopolitanism is derived from the Greek, *Kosmos* and *Politês*. *Kosmos* refers to "the universe" or "the world" while *Politês* refers to "citizen." Simply, cosmopolitan can be defined as the citizen of the world. It is in line with what was written in the memoir of Diogenes: when he was asked "Where are you from?" he answered, "I am a citizen of the world (*kosmopolitês*)" (Diogenes, 1925, p. 71 in Alunaza, 2017). In general, cosmopolitanism can be defined as an idea to revive global citizenship and to promote an identity without territorial limitations (Breckenridge, 2002, p. 2).

On the relationship between market and lifestyle in the fashion industry, Sandy Black says, "Consumers now have the income to influence fashion by their buying decisions" (Black, 2005, p. 32). It can be seen, for instance, that the fashion and culinary industries are an implementation of human basic needs of clothing and food. Most fashion items and food offered are the ones preferred by tourists, or in other words, the products offered should meet the consumers' lifestyle, which may be different from the fashion items and food that belong to Balinese or Indonesian culture. Thus, only products that meet market requirements and preferences are found in the society.

The scope of the fashion industry is also explained by Achwan (2014), wherein the fashion industry can be classified as an aesthetic economy. This type of economy covers all aesthetic markets marked by constantly changing products that emphasize the importance of taste. Creative workers or actors in this aesthetic industry play an important role as they design a product to trade.

In order to map the development of the fashion industry in Bali, the fashion industry involving some fashion brands should be classified based on the types of items. The classification of the fashion product is in accordance with the exclusiveness or the product level, the user, the garment type, and the functional specifications of the product.

The classification of fashion products above, as well as field observations, shows that the development of the fashion industry in Bali is very unique and complex because the condition of the industry, along with the variation and the types of products in Bali, cannot be found in other places in Indonesia. It can be seen from Table 2 that classification of the types of fashion products found in Bali is rich and varied. It is a result of the cosmopolitan lifestyle in Bali and the large number of people from different countries settling in Bali. Therefore, the fashion product as the implementation of the lifestyle may also be varied in line with the lifestyle of the target market.

Table 2. Classification in fashion industry in Bali based in the observation.

Types of Fashion Industry	Classification	Samples of Brands
Fashion industry based on the type of product	Clothes	Biasa, Body & Soul
	Clothes and accessories	Volcom, Rip Curl, Quicksilver, Surfer Girl, Rumble, Electrohell,
	Footwear	Pithecantropus, Ika Butoni
	Bags	Niluh Djelantik, Indosole
	Jewelry	Sabbatha
	Fabrics	Tulola Jewelry, Milk &Happines,
	Eyewear	Thread of Life
	Other lifestyle product	Olenka swimwear
Fashion industry based on the Owner	Balinese	Rumble by I Gede Ari Astina, Tjokorda Istri Ratna Cora, Paulina Katarina, IkaButoni
	Indonesian besides Balinese	Pithecantropus, Sabbatha, Olenka
	Foreign citizen	Biasa, Indosole
	Collaboration of Indonesian and foreign citizen	Thread of Life
Based on local content (tradition & culture)	Having Balinese local content	Tulola Jewelry, Tarum Bali, Ikat Batik
	Having local content from other places in Indonesia	Threads of Life, Pithecantropus, Sabbatha, Ika Butoni
	Having no local content, either from Bali or other places in Indonesia	Olenka, Paulina Katarina, Vol-com, Rip Curl, Quicksilver, Surfer Girl, Indosole

Based on the data and analysis above, it can be concluded that the tourism sector in Bali has a substantial effect on other sectors of life of the Balinese people, including fashion. From the study it can be inferred as well that the area that is considered the center of fashion is tourism. The fashion industry can grow in the area where the tourism also grows. Kuta, Legian, Seminyak, Ubud, Nusadua, and Sukowati are some examples of fashion centers that also serve as the tourism centers in Bali.

REFERENCES

Achwan, R. 2014. *Dua Dunia Seni: Industri Kreatif Fesyen di Bandung dan Bali. Jurnal Sosiologi MASYARAKAT, 19*(1, Januari), 57–75.

Alunaza, H. S. D. 2017. Globalisasi sebagai Katalis Kosmopolitanisme dan Multikulturalisme: Studi Kasus Resistensi Cina terhadap Kosmopolitanisme Intermestic. *Journal of International Studies*,1(2, Mei), 177–189. doi:10.24198/intermestic.v1n2.7

Black, S. 2012. *The Sustainable Fashion Handbook.* London: Thames & Hudson Ltd.

Breckenridge, C. A. 2002. *Cosmopolitanism.* Duke University Press.

Hadijah, I. 2014. *Upaya Peningkatan Export Drive Industri Fashion di Era Globalisasi*, Jurnal Teknologi dan Kejuruan, 37(1, Februari), 95–108.

Mantra, I. B. 1996. *Landasan Kebudayaan Bali.* Denpasar: Dharma Sastra.

Paeni, M. 2009. *Sejarah Kebudayaan Indonesia: Religi dan Falsafah.* Jakarta, Indonesia: Rajawali Pers.

Raharjo, S. et al. 1998. *Sejarah Kebudayaan Bali: Kajian Perkembangan dan Dampak Pari- wisata.* Jakarta: CV Eka Dharma.

Reichle, N. 2011: *Bali: Art, Ritual, Performance.* San Francisco: Asian Art Museum.

Sudharsana, T. I. R. C. 2016. Global Fashion and Clothing Discourse in Cosmopolitan Kuta. *International Journal of Multidisciplinary Educational Research*, 5(8, 1), 1–8.

Soekmono, R. 1991. *Pengantar Sejarah Kebudayaan Indonesia II.* Yogyakarta: Kanisius.

Sumadio, B. 1993. *Sejarah Nasional Indonesia II.* Jakarta: Balai Pustaka.

Suryawan, I. N. 2010. *Genealogi Kekerasan dan Pergolakan Subaltern: Bara di Bali Utara.* Jakarta: Prenada Media Group.

Pulp and paper companies and their fair value: Indonesian stock market evidence

M.I. Miala & F.T. Kristanti
Telkom University, Bandung, Indonesia

ABSTRACT: Investment in the pulp and paper sector is still promising due to high state revenue. Stock investment requires a stock valuation analysis to estimate fundamental data-based intrinsic value or fair price for stock. This study used a discounted cash flow (DCF) – free cash flow to firm (FCFF) method with price-to-book value (PBV) and price-earnings ratio (PER) approaches. Three scenarios, namely pessimistic, moderated, and optimistic, were used in historical data from 2014 to 2018 as the basis for yearly projections from 2019 to 2023. The result using DCF method revealed that INKP's, FASW's, and TKIM's fair price was in overvalued conditions. The result using relative valuation (RV) revealed that there was a significant difference between market value and fair value using FCFF and RV methods with PER approach. However, there was no significant difference between market value and fair value using RV method with PBV approach.

1 INTRODUCTION

Industrial companies engaged in the pulp and paper sector are still one of the biggest contributors to state revenues. The pulp and paper industry in Indonesia obtains some benefits from Indonesia's geographical position on the equator with many types of trees growing on average three times faster than in cold-region countries. As a result, they form large forests, which are the main raw material for the paper industry. The pulp and paper industry in Indonesia has considerable potential and the Indonesian Composite Index (IHSG/*Indeks Harga Saham Gabungan*) of the pulp and paper industry continues to move steadily downward in the past 10 years' observations.

The phenomenon that occurred was a significant increase of share price in 2017 towards three companies, namely Indah Kiat Pulp & Paper Tbk. (INKP), Fajar Surya Wisesa Tbk. (FASW), and Tjiwi Kimia Tbk. (TKIM). Valuation by comparing market value and fair value becomes essential for investors in making decision to buy a stock. An exact decision can make investors gain profits in the form of capital gain or dividends or experience losses due to falling stock prices bought.

2 LITERATURE REVIEW

The company's stock price in IHSG will always change as stock prices are determined through the mechanism of buying and selling (supply and demand). The more shares that are sold, the cheaper the stock price will be, whereas the more shares that are bought, the more expensive the stock price will be. Information and sentiments circulating in the stock market can quickly affect the movement of stock prices. The company's value is an investor's perception of the level of the company's success in managing the exant resources frequently associated with its stock price. High stock prices increase the company's value and increase market confidence in the company's cash flow performance and future prospects.

According to Damodaran (2012), intrinsic value (fair value) is defined as the value of cash inflows obtained throughout the company's life. Stock valuation is the process of determining fair value for a stock. Although stock prices continually change, by knowing their fair value, it will be easier to face market turmoil. Three approaches were used in conducting valuations, namely DCF, contingent claim valuation, and RV. Hutapea, Putri, and Sihombing (2012) examined the intrinsic value of PT Adaro Energi Tbk. using the DCF method. Ivanovski, Ivanovska, and Narasanov (2015) studied the effectiveness of the RV model for the stock valuation in the Macedonian Stock Exchange (MSE). Santos (2017) calculated the valuation of Adidas Group's stock using DCF. Saptono & Kristanti (2018) studied the valuation of Indonesian Provider Tower Industries using DCF with the FCFF approach and RV methods. Based on the study objectives and previous studies, this study used DCF with the FCFF approach, RV with PER, and PBV approaches.

3 METHODOLOGY

The study was conducted with regard to the valuation of pulp and paper–sector companies on the IDX, namely INKP, TKIM, and FASW. The variable used was the fundamental value-based intrinsic value of company shares. Then, those variables were calculated using DCF and FCFF approach and using RV method with PER and PBV approaches. The sampling technique used purposive sampling with criteria: shares on IDX in the pulp and paper sector, active transaction volumes on IDX and not suspended until 2018, and shares experienced significant growth in 2016–2018.

This study was conducted using a Kolomogorov-Smirnov test to compare data distribution (to test its normality) with standard normal distribution. If the significance is below 0.05, its data is abnormal. Then, a t-test was conducted to compare fair value valued by FCFF and market price; and compare fair value of PER and PBV valued by RV market value of PER and PBV, to determine whether it has a significant difference. By using hypothesis test, if P-value > 0.05, there is no significant difference. If stock market prices are smaller than stock fair prices with FCFF and RV, it is called an undervalued condition. If stock market prices are greater than stock fair prices with FCFF and RV, it is called an overvalued condition (expensive).

4 FINDINGS AND DISCUSSIONS

4.1 *Revenue growth and sales projection*

Revenue growth is an important indicator of market acceptance of the company's products or services calculated from year to year. Consistent revenue growth and profit growth are also considered important for companies that are sold to the public through shares. FASW experienced the highest revenue growth of 53.59%, calculated from revenue in 2018 of Rp164,344 billion, up from Rp107,000 billion in 2017. Average revenue growth from 2014 to 2018 of INKP was 8.07%, while TKIM had average revenue growth of 3.56%. The sales projection will be taken in the next five years from 2019 until 2023. The projected revenue growth in 2019–2023 for the three companies – INKP, FASW and TKIM – is by following the industry growth projection of 7.40%. This sector, according to Kristanti and Isynuwardhana (2018) is one of the sectors that has the lowest possible distress time, which is 10.5 years.

The calculation of the company's growth projection in the optimistic scenario is the sum of the value of industry growth projection plus the historical growth spread, and subsequently plus another half the historical growth spread. The value of calculation results for INKP was 13.46%, FASW was 60.24%, and TKIM was 19.13%. In the moderated scenario, the calculation of the company's growth projection was obtained from the sum of the value of industry growth projections plus the historical growth spread only, so that the value of INKP was 11.44%, FASW was 42.63%, and TKIM was 15.22%. In the pessimistic scenario, the calculation of company's growth projection was assumed to be the same as the value of industry growth projection, so that the growth projection of INKP, FASW, and TKIM was 7.40%.

4.2 Value of intrinsic shares condition DCF–FCFF method

Through the results of processing and analyzing the overall data of stock valuation using DCF method, the stock fair value for each company was obtained as presented in Table 1. The data show that in all calculation scenarios, the stock price for the three companies on January 1, 2019 was overvalued compared to their intrinsic value. The low intrinsic value in all scenarios with market value was due to the estimated growth of average revenue growth for the next five years of 7.40%.

4.3 Intrinsic value using relative valuation - PER and PBV

The calculation results using the RV method with PER and PBV approach are shown in Table 2. In the pessimistic scenario with PER approach, TKIM's stock price was cheaper than INKP's and FASW's, which means that, if investing in TKIM shares, the payback period was much faster compared to FASW and INKP shares. Whereas, if using the PBV approach, FASW's stock price was cheaper than INKP and TKIM shares because PBV value of FASW was smaller than INKP and TKIM which was equal to 0.75 meaning that the cash flow of FASW's stock price was 0.75 times its intrinsic value. Thus, it is recommended for investors to choose FASW shares.

Furthermore, the calculation result in the moderated scenario show that PER value of TKIM was lower than FASW and INKP which were 18.23. Therefore, it is recommended investors choose TKIM shares. PBV value of FASW was smaller than INKP and TKIM which were 1.74 times, which means the FASW's stock price was 1.74 times the intrinsic value. Thus, it is recommended investors choose FASW's stock price. Moreover, the result of study in the optimistic scenario show that PER value of INKP was lower than FASW and TKIM. In addition, PBV value of FASW was smaller than TKIM and INKP.

4.4 Analysis of Kolmogorov-Smirnov test and difference test

The normality test result of comparative data between the market value of stock price and intrinsic price of FCFF method was normal in that the significance was 0.136. Besides that, the normality test result between the market value of PER and the intrinsic value of PER was

Table 1. Intrinsic value shares condition using DCF–FCFF Method.

[INKP] PT. Indah Kiat Pulp & Paper Tbk			
Scenario	Market Share Value	Intrinsic Stock Price	Condition
Pessimistic	11,550	2,391	Overvalued
Moderated	11,550	2,731	Overvalued
Optimistic	11,550	2,910	Overvalued
[FASW] PT. Fajar Surya Wisesa.Tbk			
Scenario	Market Share Value	Intrinsic Stock Price	Condition
Pessimistic	7,775	1,876	Overvalued
Moderated	7,775	4,360	Overvalued
Optimistic	7,775	6,158	Undervalued
[TKIM] PT. Pabrik Kertas Tjiwi Kimia Tbk			
Scenario	Market Share Value	Intrinsic Stock Price	Condition
Pessimistic	11,100	3,890	Overvalued
Moderated	11,100	7,346	Overvalued
Optimistic	11,100	8,394	Overvalued

Table 2. Relative valuation - PER and PBV.

[INKP] PT. Indah Kiat Pulp & Paper Tbk

Scenario	Intrinsic PER	Industry PER	Intrinsic PBV	Industry PBV
Pessimistic	16.48	36.52	2.39	1.60
Moderated	18.82	36.52	2.73	1.60
Optimistic	20.06	36.52	2.91	1.60

[FASW] PT. Fajar Surya Wisesa Tbk

Scenario	Intrinsic PER	Industry PER	Intrinsic PBV	Industry PBV
Pessimistic	12.72	36.52	0.75	1.60
Moderated	29.57	36.52	1.74	1.60
Optimistic	41.77	36.52	2.46	1.60

[TKIM] PT. Pabrik Kertas Tjiwi Kimia Tbk

Scenario	Intrinsic PER	Industry PER	Intrinsic PBV	Industry PBV
Pessimistic	9.65	36.52	1.26	1.60
Moderated	18.23	36.52	2.37	1.60
Optimistic	20.82	36.52	2.71	1.60

normal in that the significance was 0.170. Additionally, the normality test result between the market value of PBV and intrinsic value of PBV was normal in that the significance was 0.175. The t-test for FCFF sample shows that P-value < 0.05, which means that there was a significant difference between fair stock price and market stock price. The t-test for the PER sample shows that P-value < 0.05, which means there was a significant difference between fair value of PER and market value of PER, while t-test for PBV shows contradictory result in that P-value > 0.05, which means there was no significant difference between fair value of PBV and market value of PBV.

5 CONCLUSION AND RECOMMENDATION

The result of intrinsic value revealed that since the stock prices are in overvalued conditions, investors at INKP, FASW, and TKIM companies are required to sell or not to buy the stock. The calculation result using RV method with PER approach, in the pessimistic and moderated scenarios, TKIM's stock price was cheaper than FASW's and INKP's stock price, whereas in the optimistic scenario INKP's stock price was cheaper than TKIM's and FASW's stock price. Meanwhile, using RV method with PBV approach, in the pessimistic, moderated, and optimistic scenarios, FASW's stock price was cheaper than TKIM's and INKP's stock price. Statistical test results show that there is a significant difference between the intrinsic value of FCFF and the intrinsic value of PER compared to the market stock price and the market value of PER, while the comparison of PBV's intrinsic value to market value shows that there is no significant difference.

For further study, it is recommended to try using longer historical data (i.e., more than five years) to reduce the risk of abnormal data because one or more historical sample data is significantly different to other samples resulting in different market values. Additionally, it is worth to try other valuation methods such as the contingent claims method. Since stock valuation is very dependent on the data and assumptions used, cleansing the right data needs to be conducted as in other revenue data.

REFERENCES

Damodaran, A. 2012. *Investment Valuation: Tools and Technique for Determining The Value of Any Asset*. New York: Publisher John Wiley and Sons Inc.

Hutapea, E.C., Putri, T.P.P., & Sihombing, P. 2012. *Evaluation of the Fair Price of PT Adaro Energi Shares Using the Free Cash Flow to Firm Method*. Jakarta: Al Azhar University.

Ivanovska, N., Zoran I., & Zoran, N. 2014. Fundamental analysis and discounted free cash flow valuation of stock at MSE. *UTMS Journal of Economics*, 5(1), 11–24.

Kristanti, F.T., & Isynuwardhana, D. 2018. How long are the survival time in the industrial sector of Indonesian companies. *International Journal of Engineering & Technologies*, 7(4.38). 856–860.

Santos, D.F. 2017. *Adidas Group Equity Valuation*. Lisbon: Universiade Catolica Portuguesa. Unpublished.

Saptono & Kristanti, F.T. 2018. Does the stock of Indonesian Provider Tower Industry have a fair value? *Global Journal of Business and Social Science Review*, 6(4), 130–139.

Sunda culture values at Sunda restaurant design in Bandung

T. Cardiah, R. Wulandari & T. Sarihati
Telkom University, Bandung, West Java, Indonesia

ABSTRACT: Restaurants as a public space are the most popular choices for urban residents today because restaurants are not only places to enjoy food but also a gathering place, with family, friends, and colleagues, and some restaurants are even tourist destinations. The attractions of Sudanese restaurants are the atmosphere and values of Sundanese culture, through interior elements both physically and non-physically. Sundanese cultural values do not always appear as an architectural statement or expression but appear in the form of aesthetic values. The research method used is a combination of qualitative and quantitative methods to obtain measurable data, while data collection is through questionnaires and focus group discussion (FGD). The results of the study are expected to give guidance and review/evaluate the implementation of aesthetic elements and Sundanese cultural values in the context of Sundanese culture education and preservation both for designers and the general public.

1 INTRODUCTION

Urban communities have a lifestyle tendency toward always looking for entertainment, recreation, and tourism, and especially culinary tourism. Well-designed interiors are a vital element for new restaurants, as dining out has become a form of entertainment. Restaurants are chosen for ambience as well as cuisine, and critics often review the design as well as the food (Pile, p. 459). Owners of capital have responded by providing various types of restaurant trademark choices, from traditional to modern in a variety of themes. Thematic design has an important role as an attraction and selling point of a restaurant: one theme is Sundanese culture. Elements and cultural values, both physical and non-physical, serve as a form of communication and trademark information on the restaurant in question. By presenting elements and values of Sundanese culture, restaurants can provide a unique experience for visitors. Theme restaurants are created around a selected image or idea, or "branding" that appeals to our visual and intellectual senses of place, persons, things, events, experiences, or the restaurant's brand (Rosemary, p. 302).

Interior designers can determine every visual aspect of a restaurant, from décor to seating configuration, tableware, graphics, and menu – sometimes even to the plating and presentation of the food – and will direct the sound level and choice of music to create the desired atmosphere (Pile, p. 459). At present, the restaurant's brand is a lifestyle commodity so restaurants must be packaged more attractively: one of them is by adopting Sundanese cultural values. Trademark restaurants must be able to provide satisfaction for visitors, in terms of comfort, security, owned experience, and even in the form of recreation and education. Restaurant atmosphere used to be developed strictly by offering diners a change of pace from their normal environment, but increasingly, today's guests use the experience of dining out as part of their everyday life (Katsigris, p. 80).

2 RESEARCH METHODS

The method of data collection used is a combination of qualitative and quantitative methods to obtain measurable data related to the assessment of the atmosphere and elements of

Sundanese culture in restaurants. Questionnaires and focus group discussions were conducted to obtain data on Sundanese aesthetics on interior elements and cultural values such as a way of eating or eating "ngariung." The questionnaire was compiled with specific parameters summarized from the findings of previous research entitled "Designing the Typology of Sundanese Restaurants."

3 RESULTS AND DISCUSSION

Restaurants and dining out is a big business today, and restaurant design is probably one of the most innovative arenas for creating interior environments (Rosemary, p. 302). Sindang Reret Restaurant features traditional elements such instrumental Sundanese music, fish ponds, a lesehan area that resembles a "saung" near a fish pond, waitress uniforms with kebaya, and wooden architecture that can still be recognized as Sundanese. The Sajian Sambara restaurant features contemporary architecture with the use of bamboo or material resembling bamboo, the waitress uniforms "nyi iteung and kabayan," Sundanese pop music, and no lesehan dining area. The Sunda Pavilion restaurant's interior has the feel of a Dutch villa house. However, this restaurant provides a lesehan dining area, because of the restaurant manager's vision that one of the representations of Sundanese restaurants is the lesehan area, which is good in the form of "saung."

The type of service differs in all three restaurants. Sindang Reret provides à la carte services so visitors choose food from the menu provided, although this restaurant offers an open kitchen to show the cleanliness of the food serving process. The other two restaurants, the Sunda Pavilion and Sajian Sambara, offer buffet services or a buffet with an open kitchen. This will also be a parameter of visitors' assessment of Sundanese restaurants. For the design of interiors, technology has added a new challenge: in addition to planning spaces to accommodate conventional furniture and furnishings, designers must now plan rooms, in homes as well as in every type of commercial or institutional space, to accommodate the equipment that has become an integral part of modern life (Pile, p. 446).

3.1 *Sindang reret restaurant*

The main reason for choosing a lesehan area is that eating is more relaxed, intimate, and comfortable, allowing for following friends, landscape elements, being close to the pool and able to see the scenery. Table chairs are considered more practical because visitors do not need to take off their shoes even though the atmosphere is more formal.

Figure 1. Interior of Sindang Reret restaurant.

The key to good dining-space design is to find the right balance between comfort, security, and the guests' tolerance for stimulation, and the target market is key (Katsigris, p.48). The Sundanese atmosphere in Sindang Reret restaurant is dominated by smells; visual elements in the form of decorations, materials, and waitress uniforms; and audial elements in the form of music and sounds of gurgling water. Sindang Reret restaurant feels like a home, suitable for use with family because of its relaxed atmosphere, and support for intimacy and gathering. This is supported by design elements such as structuring group chairs, natural elements (sound of water), and lesehan eating places. The results of questionnaires, observations, and interviews, indicate that the dominant cultural elements in Sindang Reret restaurant are non-objects or intangible elements in the form of music and the sound of water, and an atmosphere conducive to gathering and relaxing. This element appears in the form of gurgling sounds, Sundanese music, lesehan, and dining table layouts for large groups. Themes can be made rather subtle in their design, leaving one's imagination to make strong thematic connections. Contextual restaurant and café design concepts are based on good design principles and materials. The interior spaces and finishes are supportive of the dining experience and do not create an artificial or theatrical atmosphere (Rosemary, p. 302).

3.2 *Sajian sambara restaurant*

After all, the most important elements in a restaurant are food and service, However, it is the look and feel of the restaurant that invites people to go in (Katsigris, p. 51). In the question related to the creator element of the atmosphere in the Sambara restaurant, the dominant answer referred to the decoration of the room. The next thing most popular answer was music, followed by posters and photos. There are no lesehan areas or landscape elements, but the Sambara restaurant has characteristics of Sundanese culture in the form of a buffet. The components of a well-designed restaurant do not go unnoticed by its customers. Although they may not be able to pinpoint specifics, any person will tell you that attractive surroundings seem to make a meal better (Katsigris, p. 48).

The results showed that participants rated Sambara restaurant more formal because there was no dining area, furniture layout, and high chair models, and the modern design and ordering of food by taking it for themselves was the reason that Sambara's atmosphere was more relaxed than Sindang Reret. The successful layout of a dining area includes balancing multiple components: the safety of guests and employees, efficiency of service, aesthetics, and financial implications (Katsigris, p. 72). The authentic value of food served in Sundanese restaurants, that Sundanese cuisine should prioritize authentic cuisine and not aesthetic cuisine, and that food and service are the main elements are things that must be considered. Efficiency, value, and convenience are the hallmarks of modern restaurants, and all of these can be reflected in their design.

Figure 2. Atmosphere of the food serving room in Sambara Restaurant.

3.3 *Paviliun Sunda restaurant*

There are a number of reasons for choosing the lesehan area as a more relaxed area: "close to the water"; "close to the open area"; "there is a fish pond." In this very competitive industry, most merchants understand that good design is a way to set a restaurant apart from its competition, bringing the theme and concept to life (Katsigris, p. 48). In comparison between restaurants, participants considered the Sundanese Pavilion to be more intimate, family-friendly and relaxed. The assessment was based on the atmosphere of the room that resembled a house and more contemporary design. When people enter your new restaurant, how do you want them to feel? All the physical surroundings and decorative details of a foodservice establishment combine to create its atmosphere, the overall mood that also may be referred to as the ambience or energy of the space (Katsigris, p. 48). The majority of participants considered that buffet food presentation was more appropriate for Sundanese restaurants, as was the open kitchen model.

Figure 3. Lesehan and table chairs, decorative elements in the form of bamboo material in Paviliun Sunda restaurant.

Regarding the offer of authentic Sundanese menus, the majority of participants considered that the Sambara restaurant offered a more authentic menu than the other two restaurants. Interior designers can determine every visual aspect of a restaurant, from décor to seating configuration, tableware, graphics, and menu – sometimes even to the plating and presentation of the food – and will direct the sound level and choice of music to create the desired atmosphere (Pile, p. 459).

4 CONCLUSIONS AND RECOMMENDATIONS

From the experimental results of the experience of the atmosphere in Sundanese restaurants – Sindang Reret Restaurant, Sambara Serving Restaurant, and Pavilion Sunda Restaurant – it was found that intangible elements or non-tangible elements in the form of sound and creating a relaxed atmosphere were important to creating Sundanese atmosphere. The distinctive sound of the Sundanese atmosphere appears in the form of Sundanese instrument music or Sundanese pop music and the sound of water via a fish pond or fountain. This sound element is quite strong and is characteristic of Sundanese restaurants.

Meanwhile, to present an atmosphere of fun, relaxation, and togetherness, the restaurant manager presents a lesehan sitting area and structures the dining area in the form of a table for groups. The arrangement of furniture and the choice of dining furniture are secondary elements that arise as a result of these intangible elements. The next intangible element is serving food through a buffet model and an open kitchen. The open kitchen in Sundanese restaurants is part of the presentation of food that offers aroma in the restaurant. Notwithstanding its drawbacks, this celebrity culture generates media coverage that raises awareness of, and appreciation for, good design. The result is a greater demand for well-designed spaces, both public and private. Design has thus assumed the role of essential marketing tool (Pile, p. 447).

REFERENCES

Almanza, B.A., & Kotschevar L.H. 1985. *Food Service Planning Layout, Design and Equipment*. New Jersey: John Wiley & Sons.
Beriss, D., & Sutton, D. 2007. *The Restaurants Book: Ethnographies of Where we Eat*. New York: Berg.
Guest, K.J. 2016. *Essentials of Cultural Anthropology: A Toolkit for a Global Age*. Baruch College, The City University of New York.
Im, Y.S. 2002. *Interior World, Commercial Space II, Cafe, Restaurat, Shop, Designer & Works*. Korea: Archiworld Co.
Katsigris, C.T. 2009. *Design and Equipment for Restaurants and Foodservice: A Management View*, third edition. Costas.
Kaplan, M. 1997. *Theme Restaurants*. New York: PBC International, Inc.
Kilmer, R., &Kilmer, W.O. 2014. *Designing Interiors*, second edition. John Wiley & Sons.
Pile, J.F., & Gura, J. 2000. *History of Interior Design*, fourth edition. London: Laurence King Publishing. www.laurenceking.com.
Quinn, T. 1981. *Atmosphere in the Restaurant*. Michigan: Michigan State University.

Post acquisition analysis and evaluation with regard to diversification

K. Rosby & D.A. Chalid
Universitas of Indonesia, Jakarta, Indonesia

ABSTRACT: The purpose of this study was to evaluate an acquisition that had occurred four years earlier and to analyze whether it was successful or could add value to the acquiring company, considering that there were many acquisitions that failed (the "winner's curse"). After analysis and evaluation of the actual achievements and forecasts of the acquisition, it was found that the acquiring company suffered a loss in this acquisition, where the acquired company could not provide added value to the acquiring company.

1 INTRODUCTION

1.1 Background

An acquisition is expected to be able to provide added value to the acquiring company so it can increase the acquiring company's shareholder value. Steger and Kummer (2007) found that overconfidence based on the success of previous mergers and acquisitions is one factor in failure in mergers and acquisitions. Depamphilis (2017) states that overpayment leads to higher profitability expectations and results in excessive goodwill arising from the overpayment that must be written off, thereby reducing the company's profitability.

In 2015 Microsoft Corporation wrote off 96% of the value of the cellphone business it obtained from Nokia for $7.9 billion. Google has written off $2.9 billion in the cellphone business it bought from Motorola for $12.5 billion in 2012; HP has recorded a write-off of $8.8 billion from the acquisition of Autonomy for $11.1 billion; and in 2011 News Corporation sold MySpace for just $35 million after acquiring it for $580 million just six years earlier.

The acquiring company studied in this research acquired with the intent of diversifying its business, especially in the commodity industry. As the acquiring company had a reserve of excess cash, it looked for the most effective way to obtain additional net income or free cash flow in a short time.

The construction industry is an option for diversifying the acquiring company's business because, as a result of the new president's focus on infrastructure development, the focus of the state budget, which is usually on subsidies to the community, has moved to infrastructure development, which is considered to have lagged behind that of other countries.

Previous research found several different results for the performance of the acquiring company after mergers and acquisitions. Some studies suggest that there is a " winner's curse" effect because, even though the acquirer managed to acquire a company, the acquirer will also suffer losses as a result of the acquisition.

1.2 Research framework

To answer the problem statement, the following framework was used for this research:

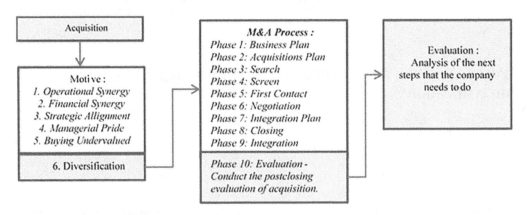

Figure 1. Research framework.

2 LITERATURE REVIEW

2.1 Mergers and acquisitions

Depamphilis (2017) outlines the process of mergers and acquisitions as follows:

- Phase 1: Business Plan—Develop a strategic plan for the entire business.
- Phase 2: Acquisitions Plan—Develop the acquisition plan supporting the business plan.
- sPhase 3: Search—Search actively for acquisition candidates.
- Phase 4: Screen—Screen and prioritize potential candidates.
- Phase 5: First Contact—Initiate contact with the target.
- Phase 6: Negotiation—Refine valuation, structure the deal, perform due diligence, and develop the financing plan.
- Phase 7: Integration Plan—Develop a plan for integrating the acquired business.
- Phase 8: Closing—Obtain the necessary approvals, resolve postclosing issues, and execute the closing.
- Phase 9: Integration—Implement the postclosing integration.
- Phase 10: Evaluation—Conduct the postclosing evaluation of acquisition.

2.2 Failures in mergers and acquisitions

Gadiesh and Ormiston (1999) conclude that there are five main causes of failure of mergers and acquisitions:

1. Bad strategic plan development
2. Cultural incompatibility
3. Difficulties in communicating and leading organizations
4. Poor integration planning and implementation
5. Paying too much for the acquired company

3 RESEARCH METHODOLOGY

3.1 Research steps

This study considered the difference in value when a company was analyzed for acquisition compared to the current value at the time of acquisition and outlined the next steps that must be taken by the company after the 2014 acquisition. The data used are those of the acquired company for 2014–2018. This research valued both the acquiring and the acquired company with free cash flow to the firm valuation.

3.2 Company profile

The acquiring company operates in Indonesia and is engaged in the commodity industry. It has about 51 subsidiaries that have a value chain in the commodity industry that is integrated from upstream to downstream.

Until 2014 business in the commodity industry still showed weakness, and at the end of 2014 the acquiring company decided to diversify into the construction industry, which was estimated to have better prospects in the future.

The acquired company operates in Indonesia and is engaged in the construction industry. The company was acquired by an acquiring company for cash payments. The company has five subsidiaries that are also engaged in the construction industry. The source of the majority of the revenue from the acquired company is contracts with state-owned enterprises or government funding fromthe state budget and expenditures that focus on infrastructure development.

4 ANALYSIS

The analysis was performed by comparing the equity value and sell value. After calculating using the steps outlined in the preceding chapter, the equity value of the acquired company was as shown in Table 1.

Sale value can also be calculated from the company's stock price of Rp1,310 per share times 700 million shares, so a sale value of Rp917 billion is obtained, with the calculations shown in Table 2.

The two calculations in Tables 1 and 2 show that the equity value of the acquired company is negative while the sale value is positive and is greater than the equity value. This means that, based on the calculation, the company acquired is better sold than maintained as part of the acquiring company.

In divesting, several qualitative factors need to be considered because they will greatly affect the survival of the acquiring company in the future. Some of the advantages and disadvantages to divesting are shown in Table 3.

Table 1. Equity value.

Year	2019	2020	2021	2022	2023	Equity
Free cash flow	−212,717	−693,998	−722,680	−783,059	−822,647	
Year	2024	2025	2026	2027	2028 & TV	
Free cash flow	−873,804	−932,351	−973,767	−1,012,364	−11,164,540	−18,191,928

Table 2. Sale value.

Description	Jumlah
Market value (rupiah per lembar)	1,310
Shares (juta lembar)	700
Sale value (juta rupiah)	917,000

Table 3. Advantage and disadvantage for the divesting company.

Advantages
1. Can increase value for the acquirer company, by being used to invest in existing business working capital or pay off debt.
2. Provide additional income or reduce expense to the acquirer.
3. The acquirer company can focus more on developing the acquirer's main business.

Disadvantages
1. Can give a bad impression to investors that the acquirer company made an acquisition error.
2. When an acquisition is made in the future, investors will question whether the acquisition is the right step or not.
3. Gives a negative impression that the management group of the acquiring company is not able to manage the business that has been acquired.

5 CONCLUSION

The company has the option to

a. Retain, provided that the acquired company must be able to achieve several operational indicators in order to get more value for its shareholders consisting of GP margin of 82%, or AR days of 282 days, or revenue growth of 8%, or operating expense to revenue ratio of 5%, or AP days of 202 days with other assumptions considered permanent.
b. Sell, noting that the acquirer company risks decreasing investor confidence in future acquisitions, which may experience the same failure as in the case of this company. The sale of all shares of the company will provide additional cash of Rp917 billion that can be used for working capital needs or payment of existing debt. The sale of shares of a company that is acquired is considered capable of providing additional value to shareholders compared to retaining the company.

REFERENCES

Depamphilis, D. M. 2017. *Mergers, Acquisitions, and Other Restructuring Activities*, 9th ed. London: Elsevier.
Gadiesh, O., & Ormiston, C. 2002. Six rationales to guide merger success. *Strategy and Leadership*, *30*(4), 38–40.
Steger, U., & Kummer, C. 2007. Why merger and acquisition (M & A) waves reoccur: The vicious circle from pressure to failure.

Financial efficiency of metal and mineral mining company in Indonesia

N.F. Sembung & F.T. Kristanti
Telkom University, Bandung, Indonesia

ABSTRACT: This study seeks to analyze the efficiency level of seven metal and mineral mining companies in Indonesia during the period of 2013-2017. Stochastic Frontier Analysis (SFA) method was employed in this study to estimate the technical efficiency score and their relationship with input, explanatory, and output variables. The results showed that overall, metal and mineral mining companies in Indonesia had been efficient with an average technical efficiency score of 0.8635, and the most prominent variables were cost of revenue, operating expense, and total equity. Based on the ownership concentration of the companies, metal and mineral mining companies owned by private sector were more efficient than metal and mineral mining companies owned by the Indonesian Government.

1 INTRODUCTION

As unsustainable materials, metal and mineral need to be utilized wisely in order to meet the needs of people in the world. Thus, their utilization must be executed properly by the metal and mineral mining companies. In 2012-2017, the metal and mineral mining business in Indonesia had a good business performance, which was indicated by the increasing growth of mining company revenue in Indonesia from 2012-2017. However, this increase was not accompanied by the growth in financial ratio growth, Net Profit Margin, EBITDA and EBIT. This issue could unfortunately lead to decreasing performance of this strategic sector of the Republic of Indonesia. This is supported by the study of Kristanti and Isynuwardhana (2018a; 2018b) who found evidence that profitability is an important determinant for the survival of this sector. Kristanti et al (2016), early identification of potential failures will help the company solve the problem. Therefore, a study focusing on the metal and mineral mining company's efficiency level is significant to be conducted.

Efficiency is a parameter that theoretically underlying the company's performance. Considering the importance of company's efficiency, some previous studies on this field have been conducted. Samal, Mohanty and Sharma (2005) studied the technical efficiency of nine mining companies in Illinois, the United State of America, using Stochastic Frontier Approach (SFA) method using 1989-2001 data. Then, Reddy, Sudhakar, and Krisna (2013) conducted a study on the efficiency to 15 open cast mines of a coal company in India using Data Envelopment Analysis (DEA). Similarly, Honglan, Ruyun, and Xiaona (2014) analyzed the production efficiency of five mining locations of a mining company in China using Data Envelopment Analysis (DEA) using data obtained in July-December 2012. Then, Hosseinzadeh, Smyth, Valadkhani, and Le (2016) analyzed the efficiency of 33 mining companies in Australia in 2008-2014 using Bootstrap DEA method.

A study on company financial efficiency analyzed using Stochastic Frontier Approach (SFA) has not been found. Hence, this study seeks address the gap by analyzing the financial efficiency level of metal and mining companies listed on Indonesia Stock Exchange using SFA, and based on its ownership concentration. This study involved two state-owned and five metal and mineral mining companies in Indonesia. By knowing the level of efficiency of the company and the most influential main variables, the management are expected to make better decisions in increasing the competitiveness and profitability of the company.

2 LITERATURE REVIEW

Financial efficiency is part of financial management that focuses on the company's function in performing effective management. Its application does not stand alone but requires various other disciplines, such as marketing management, production management, accounting, microeconomics, macroeconomics, quantitative methods, and so on. Almost all businesses, both large and small scales, and those that have profit motives and non-profit motives, will have great attention in the financial sector. As a result, an efficient financial sector in a company will greatly support the progress of the company.

Several approaches have been developed in measuring and comparing the efficiency of a company. Saxena, Thakur and Singh (2009) categorize efficiency comparisons with two methods: average method and the frontier method. The average method compares the target variable with the average performance, while the frontier method compares the variable by taking the best value of the compared variable. Furthermore, Burger and Humprey (1992) state that measuring company's efficiency can be done using traditional approach and frontier approach. The Stochastic Frontier Approach (SFA) can be used to measure the cost, profit or production relationship between input and output. The advantage of the SFA method compared to other parametric methods is that SFA allows errors, namely random error and cost inefficiency. Random error is assumed to follow symmetrical standard distribution. Whereas, cost inefficiency is assumed to follow asymmetrical distribution. In specific, this method is used to know efficiency level from time to time, hence it resulted in clear efficiency growth can be easily assessed.

3 METHODOLOGY

This study employed quantitative method since it measures certain variables using numerical data. In specific, it is also a cross-sectional study and uses time series since this study only used data from the financial statements of metal and mineral mining companies in Indonesia during the period of 2013-2017. Furthermore, the unit analysis of this study is metal and mining companies listed on Indonesia Stock Exchange (IDX). Out of nine companies listed on IDX, there were seven companies selected using purposive sampling with following criteria: (1) listed during period of 2013-2017, and (2) Had published the financial reports in period of 2013-2017. Hence, based on the ownership concentration, the selected companies comprise two state-owned companies: PT. Aneka Tambang and PT. Timah; and five private-owned companies: PT. Cita Mineral Investindo, PT. Cakra Mineral, PT. Central Omega Resources, PT. Vale Indonesia, and PT. J Resources Asia Pacific.

The input variables used are cost of revenue, operating expense, inventories, net fixed assets, personal expense and total equity. Meanwhile, revenue amount is the only output variables used. The data were obtained using document study as it provides information on theories, concept, relationship between variables obtained from secondary research, namely financial reports published in the official websites of Indonesia Stock Exchange and the companies involved in this study.

The output parameters that is representing profit function in this study, are evaluating how close a company obtains profit as achieved by the best company within same exigent condition. The profit is representing a function of input, output, and environment variables as following:

$$\ln[(\pi) = f_\pi(y,w,v) + \ln[u_\pi] + \ln[\varepsilon_\pi]] \qquad (1)$$

Where π representing profit variable, y representing output variables, w representing input variables, and v representing environmental variables that might influence corporate performance. In addition, u representing controllable factors that might influence efficiency, while ε representing uncontrollable factors or random errors. Efficiency score will be measured and compared following the SFA approach, and then regression will be performed to identify the significance level of each variables impact to the efficiency result.

Stochastic Frontier Analysis is expected to be more adequate than other methods, as it does not specifically use one of the companies as a benchmark. In its execution, Frontier 4.1 software was used to apply SFA model 2 proposed by Battese and Coelli (1995) using the effects of technical inefficiencies in the translog production function model in estimating technical efficiency. The basic SFA function model is presented by Equation 1. Specifically, this study applied the stochastic translog production function model presented by Equation 2, and inefficiency function model presented by Equation 3 below:

$$Y_{it} = \exp(X_{it}\beta + V_{it} - U_{it}) \qquad (2)$$

$$\begin{aligned}\ln(Y_{it}) = \\ \beta_0 + \beta_1 lnK_{it} + \beta_2 lnL_{it} + \beta_3 lnO_{it} + \beta_4 t + 0.5\beta_5 ln(K_{it})^2 + \beta_6 lnK_{it}(lnL_{it}) + \\ \beta_7 lnK_{it}(lnO_{it}) + \beta_8 lnK_{it}(t) + 0.5\beta_9 ln(L_{it})^2 + \beta_{10} lnL_{it}(lnO_{it}) + \beta_{11} lnL_{it}(t) + \\ 0.5\beta_{12} ln(O_{it})^2 + \beta_{13} lnO_{it}(t) + \beta_{14} t^2 + V_{it} - U_{it}\end{aligned} \qquad (3)$$

$$U_{it} = \delta_0 + \delta_1(FS_{it}) + \delta_2(A_{it}) + \delta_3(G_{it}) + \delta_4(t) \qquad (4)$$

Information: K = capital expense; Y = revenue of the company; O = other inputs; L = personnel expenses; β = natural log of vector of each input variables; t = time period of observation; V_{it} = random error which assumed to be independently distributed $N(0,\sigma^2)$; FS = total assets; U = non-negative random variables which assumed to be independently distributed $N(0,\sigma^2)$; G = GDP per capita; δ = natural log of vector of each explanatory variables, and i = number of companies.

4 FINDINGS AND DISCUSSION

The results of this study revealed that there were three variables that had the most significant impact on efficiency: cost of revenue, operating expense, and total equity. This finding showed that technical efficiency in metal and mineral mining companies would increase along with cost of revenue downsizing, operating expense restructuring and maximizing total equity resources. While revenue as the only output variable, had a significant effect on the value of efficiency. This output variable in the SFA measurement method had a positive influence on the efficiency of metal and mineral mining companies. This findings are in contrary with De Araujo et al. (2012) revealing that the revenue, as output variable, influenced the decrease in technical efficiency.

Table 1 presents the descriptive statistics of the input and output variables in the period of 2013-2017.

PT. J Resources Tbk. had the highest level of efficiency in Indonesia with an efficiency score of 0.9571. PT. Central Omega Resources was in the second position with an efficiency score of 0.9309, followed by PT. Vale Indonesia Tbk. with an efficiency score of 0.9103. In addition, the two state-owned metal and mineral mining companies, PT. Aneka Tambang Tbk. and PT. Timah Tbk., apparently occupied the last two positions. PT. Aneka Tambang Tbk occupied the sixth position with an efficiency score of 0.7349 and followed by PT. Timah Tbk. in seventh position with an efficiency score of 0.7265. There were seven mining companies in the metal and mineral sector in Indonesia that had efficiency scores above the average (above 0.8635), namely PT. J Resources Asia Pacific Tbk., PT. Central Omega Resources

Table 1. Descriptive Statistics of Metal and Mineral Mining Sector 2012-2017.

Variable Category	Variable	N	Max	Min	Mean	St.Dev
Output	Revenue*	42	12,913,740	0	4,607,627	4,463,855
Input	Cost of Revenue*	42	11,009,727	0	3,745,794	3,922,977
	OPEX*	42	1,259,629	7,904	415,028	387,208
	Inventories*	42	3,384,026	6,402	1,005,879	939,352
	Net Fixed Assets*	42	22,119,154	6,952	5,128,540	7,016,870
	Total Equity*	42	25,301,271	554,346	7,170,673	8,070,793
	Personal Expenses*	42	5,028,244	10,585	661,499	1,328,106

* in millions of Rupiah.

Tbk., PT. Vale Indonesia Tbk., PT. Cakra Mineral Tbk., and PT. Citra Mineral Investindo Tbk. Lastly, there were two companies with efficiency scores below the average efficiency in Indonesia: PT. Aneka Tambang Tbk. and PT. Timah Tbk.

Out of the seven companies, two state-owned companies, PT. Aneka Tambang Tbk. and PT. Timah Tbk., were positioned in the two last positions, with efficiency scores of 0.7349 and 0.7365 accordingly. In contrary, the five private-owned companies, PT. J Resources Asia Pacific Tbk., PT. Central Omega Resources Tbk., PT. Vale Indonesia Tbk., and PT. Cakra Mineral Tbk., were positioned in the first top five, with efficiency scores of 0.9571, 0.9309, 0.9103. 0.9023, and 0.8825 accordingly. Hence, it can be inferred that private-owned metal and mining companies were more financially efficient compared to state-owned companies.

5 CONCLUSIONS AND RECOMMENDATIONS

The findings revealed that input variables consisting of cost of revenue, operating expense and total equity had a significant influence on the efficiency score of metal and mineral mining companies. The metal and mineral mining companies in Indonesia had already moderately efficient, but there were still room for metal and mineral mining companies to improve their technical efficiency through the improvement of cost of revenue, operating expense and total equity. Seen from the technical efficiency scores, PT. J Resources was the most efficient metal and mineral mining company during the period, while PT. Timah was the most inefficient metal and mineral mining company during the period. Hence, it can be inferred that based on its ownership concentration, private-owned metal and mineral mining companies were more efficient than state-owned companies.

As the final remark, the policymakers are recommended to supervise the company's financial efficiency, through Indonesian Stock Exchange, in order to maintain the good performance of companies executing metal and mineral products, since their existence are not sustainable and cannot be recycled.

REFERENCES

Berger, A., & Humprey, D. (1992). Efficiency of Financial Institution: International Survey and Directions for Future Research. European Journal of Operational Research, 97–05.

Honglan, L., Ruyun, Y., Xiaona, Q. (2014). Productivity Analysis and Benchmark Selection of X Mining Company by DEA, JOPCR, ISSN: 0975-7384.

Hosseinzadeh, A., Smyth, R., Valadkhani, A. and Le. (2016). Analyzing the Efficiency Performance of Major Australian Mining Companies using Bootstrap Data Envelopment Analysis, Research Gate.

Kristanti, F.T & Isynuwardhana, D. (2008a). Survival Analysis of Industrial Sectors in Indonesia Companies. Jurnal Keuangan dan Perbankan. 22(1). 23–36.

Kristanti, F.T. & Isynuwardhana, D. (2018b). How long are The Survival Time in The Industrial Sector of Indonesian Companies. Internationa Journal of Engieering & Technologies. 7(4.38). 856–860.

Kristanti, F.T., Effendi, N., Herwany, A. & Febrian, E. Does Corporate Governance Affect the Distress of Indonesian Company? A Survival Analysis Using Cox Hazard Model with Time-Dependent Covariates. Advanced Science Letters. 22. 4326–4329.

Reddy, G.T., Shudakar, K., Khrisna, S.J. (2013). Benchmarking of Coal Mines using Data Envelopment Analysis, International Journal of Advanced Trends in Computer Science and Engineering. 2(1).159–164.

Samal, A.R., Mohanty, M.K., Sharma, M.C. (2005). Technical efficiency Analysis in Illinois Coal Mining Sector, Taylor & Francis Group, London, ISBN 04 1537 4499.

Saxena, V., Thakur, T., and Singh., R.P., (2009). Evaluating the Performance of Mobile Telecom Operators in India. International Journal of Simulation System, Science, & Technology IJSSST, 10(4).

Battese, G., & Coelli, T. (1995). A Model for Technical Inefficiency Effects in a Stochastic Frontier Production Function for Panel Data. Empirical Economics, 325–332.

Identification of tourism destination preferences based on geotag feature on Instagram using data analytics and topic modeling

H. Irawan, R.S. Widyawati & A. Alamsyah
Telkom University, Bandung, West Java, Indonesia

ABSTRACT: Tourism is one of the fastest growing sectors in the world and has become a leading sector in Indonesia. The Indonesian government has determined four superpriority tourist destinations in order to achieve visits by 20 million foreign tourists and 275 million domestic tourists in 2019. Moreover, a social network provides essential information that raises a new paradigm known as smart tourism. The geotag social network can identify tourists' preferences, perceptions, and also attitudes toward tourist destinations from their posts on Instagram. It is important for the government and tourism management to understand tourists' preferences and interests to develop a better strategy and policy. This study provides an alternative method to evaluate the superpriority tourist destinations using data mining shared on social media. It identifies the pattern of tourist visits and tourists' interest related to their opinions of the tourist destinations using topic modeling, which generally describes the scenery, favorite spots, and events. In conclusion, this article provides recommendations for determining favorite tourist destinations based on tourists' preferences as expressed on social media.

1 INTRODUCTION

The tourism sector is one of the largest economic sectors and has the fastest growth rateworldwide. It is considered the leading sector in Indonesia because it has positive growth and also an important role in the economy. The Indonesian government has set a target of 275 million domestic tourists and 20 million foreign tourists to visit tourism destinations in Indonesia in 2019. To achieve this goal, the government is supporting the acceleration of the development of the tourism sector in four superpriority destinations: Mandalika in West Nusa Tenggara, Labuan Bajo in East Nusa Tenggara, Borobudur in Central Java, and Toba lake in North Sumatra.

In general, both tourists and locals have different motivations and activities when they visit tourist destinations, so the management of the tourist destinations must have a good understanding of the behavior and preferences of the visitors (Vu et al., 2016). The internet and social media have been widely used by tourists to find, organize, and share travel stories and their experiences (Leung et al., 2013). Social media has the potential to generate different information compared with other data sources, provide massive and real-time insight, and is also very widely available. Currently, Instagram is a popular form of social media to share photos supplemented with information about geographic locations (geotags). One of the common Instagram content types is a photo or video post when someone visits historic buildings and interesting locations (Mukhina et al., 2017). This study produced analytic data derived from Instagram users' posts on travel destinations by using a time window strategy to separate the posts between tourists and locals. This study also used topic modeling to identify topics that are often discussed by tourists related to the tourist destinations on Instagram in order to make it easier to understand the preferences and opinions of the tourists as feedback from their experiences during their visit. Tourist opinions can be utilized to gain more insight into the strengths and weaknesses of destinations, and the management of the destination can thereby increase its quality (Alamsyah et al., 2015). This study aimed to identify tourists' preferences based on their experiences posted on Instagram and reflected through the posting patterns and the topics often discussed regarding the four superpriority destinations in Indonesia. This study is expected to provide insights for the government and tourist destination management in identifying the preferences and interests of tourists through social media to facilitate better management policy and strategy decision-making related to the development of tourism in Indonesia.

2 LITERATURE REVIEW

The development of tourist destinations is accompanied by a high usage of social media that includes the ability to upload photos with ease along with information on the location (Okuyama & Yanai, 2013). Some studies aiming to determine the perception of tourists related to tourism destinations have been conducted before, such as one that took advantage of geotagged photos posted on social mediato understand the preferences and activity of tourists when visiting urban parks in Hong Kong (Vu et al., 2016).

One study utilized geotagged photos on social media to identify and analyze the main tourist attraction in Europe by separating tourists and locals so that the areas visited by tourists and locals could be distinguished (Garcia-Palomares et al., 2015). There are several methods that can be used to identify tourists and surrounding locals, as in the study of Mukhina et al. (2017), which used the time window approach to distinguish tourists and locals to identify differences in each of their favorite places.

In analyzing large amounts of text data or reviews, topic modeling can be used to determine the topics that are often discussed in the data. Topic modeling produces a set of topics that are a collection of words that appear together in the document (Hunt et al., 2014). The topic modeling uses Latent Dirichlet Allocation (LDA), which produces content in the form of topic distribution consisting of a collection of words (Ko et al., 2018). Results of topic modeling in this study provide suggestions to managers of the tourist destinations for understanding the interest and attraction of tourists toward particular destinations.

3 METHODOLOGY

This study consisted of several steps, starting from data collection to data preprocessing, the main process, and analysis results. Data collection was performed by streaming data using cloud-based applications that provide access to retrieve data based on user-generated content (UGC) on Instagram. Geographic information on the destinations comprises the keywords that are used to collect data for two months. In the data preprocessing stage, data cleaning is conducted manually in order to remove posts or accounts that are irrelevant (noises). The next stage is to separate the posts uploaded by tourists from the posts by locals. This study used the time window strategy to identify the posts uploaded by tourists and locals by checking the time periods of the posted photos. If an account uploaded photos at the target destination at a certain time then the account was considered as tourists but if the time span of the post was longer then it was considered as locals (Mukhina et al., 2017). Based on the Kajian Data Pasar Wisatawan Nusantara 2017, tourists have an average travel time of four days.

At the main process, posted data uploaded by tourists who were considered unique visitors were processed to produce dynamic patterns on a daily basis and daily aggregation to discover tourist visiting patterns and trends for each destination. In the next stage text data were processed based on captions or reviews of posts in Bahasa Indonesia using topic modeling. Topic modeling was conducted on each dataset in every tourist destination to bring up some topics that contain sets of words for each destination.

4 RESULTS AND DISCUSSION

Data collected from the process of streaming data amounted to 99,912 posts from 58,398 Instagram user accounts. From these data, we can identify the intensity of posted photos or videos related to the four superpriority tourist destinations on Instagram. From the results, Borobudur has the highest intensity posts and Toba Lake has the lowest. To provide further insights, pattern analysis was conducted in order to generate tourist traffic patterns on each tourist destination shown in Figure 1.

The graphs in Figure 1 represent the daily patterns of dynamic tourist visits in two months. In general, the pattern of tourist visits tended to fluctuate and have the same variations. A significant increase often occurred on weekends, holidays, or when an event was taking place, and a decrease often occurred during working days, the end of the month, or due to bad weather. The pattern of daily tourist visits used to identify changes in the number of tourists visits thus makes it easier to determine the peaks and valleys of the visits in the tourist destination within a certain period.

Based on daily aggregation tourist visit results in Borobudur, Mandalika and Toba Lake tended to fluctuate, which decreased during working days (weekdays) and increased on weekends. Conversely, Labuan Bajo has a different pattern among other destinations, tending to fluctuate both on weekdays and over the weekend. These patterns can be used to identify tourists' preferences related to the day when many tourists visit and when a few tourists visit in a certain period. Figure 2 shows the results of topic modeling.

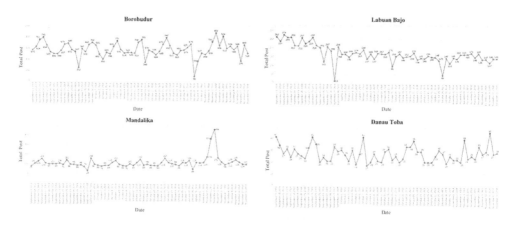

Figure 1. The daily pattern of dynamic tourists visits in four superpriority tourist destinations.

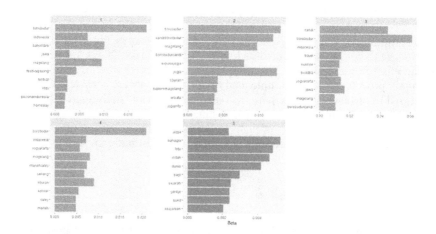

Figure 2. Topic modeling result for Borobudur.

The five topics contain a collection of words forming a topic that is often discussed by tourists related to the Borobudur. Topic 1 describes the umbrella and coffee festival in Borobudur. Topic 2 describes Borobudur as a famous vacation destination. Topic 3 describes Borobudur as having the largest Buddhist temple and a wonderful view of the sunrise. Topic 4 describes the international music event of Borobudur Symphony 2018, Mariah Carey Live in Concert. Topic 5 describes the Borobudur temple, which is one of the wonders of the world. In general, the topics that emerged from superpriority tourist destinations describe the things that became icons of tourist destinations, events, and conditions of the destinations.

5 CONCLUSION

This study produced meaningful insights related to the tourism sector, especially the superpriority tourist destinations in Indonesia, by utilizing big data from social media using user-generated content from Instagram to identify the preferences and interests of tourists toward tourist destinations. The time window strategy is an approach that is suitable to separate the posts of tourists and surrounding locals in order to obtain the posting pattern on each tourist destination. The pattern helps the destination's manager to understand an increase or decrease of tourist visits at a specific time to assist in making decisions related to service management in the destinations such as holding an event to increase tourist visits, increasing the availability of services (lodging), and performing scheduling in anticipation of the high number of tourist visits.

Identification of the favorite tourist destinations based on social media posts provides a strong recommendation to the Indonesia government to describe tourists' preferences and is used as an approach in determining the leading destinations for the increase of tourist visits and development of national tourism. Topic modeling using LDA is a suitable method for processing text data in order to discover the topic preferences that are often discussed by tourists in social media as well as what customers like the most. These insights provide suggestions to tourism managers in improving and developing the quality of services or facilities related to the tourist destinations. The scope of this study still had limits, and future studies are therefore expected to explore big data to a greater extent for the service industry, especially the tourism sector, utilizing other social media and longer data collection periods as well as other analysis methods (sentiment analysis, classification, text network analysis) so that they can make further contributions to the service industry, especially the tourism sector.

REFERENCES

Alamsyah, A., Rahmah, W., & Irawan, H. 2015. Sentiment analysis based on appraisal theory for marketing intelligence in Indonesia's mobile phone market. *Journal of Theoretical and Applied Information Technology*, *82*(2).

Asuncion, A., Welling, M., Smyth, P., & Teh, Y.W. 2009. On smoothing and inference for topic models. In *Proceedings of the Twenty-Fifth Conference on Uncertainty in Artificial Intelligence* (pp. 27–34). AUAI Press.

García-Palomares, J.C., Gutiérrez, J., & Mínguez, C. 2015. Identification of tourist hot spots based on social networks: A comparative analysis of European metropolises using photo-sharing services and GIS. *Applied Geography*, *63*, 408–417.

Holloway, J.C. 2009. *The Business of Tourism*. London: Pearson Education.

Hunt, C.A., Gao, J., & Xue, L. 2014. A visual analysis of trends in the titles and keywords of top-ranked tourism journals. *Current Issues in Tourism*, *17*(10), 849–855.

Ko, N., Jeong, B., Choi, S., & Yoon, J. 2018. Identifying Product opportunities using social media mining: Application of topic modeling and chance discovery theory. *IEEE Access*, *6*, 1680–1693.

Leung, et al. 2013. Social media in tourism and hospitality: A literature review. *Journal of Travel & Tourism Marketing*, *30*(1–2), 3–22.

Mukhina, et al. 2017. Detection of tourists attraction points using Instagram profiles. *Procedia Computer Science*, *108C*(2017), 2378–2382. Retrieved from ScienceDirect.

Okuyama, K., & Yanai, K. 2013, A travel planning system based on travel trajectories extracted from a large number of geotagged photos on the web. In *Proceedings of the Era of Interactive Media* (pp. 657–670). New York: Springer Science+Business Media.

Vu, H.Q., Leung, R., Rong, J., & Miao, Y. 2016. Exploring park visitors' activities in Hong Kong using geotagged photos. In *Information and Communication Technologies in Tourism 2016* (pp. 183–196). Cham, Switzerland: Springer.

Social network analysis of the information dissemination patterns and stakeholders' roles at superpriority tourism destinations in Indonesia

H. Irawan, D.A. Digpasari & A. Alamsyah
Telkom University, Bandung, Indonesia

ABSTRACT: Indonesia's tourism is one of the leading sectors in its economic, social, and cultural development. In an effort to increase the number of tourists, tourism stakeholders use online social media as a strategy to share information to promote tourism destinations. This research aimed to discover the patterns of dissemination of information and to uncover the role of tourism stakeholders in disseminating information at superpriority tourism destinations in Indonesia. This research used social network analysis to process data sources from Twitter and found new insights into a similar pattern at superpriority tourism destinations. The visualization shows that most Twitter users have less of a contribution to social interaction, such as events, hotels, and business accounts; instead, individual stakeholders have a more significant role in disseminating information on tourism activities in those destinations. The results of this study can be implemented as a strategy to carry out effective information dissemination in the tourism sector.

1 INTRODUCTION

Indonesian tourism is a competitive and progressive sector that is growing rapidly and provides a large contribution to the country's economy. In 2019, the government targeted a 10% growth in tourist visits or 20 million foreign tourists and 275 million domestic tourists. In order to achieve this target, the government created four superpriority destinations consisting of Borobudur Temple, Lake Toba, Labuan Bajo, and Mandalika. In addition, tourism stakeholders were needed as supporters and coordinators in each tourism sector policy. In its strategy, the government utilized technological advancements for the growth and development of sustainable destinations by leading to smart tourism to gain wider market access. This takes advantage of social media as a platform for generating and disseminating information to potential tourists (Tran et al., 2014). A large amount of information in the form of user-generated content on social media can be used to strengthen references in planning a trip (Fotis et al., 2012).

Information dissemination on social media will create interactions that produce new knowledge and enhance the sustainability of the organization (Yadav & Rahman, 2017). The interactions will result in a complex pattern or network that can be modeled with social networks. In addition, there will be actors who influence a network more than other actors. Social network analysis is an approach that utilizes big data that can help analyze phenomena about tourism activities including stakeholder relationships, the influence of relationships between companies and tourists, information dissemination, and tourism trends (Casanueva et al., 2014).

The purpose of this study was to discover the use of social network analysis in identifying patterns of information dissemination and the role of tourism stakeholders in the dissemination of information on Indonesia's superpriority tourism destinations based on Twitter's social media data as a form of tourism analytics. Identifying social networks on Twitter is an effective way to produce more data compared to conventional methodologies such as questionnaires and interviews (Alamsyah et al., 2015). Using social network analysis can influence tourist decisions; therefore tourism stakeholder involvement is needed in considering this approach as a more efficient digital marketing campaign development.

2 LITERATURE REVIEW

Social network analysis is a method to identify valuable information on large-scale structured data (Alamsyah et al., 2017). Social network analysis is a social interaction activity carried out by individuals or groups of individuals (Liu et al., 2017). Previous research explained that it is important to identify interactions in the network and roles of the actors involved (Bokunewicz and Shulman, 2017; Casanueva et al., 2014; Deborah et al., 2019; Tran et al., 2014; William et al., 2015). Interaction in dissemination of information involves parties who have no interest in only one destination but help to disseminate information together to create viral marketing and E-WOM in the digital world to create an impact on the brand image of a destination.

Therefore a social network analysis approach is needed in the tourism sector. The analysis in this study used measurement of network properties to identify a pattern of information dissemination and describe the character of a network regarding management insights in the organization based on the value of nodes, edges, average degree, diameter, and average path length. Centrality measurement was used to identify the role of each tourism stakeholder in a social network based on the degree of centrality consisting of indegree and outdegree, betweenness centrality, closeness centrality, and eigenvector centrality.

3 METHODOLOGY

The framework in this study consisted of data collection, data processing, data visualization, and measurement properties based on social network analysis. The data obtained were user-generated content (UGC) based on interactions that form Twitter users' network, namely replies, mentions, and retweets in the dissemination of information related to superpriority tourism destinations. Data were collected by streaming data using Python, which already has an API for permission to access and retrieve tweets on Twitter. Data retrieval was carried out from September 13, 2018, to November 13, 2018. The next stage was preprocessing data, which consisted of removing irrelevant duplicates, spam, bots, and tweets from the collected data. Prior to social network analysis, Twitter users from the clean data were classified into tourism stakeholder types based on Twitter user bio profiles.

After preprocessing data, social network analysis was used for social network modeling, followed by measurement of social network properties in each superpriority tourism destination. It aimed to analyze the patterns of information dissemination from the interaction of tourism stakeholders. The final step was to measure centrality in order to identify the role of each tourism stakeholder.

4 RESULTS

The social network is a representation of interaction in the pattern of information dissemination carried out and influenced in a network of tourism stakeholders. In the visualization, there are nodes, which are a representation of tourism stakeholder actors and edges, namely interactions conducted by nodes. The processed product is visualized using Gephi. Figure 1 is a formed pattern of information dissemination.

The characteristics of information dissemination patterns on social networks are depicted in Table 1. A node is a representation of tourism stakeholders symbolized by a point. The higher the value of a node, the higher the level of awareness in information dissemination. Borobudur Temple has the highest node value of 20,896 tourism stakeholders. Edges is a representation of the

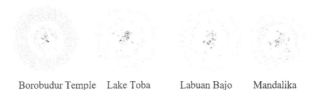

Borobudur Temple   Lake Toba   Labuan Bajo   Mandalika

Figure 1. Social network analysis.

Table 1. Network properties.

Superpriority tourism destinations	Nodes	Edges degree	Average	Diameter path length	Average
Borobudur Temple	20,896	27,131	1,081	8	3,049
Lake Toba	3,772	5,129	1,354	10	2,955
Labuan Bajo	2,600	3,213	1,234	12	3,175
Mandalika	1,192	1,568	1,069	7	2,493

amount of interaction in the information dissemination symbolized by lines. Borobudur Temple has the highest edge value of 27,131 information dissemination tweets.

Average degree is the average number of relationships that occur in a social network owned by one node to another node (Barabasi, 2012). The higher the average degree, the more complete the information disseminated. Lake Toba has the highest average degree value of 1,354. Diameter is the shortest and longest path length between nodes that share information in the network (Arnaboldi et al., 2015). The smaller the diameter value of a network, the greater the distribution of information on social networks. Mandalika has the smallest value of 7. Therefore, the social network in the information distribution is denser than in the other destinations.

The last measurement is average path length, which calculates information dissemination speed based on the average distance between all pairs of nodes that interact with each other in the distribution of information on the network (Barabási, 2012). Mandalika has the smallest value of 2,493, meaning the path for information dissemination is faster.

Furthermore, the authors take measurements of centrality to identify the role of tourism stakeholders in the dissemination of information on Twitter. Analyzing the value of degree centrality is measured by identifying the value of indegree and outdegree. The indegree result indicates that the individual stakeholders have the highest value. That is because the role of individuals tends to connect more nodes that direct their arrows at individual stakeholders. On the other hand, business stakeholders have the lowest indegree value because business stakeholders have few arrows directed at the business account. Furthermore, individual stakeholders also have the highest outdegree value, which can be interpreted as indicating that individual stakeholders often mention, reply, or retweet to other nodes. On the contrary, event stakeholders have the lowest outdegree value, meaning that event stakeholders interact little with other nodes on the social network. The next measurement, betweenness centrality, shows that individual stakeholders have the highest value. Therefore, individual stakeholders have a great opportunity as gatekeepers in the social network. In contrast, business stakeholders have the lowest value, which means a lack of involvement of business stakeholders in dissemination of information in each tourism destination.

Closeness centrality results indicate that individual stakeholders have the highest value in the social network. Thus, individual stakeholders have more immediate potential in spreading information to nodes in social networks, whereas business stakeholders have the lowest closeness value. Therefore, information dissemination is more quickly spread by individual than by business stakeholders. For eigenvector centrality, individual stakeholders are the node with the highest value and hotel is the node with the lowest value. Therefore, individual stakeholders have a greater influence than hotel stakeholders. This is in accordance with previous research that the nodes involved in social networks tend to lead to nodes that have large eigenvector centrality values (Newman, 2010).

5 CONCLUSIONS AND RECOMMENDATIONS

In this study, the authors managed to find new insights for the tourism sector, especially related to superpriority tourism destinations. Based on the analysis, it can be concluded that social network analysis is an approach to analyze patterns of information dissemination and the role of tourism stakeholders. In the pattern of information dissemination, network property measurement is used and shows a similarity of patterns, namely the existence of a circle gap between forming groups. This indicates that tourism stakeholders lack interaction in spreading information. There is also a gap supported by the average degree value, which has an average interaction between one and two Twitter users only. Furthermore, in the process of identifying the role of stakeholders, individual stakeholders have an important role in information dissemination based on the centrality measurement. Tourism stakeholders such as government, media, community, and travel play a good role in disseminating information. In contrast, tourism stakeholders such as events, hotels, and businesses do not play a good role because of the small number of nodes involved and the lack of interaction in the distribution of information.

In identifying patterns of information dissemination, a creative way to share tourism destination content is needed to involve more actors so that information can be more widely spread and thereby have an impact on visits to tourism destinations. Furthermore, identifying the role of tourism stakeholders becomes a parameter of the extent to which tourism stakeholders are involved in disseminating information on social media. Therefore, it may serve as a feedback for the government to maximize the role of tourism stakeholders.

Further research could analyze information disseminated by tourism stakeholders using other methods such as sentiment analysis, topic modeling, and text network analysis using data sources from various social media outlets such as Instagram and Facebook.

REFERENCES

Alamsyah, A., Bratawisnu, M. K., & Sanjani, P. H. 2017. Finding pattern in dynamic network analysis. In *ICOICT Conferences.*
Alamsyah, A., Rahmah, W., & Irawan, H. 2015. Sentiment analysis based on appraisal theory for marketing intelligence in Indonesia's mobile phone market. *Journal of Theoretical and Applied Information Technology, 82*(2).
Arnaboldi, V., Passarella, A., Conti, M., & Dunbar, R. I. 2015. *Online Social Networks: Human Cognitive Constraints in Facebook and Twitter Personal Graphs.* UK: Elsevier.
Barabasi, A.-L. 2012. Network Science: Luck or reason. *Nature, 489*, 507–508.
Bokunewicz, J. F., & Shulman, J. 2017. Influencer identification in Twitter networks of destination marketing organizations. *Journal of Hospitality and Tourism Technology, 8*(27).
Casanueva, C., Gallego, A., & Garcia-Sanchez, M. R. 2014. Social network Analysis in Tourism. *Current Issues in Tourism.*
Deborah, A., Michela, A., & Anna, C. 2019. How to quantify social media influencers: An empirical application at the Teatro Alla Scala. *Journal of Heliyon.*
Fotis, J., Rossides, N., & Buhalis, D. 2012. Social media use and impact during the holiday travel planning process. *Information and Communication Technologies in Tourism 2012.* doi: 10.1007/978-3-7091-1142-0_2.
Liu, Z., Dong, Y., Zhao, X., & Zhang, B. 2017. A dynamic social network data publishing algorithm based on differential privacy. *Journal of Information Security.*
Newman, M. E. 2010. *Networks: An Introduction.* New York: Oxford University Press.
Tran, Mai. T. T., Jeeva, A. S., & Pourabedin, Z. 2016. Social Network Analysis in Tourism Services Distribution Channels. *Journal of Tourism Management Perspectives.*
Williams, N. L., Inversini, A., Buhalis, D., & Ferdinand, N. 2015. Community crosstalk: An exploratory analysis of destination and festival eWOM on Twitter. *Journal of Marketing Management, 31* (9).
Yadav, M., & Rahman, Z. 2017. Measuring consumer perception of social media marketing activities in e-commerce industry: Scale development & validation. *Journal Telematics and Informatics.*

The effect analysis of financial performance of companies before and after M&As

A.R. Bionda, I. Gandakusuma & Z. Dalimunthe
Universitas Indonesia, Jakarta, Indonesia

ABSTRACT: This research aimed to describe the analysis effect of financial performance on Cumulative Abnormal Returns of companies listed in Indonesia Stock Exchange for the 2014-2018 period using the event study method. The results of this research indicate that there is a significant difference in Return on Equity (ROE) while Tobin's Q Ratio and Cumulative Abnormal Returns have no significant differences before and after merger and acquisitions. Moreover, Return on Equity (ROE) and Tobin's Q Ratio do not have a statistically significant effect on Cumulative Abnormal Returns before and after merger and acquisitions. Since Return on Equity (ROE) has a significant difference before and after merger and acquisitions, it can be meant that merger and acquisitions have the implications to the synergy and value-added to the companies before and after can be shown by the profit margin and shareholders' equity of the companies as the components of Return on Equity (ROE).

1 INTRODUCTION

Merger and acquisitions is a business combination conducted by two or more companies. In general, the goal of completing mergers and acquisitions is for the company to obtain synergy or added value (Aulina, 2012). Most companies do merger and acquisitions as their strategy as a short way for them to provide new products without having to proceed from the beginning and the results which could be obtained from mergers and acquisitions are companies becoming bigger, more economical, moreover companies can control assets, markets, and also their business potential. In Indonesia, there were companies conducted merger and acquisitions such as the merger of 4 government banks into Bank Mandiri on 1999, the merger of Bank Niaga and Lippo-Bank into CIMB NIAGA on 2007, the acquisitions of AXIS by XL AXIATA on 2013, and the acquisitions of Bank BTPN by SMBC on 2019. The results of the previous merger and acquisitions are the companies are getting bigger and have reached the broader market industry. Earlier researches have been conducted on the effect of merger and acquisitions on the financial performance of companies. Sihombing and Kamal (2016) concluded that there is a partial significance in Current Ratio, Total Assets Turnover (TATO), Debt to Equity Ratio (DER), Return on Investment (ROI), and Earning per Share (EPS) while there is no difference in the ratio of Abnormal Returns and Cumulative Abnormal Returns before and after mergers and acquisitions. Pervan, Visic, and Barnjak (2015) concluded that merger and acquisitions did not have a significant effect between on the profitability of companies in Croatia.

Based on the results of earlier researches, on the company's financial performance before and after conducting mergers and acquisitions, this research was conducted in order to provide an overview of whether the company's financial performance has an influence on Cumulative Abnormal Returns on companies listed on the Indonesia Stock Exchange in the period 2014-2018 before and after mergers and acquisitions. The research is conducted by calculating the Return on Equity (ROE) and Tobin's Q Ratio with the quarterly financial statements before and after the date of notification of mergers and acquisitions while for the calculating of the company's Cumulative Abnormal Returns the period which being used is 30 days before and after the date of notification of mergers and acquisitions.

2 LITERATURE REVIEW

2.1 Merger and acquisitions

Article 1 sub 9 of Company Law No. 9 of 2007 defines Merger as a legal act by one or more companies to merge with other companies while based on Article 1 number 3 PP Number 27 of 1998 concerning Merger, Consolidation, and Takeover of Limited Liability Companies, acquisitions are legal actions carried out by legal entities or individuals to take over either all or most of the company's shares which may result in the transfer of control over the company.

2.2 Profitability ratio

Profitability ratio measures the efficiency of a company in using its assets and managing its operations (Ross, Westerfield, & Jordan (2013). Profitability ratio used in this study are Return on Equity (ROE), which can also be referred to as Own Capital Profitability and one of indicators used by shareholders to measure the success of the business being undertaken.

2.3 Tobin's ratio

Tobin's Q ratio is a parameter measuring company performance that shows management's performance in managing company assets (Sudiyatno & Puspitasari, 2010).

2.4 Cumulative Abnormal Returns

Cumulative Abnormal Returns are accumulations of Abnormal Returns from period to period. Cumulative Abnormal Returns during the period before an event occurs will be compared with Cumulative Abnormal Returns during the period after an event occurs. The comparison of Cumulative Abnormal Returns will provide information on what stocks are affected by an event, both positive and negative effects (Jogiyanto, 2009).

3 DATA AND METHODOLOGY

The source of the data in this research is from the quarterly financial statements and closing prices of shares of companies that did mergers and acquisitions and were listed on the Indonesia Stock Exchange in the 2014-2018 period. The data was obtained from the Indonesia Stock Exchange website, related company website, IDN Financials website, Yahoo! Finance, and Bloomberg. The population in this research are companies that conduct mergers and acquisitions and were listed on the Indonesia Stock Exchange in the 2014-2018 period. The sample selection uses a purposive sampling method, where the sample is chosen based on certain criteria, such as; nonbanking companies, active companies, and a complete financial statements and stock price data during the research period. The variables which are being used in this research described in the Table 1.

Table 1. Research variables.

Research Variables	Measure	Sources
Dependent Variable		
Cumulative Abnormal Returns	$\sum_{t3}^{t} RTN_{i,t}$	Jogiyanto (2009)
Independent Variables		
Return on Equity	$\frac{Net\ Income}{Total\ Equity}$	Ross, Westerfield, & Jordan (2013)
Tobin's Q Ratio	$\frac{(MVS+D)}{Total\ Assets}$	Wolfe & Sauaia (2003)

The research model which are used in this research are denoted below:

$$CARB_i = \alpha + \beta_1 ROEB_i + \beta_2 TOBINB_i + \varepsilon_i \quad (1)$$

$$CARA_i = \alpha + \beta_1 ROEA_i + \beta_2 TOBINA_i + \varepsilon_i \quad (2)$$

4 RESULT AND DISCUSSION

During 2014-2018 there were 335 number of merger notifications included in the merger notifications list at KPPU. The number of merger notifications came from all companies both listed on the Indonesia Stock Exchange and those not listed on the Indonesia Stock Exchange. There were 79 companies listed on the Indonesia Stock Exchange which were included in the merger notification list at KPPU during 2014-2018. Since this research choose to use non-banking companies, there are 66 non-banking companies, 1 of them is suspended, and 4 of them do not have completed financial statements, therefore, there are 61 companies fit the criteria of the research. This research is using 2 periods which are before and after merger and acquisitions therefore there are 122 data to be observed in this research.

Based on Table 2, the regression significance of F Test before and after merger and acquisitions results 0.624 and 0.893 which are greater than 0.05. It is concluded that the independent variables simultaneously do not affect the dependent variable before and after mergers and acquisitions. For the regression significance of T Test, Return on Equity (ROE) results 0.671 and 0.689 while Tobin's Q Ratio results 0.397 and 0.815. Both ratios result the regression significance of T Test greater than 0.05. It is concluded that each of the independent variables do not affect the dependent variable before and after mergers and acquisitions.

Based on Table 3, the regression significance of Paired Sample T-Test before and after merger and acquisitions for each variable in this research results that only Return on Equity (ROE) has the regression significance 0.020 less than 0.05 while Tobin's Q Ratio and Cumulative Abnormal Returns have the regression significance greater than 0.05. It is concluded that Return on Equity (ROE) has a significant difference before and after mergers and acquisitions.

Table 2. F and T Test.

Test	Sig. Before	Sig. After
F	0.624	0.893
T		
Return on Equity (ROE)	0.671	0.689
Tobin's Q Ratio	0.397	0.815

Table 3. Paired Sample T-Test.

Variable	Sig. (2-tailed)
Cumulative Abnormal Returns	0.781
Return on Equity (ROE)	0.020
Tobin's Q Ratio	0.257

5 CONCLUSION

Based on the results of the tests and the analyzes that have been done, it can be concluded that there is a significant difference of Return on Equity (ROE) ratio before and after mergers and acquisitions while Tobin's Q Ratio and the Cumulative Abnormal Returns ratio have no significant differences before and after mergers and acquisitions. Moreover, Return on Equity (ROE) and Tobin's Q Ratio do not have a statistically significant effects on Cumulative Abnormal Returns before and after mergers and acquisitions. Return on Equity (ROE) consists of profit margin and shareholders' equity of the companies, so it can be meant that merger and acquisitions have the implications to the synergy and value-added to the companies before and after merger and acquisitions which can be shown by Return on Equity (ROE). Tobin's Q Ratio and Cumulative Abnormal Returns consist of the stock price of each company that done mergers and acquisitions, it means the announcement of the mergers and acquisitions of the companies is not a shocking news for the investors since the return of each stocks is not overly abnormal.

REFERENCES

Aulina, S. (2012). Analysis of Company Financial Performance Before and After the Acquisition of PT Hanjaya Mandala Sampoerna Tbk. Jurnal Ilmu & Riset Manajemen, 1(14),1–18.
https://docplayer.info/52874437-Jurnal-ilmu-riset-manajemen-vol-1-no-14-2012.html

Jogiyanto. (2009). Portfolio Theory and Investment Analysis (6 ed.). Yogyakarta: PT BPEE Yogyakarta.

Pervan, M., Višić, J., & Barnjak, K. (2015). The Impact of M&A on Company Performance: Evidence from Croatia. Procedia Economics and Finance, 23, 1451–1456.
https://www.sciencedirect.com/science/article/pii/S2212567115003512

Ross, S.A., Westerfield, R. W., & Jordan, B. D. (2013). Fundamentals of Corporate Finance. New York: McGraw-Hill Irwin.

Sihombing, N., & Kamal, M. (2016). Analysis of the Effect of Merger and Acquisition Announcements on Abnormal Stock Returns and Corporate Financial Performance. Diponegoro Journal of Management, 5(3),1–15.
https://ejournal3.undip.ac.id/index.php/djom/article/viewFile/14882/14398

Sudiyatno, B., & Puspitasari, E. (2010). Tobin's Q and Altman Z-Score as Indicators of Firm Performance Measurement. Kajian Akuntansi, 2(1), 9–21
https://www.unisbank.ac.id/ojs/index.php/fe4/article/download/223/162

Effect of transformational leadership and culture on creativity and innovation

A. Sulthoni & J. Sadeli
University of Indonesia, Jakarta, Indonesia

ABSTRACT: The global market demands a skilled workforce of individuals who are intellectually active, creative, innovative, and capable of critical thinking. This study examined the factors determining employee creative and innovative undertakings at work in a power generation company with a focus on transformational leadership and organizational culture. The study adopted a survey design and simple and moderated regression analyses to test three hypotheses. Data were collected from a sample of 158 randomly selected staff. Employees completed questionnaires consisting of validated scales of each variable in the study. It was concluded that transformational leadership (t: 7,768 dan sig: 0,00) and appropriate organizational culture (t: 3,635 dan sig: 0,00) are important factors in facilitating employee creativity and innovation in the PT PJB UP Muara Tawar. It was recommended that companies rapidly increase employee creativity and innovation by promoting and investing in transformational leadership training, spreading the transformational leadership way to work, as well as continuing to promote a creative and innovative organizational culture.

1 INTRODUCTION

The business environment in the age of technology 4.0 is changing rapidly as a result of rapid technological changes, shorter product turnover cycles, dynamic service delivery, and globalization. Employee creativity and innovation are important components that will improve organizational performance and core competencies that organizations must develop to survive, develop, and maintain competitiveness in today's turbulent business environment (Alsop, 2003; Nussbaum, 2005). Through creative and innovative efforts, employees in organizations try to change certain aspects of their work or work products to get some benefits that they value for themselves or for the organization. Some of these benefits are increased employee performance; higher productivity; and improved product quality, production routines, and service delivery as well as the development of new markets to meet organizational goals and achieve a competitive advantage (Odetunde & Ufodiama, 2012).

The important factor that will facilitate employee innovative efforts and organizational innovation is the quality of leadership and organizational culture inherent in the organization. Urbancová (2013) argues that creative and innovative organizations do not come about accidentally but rather from the role of leaders encouraging and controlling intentional changes in structure, culture, and processes to transform such organizations into creative, effective, and productive organizations.

In particular, transformational leadership has been raised to promote more effective activities in increasing organizational innovation (Gardner & Avolio, 1998; Howell & Avolio, 1993). Yukl (1989) revealed that a leadership style that can be a good role model, and one that is able to solve problems and motivate other employees to work harder, is a transformational leadership style. Organizational culture, defined as the norms and organizational vision of how people behave and how things are done in an organization (Glisson & James, 2002), has been considered to have a significant influence on innovation in organizations (Carmeli, 2005).

Based on the background and results of previous studies, the main issues to be discussed in this article are the following:

1. Does transformational leadership have a significant and positive influence on employee creativity and innovation?
2. Does the organizational culture have a significant and positive influence on employee creativity and innovation?
3. Does transformational leadership and corporate culture have a significant and positive simultaneous effect on employee creativity and innovation?

2 LITERATURE REVIEW

Although closely correlated and often used interchangeably, many researchers have shown that creativity and innovation individually refer to two different activities. For example, many researchers conceptualize innovation as two different processes of the generation of new ideas and the implementation of these ideas. Creativity is defined as the development of novel (i.e., original, unexpected) and useful ideas, products, or problem solutions (Amabile et al., 2005), while innovation is the translation of creative ideas into relevant and applicable solutions or fill the identified problem or gap. Amabile et al. (2004) further state that no innovation is possible without a creative process that marks the initial stages of the process: identifying important problems and opportunities, gathering information, generating new ideas, and exploring the validity of those ideas.

According to this theory transformational leadership can lead to unusual commitments to its followers to produce extraordinary performance (Bass & Riggio, 2006). Many studies also prove that the transformational leadership paradigm is promising. Bryman (1992) cites several studies on organizations clearly showing that transformational leadership styles have a positive relationship with employee satisfaction and performance. Denison (1990) in his research on culture created a model known as the Denison Model.

3 METHODOLOGY

This research started with establishing the variables that affect creativity and innovation at PT. PJB UP MTW. These variables were decided after a literature review. The variables formed from the previous research needed to be tested for significance to determine the relationship of each independent variable to the dependent variable. If certain variables were known to have no significant effect on the dependent variable, they were not included in the next test. Furthermore, the initial model was built from the selected variables.

Then, the next step was to build a model with variables that passed the multivariate test and then test the model formed with various statistical tests. These tests were carried out to see the performance of the model in terms of the accuracy of the results and its ability to represent real conditions.

From the Slovin equation, this research used data from a sample of 158 randomly selected staff because the main purpose was to determine effects on employee behavior. A questionnaire developed by Podsakoff et al. (1990) was used to measure transformational leadership behavior in this study. This measuring instrument consists of six components: articulating the vision, providing an appropriate model, accepting group goals, having high performance expectations, providing individual support, and providing intellectual stimulation. Organizational culture variables were measured using a questionnaire developed by D. R. Denison (1990). Employee creativity and innovation were measured using a questionnaire developed by Zhou and George (2001) and by De Jong and Den Hartog (2010).

4 RESULTS AND DISCUSSION

After we pretested the variables (validity and reliability test) we performed the normality, heteroskedasticity, and multicolinearity tests. We validated that the data are normal and show no heteroskedasticity and no multicolinearity.

Coefficient values can be seen in the Table 1 and entered into the multiple linear regression equation, producing following equation, where Y is creativity and innovation, $X1$ is transformational leadership, and $X2$ is organizational culture:

$$Y = 26.855 + 0.477\ X1 + 0.233\ X2 \tag{1}$$

The t-test is done to see the effect of independent variables with partial on the dependent variable. Table 2 shows the results of the t-test for each variable. Decision making is based on the comparison between the t arithmetic and t-table.

According to the results in Table 2 the value of the t-arithmetic transformational Leadership variable is worth 7.768 more than the t-table value of 1.975 and Sig. value of 0.000 less than 0.05 The t-value of the Organizational Culture variable in Table 2 is 3.635 more than the t-table value of 1.975 and the Sig. value of 0.000 is less than 0.05, so the conclusion can be drawn that the Transformational Leadership and Organizational Culture variables have a strong positive effect on the Creativity and Innovation variables.

Based on research conducted at this time in the PJB UP Muara Tawar plant which has nearly 300 employees, it was proven that transformational leadership significantly influenced creativity and innovation. This is in accordance with previous studies conducted by Sosik et al. (1999) and Avolio et al. (1999), who found that transformational leadership will increase employee creativity and innovation.

Organizational culture variables proved to have a significant and positive effect also on employee creativity and innovation. This is in accordance with previous research conducted by Amabile (1988) and Gadomska-Lila (2010), who found that corporate culture that supports the values of creativity and innovation will increase the creativity and innovation of its employees.

Transformational leadership and organizational culture have proven to have a significant effect on employee creativity and innovation. This is in accordance with previous research conducted by Jung et al. (2003) and Zairi and Al-Mashari (2005), who found that leaders who have transformational leadership with a corporate culture that supports innovation will increase the creativity and innovation of its employees.

Table 1. Result of multiple linear regression equation.

	Unstandardized coefficients B	Std. error	Standardized coefficients beta	t	Sig.
(constant)	26.855	4.883		5.500	0.000
Transformational Leadership	0.477	0.061	0.532	7.768	0.000
Organizational Culture	0.233	0.064	0.249	3.635	0.000

Table 2. Results of t-test.

Model	t	Sig.
Transformational Leadership	7.768	0.000
Organizational Culture	3.635	0.000

REFERENCES

Alsop, S., & Watts, M. 2003. Science education and affect. *International Journal of Science Education, 25* (9), 1043–1047.

Amabile, T. M. 1988. A model of creativity and innovation in organizations. *Research in Organizational Behavior, 10*(1), 123–167.

Amabile, T. M., Barsade, S. G., Mueller, J. S., & Staw, B. M. 2005. Affect and creativity at work. *Administrative Science Quarterly, 50*(3), 367–403.

Amabile, T. M., Schatzel, E. A., Moneta, G. B., & Kramer, S. J. 2004. Leader behaviors and the work environment for creativity: Perceived leader support. *The Leadership Quarterly, 15*(1), 5–32.

Avolio, B. J., Bass, B. M., & Jung, D. I. 1999. Re-examining the components of transformational and transactional leadership using the Multifactor Leadership. *Journal of Occupational and Organizational Psychology, 72*(4), 441–462.

Bass, B. M., & Riggio, R. E. 2006. *Transformational Leadership*. London: Psychology Press.

Bryman, A. 1992. *Charisma and Leadership in organizations*. Thousand Oaks, CA: SAGE.

Carmeli, A. 2005. The relationship between organizational culture and withdrawal intentions and behavior. *International Journal of Manpower, 26*(2), 177–195.

De Jong, J., & Den Hartog, D. 2010. Measuring innovative work behaviour. *Creativity and Innovation Management, 19*(1), 23–36.

Denison, D. R. 1990. *Corporate Culture and Organizational Effectiveness*. New York: John Wiley & Sons.

Gadomska-Lila, K. 2010. Charakterystyka i uwarunkowania proinnowacyjnej kultury organizacyjnej-wyniki badań. *Przegląd Organizacji, 2*, 12–15.

Gardner, W. L., & Avolio, B. J. 1998. The charismatic relationship: A dramaturgical perspective. *Academy of Management Review, 23*(1), 32–58.

Glisson, C., & James, L. R. 2002. The cross-level effects of culture and climate in human service teams. *Journal of Organizational Behavior, 23*(6), 767–794.

Hunt, J. G. J., Stelluto, G. E., & Hooijberg, R. 2004. Toward new-wave organization creativity: Beyond romance and analogy in the relationship between orchestra-conductor leadership and musician creativity. *The Leadership Quarterly, 15*(1), 145–162.

Jung, D. I., Chow, C., & Wu, A. 2003. The role of transformational leadership in enhancing organizational innovation: Hypotheses and some preliminary findings. *The Leadership Quarterly, 14*(4–5), 525–544.

Sosik, J. J., Kahai, S. S., & Avolio, B. J. 1999. Leadership style, anonymity, and creativity in group decision support systems: The mediating role of optimal flow. *The Journal of Creative Behavior, 33* (4),227–256.

Nussbaum, B. 2005. Getting schooled in innovation. *Business Week*.

Odetunde, O. J., & Ufodiama, N. M. 2017. Transformational leadership and organisational culture as predictors of employee creativity and innovation in the Nigerian oil and gas service industry. *IFE PsychologIA, 25*(2), 325–349.

Podsakoff, P. M., & Mackenzie, S. B., & Morrman, R. H., & Fetter, R. 1990. Transformational leader behaviours and their effects on follower's trust in leader, satisfaction, and organizational citizenship behaviours. *Leadership Quarterly, 1*, 107–142.

Yukl, G. 1989. Managerial leadership: A review of theory and research. *Journal of Management, 15*(2), 251–289.

Zairi, M., & Al-Mashari, M. 2005. Developing a sustainable culture of innovation management: A prescriptive approach. *Knowledge and Process Management, 12*(3), 190–202.

Zhou, J., & George, J. M. 2001. When job dissatisfaction leads to creativity: Encouraging the expression of voice. *Academy of Management Journal, 44*(4), 682–696.

Essence and implementation of enterprise resource planning in the textile industry: Critical success factor

A.R. Muafah, R.W. Witjaksono & M. Lubis
Telkom University, Bandung, Indonesia

ABSTRACT: The Enterprise Resource Planning (ERP) system is an integrated system consisting of several modules, where data flow and information are integrated into each module. One of the objectives of the ERP system is the automation and integration of corporate business processes. However, in practice, not all companies wishing to implement ERP systems succeed in their implementation process, and many companies fail in this effort, as disclosed by Standish Group, which found that only 10% of companies are successfully implementing ERP. Given this paradoxical situation, why there are still many companies, in particular those in the textile industry, implementing ERP? Consequently, the purpose of the qualitative research described in this article, based on data collection using interviews and observations, was to analyze and determine critical success factors of ERP implementation in the textile industry, formulate why companies in the textile industry want to implement ERP, and identify the essence of ERP implementation in companies in the textile industry.

1 INTRODUCTION

Not all companies successfully implement the Enterprise Resource Planning (ERP) system. Reasons for failures include environmental factors and process failures. The Standish Group found that only 10% of companies successfully implemented ERP, 35% of projects were canceled, and 55% had delays (Widiyanti, 2013). Based on the results of the survey it can be concluded that not all companies will be matched using the same ERP system. It is therefore paradoxical that although quantitatively, many companies are experiencing failure in implementing ERP, at present there are still many companies that want to implement an ERP system. Companies in the textile industry, which manufacture raw textile materials into products, implement ERP systems mostly to assist their business processes. But some textile companies have not managed to implement ERP because of problems regarding outside consultants, as experienced by W. L. Gore, one of the companies that produce fluoropolymer. In a lawsuit filed by W. L. Gore against PeopleSoft and Deloitte & Touche, the company accused PeopleSoft of sending consultants who were not qualified to do the job and forced the company to rely solely on the customer service hotline to remedy a major problem in the application of the program after a the system went live. Deloitte & Touche paid a referral fee to PeopleSoft, encouraging PeopleSoft to recommend them as consultants even though they did not have the skills needed to implement the application. Then Gore was asked to hire another group of consultants to repair the damages, which cost hundreds of thousands of dollars more (Barton, 2014). The purpose of the research described in this article is to identify the critical success factors of ERP implementation in the textile industry, formulating the reasons why companies in the textile industry want to implement ERP, and analyzing the essence of ERP implementation in the textile industry.

2 LITERATURE REVIEW

2.1 *Enterprise resource planning*

The following are some definitions of Enterprise Resource Planning (ERP) by some experts (Rashid et al., 2002):

a. "ERP (Enterprise Resource Planning system) consists of a commercial software package that promises the integration of all information flows through the company's finances, accounting, human resources, supply chains, and customer Information" (Davenport, 1998).
b. "ERP systems are computer-based systems designed for process transactions within the organization and facilitate integration, and planning, production, and responding to customers in real-time" (O'Leary, 2001).

ERP is an integrated system consisting of several modules, where data flow and information are integrated into each module.

2.2 *Diffusion on innovations*

According to Rogers (1983) diffusion is the process by which an innovation is communicated through certain channels over time among the members of a social system, while innovation is an idea, practice, or object that is considered new by an individual or other adoption units. Diffusion of innovation is the process by which an idea, practice, or object is communicated through a particular channel over time among members of the social system. The following are variables determining the rate of adoption according to Rogers (1983).

a. Attributes of innovations (relative advantage, compatibility, complexity, trialability, observability)
b. Type of innovation-decision (optional, collective, authority)
c. Communication channels
d. Nature of the social system
e. Promotion efforts of agent change rates

2.3 *Textile industry*

Textile manufacturing is based on the conversion of fibers derived from cotton into yarn, then into fabrics and finally higher-valued textiles in a process consisting of weaving, fabric forming, and finishing and staining complexity. These processes are capable of producing a wide range of products.

3 RESEARCH METHODOLOGY

The qualitative research described in this article used a case studies approach that aimed to investigate fully and in-depth ERP failures as a paradox. Data collection techniques used observation methods and interviews. The sampling techniques used in this study were purposive sampling to determine the criteria for selecting respondents. Several criteria for selecting the respondents included the following: the minimum position is the staff that operating ERP system, minimum education is high school/equivalent, and length of employment is at least 1 year in the company. The companies studied in this research are engaged in the textile sector: CV Kenari, CV Terang Mulia, and Nirwana. These companies have implemented ERP. The respondents in this study selected two speakers in each company who were experienced in using the ERP system. The speakers in this study were Mr. Yosef Heru, Head of HR & Operational at CV Kenari; Mr. Anto, Coordinator of Sales & Marketing at CV Kenari; Mr. Dede Haryono, Supervisor of Operations at CV Terang Mulia; Mrs. Sancia, Supervisor of Finance &

Accounting at CV Terang Mulia; Mrs. Susi, Marketing at Nirwana; and Mr. Yokky, Head of shopat Nirwana.

4 PROBLEM CLASSIFICATION

Table 1. Problem classification.

Object	Problem	Problem classification
CV Kenari	The systematic use of programs	Behavior
	Internet connection Interruption	Technical
	Data not saved	Administration
	Data access restriction	Managerial
CV Terang Mulia	No detailed modules	Behavior
	Data is out of sync	Technical
	Stock data difference	Administration
Nirwana	Network interference	Technical
	Program updates	Administration

5 DEPENDABILITY ANALYSIS

Dependability analysis was performed to determine data dependence using the Rogers model to avoid an error. The dependability analysis is interpreted so as to develop the interviewee's answers. The results obtained from the dependability analysis are clarity of the attachment/importance of the speaker in this study. Dependability corresponds to data validity.

Table 2. Dependability analysis.

Rogers's attribute	Means	Thread	Consideration	Weakness	Opportunities
Relative advantage	Compliance with operational processes Good data management Data integration of all divisions Operational time Efficiency	Data are crucial, so if the system inhibits data flow, both due to programs and infrastructure, the unsaved data will cause a problem in the final report.	The conformity of the company's operating system is the final result of the system's final data, so if the report data are already good the system used can also be considered successful.	Data management is very influential in the operational process so that when there is an incident in the data flow, the operational process is also hampered.	System implementation must comply with the company's opera-tional process. Once appropriate data flows and data integration can be oper-ation-ally time-efficient.
Compatibility	Complies with operational flows Modules for users	In the development process for updating the program, if it is not able to analyze the properly then cannot find the right solution	The system must com-ply with the needs of the company, so in the errors of the system in the field should be analyzed well.	To customize operational process the system must always be developed to know whether the system is appropriate or not for the case.	With suitability the system with the operational flow of work of the system becomes more effective and efficient. Guidebooks are very useful for novice users.

(Continued)

Table 2. (Continued)

Complex-ity	Set up user authority	With author-ity, data change be done for a certain position but will take time when there are many input errors.	The system usage step is quite complex, so if there is a system input error in a certain position it will appear in the final result of the report.	To keep the data safe, the data already input are difficult to modify, so only certain positions can change the data.	With authority for every position, data in the system will not be easily manipu-lated.
Trialability	Data integration Easy recap of data	In the case of technical errors, the data in the process are not synchronized with the system, and help from the consul-tant is needed.	Time for data synchronization; if there is an error it should not impede ongoing operational activities.	With the integra-tion of data, in case data error occurs in one section, it will impede performance in other parts.	With the integration for each section, the data flow becomes easier, so in data recording, users can more easily respond to each report's final data.
Observa-bility	Ease of operation Simplify logging Simplify reports	Not synchronizing data extends system administration time, because every day the company must have complet-ed the sales day report.	When exceeding one program, vendors should consider the effectiveness of data synchronization between programs.	Data retrieval from asynchron-ous program serv-ers inhibits daily reporting.	All data are integrated so that data management and user report generation are facilitated by the automatic logging of the system of stock data.
Communication channel	System introduction Simulation consultation	The occurrence of system errors at the time of operation inhibits operational and administrative processes.	Distance consultants with users cause communication and repair to be time consuming	The distance bet-ween consutants and users does not allow them to repair the system directly, so it is done remotely.	The system can continue to be developed, as many problems exist in the field.

6 CONCLUSION

Based on the research results, it can be concluded that several factors determine the success of the ERP in the textile industry. One factor is conformity between systems with business processes so that the company achieves benefits such as ease of operation, ease of processing data, and efficiency of time and administration costs. Another factor is the compatibility of system customization with business needs; with an appropriate system the company will achieve the results that can be used for long-term business continuity. The last factor is communication media used by consultants for the introduction of new systems to users; with more efficient communication users will get more information and experience in overcoming the operational constraints of the system. The reason why companies in the textile industry implement ERP is that the most complex element in textile companies is stock data of goods and final reports, and by using an integrated system, the data management of goods will be easier.

With an integrated system the data flow of goods can also be more easily monitored so that the data on the final report can also be more easily compiled. The essence of the ERP system in the textile industry is how the system can customize the company's business process flow, so the system needs to be flexible for all corporate business processes in the textile field.

REFERENCES

Barton, P. 2014. Enterprise Resource Planning: Faktor yang Mempengaruhi Keberhasilan dan Kegagalan (Part 1). http://dotsystem.co/blog/?p=26 (online) (accessed July 2, 2019).
Rashid, M. A., Hoassain, L., & Patrick, J. D. 2002. *The Evolution of ERP Systems: A Historical Perspective*. Idea Group Publishing, *1*, 1–16.
Rogers, E. M. 1983. *Diffusion of Innovations*. New York: Collier Macmillan.
Widiyanti, S. 2013. Kesuksesan dan Kegagalan Implementasi Enterprise Resource Planning (ERP) Dan Contoh Studi Kasus PT. Semen Gresik dan Fox Meyer. Thesis, MB-IPB..

Innovation of motif design for traditional batik craftsmen

F. Ciptandi
Telkom University, Bandung, Indonesia

ABSTRACT: The traditional batik motif is moving in a direction in which the market demands new designs without diminishing its distinctiveness. The main obstacle is the limitations of craftsmen in developing new ideas, as they still tend to be conservative both in thinking and competency. As a solution, technology can be used to form a design using j-Batik software that is able to produce many motif variations more quickly and easily. However, it raises other problems, that is, how to persuade the craftsmen to make batik with a new motif that had never been made before. The experimental approach used here was based on diffusion innovation theory, and the study concluded it can appropriately be applied to a group of traditional batik craftsmen in responding to the innovation. In addition, the craftsmen can engage in more dynamic thought to move out of their own conventional customary habits.

1 INTRODUCTION

The distinctive style of Tuban is an expression of a traditional society in interpreting the values and meanings associated with its philosophy of life, with the five senses playing an important role in selecting the objects that inspire it. These expressions survived for years, until finally they started to fade and were replaced by dynamic forms of creative expression. Batik no longer has to carry the value and meaning of the traditional philosophy; it can be freely created in accordance with modern trends and lifestyles. There should be no conflict because both have been able to coexist synergistically (Ciptandi, 2016).

In realizing innovation of the traditional motif, there are still some obstacles, especially limitations of the craftsmen, who tend to be conservative in thinking and skills, in developing ideas for motifs. As a solution to overcome such limitations, technological assistance can be used in the form of j-Batik design software that can produce a variety of motifs more quickly and easily. However, the challenge lies in the ability of the new appearance to still carry a strong spirit of traditional expression.

An appropriate approach to the craftsmen community is needed in applying innovation in the form of j-Batik software technology combined with traditional knowledge and skills about batik possessed by the craftsmen to create synergy between traditional and modern elements.

2 LITERATURE REVIEWS

According to Hariadi et al. (2013, pp. 84–85), the application of fractals to batik by means of jBatik software is not unintentional, but has gone through a series of research processes by calculating fractal dimensions until finally it could be proven that there are fractal elements in batik. The existence of fractal properties in batik can be explained by first understanding the nature of the fractal itself, namely self-similarity (similarity with the shape itself), which means there is a detailed geometry on a smaller scale.

Making batik using jBatik software places it as a modern product oriented to innovation and novelty. Through fractal batik, tradition is able to stand collaboratively side by side with modernity (technology) (Ciptandi, 2018).

3 RESEARCH METHODOLOGY: DIFFUSION OF INNOVATIONS

According to Rogers (1983, pp. 10–27), the diffusion of innovation is a communication theory needed to explain how a new idea and technology can be disseminated to a social system.

Rogers mentioned that there are at least four main elements that become the conditions for this theory to be carried out: (1) innovation, (2) communication media and ways of communicating, (3) time constraints, and (4) social systems.

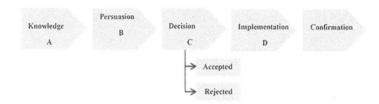

Figure 1. Level of adaptation process in innovation diffusion theory. (Source: Roger, 1983).

The levels of the adaptation process, as shown in Figure 1, are as follows:

A: The level of knowledge: A person is exposed to innovation for the first time but lacks information about the innovation.

B: Persuasive level: Someone starts to be interested in innovation and looks for information.

C: Decision level: A person accepts an innovation concept and starts to consider it.

D: Implementation level: Someone starts to use innovation depending on the situation.

E: Confirmation level: A person finally decides to make a decision to continue using the innovation.

In this method the adoption rate is also used, including:

– Innovators: Are willing to take risks, have the highest social status and financial security, are sociable, and have a close relationship with other innovators.
– Early adopters: Have a good level of leadership, a higher social status, financial ability, and better education. In terms of taking decisions early adopters are more careful and prudent to maintain their position in central communication.

The strategy to implement this innovation for faster adoption using innovation diffusion theory is to present the innovation to a group of individuals who are in the category of innovators and early adopters, so the influence of leadership opinions could affect the receptivity of other communities to such innovations.

4 RESULTS AND DISCUSSION

Based on the aesthetic experiment carried out using jBatik software, two patterns of motifs were developed, as shown in Figure 2.

Figure 2. Batik motif developed using jBatik software: Traditional motif inspiration of *Ganggeng* and *Srigunting*.

Furthermore, the design was given to the innovator group to measure the level of possibility of its realization by the craftsmen in Tuban. At this stage, the innovators, consisting of two people (Mrs. Sri Lestari and Mrs. Rukayah), stated that the newly created motif using j-Batik software had the potential to be applied by batik craftsmen in Tuban.

Next, supported by good knowledge and their authority, these two innovators would transfer knowledge and skills to craftsmen at the level of early adopters to be realized on sheets of clothes.

Table 1. Experimental assignment participants from the innovator category.

Name	Photo-graph	Description
Rukayah (57 years old)		Not in the highest social and economic status, but having experience and opinion leadership by having been a coordinator for all fabric craftsmen in the Tuban region. In addition, Rukayah also has a large role in popularizing the traditional cloth of Tuban by promoting it overseas, and was also awarded Upakarta by the president.
Sri Lestari (45 years old)		In addition to being experienced and having an opinion leadership by employing quite a number of craftsmen in Tuban, Sri Lestari is also starting her own business. She was very open to various innovations in developing traditional fabrics, leading her to win the provincial batik design competition several times.

Table 2. Results of the innovators' methods of communication to early adopters.

Early adopters

Craftsman 1 Description
The assignment of experiments was carried out by Mrs. Sri Lestari. The innovations given can be well received, shown as an attitude of carrying out the process of experimentation with diligence and discipline. Craftsman 1 has the ability to absorb information well; it can be seen from a number of directions given that he has been able to directly carry out the experimental process correctly as expected. The resulting batik cloth has very good quality and does not require a long time to produce although it is still in the adjustment stage.

Result analysis
The batik cloth produced has very good quality. Based on the level of similarity with the given reference image, it is able to reach 80%. The details of the form in the motif can be replicated well, although the composition is still not in accordance with the given picture.

(*Continued*)

Table 2. (*Continued*)

Early adopters

Craftsman 2	Description The assignment of the experiment was carried out by Mrs. Rukayah. The innovations given were well received, although at the beginning there had been some doubtful expressions. Craftsman 2 has the ability to absorb information well; it can be seen from several directions given that he can carry out the experimental process correctly. The resulted batik cloth has very good quality and does not require a long time to produce although it is still in the adjustment stage

	Result Analysis In the second cloth, the quality of the cloth produced is also consistently having very good quality. Based on the level of similarity with the reference images given, it is able to reach 85%. In general, the motifs that have been produced still have a very strong impression representing the visual characteristics of the traditional batik cloth typical of Tuban.

The results of the innovators' methods of communication to early adopters are shown in Table 2.

5 CONCLUSIONS

After applying the innovation diffusion method to design innovation of the typical Tuban batik motif using J-Batik software, it can be concluded that the innovation could be achieved while still displaying the characteristics of the typical Tuban fabric. Therefore, it can be concluded that in addition to being able to adapt to technological interventions, through the right approach, the traditional Tuban cloth can still strongly exhibit its identity.

It is also known that the traditional elements in the typical Tuban motif have a dominant characteristic so that the development carried out on them also remains in the shadow of traditional visual characteristics. This cannot be separated from the mentality of the craftsmen in Tuban who still carry their traditional attitudes and behavior in making fabrics even though they have shown an attitude of acceptance of innovation.

Based on observations of the innovation diffusion method that has been applied, by involving the role of the innovator and also the early adopters, innovation has reached the persuasive/persuasion level and is in the process of strengthening to arrive fully at the level of decision, where someone who receives a concept of innovation and begins to consider it. Further research is needed to be able to provide evidence of the impact of the benefits of the application of this innovation to the community, so that it will reach a higher level, that is, the level of implementation.

REFERENCES

Ciptandi, F. 2018. Transformasi Desain Struktur Tenun Gedog dan Ragam Hias Batik Tradisional Khas Tuban Melalui Eksperimen Karakteristik Visual (Transformation on Design of Gedog Weaving and Traditional 'Tuban' Batik Decoration Through Visual Characteristic Experiment), Disertasi Program Doktor, Institut Teknologi Bandung.

Ciptandi, F., Sachari, A., & Haldani, A. 2016. Fungsi dan Nilai pada Kain Batik Tulis Gedhog Khas Masyarakat di Kecamatan Tuban, Kabupaten Tuban, Jawa Timur. *Panggung, 26*(3).

Fajar, D. C., & Agus, S. 2018. "Mancapat" concept on traditional cloth cosmology of Tuban community, East Java, Indonesia. *Advanced Science Letters, 24*(4), 2243–2246.

Hariadi, Y., Lukman, M., & Destiarmand, A. H. 2013. Batik fractal: Marriage of art and science. *Journal Visual Art and Design, 4,* 84–85.

Rogers, E. M. 1983. *Diffusion of Innovations* (pp. 10–27). New York: The Free Press.

Design documentation of software specifications for a maternal and child health system

G.S. Hana, T.L.R. Mengko & B. Rahardjo
Institute of Technology Bandung, Bandung, Indonesia

ABSTRACT: Babakan Surabaya Community Health Center is a Community Health Center under the Bandung city government that organizes Maternal and Child Health (MCH) services. The current MCH history is recorded in the Cohort Book, but individual patient records are held in the form of an MCH Handbook. The process of recording and reporting MCH is still done manually, and the possibility of data entry errors, problems with data accuracy, and data storage that cannot be accessed at any time can present complications. To be able to produce a good MCH information system, the system needs to undergo functional and nonfunctional analysis. The system requirements are then poured into a Software Requirement Specification (SRS) document. In this study, researchers also measured SRS documents that were produced to determine the quality of the documents.

1 INTRODUCTION

Babakan Surabaya Community Health Center is a Community Health Center under the Bandung city government that organizes Maternal and Child Health (MCH) services. MCH services include examining pregnant women, childbirth, infant immunization, bride and groom immunization, Integrated Visual and Auditory (IVA) testing, and family planning.

The current MCH history is recorded in the Cohort Book, while individual patient records are held in the form of an MCH Handbook. The process of recording and reporting MCH is still done manually. The cohort book is the only source for monitoring and reporting on MCH development in Babakan Surabaya Community Health Center.

Because it still done manually, there are problems related to recording and making MCH reports. Apart from being sourced from MCH services carried out in Puskesmas, the MCH history can also come from services performed outside the building (Posyandu). Constraints arise when MCH services are carried out simultaneously while only one Cohort Book is available. In addition, the possibility of data entry errors, problems with data accuracy, and data storage that cannot be accessed at any time can present complications. In order to produce a good information system and provide maximum benefits, the system must really answer the needs of the user organization. The same goes for the MCH information system. To be able to produce a good MCH information system, the system needs to undergo functional and nonfunctional analysis. Availability of system requirements can also be used as material to cross check whether the application development at the design and implementation stage is in accordance with the plan. The system requirements are then poured into a Software Requirement Specification (SRS) document. In this study, researchers also measured SRS documents that were produced to determine the quality of the documents.

2 RESEARCH OBJECTIVES AND QUESTIONS

The purpose of this research was to answer the following questions:

1. How is the SRS for the MCH system at Babakan Surabaya Community Health Center measured?
2. Has the SRS already produced described the MCH data management needs in the Community Health Center based on the opinions of health workers at Babakan Surabaya Community Health Center?
3. How is the quality of the SRS evaluated?

3 THEORETICAL FOUNDATION

Many parties are concerned about the implementation of MCH services. Some parties have applied the concept of MCH services using a computer-based Information System that is already an application. From several examples of the application of eHealth for MCH in Indonesia, it can be concluded that the use of eHealth can improve the following:

1. Access to services
2. Efficiency in the provision of health services
3. Quality and safety of care
4. Access to health knowledge and education
5. Innovation and growth

4 RESEARCH METHODOLOGY

This research followed a mixed method approach, which combines qualitative and quantitative methods. Qualitative methods are used to produce SRS documents; to produce these documents the researcher must explore the problems that occur at the Community Health Center and find out what the Community Health Center needs for the MCH information system to be built. The quantitative method is used to measure the quality of the SRS produced.

The stages of this research can be seen in Figure 1.

Figure 1. Research stages.

The Requirement Elicitation (RE) is the process of gathering needs from users, stakeholders, and customers in building a system (Swathine & Lakshmi, 2014). RE techniques used are interviews, document collections, and brainstorming analysis. Requirement analysis is an analysis method software requirements collected in the requirements process, placing the discussion in a broader perspective. Confirmation and validation test the SRS to determine its quality. The researchers solicited opinions from 33 individuals, consisting of 7 end-users, 22 people from the software developer group, and 4 people from the academic group.

5 SRS QUALITY MODEL

The SRS quality model used in this study refers to the SRS quality standards issued by Institute of Electrical and Electronics Engineers (IEEE) and can be seen in Figure 2.

6 PROPOSED SYSTEM

This application is designed to be used in Babakan Surabaya Health Center in conducting MCH services. The final result of processing the data is the production of reports in accordance with the standards of maternal and child health services.

This application consists of two subsystems:

1. The web apps subsystem (desktop application) is intended for doctors and midwives, who use the application to record the results of MCH services and to produce reports.
2. The mobile apps subsystem is intended for mothers/patients to view medical history records and get notifications or information about MCH services provided by Puskesmas.

7 SRS QUALITY

SRS quality assessment is carried out on eight variables in accordance with the specified model, except for resource persons from the user group, assessing only four variables according to their domain and understanding. The system has functional requirements (25 codes for web and 14 for mobile system) and 22 nonfunctional requirements for people without technical knowledge. The form of assessment given by the resource person is the answer to the questionnaire with the following alternatives: Strongly Agree, given a value of 5 until Strongly Disagree, given a value of 1. Based on the alternative and weighting of the answers, the maximum value expected in the SRS

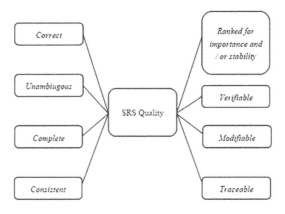

Figure 2. SRS quality model. (Source: IEEE, 2011).

quality testing is $(8 \times 33 \times 5) - (4 \times 7 \times 5)$, which is equal to 1,180. The following are the results of the calculation of the ratings given by the resource person:

1. The total number of weights given by the resource person is 1,046.
2. Value interval:

Number of statements	= 236
Highest value	= 236 × 5 = 1,180
Lowest value	= 236 × 1 = 236
Class interval	= (1,180-236)/5 = 188.8

   The SRS quality assessment intervals are as follows:

236 × 424.8	= Very Bad
424.8 <× 613,6	= Not Good
613.6 <× 802.4	= Doubtful
802.4 <× 991,2	= Good
991.2 <× 1180	= Very Good

Based on the final weight of 1,046 and using predetermined intervals, the result for SRS quality is Very Good.

8 CONCLUSION

The conclusions of this study are as follows:

1. The SRS document for the MCH information system in Babakan Surabaya Community Health Center was produced using the IEEE standard 830-1998. The analysis results show that in the SRS document there are two subsystems: web apps and mobile apps.
2. The SRS has already described the MCH data management needs. All the requirements for MCH data and reporting management at Babakan Surabaya Community Health Center are contained in the SRS document. The web apps subsystem consists of 25 functional requirements and the mobile apps subsystem consists of 14 requirements. There are also 22 nonfunctional requirements.
3. Based on the assessment of 33 sources, the quality of the SRS documents that have been produced is very good.

REFERENCES

IEEE. 2011. International Standard ISO/IEC/IEEE 29148 — Systems and software engineering — Life cycle processes — Requirements engineering.

Swathine, K., & Lakshmi, J. K. 2014. Requirement elicitation for requirement in software engineering. *International Journal of Engineering, Science, and Technolnology*, 3(12).

Design strategies in the market competition of capsule hostels

S. Rahardjo
Telkom University, Bandung, Indonesia

ABSTRACT: As a favorite travel destination, Bandung is always visited by tourists, especially on weekends. Thus the business opportunity and competition in the accommodation sector have increased. There are broad accommodation choices including hotels, guest houses, and hostels, all of which have their own segmented markets, such as high-class tourists, low-budget tourists, and backpackers. The markets are becoming predisposed toward hostels, many of which provide capsule bedrooms with adequate privacy and comfort for their guests. With several emerging capsule hostels in this city, this article aimed to find the strategies hostels use to win market competition through their capsule pod designs. A comparative method was used for the research through a survey at INAP at Capsule Hostel, Pinisi Backpacker Hostel, and Tokyo Cubo. The finding shows that each hostel has a unique strategy, which is shown respectively in compact design, fancy LED lighting, and thematic design.

1 INTRODUCTION

1.1 *The rising number of backpackers as the foundation of the hostel market*

Several years ago, there were only a few hostels listed in Bandung because most tourists, especially domestic tourists, prefered to choose a hotel or a guest house to meet their accommodation needs. At that time, the hostels were segmented to a smaller group of tourists who called themselves backpackers and usually came from foreign countries.

With the rise of low-cost airlines and the development of communication technology, travel has become easier and cheaper, and today, it has become a lifestyle for many Indonesians (Hermawan & Hendrastomo, 2017). Cohen (2010) even stated that backpacking as a form of travel is "a way of life." As seen in several popular blogs and Instagram accounts, many young Indonesian people claim to be independent travelers or backpackers who are proud to show that they are able to travel far on a shoestring budget.

According to Paris (2010), backpacking is not something new. Instead, he suggested backpacking as a mainstream phenomenon in tourism that symbolizes the increasingly mobile world. In line with it, Hyde and Lawson (2003) also mentioned that independent travel has also become an important and growing sector of worldwide tourism. So, with greater numbers of independent travelers or backpackers in a city, the demand for accommodation facilities has also increased, especially for hostels as budget accommodations for backpackers.

1.2 *A shift in the hostel market*

The data from Badan Pusat Statistik Jawa Barat (2018) show that until 2016, Bandung received the largest number of visits compared to other cities in West Java, and according to Badan Pusat Statistik Kota Bandung (2017), those visitors were mostly domestic tourists. As a result, today we can easily find many more hostels in Bandung with different styles, designs, and prices. There is a demand for hostels among domestic tourists, who outnumber foreign tourists,, regardless of their tendencies to travel with a sense of prestige or to avoid the fear of missing out (Hermawan & Hendrastomo, 2017). According to Hao and Har (2014), "As part of the process of rapid growth, the tourism industry has become more diversified, targeting different consumers with more sophisticated products." Today, there are broad choices for tourist accommodations, including hotels, guest houses, and hostels. Previously, these markets were classified into medium- to high-budget tourists, low-budget tourists, and backpackers respectively, but now the markets are predisposed to hostels, as many provide capsule bedrooms that meet their guests' expectations for privacy and comfort.

Hence, the hostel market has shifted from segmented to almost borderless, especially when travelers are budget conscious (Cho, 2005). Moreover, hostels also enable tourists to stay at a strategic location at a reasonable price (Nikova, 2019). With various choices of hostel designs, people can choose from among different hostels. Hostels then become a common typology of travel accommodation not only for backpackers but also for other groups of tourists with higher budgets, such as flashpackers (Paris, 2012), a group of friends and family, or young people. Therefore, with several emerging capsule hostels in this city, this article aimed to find the strategies of hostels to win market competition through their capsule pod designs.

2 METHOD

2.1 Method

This study used a comparative method through observation in three selected hostels. From the large number of hostels and several choices of hostel styles, all of the selected objects had to meet the following criteria: the hostels are good examples in the market; the studied hostels have a good degree of popularity in booking websites; the hostels adhere to the basic value of a hostel in sharing public facilities, the selected objects are hostels that offer only beds in dormitories and not a private room; and to ensure that the studied objects meet the hostel trend in today's era, all hostels must have beds designed as capsule pods that are built in accordingly to the interior space, not mass produced capsules arranged in a stack system.

2.2 Study objects

From names of hostels listed on Hostelworld.com and Booking.com, three hostels were found to meet the aforementioned requirements: Pinisi Backpacker Hostel, Tokyo Cubo, and INAP at Capsule Hostel. Pinisi Backpacker Hostel was the oldest hostel among the three that were studied. Although it does not have the highest review score, it still pops up as first on the popularity list on Hostelworld.com. In contrast, Tokyo Cubo was the newest hostel, with a very good review, scoring 9.3 out of 10 on Booking.com and also was first on the popularity list on the same website. The last one is INAP at Capsule Hostel. Although INAP did not appear first on the list or gained the best review, many personal blogs post about the writers' positive experiences about this hostel and those are considered as a good measure of popularity.

3 DISCUSSION

3.1 Pinisi backpacker hostels

Located very near the train station, this hostel already has a great opportunity to win market share based on its location. This hostel operated before the era of capsule pods and still keeps half of its dormitory rooms with the conventional bunk beds, but it also offers another half of the space as capsule pods to attract customers who prefer more privacy in the bedroom. To transform the dormitory into capsule pods, the hostel only adds several wall partitions and finishes them with paint. Counting only on paint and the efficient use of plywood material, the design appears very simple and far from an impression of being costly, yet is still well decorated with colorful LED stripes that trick the eyes to not think that the room is actually painted in plain white. This hostel also benefit from the exsisting building, which was designed according to the principles of tropical houses. Natural air circulation is possible in and out of the building, so the rooms do not urgently need air conditioning. The air flow in the capsules can still be managed with fans and exhausts.

On the other hand, there are some design weaknesses that may diminish the value of this hostel. First of all, the appearance is not as trendy as that of other, newer hostels. The plain white paint on plywood also creates an impression that the design is cheap. Obviously, those will filter their market to tourists whose budgets are limited. Thus, the hostel will face difficulty in winning the market if it raises its price. Nevertheless, if the market of this hostel is already filtered, travelers with limited budgets or the real backpackers will more likely choose this hostel compared to the newer ones when they make reservations from booking websites.

3.2 Tokyo Cubo

Even though this hostel is still brand new and has very good customer reviews, it does not mean that Tokyo Cubo can easily rule the hostel market. This hostel has a premium location very close to the Bandung train station. One of its competitors is Pinisi Backpacker Hostel, which is only a few meters away. Although the market for Pinisi Backpacker Hostel is already segmented based on the impression of its design, there is another popular accommodation nearby that is almost as new as this hostel, named Bobobox. Therefore, this hostel needs a unique concept to compete with similar accommodations nearby.

The uniqueness of this hostel comes in the form of a strong Japanese thematic concept. Every interior element inside the dormitory is arranged in a grid pattern with clean and neat geometrical shapes. The ratios and proportions that are used in the visual composition of the interior elements are similar to those usually found at Japanese houses. As Japan is also one of Indonesian tourists' dream countries for traveling, there are many domestic travelers who can relate the Japanese theme to their traveling lifestyle. Other than that, the design of this hostel also gives a clean, stylish, and comfortable impression. As the hostel is still new, those impressions create another impression that this is an expensive or luxurious hostel. In contrast to the case at Pinisi Backpacker Hostel, backpackers with limited budgets may feel hesitant to stay at this place. The market for this hostel then is also segmented in the opposite direction, which is to cater to those who can actually afford a higher price.

3.3 INAP at Capsule Hostel

Where other hostels in Bandung are placed inside shop houses or become an adaptive reuse of a residential house, INAP at Capsule Hostel uses a unit of an apartment at the center of the city. The original shape of the apartment unit is very small, which makes the design of each capsule very compact, yet very interesting.

It is found that this hostel has the largest number of facilities inside each capsule unit compared to any other hostel in Bandung. Whereas a capsule in other hostels generally provides an electric socket, personal reading lamp, and an area to put the guests' personal belongings, a capsule at INAP provides extra entertainment facilities, such as a TV and aux jack, a private mirror, and a real aquarium, all of which are placed inside the capsule unit and are dedicated for single-person use. With such facilities, this hostel is fit for travelers who are keen on entertainment and a lifestyle with electronic devices. Guests do not need to have any other activities inside the hostel if they are already able to entertain themselves with all kinds of facilities provided right from their beds.

As a consequence, this capsule cannot achieve total darkness. With a real aquarium with real fishes, the capsule cannot achieve total silence or total darkness because the aquarium has a light that is always turned on and the fish cannot be instructed not to make any sound at all.

3.4 Comparison of the studied objects

Table 1. Strengths and weaknesses of the studied objects.

Strengths of hostels	
Pinisi	Provides choices of conventional bunk beds or capsule pods
	Simple yet colorful designs through applications of paint and LED stripes.
	Capsule pods are built inside the room with natural air flow, which does not require air conditioning.
Tokyo Cubo	Strong Japanese thematic design.
	Every element is arranged in a grid pattern with clean geometrical shapes
	Gives a clean, stylish, and comfortable impression
INAP	Sophisticated facilities
	Extra entertainment facilities compared to other hostels
	Compact design: all facilities are accessed from the bed inside each capsule unit
Weaknesses of hostels	
Pinisi	Less attractive design
Tokyo Cubo	An expensive impression creates hesitation on the part of backpackers
Sophisticated facilities	Guests cannot sleep in total darkness or total silence

4 CONCLUSIONS

Three design approaches were found among the three studied hostels. Pinisi Backpacker Hostel, as the simplest one, applies colorful LED stripes on their white painted plywood that looks simple but still shows an effort to decorate the capsule unit in a budget- conscious way. Thus, this hotel may aim for the market of budget travelers or real backpackers. In contrast, Tokyo Cubo applies a strong thematic approach based on a popular travel destination country. Thus, this hostel may target domestic travelers who seek comfort and lifestyle, yet still wish to experience staying at a hostel. The last and the unique one is INAP at Capsule Hostel, which provides extravagant entertainment embedded in electric devices attached together inside the capsule as a compact design. Thus, this hostel may attract the group of flashpackers.

REFERENCES

Cho, M. 2005. Budget traveler accommodation satisfaction: The case of Yogwans during the 2002 FIFA World Cup Korea/Japan. *Asia Pacific Journal of Tourism Research*, *10*(3), 275–286.

Cohen, S.A. 2011. Lifestyle travellers: Backpacking as a way of life. *Annals of Tourism Research*, *38*(4), 1535–1555.

Hao, J.S.C., & Har, C.O.S. 2014. A study of preferences of business female travelers on the selection of accommodation. *Procedia: Social and Behavioral Sciences*, *144*, 176–186.

Hermawan, H., & Hendrastomo, G. 2017. Traveling sebagai gaya hidup mahasiswa Yogyakarta. *Jurnal Sosiologi Universitas Negeri Yogyakarta*.

Nikova, T.V., et al. 2019. Features and factors of the hostel market development in Russia. In *The European Proceedings of Social & Behavioural Sciences EpSBS. GCPMED 2018 International Scientific Conference*, December 6–8, 2018. Russia.

Paris, C. M. 2010. The virtualization of backpacker culture: virtual moorings, sustained interactions, and enhanced mobilities. In K. Hannam & A. Diekmann (eds.), *Backpacker Tourists: Experiences and Mobilities* (p. 406). Clevedon: Channel View Publications.

Paris, C.M. 2012. Flashpackers: An emerging sub-culture? *Annals of Tourism Research*, *389*(2), 1094–1155.

https://jabar.bps.go.id/statictable/2018/03/23/475/jumlah-kunjungan-wisatawan-ke-obyek-wisata-menurut.html

https://bandungkota.bps.go.id/statictable/2017/08/29/120/jumlah-wisatawan-mancanegara-dan-domestik-di-kota-bandung-2016.html

The link of e-servqual & perceived justice of e-recovery to satisfaction & loyalty

S. Syafrizal & S.L. Geni
Andalas University, Padang, Indonesia

ABSTRACT: The rapid growth of internet users has facilitated the growth of e-commerce business in Indonesia. This study aims to analyze the effect of electronic service quality and perceived justice of electronic service recovery on the satisfaction and loyalty of online shoppers. Survey questionnaire was conducted on 338 online shoppers in Padang, Indonesia. Research instrument of this study was adapted from previous studies. Research data were analyzed through structural equation modelling with the help of Smart PLS 3.0 software. The results of this study show that electronic service quality has positive significant effect on satisfaction and loyalty. In addition, perceived justice of electronic service recovery also has positive significant influence on satisfaction and loyalty. Furthermore, it is found that satisfaction has positive significant effect on loyalty.

1 RESEARCH BACKGROUND

The development of information communication technology has spurred the growth of e-commerce industry over the last decade. Kuan, Bock and Vathanopas (2005) had stated that the growth of online retail selling has surpassed the growth of offline retail selling. Statistical data show that e-commerce contributed as much as 5.9% ($1.3 trillion) to total global retail sales in 2014 (Renjith, 2015). It has been predicted that this contribution will increase to 8.8% ($2.5 trillion) of global retail sales (Renjith, 2015).

Even though the e-commerce industry has a good prospect, it still has many challenges to grow appropriately. According to Marimon (2011) online customers are difficult to be acquired and maintained. In addition, intense competition results in the difficulty to establish and develop loyal customers in the e-commerce industry (Renjith, 2015). Besides competition factor, the introduction of new products with shorter life cycle and the increase in customer expectation are also challenging the e-commerce businesses in maintaining and increasing their customers' loyalty (Barani, Menon and Ajay, 2017; Romano and Fjermestad, 2001). Online business has a unique characteristic that makes it difficult to avoid service issues such as late merchandise delivery, problem in transaction system, etc. (Qin, Chen, Wan 2012). These service issues will consequently trigger customers' complaint. According to an NGO that provides support to consumers (YLKI), it reported that complaints related to e-commerce ranked the top of consumers' complaint in 2017. Marketing Charts Staff (2013) had found that as high as 66% of online customers had lodged complaints against certain e-commerce firms. Consumers have been advised to stop purchasing from these e-commerce businesses if they fail to conduct service recovery. According to Collier (2006) service recovery is one of key success factors for an e-commerce firm.

Besides service recovery, the literature has also revealed that e-servqual has an important role in establishing and developing customer loyalty. Many previous studies conducted in various industries and countries have shown that e-servqual dimensions such as *system availability, fulfillment, responsiveness, efficiency, privacy, web design have positive and significant effect on customer loyalty* (Jiang, Jun & Yang, 2016; Puriwat & Tripopsakul, 2017).

Even though e-servqual and e-service recovery have been proven to play important roles in developing and maintaining customer loyalty, there is still very few being conducted to investigate the effect of these variables on customer loyalty simultaneously, especially in e-commerce industry. Therefore, this study aims to fulfill this research gap by choosing the research topic: "The effects of e-servqual and perceived justice of e-service recovery on customer loyalty"

2 LITERATURE REVIEW

2.1 Electronic service quality, satisfaction and customer loyalty

Research on e-servqual has been conducted over last two decades. Scholars and researchers are yet to agree on what are the permanent dimensions of e-servqual. Currently, there are three dominant approaches to e-servqual research. Those approaches are: marketing approach, represented by e-servqual model by Parasuraman (2005); information communication technology approach, represented by TAM and UTAUT models; and mixed model which is a combination of the two previous models.

Online shopping is becoming a life style for the millennial generation. Delivering electronic service quality that equal or exceed customers' expectation will improve their satisfaction. In other words, perceived high electronic service quality dimensions such as reliability, privacy, responsiveness, fulfillment and website design by customers will improve their satisfaction. Previous studies of various industries (retail, hotel, etc.) had found that electronic service quality has positive and significant effect on customer satisfaction (Pool, Deegan, Jamkhaneh, Jaberi and Sharifkhani, 2018; Jeon & Jeong, 2017; Sundaram, Ramkumar, Shankar, 2017; Puriwat, Tripopsakul, 2017).

A good quality electronic service will result in positive attitude and experience, which in turn will increase the customers' intention to repeat their purchases. Previous studies have found that electronic service quality has positive significant effect on return intention (Jeon & Jeong, 2017) and customer loyalty (Puriwat, Tripopsakul, 2017). In addition, the literature also reveals that the dimensions of electronic service quality such as *availability, fulfillment, efficiency, privacy, responsiveness dan web design* have positive significant effect on customer loyalty (Sundaram, Ramkumar, Shankar, 2017; Pool, Dehghan, Jamkhaneh, Jaberi and Sharifkhani, 2018; Bai, Cui, & Ye, 2014).

2.2 Perceive justice of electronic service recovery, satisfaction and customer loyalty

Previous studies revealed that e-service recovery strategy has an important role in improving customer satisfaction and customer loyalty. For example, Marimon, Jaya and Casadesus (2011) in their studies had found that e-service recovery has significant effect on customers' satisfaction and loyalty. Furthermore, the literature also argued that perceived justice of e-service recovery has significant effect on customer satisfaction (Jung and Seock, 2017). Zehir and Narcikara (2016) in their research found that perceived procedural justice and interactional justice of e-service recovery have significant effect on customer loyalty.

3 RESEARCH METHOD

This study used quantitative research design. Survey questionnaire was administered on 338 respondents (online-store customers) in Padang city, Indonesia. Purposive sampling technique was used to select the respondents. Research instrument was adapted from previous research. Structural equation modelling technique with software Smart PLS 3.0 had been used to analyze research data.

4 RESULT AND DISCUSSION

All of the components in the mesurement model such as main loading indicator, average variance extracted (AVE) and discriminant validity have fulfilled the PLS requirement. Then, the structural model of Smart PLS.3.0 results indicated a positive significant effect of electronic service quality on satisfaction ($\beta = 0.456$, $t = 3.384$; $p < .01$), perceived justice of service recovery on satisfaction ($\beta = 0.323$, $t = 2.708$; $p < .05$), electronic service quality on loyalty ($\beta = 0.410$, $t = 3.241$; $p < .01$), and perceived justice of service recovery on loyalty ($\beta = 0.329$, $t = 2.531$; $p < .05$).

4.1 The effect of electronic service quality on customer satisfaction and customer loyalty

In this study, it is found that e-servqual has significant effect on satisfaction and loyalty. This means that the higher the e-servqual, the higher customers' satisfaction and loyalty will be. This is in line with previous studies that found e-servqual has significant effect on customer satisfaction (Jeon and Jeong, 2017; Puriwat dan Tripopsakul 2017; Dehghan, Jamkhaneh, Jaberi and Sharifkhani, 2018) and customer loyalty (Puriwat and Tripopsakul, 2017; Dehghan, Jamkhaneh, Jaberi and Sharifkhani, 2018). Majority of this study's respondents were the millennial generation, this means that electronics service quality is an important predictor to their online purchase decisions. This makes sense since online purchase is a daily activity and has become a lifestyle for these millennials. Based on the literature, electronic service quality is generally consisted of five dimensions: reliability, privacy, website design, fulfillment and efficiency (Kayabaşı, Çelik, Büyükarslan, 2013; Amin, 2016). Therefore, improving electronic service quality dimension will improve customers' satisfaction and loyalty.

4.2 The effects of perceived justice of electronic service recovery on customer satisfaction and customer loyalty

This study had found that perceived justice of electronic service recovery has positive and significant effect on customers' satisfaction and loyalty. This is in line with previous studies that found perceived justice of electronic service recovery has significant effect on satisfaction (Jung and Seock, 2017; Marimon, Jaya and Casadesus, 2011) and loyalty (Abolfathi and Feizi, 2015; Shirabad and Gilaninia, 2015). Marketing literature revealed that customer satisfaction to service recovery action is determined by the degree of perceived justice in service recovery effort by the service firm (Jung and Seock, 2017). According to theory, customer complaints that are appropriately handled by service firm will resolve the customers' issues. In fact, it will result in an even higher satisfaction than customers who have never complaint before (McCollough, Michael, and Bharadwaj, 1992).

5 CONCLUSION AND IMPLICATION

This study has found that electronic service quality has a more important role than perceived justice of service recovery to predict customers' satisfaction and loyalty. Furthermore, this study explains that the dimension of electronic service quality has an important role in improving customers' satisfaction and loyalty. In addition, this study also shows that perceived interactional and procedural justice have important roles in improving satisfaction and loyalty.

Regarding the significant effect of electronic service quality on customers' satisfaction and loyalty, it implies that online store should maintain the high electronic service quality indicator value and put in more efforts to improve their low electronic service quality indicator value. Furthermore, online stores also need to put in more efforts in improving their perceived interactional and procedural justice, which will improve their customers' satisfaction and loyalty. This is primarily important for millennial generation who conduct online transactions daily.

REFERENCES

Abolfathi, Y., & Feizi, M. 2015. To explain the role of service recovery strategy on trust and behavioral intentions of TOSEH insurance's customers of Ardabil. *SUSUREA* 3(1): 24–34.

Amin, Muslim. 2016. Internet banking service quality and its implication on e-customer satisfaction and e-customer loyalty. *International Journal of Bank Marketing* 34(3): 280–306.

Barani, Menon & Ajay. 2017. A study on the customer retention strategies used by the e-commerce companies. *International Journal of Engineering Technology, Management and Applied Sciences* 5(5): 527–535.

Jeon, M. M., & Jeong, M. 2017. Customers' perceived website service quality and its effects on e-loyalty. *International Journal of Contemporary Hospitality Management* 29(1): 438–457.

Jiang, L., Jun, M., & Yang, Z. 2016. Customer-perceived value and loyalty: how do key service quality dimensions matter in the context of B2C e-commerce?. *Service Business* 10(2): 301–317.

Kayabaşı, A., Çelik, B., & Büyükarslan, A. 2013. The analysis of the relationship among perceived electronic service quality, total service quality and total satisfaction in banking sector. *Journal of Human Sciences* 10(2): 304–325.

Kuan, H. H., Bock, G. W., & Vathanophas, V. 2005. Comparing the effects of usability on customer conversion and retention at e-commerce websites. *In Proceedings of the 38th annual Hawaii international conference on system sciences, Hawaii, January 2005*. IEEE.

Parasuraman, A., Zeithaml, V. A., & Malhotra, A. (2005). ES-QUAL: a multiple-item scale for assessing electronic service quality. *Journal of service research* 7(3): 213–233.

Pool, J. K., Dehghan, A., Jamkhaneh, H. B., Jaberi, A., & Sharifkhani, M. 2018. The effect of e-service quality on football fan satisfaction and fan loyalty toward the websites of their favorable football teams. In *Sports Media, Marketing, and Management: Breakthroughs in Research and Practice*: 470–485.

Puriwat, W., & Tripopsakul, S. 2017. The impact of e-service quality on customer satisfaction and loyalty in mobile banking usage: case study of Thailand. *Polish Journal of Management Studies* 15(2): 183–193.

Renjith, S. 2015. An Integrated framework to recommend personalized retention actions to control B2C e-commerce customer churn. *International Journal of Engineering Trends and Technology (IJETT)* 27(3): 152–157.

Zehir, C., & Narcıkara, E. 2016. E-service quality and e-recovery service quality: effects on value perceptions and loyalty intentions. *Procedia-Social and Behavioral Sciences* 229: 427–443.

Technical efficiency of information and communications technology companies in East Asia and Southeast Asia

Y. Yuliansyah & P.M. Sitorus
Telkom University, Bandung, Indonesia

ABSTRACT: This research aimed to analyze the efficiency level of 30 information and communications technology (ICT) companies in Southeast Asia and East Asia for the period 2013–2017. This research used the Stochastic Frontier Analysis (SFA) method with the Translog Production model to estimate the technical efficiency score and its correlation with input variables and explanatory variables. The results revealed that overall, ICT companies in Southeast Asia and East Asia have been efficient, with an average technical efficiency value of 0.8361, and the most influential variables are labor, company's age, and company's size. Telkom, China Mobile, and NTT were the most efficient ICT companies, while Time dotCom, Smartfren, and Samart Telcoms were the most inefficient ICT companies. Based on their operating location, ICT companies in the East Asia region were more efficient than those in the Southeast Asia region, and ICT companies in high-income countries were more efficient than those in middle-income countries.

1 INTRODUCTION

The information and communications technology (ICT) industry is still an attractive growth business, while companies' financial performance in terms of profits shows a downward trend. The growth of the technology market in the Asia Pacific region was 2.1% in 2013–2017 and predicted to grow around 5.5%–6.8% in 2018 (Bartels & Giron, 2017). On the other hand, companies' financial performance in terms of profits continues to decline. Based on 2013–2017 financial data, average income showed a 2% growth, while EBITDA margins tended to decline 0.8%. Based on these phenomena, technical efficiency analysis is required to help management to evaluate and improve company performance. By knowing the efficiency level and the most influential variables, management can make better decisions to increase the company's competitiveness and profitability (Masson et al., 2016).

2 LITERATURE REVIEW

Efficiency is one of the parameters of company performance related to resource management. (Porcelli, 2009). Efficiency can be measured by several methods, such as the Stochastic Frontier Approach (SFA), Distribution Free Approach (DFA), Thick Frontier Approach (TFA), Data Envelopment Analysis (DEA), and Free Disposal Hull (FDH) (Berger & Humprey, 1997, p. 5).

According to the research objectives and the previous research, this research used the SFA method developed in 1977 by Aigner et al. Hjalmarsson et al. (1996) found that the SFA model is the most preferred model compared to DEA and DFA because it shows a more constant scale of returns. Similarly, Coelli et al. (2005) stated that SFA has a major advantage over other methods that consider measurement errors and other sources of statistical noise that are considered as technical inefficiencies.

3 METHODOLOGY

3.1 Data and variables

This research used panel data from 30 ICT companies in the 2013–2017 period. The samples were selected using a purposive sampling technique based on the criteria of ICT companies listed in the Top 300 Most Valuable Telecommunications 2018 categories operating in the East Asia and Southeast Asia regions and listed on the stock exchange since 2013 with financial accounts detailed in accordance with the research variables.

Nine variables were used in this research, which were divided into three categories. Input variables consisted of capital expenditure (Hendrawan & Nugroho, 2018); personnel expenses (Prabowo & Cabanda, 2011; Hendrawan & Nugroho, 2018); other input, such as the total of cost of goods sold/services, selling & marketing expenses, general and administrative expenses (Battese & Coelli, 1995; Leurcharusmee et al., 2016; Hendrawan & Nugroho, 2018); and time period of observation (Battese & Coelli, 1995; Hjalmarsson et al., 1996; Moriwaki et al., 2010). Revenue was used as an output variable (Torres & Bachiller, 2013; Hendrawan & Nugroho, 2018). Explanatory variables consisted of total assets (Prabowo & Cabanda, 2011; Torres & Bachiller, 2013; Hendrawan & Nugroho, 2018), company's age (Prabowo & Cabanda, 2011), gross domestic product (GDP) per capita (Moriwaki et al., 2010; Torres & Bachiller, 2013), and time trend (Hjalmarsson et al., 1996; Prabowo & Cabanda, 2011).

3.2 Stochastic frontier analysis model

This research used Frontier 4.1 software to apply SFA model 2 proposed by Battese and Coelli (1995) using the effects of technical inefficiencies in the Translog Production function model in estimating technical efficiency. The basic SFA function model is presented in Equation (1). Specifically, this research applied the stochastic Translog Production function model presented in Equation 2 and the inefficiency function model presented in Equation 3:

$$Y_{it} = \exp(X_{it}\beta + V_{it} - U_{it}) \tag{1}$$

$$\begin{aligned}\ln(Y_{it}) =\ & \beta_0 + \beta_1 \ln K_{it} + \beta_2 \ln L_{it} + \beta_3 \ln O_{it} + \beta_4 t + 0.5\beta_5 \ln(K_{it})^2 + \beta_6 \ln K_{it}(\ln L_{it}) \\ & + \beta_7 \ln K_{it}(\ln O_{it}) + \beta_8 \ln K_{it}(t) + 0.5\beta_9 \ln(L_{it})^2 + \beta_{10} \ln L_{it}(\ln O_{it}) \\ & + \beta_{11} \ln L_{it}(t) + 0.5\beta_{12} \ln(O_{it})^2 + \beta_{13} \ln O_{it}(t) + \beta_{14} t^2 + V_{it} - U_{it}\end{aligned} \tag{2}$$

$$U_{it} = \delta_0 + \delta_1(FS_{it}) + \delta_2(A_{it}) + \delta_3(G_{it}) + \delta_4(t) \tag{3}$$

where Y = revenue of the company; K = capital expense; L = personnel expenses; O = other inputs; t = time period of observation; β = natural log of vector of each input variables; V_{it} = random error, which is assumed to be independently distributed $N(0, \sigma^2)$; U = nonnegative random variables which assumed to be independently distributed $N(0, \sigma^2)$; FS = total assets; A = company's age; G = GDP per capita; δ = natural log of vector of each explanatory variables; and i = number of companies.

4 FINDINGS AND DISCUSSION

4.1 Correlation of input variables and explanatory variables on technical efficiency

The research results revealed that three variables—labor, company's size, and company's age—had the most significant impact on efficiency. This finding indicated that technical efficiency in ICT companies would increase along with labor utilization and an increase in company's size and company's age. This result was supported by the findings of Leurcharusmee et al. (2016), Prabowo and Cabanda (2011), and Hjalmarsson et al. (1996), which stated that labor had a positive effect on a company's efficiency, while Torres and Bachiller (2013) revealed that large companies were more technically efficient than small companies, and Prabowo and Cabanda (2011), who stated that a company's age or operational experience was able to improve technical efficiency.

4.2 *Technical efficiency analysis*

The value of the technical efficiency of the average ICT companies in the East and Southeast Asia was 0.8361 (Table 1). These results indicated that overall the level of efficiency of ICT companies in the East and Southeast Asia was relatively high (greater than 0.7). The highest technical efficiency score was obtained by Telkom in 2016 (0.9850), meaning that on average, there was still room for ICT companies to improve their technical efficiency scores by 0.1414.

There were 23 out of 30 companies that had carried out production activities efficiently with a technical score of more than 0.7. On average, Telkom (0.9836), China Mobile (0.9762), and NTT (0.9681) were the most efficient ICT companies, while Time dotCom (0.4062), Smartfren (0.4708), and Samart Telcoms (0.5060) were the most inefficient ICT companies.

Based on the region (Table 2), ICT companies operating both in East Asia and Southeast Asia had already been efficient. However, ICT companies operating in the East Asia region had a much higher average technical efficiency score than those operating in the Southeast Asia region, which was 0.9128 compared to 0.7850.

Based on the income status of the region (Table 3), both groups had already been efficient, with ICT companies in high-income countries still much more efficient than those in middle-income countries. This finding was in line with Moriwaki et al. (2010) stating that the higher the state income, the higher the level of efficiency of companies in the region.

Table 1. Technical efficiency scores of ICT companies.

Companies	2013	2014	2015	2016	2016	Average
AIS	0.9514	0.9496	0.9580	0.9207	0.9484	0.9456
Axiata	0.9559	0.9519	0.9461	0.9501	0.9594	0.9527
China Mobile	0.9748	0.9764	0.9759	0.9767	0.9773	0.9762
China Telecom	0.9709	0.9662	0.9566	0.9552	0.9604	0.9619
China Unicom	0.9396	0.9513	0.9179	0.9424	0.9635	0.9429
Chunghwa	0.9423	0.9454	0.9624	0.9618	0.9551	0.9534
Digi.com	0.9196	0.9258	0.9163	0.9303	0.9240	0.9232
FarEasTone	0.8561	0.8402	0.8464	0.8628	0.8730	0.8557
Globe Telecom	0.9025	0.9264	0.9263	0.9357	0.9297	0.9241
HKT	0.9450	0.9591	0.9622	0.9688	0.9719	0.9614
Indosat	0.9140	0.9228	0.9355	0.9475	0.9492	0.9338
Jasmine	0.6170	0.6188	0.5500	0.5601	0.6011	0.5894
KT Corp	0.9626	0.9584	0.9697	0.9700	0.9704	0.9662
LG U+	0.9089	0.8837	0.8778	0.9046	0.9154	0.8981
Link Net	0.4468	0.5463	0.5947	0.6668	0.6974	0.5904
M1	0.7146	0.7287	0.7350	0.7220	0.7174	0.7235
Maxis	0.9668	0.9704	0.9698	0.9745	0.8942	0.9551
NTT	0.9760	0.9699	0.9702	0.9619	0.9623	0.9681
PLDT	0.9669	0.9656	0.9265	0.9340	0.9381	0.9462
Samart Telcoms	0.5531	0.5129	0.4607	0.4756	0.5276	0.5060
Singtel	0.9646	0.9589	0.9616	0.9558	0.9512	0.9584
SK Telecom	0.9640	0.9619	0.9649	0.9623	0.9633	0.9633
Smartfren	0.4080	0.4482	0.4724	0.4778	0.5477	0.4708
Smartone	0.7253	0.6832	0.4297	0.9572	0.6434	0.6878
Starhub	0.7894	0.7976	0.7869	0.7835	0.7836	0.7882
Taiwan Mobile	0.7699	0.7293	0.8529	0.8854	0.8552	0.8186
Telkom	0.9821	0.9829	0.9831	0.9850	0.9849	0.9836
Thaicom	0.6242	0.6894	0.6990	0.6414	0.6309	0.6570
Time dotCom	0.3850	0.3942	0.3886	0.4122	0.4507	0.4062
XL Axiata	0.8838	0.8702	0.8835	0.8760	0.8615	0.8750
Mean	0.8294	0.8329	0.8260	0.8486	0.8436	0.8361

Table 2. Technical efficiency scores of ICT companies by region.

Region	2013	2014	2015	2016	2017	Average
Southeast Asia	0.7748	0.7867	0.7830	0.7861	0.7943	0.7850
East Asia	0.9113	0.9021	0.8906	0.9424	0.9176	0.9128

Table 3. Technical efficiency scores of ICT companies by income status.

Region	2013	2014	2015	2016	2017	Average
Middle Income	0.7748	0.7867	0.7830	0.7861	0.7943	0.7850
High Income	0.9113	0.9021	0.8906	0.9424	0.9176	0.9128

5 CONCLUSIONS AND RECOMMENDATIONS

This research provided relevant results in helping management to evaluate and make better decisions in terms of operational or production activities. The research result revealed that three main variables—labor, company's size, and company's age—had a significant effect on efficiency. During the five-year period of 2013–2017, ICT companies in East Asia and Southeast Asia were already relatively very efficient in terms of technical efficiency, but there was still room for ICT companies to improve their technical efficiency through the labor utilization, company size increase, and company's operational experience.

Based on the technical efficiency scores, it was found that Telkom was the most efficient ICT company during the period 2013–2017, while Time dotCom was the most inefficient ICT company. In addition, based on the group and its region and income status, ICT companies operating in the East Asia region were more efficient compared with those in the Southeast Asia region, and ICT companies in high-income countries were more efficient compared with those in middle-income countries.

REFERENCES

Bartels, A., & Giron, F. 2017. *Asia Pacific Tech Market Outlook for 2017 To 2018*. Cambridge, MA: Forrester Research.

Battese, G., & Coelli, T. 1995. A model for technical inefficiency effects in a Stochastic Frontier Production function for panel data. *Empirical Economics*, 325–332.

Berger, A., & Humprey, D. 1997. Efficiency of financial Institution: International Survey and Directions for Future Research. *European Journal of Operational Research*, 97–105.

Coelli, T., Rao, D., O'Donnell, C., & Battese, G. 2005. *An Introduction to Efficiency and Productivity Analysis*, 2nd ed. New York: Springer Science+Business Media.

Hendrawan, R., & Nugroho, K. 2018. Telecommunication sector reform in Southeast Asia: A new rationality. *Global Journal of Business and Social Science Review*, 6(4), 147–154.

Hjalmarsson, L., Kumbhakar, S., & Heshmati, A. 1996. DEA, DFA, and SFA: A comparison. *The Journal of Productivity Analysis*, 7, 303–327.

Leurcharusmee, S., Jirakom, S., & Pruekruedee, S. 2016. Firm efficiency in Thailand's telecommunication industry: Application of the Stochastic Frontier model with dependence in time and error components. *Studies in Computational Intelligence*, 622, 495–505.

Masson, S., Jain, R., Ganesh, N., & George, S. 2016. Operational efficiency and service delivery performance: A comparative analysis of Indian telecom service providers. *Benchmarking*, 23 (4).

Moriwaki, S., Era, A., Osajima, M., & Umino, A. 2010. *Productivity and Efficiency Analysis of Telecommunications Industries: The Case of Asia-Pacific Countries*. Tokyo, Japan: Ministry of Internal Affairs and Communications of Japan.

Porcelli, F. 2009. *Measurement of Technical Efficiency: A Brief on Parametric and Non-Parametric Techniques*, Vol. 11. Coventry, UK: University of Warwick.

Prabowo, T., & Cabanda, E. 2011. Stochastic Frontier analysis of Indonesian firm efficiency: A note. *International Journal of Banking and Finance*, 8(2), 74–91.

Torres, L., & Bachiller, P. 2013. Efficiency of telecommunications companies in European countries. *Journal of Management & Governance*, 17, 863–886.

Examining factors influencing internal audit effectiveness

F.D. Izzuddin & H. Mohd Hanafi
Multimedia University, Cyberjaya, Malaysia

ABSTRACT: An internal audit (IA) adds value to the organization by contributing to governance and risk management. Therefore, internal auditors need to be effective in carrying out their duties. This article examines factors that may affect IA effectiveness through management support and internal auditors' competency. This article proposes utilization of an experimental approach with multiple 2 × 2 factorial designs. Management support will be explored through trainings and internal auditors' competency through their professional qualification. Studies have indicated that both management support and internal auditors' competency could lead to a positive influence on the effectiveness of the IA. The survey will be collected through a questionnaire sent to the members of The Institute of Internal Auditors. Findings should assist the management to focus on the types of resources needed to be deployed in maximizing the capability of their IA team.

1 INTRODUCTION

Internal audit (IA), as an independent, objective assurance and consulting activity, is designed to add value and improve an organization's operations (IIA, 2017). The added value for the organization includes, among others, monitoring the company's risk profile and identifying areas to improve risk management (Goodwin-Stewart & Kent, 2006). Therefore, it is important to ensure the IA function is effective in carrying out its duties to ensure the successful implementation of risk management as well as to achieve its operational, financial, and compliance objectives.

Several studies have examined the factors that influence the effectiveness of the IA. Aaron and Gabriel (2010) indicated that top management support is the main determinant of IA and provided some supporting evidence from the organizational independence of the IA. Huong (2018) found four factors contributing to IA effectiveness: (1) the independence of IA, (2) the competence of IA, (3) the management support for IA, and (4) the quality of the IA. Another study found that factors that affect IA effectiveness are management support, experienced staff, and independent IA department (Alzeban et al., 2014). Even though researchers have examined these various factors, further insights are needed. Therefore, this conceptual article examines two factors that affect IA effectiveness: management support and internal auditors' competency.

The remainder of this article is structured as follows. The next section describes the various studies that examined the measurement of IA effectiveness and the development of our hypotheses. This will be followed by the research design and finally a conclusion.

2 IA EFFECTIVENESS AND DEVELOPMENT OF HYPOTHESES

2.1 Internal audit effectiveness

The IA has become an integral part of an organization. Hence, to ensure effective IA performance, it is important to identify the determinants that would affect its effectiveness. According to the International Professional Practices Framework (IPPF), the effectiveness of IA is crucial to ensure that organizational objectives are met and to support continuous improvement.

A study by Ziegenfuss (2000) has developed a questionnaire that contains 84 criteria to measure the effectiveness of IA. These include the output of the audit, the input into the audit, and the auditing process. In a more recent study using regression analysis, Temesgen and Estifanos (2018) revealed that overall contribution of IA quality, independence of auditors, proficiency of internal auditors, organizational setting, and scope of IA work accounted for more than half of the variation in the IA effectiveness. The recommendations in this article will be measuring IA effectiveness from the perspective of the audit recommendations that will be implemented by the top management.

2.2 Management support

The literature review provides findings from several studies that have examined the relationship between management support and IA effectiveness. A study conducted by Mihret and Yismaw (2007) on a large public sector of a higher educational institution in Ethiopia indicated that IA effectiveness is greatly affected by two factors: the management support and IA quality. Alzeban and Gwilliam (2014) argued that there is a positive relationship between top management support and IA effectiveness. Dellai and Omri (2016) provide further evidence that IA effectiveness is highly influenced by management support for IA. A survey on internal auditors conducted by Onumah and Yao Krah (2012) revealed that IA effectiveness was impeded when there is a lack of management support and insufficient resources for the IA department.

As many studies had found that management support is crucial toward how IA could function, it is important to receive full support from the top management. Many forms of support are given to IA staff, which could be in the form of budget allocation, manpower, or approved trainings. This study examined the effect of management support on IA effectiveness through training given to the IA staff. It is presumed that attendance at the trainings would improve and enhance the existing skills of internal auditors. Therefore, the following hypothesis is proposed:

H1: Internal audit effectiveness is positively related to management support given to IA staff.

2.3 Internal auditors' competency

Researchers argue that to be effective, internal auditors are expected to be adequately qualified along with having professional competence. Studies by Al-Twaujry et al. (2003) and Mihret et Al. (2007) indicated that internal auditors are expected to have adequate qualification, preferably with professional competence as the "ability to perform to recognized standards." It is further posited that an audit would require professional staff with essential education, experience, and training to ensure audit activities can be performed effectively. A study by Hajiha and Farhani (2012) on the IA qualification against the Economic Value Added (EVA) as a comprehensive performance index in the Iranian context suggest there is a positive significant relationships between the experience of IA staff and the EVA. Similarly, Badara and Saidin (2014) conducted a study to provide empirical evidence of antecedents of IA effectiveness from the Nigerian perspective. The antecedents are risk management, effective internal control system, audit experience, cooperation between internal and external auditors, and performance measurement. Their study found a significant positive relationship between all antecedents and IA effectiveness. A more recent study by Nurdiono and Gamayuni (2018) on internal auditors of local government also found that their competency can affect IA quality.

Based on the findings from previous studies, it is posited that internal auditors' competency should have a positive effect on IA effectiveness. Internal auditors have various background and it is not uncommon for some of their team members in the department to possess professional qualifications. Thus, the internal auditors' competency will be examined through a comparison between those having professional qualifications and those who do not. Therefore, the following hypothesis is proposed:

H2: Internal audit effectiveness is positively related to internal auditors' competency.

2.4 Gender effect

Internal auditors come from various background and characteristics. They differ based on their gender, age, educational level, races, religion, level of experience, etc. Studies have found that gender may become a possible explanatory factor in assessing audit quality, with female auditors tending to be more risk averse (Mgbame et al., 2012). A study investigating the association between an audit partner's gender and the likelihood of issuing a going-concern opinion for a financially distressed client revealed that female auditors are less likely to award that opinion (Hossain et al., 2012). A study on Jordanian auditors by Abed and Al-badainah (2013) found no difference between audit fees charged and gender. However, based on interviews with the male auditors, the study found that females generally refuse to engage in any unethical behavior.

Based on the findings from the aforementioned studies, it could be expected that there will be difference in responses between male and female internal auditors toward management support and IA competency. Therefore, the following are the expected hypotheses:

H3: Gender will moderate the results in H1, such that women will be more likely than men to perceive IA will be more effective when management gives its support to the IA staff.

H4: Gender will moderate the results in H2, such that women will be more likely than men to perceive IA will be more effective when IA are competent.

3 RESEARCH DESIGN

This is an experimental study with multiple 2 × 2 between-subjects designs. Each design has the same common subject variable, gender, whereby responses will be compared between male and female internal auditors. The independent variable for the first design is the training granted to the internal auditors. The independent variable for the second design is the internal auditor's professional qualifications. The dependent variable that will be used to evaluate IA effectiveness is whether the top management would implement the recommendations suggested by the internal auditors in their audit report. The Institute of Internal Auditors (IIA) Malaysia will be contacted to obtain their support. The questionnaire will be distributed at the end of the identified trainings/workshops to the attending internal auditors.

4 CONCLUSION

The primary objective of this study is to examine IA effectiveness through management support and internal auditors' competency. There are three expected potential contributions from this study. The first is to gain evidence on whether providing more training would make a difference in the effectiveness of the IA and the importance of top management support in this effort. This will be important to determine whether management should approve more training for the IA team. Second, it is expected that this study would provide empirical results on the degree of influence of IA competency such as professional qualifications on IA effectiveness. This would be a great input for Human Resources during the employment of future IA staff. Finally, any results from the gender observation would provide more perspective on how male and female internal auditors may perceive things differently.

REFERENCES

Abed, S., & Al-badainah, J. 2013. The impact of auditor's gender on audit fees: Case of Jordanian auditors. *8*(14), 127–133.

Alzeban, A., & Gwilliam, D. 2014. Factors affecting the internal audit effectiveness: A survey of the Saudi public sector. *Journal of International Accounting, Auditing and Taxation, 23*(2), 74–86.

Bidara, M. S., & Saidin, S. Z. 2014. Empirical evidence of antecedents of internal audit effectiveness from Nigerian perspective. *Middle-East Journal of Scientific Research, 19*(4), 460–471.

Cohen, A., & Sayag, G. 2010. The effectiveness of internal auditing: an empirical examination of its determinants in Israeli organisations. *Australian Accounting Review*, *20*(2), 296–307.

Dellai, H., & Omri, M. A. B. 2016. Factors affecting the internal audit effectiveness in Tunisian organisations. *Research Journal of Finance and Accounting*, *7*(16), 208–221.

Goodwin-Stewart, J., & Kent, P. 2006. The use of internal audit by Australian companies. *Managerial Auditing Journal*, *21*, 81–101.

Hajiha, Z., & Farhani, M. G. 2012. An investigation on the relationship between internal audit quality and economic value added: Evidence from Iran. *Journal of Basic and Applied Scientific Research*, *2*(7), 6872–6881.

Hossain, S., Chapple, L., & Monroe, G. S. 2012. Does auditor gender affect issuing going-concern decisions for financianlly distressed clients? *Accounting and Finance*, *58*(4). 1027–1061.

Institute of Internal Auditors (IIA). 2017. *International Standards for the Professional Practice of Internal Auditing (Standards)*. Retrieved from https://na.theiia.org/standards-guidance/mandatory-guidance/Pages/Definition-of-Internal-Auditing.aspx (accessed August 5, 2019).

Mgbame, C. O., Izedonmi, F. I. O., & Enofe, A. 2012. Gender factor in audit quality: Evidence from Nigeria. *Research Journal of Finance and Accounting*, *3*(4), 81–88.

Mihret, D. G., & Yismaw A. W. 2007. Internal audit effectiveness: an Ethiopian public sector case study. *Managerial Auditing Journal*, *22*(5), 470–484.

Onumah, J. M., & Yao Krah, R. 2012. Barriers and catalysts to effective internal audit in the Ghanaian public sector. *Research in Accounting in Emerging Economics*, *12*, 177–207.

Temesgen, A., & Estifanos, L. 2018. Determinants of internal audit effectiveness: Evidence from gurage zone. *Research Journal of Finance and Accounting*, *9*(19), 15–25.

Ziegenfuss, D. E. (2000). Measuring performance. *Internal Auditor*, February, 36–40.

Implementation of cognitive multimedia learning theory on mobile apps interactive story Kisah Lutung Kasarung

D. Hidayat & M.I.P. Koesoemadinata
Telkom University, Bandung, Indonesia

M.A. Bin Mat Desa
Universiti Sains Malaysia, Penang, Malaysia

ABSTRACT: Multimedia technology has changed the way people learn and obtain information. Cognitive theory about multimedia learning states that the information processing system in humans consists of two channels, visual/pictorial processing channels and auditory/verbal processing channels. Both channels have limited capacity for processing. Therefore multimedia applications for learning purposes must adapt to the way humans learn. This study aims to analyze how the application of the principles of cognitive multimedia learning in mobile apps Interactive Story *Kisah Lutung Kasarung*. This research is a qualitative research, data collection is carried out using literature study and documentation methods. Data analysis was performed using a visual analysis method through the stages of description, analysis, interpretation, and judgment. The results showed that the cognitive theory of multimedia learning has been well applied to the mobile apps Interactive Story *Kisah Lutung Kasarung*, so that these mobile apps deserve to be used as learning media about Sundanese cultural values.

1 INTRODUCTION

1.1 *Background of study*

Indonesia has many folklore that contain moral values and education for children. But with the presence of more modern recreational and entertainment media, Indonesian folklore began to be forgotten. To reintroduce Indonesian folklore to children can be done using modern media such as mobile apps. Based on observations made, mobile apps with Indonesian folklore content for children currently available generally prioritize entertainment aspects rather than education. The aspects of education and learning should take the form of moral and cultural values must also be conveyed properly to the user.

The object of this research is the Mobile apps *Kisah Lutung Kasarung* (KLK) created by the game developer Educa Studio. KLK Mobile apps was chosen because it has a good rating as a recommended application for children users, and has received the Dicoding award that is 1K Download Club Android NDK Challenges. This mobile apps are made by Educa Studio, a game development company that produces interactive children's story series on Indonesian folklore stories and fairy tales from around the world. In the mobile app KLK interactive stories contained Sundanese cultural values delivered narratively.

The purpose of this study was to identify the application of the principles of cognitive multimedia learning from Richard E. Meyer (2016) to the mobile apps that became the study sample. The research conducted is descriptive qualitative research. Data collection was carried out using the literature study and documentation method. Data analysis was performed using a content analysis approach, which is an in-depth observation of aspects of multimedia elements and story elements contained in mobile apps. Data analysis was performed using visual analysis methods through steps of description, analysis, interpretation, and judgment (Feldmann, in Aland & Darby 1992).

1.2 Research question

Based on the background of the research, the problem in this study is how the principles of cognitive multimedia learning are implemented in mobile apps interactive story *Kisah Lutung Kasarung*?

1.3 Theoretical review

Multimedia also refers to technology for presenting material in verbal and visual form (Mayer, 2009). In the context of computers, multimedia is the use of computers to present and combine text, sound, images, animation and video with tools and connections (links) so that users can navigate, interact, work and communicate (Hofstetter, 2001). In the cognitive process and multimedia motivational process has 2 elements. First, multimedia elements, namely Visual input (text, pictures, diagrams, video, animation) and Auditory input (sound, signal/cues, music, narration, instructions). Second, story elements, namely literary text elements build "stories" (theme, character, plot, structure, settings, point of view, foregrounding) (Astleiner, 2004).

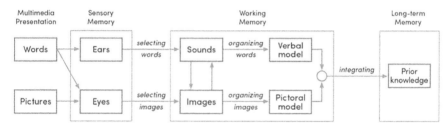

Figure 1. The principle of cognitive multimedia learning of Mayer.

In the context of learning, multimedia learning can be understood as the delivery of instructional content using multiple modes that include visual and auditory information (Elliot, 2009). Multimedia learning is instructional multimedia or multimedia instructional presentations, namely the presentation of messages involving words and images intended to enhance learning (Mayer, 2009).

The theory of cognitive multimedia learning from Mayer contains three assumptions. First, the dual-channel assumption, which states that humans have separate channels for processing information for visual and auditory material. The second assumption, humans have a limited capacity for the amount of information that can be processed in each channel at a certain time. The third assumption, humans are active processors who cognitively try to reason and understand every multimedia presentation (Mayer, 2009). Therefore, the delivery of educational content and affective values into multimedia learning requires an appropriate approach so that predetermined learning goals can be achieved optimally, by considering the channels and cognitive burdens of individuals that are limited (Mayer 2001; Chandler & Sweller 1991).

Richard E. Mayer discusses 12 principles that shape the design and organization of multimedia presentations. Some examples are included: 1) Coherence Principle, people learn better when extraneous words, pictures and sounds are excluded rather than included; 2) Signaling Principle, people learn better when cues that highlight the organization of the essential material are added; 3) Redundancy Principle, people learn better from graphics and narration than from graphics, narration and on-screen text.; 4) Spatial Contiguity Principle, people learn better when corresponding words and pictures are presented near rather than far from each other on the page or screen; 5) Temporal Contiguity Principle, people learn better when corresponding words and pictures are presented simultaneously rather than successively; 6) Segmenting

Principle, people learn better from a multimedia lesson is presented in user-paced segments rather than as a continuous unit; 7) Pre-training Principle, people learn better from a multimedia lesson when they know the names and characteristics of the main concepts; 8) Modality Principle, people learn better from graphics and narrations than from animation and on-screen text; 9) Multimedia Principle, people learn better from words and pictures than from words alone; 10) Personalization Principle, people learn better from multimedia lessons when words are in conversational style rather than formal style; 11) Voice Principle, people learn better when the narration in multimedia lessons is spoken in a friendly human voice rather than a machine voice; 12) Image Principle, people do not necessarily learn better from a multimedia lesson when the speaker's image is added to the screen (Mayer, 2009).

2 METHOD

This research is a qualitative research, carried out by steps of data reduction, data presentation, and data analysis (Miles & Huberman, 1994). Data collection was carried out using the literature study and documentation. Literature study was used to understand theories about multimedia and the principles of cognitive multimedia learning. The documentation method was used to divide the object of research into units of data that can be analyzed. Data analysis was performed using a content analysis approach, carried out through the steps of description, analysis, interpretation, and judgment (Feldmann, in Aland & Darby, Soewardikoen, 2013).

3 DISCUSSION

3.1 *Object of the research*

KLK Mobile apps combine interactive stories and interactive education in one application. In interactive story mode there are two options, which are automatic and read by yourself. Automatic mode will run the story automatically from beginning to end, accompanied by narration according to the story on the page. While with its own reading mode the user can freely choose the desired page, the narration is replaced by a panel containing writing that contains the story according to the scene on the page. In interactive education mode there are four games that can be played, namely puzzle shapes, dress ups, body wash and fruit caterpillars. This research will focus on KLK Mobile apps in an interactive story mode, in which narratively communicates Sundanese cultural values.

3.2 *Data processing techniques*

Data collection is carried out through literature and documentation studies. KLK Mobile apps are broken down into pages based on story scenes. The study focused on analyzing mobile apps in interactive story mode and produced a total of 13 pages or scenes which were then numbered sequentially to facilitate the analysis process.

Figure 2. The scenes of Mobile apps KLK.

3.3 *Analysis*

Analysis of each scene is carried out using a content analysis approach (Feldmann, in Aland & Darby 1992), carried out by the steps: 1) description, in-depth observation of aspects of multimedia elements and story elements; 2) analysis, analyzing the relationships between the elements using multimedia cognitive learning theory; 3) interpretation, interpret the results of the analysis; and 4) judgment, providing conclusions.

The following aspects is a discussion of the implementation of Mayer's cognitive learning multimedia principles in the aspects of multimedia elements and story elements found in each KLK mobile apps scene: 1) Coherence Principle, multimedia elements and story elements contained in all scenes are strongly related with the story's content; 2) Signaling Principle. KLK Mobile apps include various sound effects when the user touches images and certain parts of the scene, and activates animations on certain images; 3) Redundancy Principle, mobile apps have two options; Automatic mode that uses narration and reading mode that using the writing panel containing the story; 4) Spatial Contiguity Principle, narrative and written elements of the story placed close to the multimedia elements, which are placed in one scene or the same page; 5) Temporal Contiguity Principle, story narrative and written elements are displayed simultaneously with multimedia elements, so there is a connection between the story and multimedia elements; 6) Segmenting Principle, the stories are delivered separately and in sequence based on the scene or page - making it easy for users to understand the story and move to the next scene according to their wishes; 7) Pre-training Principle, available scenes or pages containing thumbnails of each scene contained in the KLK Mobile apps, to ease the users choosing the page they desired; 8) Modality Principle, implemented in the KLK mobile apps, with the presence of animation in certain parts of the scene, and text that can be activated in its reading mode; 9) Multimedia Principle. All scenes in the KLK mobile apps are a combination of visual and auditory multimedia elements, and story elements contain characters, plots, structures, settings, point of view, and foregrounding; 10) Personalization Principle, the stories are told in simple sentences and easily understood by children; 11) Voice Principle, the story will be told by the narrator will tell the story using human original voice; 12) Image Principle. KLK mobile apps don't display narrators in human form. Users focus more on the story without having to be interrupted by the appearance of the narrator in human form that has nothing to do with the story's contents.

4 CONCLUSION

The research shows that KLK mobile apps have implemented 12 principles of cognitive multimedia learning. These 12 principles are implemented both in multimedia elements and in story elements found in all scenes or pages of the KLK mobile apps. The application of 12 principles of cognitive multimedia learning in an interactive multimedia in the form of mobile apps interactive stories will enhance the functions and benefits of these mobile apps. Besides being attractive and functioning as a means of entertainment, the mobile apps are useful as learning media.

This research only discusses the implementation of multimedia cognitive principles on KLK mobile apps. Further research is needed to measure the effectiveness of KLK mobile apps in conveying Sundanese cultural values to users.

REFERENCES

Astleitner, H. 2004. *Multimedia, Stories, and Emotion – Integrated Model for Research and Design*, e-Journal of Instructional Science and Technology, Volume 7, No. 2, 2004. Retrieved November 15, 2018, from https://files.eric.ed.gov/fulltext/EJ850353.pdf.
Chandler, P. & Sweller, J. 1991. *Cognitive Load Theory and The Format of Instruction*. Cognitive and Instruction, 8, 293–332.
Elsom, Cook, 2001. *Principles of Interactive Multimedia*. New York: MacGrawHill.

Fred T. Hofstetter, 2001, *Multimedia Literacy*, Third Edition. New York: MacGrawHill.
Gjedde, Lisa, 2005. *Designing for Learning in Narrative Multimedia Environment*. London: Idea Group Inc.
Mayer, R., 2009. *Multimedia Learning*, Second Edition. New York: Cambridge University Press.
Miles, Mattew B & Huberman A. 2007. *Analisis Data Kualitatif Buku Sumber tentang Metode-Metode Baru.Terjemahan Tjetjep Rohendi Rohisi*. Jakarta: Universitas Indonesia.
Soewardikoen, D. W., 2013. *Metodologi Penelitian Visual, dari Seminar ke Tugas Akhir*, Bandung: Dinamika Komunika.
Sweller, J. 1999. *Instructional Design in Technical Areas*. Camberwell, Australia: ACER Press.

Efficiency of legal and regulatory framework in combating cybercrime in Malaysia

S. Khan, N. Khan & O. Tan
Multimedia University, Selangor, Malaysia

ABSTRACT: This article examines the increase of cybercrime in Malaysia and how this issue should be resolved. Cybercrime, which is also known as computer crime, involves the use of a computer as an instrument to further illegal ends, such as committing fraud, trafficking in child pornography, intellectual property violations, cyberbullying, identity theft, and privacy violations. There are several legislations covering cybercrime, but statistics show that cybercrime has been increasing in Malaysia. Thus, the objective of this article is to investigate and examine the efficiency of the existing legal and regulatory framework in combating cybercrime and to recommend whether to enact a new legislation to address all types of cybercrime including cyberbullying, which is a serious crime committed by students. Qualitative analyses were used and selected interviews were conducted with relevant agencies.

1 INTRODUCTION

Cybercrimes are criminal offences committed through the use of a computer and the internet, raising great concern about cyber fraud and identity theft through which crimes such as spamming, phishing, email spoofing, defamation, pornography, cyber terrorism, espionage, and other serious offences are perpetrated. Cyber law refers to laws related to governing and protecting the use of a computer and internet and other online communication technologies and consist of a combination of state and federal statutory, decisional, and administration laws arising from the use of the internet. As information and communication technology in this era is becoming a vital issue, the Malaysian government has enacted cyber laws to govern and protect users' computer and internet rights.

2 BACKROUND OF THE STUDY

For Malaysia to be a world leading country in information and communications technology (ICT) development and a global center and hub for communication and multimedia information and content services, a law needs to be put in place to promote a high level of consumer confidence and to protect the security of information and network reliability and integrity. Malaysia is making every effort to progress in a safe and confidential cyberspace environment with the enactment of several cyber laws that have helped to increase computer and internet usage in online transactions, and because of the rapid development of technology, review and enactment of new cyber laws are required, as hackers are becoming more adept. Cyber hacking is a criminal offence against Malaysian laws and hackers can be charged under Section 4 of the Computer Crimes Act 1997, which carries a fine of not more than RM150,000, a jail term of not more than 10 years, or both. In Malaysia the National Security Council, an agency under the prime minister's department, is responsible for managing and coordinating the implementation of policies related to national security; Cybersecurity Malaysia is the national cyber security specialist agency under the Ministry of Science, Technology and Innovation; and the Malaysia

Computer Emergency Response Team or MyCERT is concerned with the security of internet users. Yet, the rate of cyberattacks in Malaysia in on the rise.

3 OVERVIEW OF LEGAL AND REGULATORY FRAMEWORK

The important steps that were taken by the government of Malaysia to combat this new type of crime were the introduction of a new legal framework to facilitate the development of ICT systems by countering the threats and abuses related to such systems called Cyber Laws of Malaysia. The Malaysian cyber laws consist of Computer Crime Act 1997, Digital Signature Act 1997, Telemedicine Act 1997, Communication and Multimedia Act 1998, Copyright (Amendment) Act 1997, Malaysian Communication and Multimedia Commission Act 1998, and Optical Disk Act 2000. There are other existing laws that are used in conjunction with these acts: the Official Secret Act 1976, Patent Act 1983, Prison Act 1995, Akta Arkib Negara 44/146, and other relevant legislations.

This set of Acts has made Malaysia one of the first countries to enact a comprehensive set of cyber laws. The aforementioned Acts were passed for the purpose of safeguarding consumer and service providers in addition to online businesses and owners of intellectual property. In addition, there was a recent amendment to the Evidence Act that was passed in Parliament on May 19, 2012 that increased censorship on the internet. Instead of applying different sets of laws in a nation, there must be a specific legislation that prevents and protects people from these crimes. There must be enforcement officers who are highly knowledgeable in technology-related matters to testify in court where these cyber criminals could be prosecuted for their offences. In addition, people should be aware of the existence of the laws that can be used to protect them. The Penal Code often fails; the laws are either overly vague, leaving targeted behaviour undefined, or underinclusive, leaving gaps in the law when legislatures are unable to keep up with the rapid pace of technological advancement. Given the dynamic nature of the internet, victims of online crimes are often left with inadequate protection from their harassers. One particular shortcoming is the requirement of proving intent, which does not apply to those with less culpable mental states. Thus, even if the victim experiences severe emotional distress that causes him or her to withdraw from school, work, and society, the claim will fail if it cannot be proven that the perpetrator actually intended to cause distress.

4 THE CHALLENGES

Managing cybersecurity has become a real challenge, as the number of attacks is ever increasing, as are the difficulties in defending against these attacks. Enactment of specific acts in laws is essential to combat cybercrimes but it could not guarantee full protection in cyberspace because of the borderless communication involved. Enforcement officers who are well versed in technology are also one of the requirements to ensure that the law can be enforced successfully. The Malaysian Police Force established a special section under the commercial crime division to deal with cybercrimes. But to tackle cybercrimes, the enforcement officers need to be technology savvy and on par with these cyber criminals who are using sophisticated technology to commit the crimes. The resources allocated at the national or international level are just not sufficient to track down and prosecute cyber criminals. The lack of a physical presence in commission of these crimes makes it difficult to trace these cyber criminals.

5 METHODOLOGY

This study was based on both qualitative and quantitative analysis. It examined and analysed the existing legislations relating to cybercrime and investigated their efficiency. A comparative research approach was used by examining the implementation of cybercrime law in other jurisdictions, specifically Singapore, the United States, and the United Kingdom and whether these laws should be modeled upon. The information on the efficiency of existing legislation was analysed through interviews with the relevant bodies, specifically Judiciary, AG Chamber, Legal Practitioners, CyberSecurity Malaysia, Malaysian Communications and Multimedia Commission (MCMC), and enforcement agencies such as the National Cyber Security Agency (NCSA) and the Royal Police. A quantitative research approach was used to conduct a survey among the general public to determine its awareness of cybercrime in Malaysia and whether a specific legislation is needed to combat all sorts of cybercrime in Malaysia.

6 FINDINGS AND RECOMMENDATIONS

Current legislations are inadequate or ineffective and do not cover or spell out the cybercrimes such as state-sponsored cybercrime and civil liability, online impersonation, cyberbullying, cyber harassment, and cyberstalking. Thus, it is important to review the current legislation to ensure that cybercrimes can be efficiently handled in Malaysia.

Table 1. Cybercrime handled in Malaysia.

	2017	2018	2019 (until May)
Fraud	3,821	5,123	2,563
Cyber harassment	560	356	101
Total	7,962	10,699	3,743

The study reveals that the potential economic loss in Malaysia due to cybercrime incidents can hit a staggering US$12.2 billion, with the financial sector having the worst hit. Cybercrime has grown with the development of technology and these figures show that there must be something lacking either in the investigation, prosecution, or the laws. To achieve a successful prosecution in cybercrime cases, Malaysia needs a comprehensive new cyber law instead of the current piecemeal legislation and guidelines. This research found a total of 916,833 cybercrime cases in the first four months of 2019.

The Computer Crimes Act 1997 (CCA) criminalises the act of gaining unauthorised access into computers or networks; spreading malicious codes (e.g. viruses, worms and Trojan horses); unauthorised modification of any program or data on a computer; as well as wrongful communication by any means of access to a computer to an unauthorised person. The case of *Basheer Ahmad Maula Sahul Hameed v. PP* [2016]6 CLJ 422 is an example of a hacking offence under the Act. Communications and Multimedia Act 1998 (CMA), which provides for and regulates various activities carried out by licensees (i.e., network facilities providers, network service providers, application service providers, and content applications service providers) as well as those utilizing the services provided by the licensees. The Act prohibits fraudulent or improper use of network facilities or network services. In cases in which computer/internet-related crime activities are involved but do not specifically fall within the ambit of any of the aforementioned statutes they may be charged under the Penal Code, which is the main statute that deals with a wide range of criminal offences and procedures in Malaysia.

Regulators responsible for enforcing requirements are generally either sectorspecific or subject matter specific: Malaysian Communications and Multimedia Commission (MCMC) for information security/network reliability and integrity under Communications and Multimedia Act 1998, Personal Data Protection Department/Commissioner's Office for personal data under Personal Data Protection Act 2010, Royal Malaysian Police for penal offences under Penal Code and Computer Crimes Act 1997, Securities Commission Malaysia (SC) & Central Bank of Malaysia or Bank Negara Malaysia (BNM) for sector-specific regulations under Securities Commission Guidelines, and the Banking and Financial Sector Guidelines.

Countering cybercrime will be increasingly challenging as a result of the exponential growth of connected devices, especially in the Fourth Industrial Revolution (4IR). New legislation would help safeguard the interests and well-being of people because every complaint could be acted upon through legal means. There were 1,524 cyberbullying cases recorded in the past five years and most of them involved students. Although the number is low compared with the more than 5 million students nationwide, the statistics are based only on reported cases. Research revealed a shocking fact that one in four students admitted that they had experienced cyberbullying. The Penal Code often fails; laws are either overly vague, leaving targeted behaviour undefined, or underinclusive, leaving gaps in the law when legislatures are unable to keep up with the rapid pace of technological advancement. As warfare evolves, cyberattacks, cybercrime, and cyberespionage have become more prevalent and inevitable than ever before. The current legislation in Malaysia is inadequate or ineffective to counter these issues.

7 CONCLUSION

Countering cybercrime will be increasingly challenging as a result of the exponential growth of connected devices, especially in the Fourth Industrial Revolution (4IR). New legislation would help safeguard the interests and well-being of people because every complaint could be acted upon through legal means. New legislation would address all types of cybercrime including cyberbullying, which is more serious than what has been reported. In addition, current legislation does not cover or spell out cybercrimes such as state-sponsored cybercrime and subjects such states to civil liability, online impersonation, cyberbullying, cyber harassment, cyberstalking, and so forth.

REFERENCES

Beatty, D. L. 2017. Malaysia's Computer Crimes Act 1997 gets tough on cybercrime but fails to advance the development of cyberlaws. *Pacific Rim Law & Policy Journal, 7*(2), 351–376.
Blinderman, E., & Din, M. 2017. Hidden by sovereign shadows: Improving the domestic framework for deterring state-sponsored cybercrime. *Vanderbilt Journal of Transnational Law, 50*, 889.
Communications and Multimedia Act 1998 (CMA).
Computer Crimes Act 1997 (CCA).
Fehr, C., Licalzi, C. & Oates, T. 2016. Computer crimes. *American Criminal Law Review, 53*(4), 977–1026.
Holt, T. J. (2018). Regulating cybercrime through law enforcement and industry mechanismss. *Annals of the American Academy of Political and Social Science, 679*, 140.
Koch, C. M. 2017. To catch a catfish: A statutory solution for victims of online impersonation. *University of Colorado Law Review, 88*, 233.
Lafen, S. A. 2018. U.N. regulation: The best approach to effective cyber defence? *Syracuse Journal of International Law and Commerce, 45*, 249.
Mohamed, D. 2012. Investigating cybercrimes under the Malaysian Cyberlaws and the Criminal Procedure Code: Issues and challenges. *Malayan Law Journal, 6*, i.
Mohd Nor, M. W., & Ab Razak, S. 2017. Regulating hate speech on social media: Should we or shouldn't we? *Malayan Law Journal, 4*, cxxix. Penal Code.
Perloff-Giles, A. 2018. Transnational cyber offenses: Overcoming jurisdictional challenges. *Yale Journal of International Law, 43*, 191.

The case of women entrepreneurs: Comparison study between two cities in Indonesia and Zimbabwe

R.L. Nugroho & S. Mapfumo
Telkom University, Bandung, Indonesia

ABSTRACT: In Indonesia and Zimbabwe, entrepreneurship has become a very popular option among women. The high rate of entrepreneurship as a career choice among women is attributed to factors that are specific to women in these two countries, which may also possibly relate to other countries. The purpose of this article is to understand the variables and their interrelationships that lead to sector crowding in entrepreneurship among women in Harare city, Zimbabwe and Bandung city, Indonesia. A phenomenological and comparative qualitative approach to research was adopted for this study to help the researcher gain a deeper understanding of the experiences of Bandung and Harare women entrepreneurs. Little has been known in the extant literature until now concerning the sector preferences in women entrepreneurship in both of these cities. As a comparative study in the context of Indonesia and Zimbabwe, this study therefore sought to open some space for different points of view.

1 INTRODUCTION

The entrepreneurial capacity of women has been acknowledged as governments seek to hasten economic growth and to attract more women to new ventures (Butler, 2003; Singh & Belwal 2008). The Global Entrepreneurship Monitor (GEM) reported that an estimated 163 million women were starting or running new businesses in 74 economies around the world. In addition, an estimated 111 million were running established businesses (GEM, 2017). This not only shows the impact of women entrepreneurs across the globe but also highlights their contributions to the growth and well-being of their societies. A strategic approach to managing small firms by women is necessary for company growth. Globally, there is a consensus that governments can indeed play a positive role in encouraging small business or entrepreneurship, particularly those small and medium-sized enterprises (SMEs) run by women.

According to Anna et al. (2000), Coleman (1999), and Du Rietz and Henrekson (2000), women often desire to launch firms in highly competitive and low-growth industries such as personal services and retail entrepreneurship sectors that require little innovation, with less technicality, and do not require much attention so as to accommodate the social context of work–life balance. The 2018 Women, Business and Law report identified that globally, more than 2.7 billion women are legally restricted from having the same choice of jobs as men and 104 economies still have laws preventing women from working in specific jobs, which subsequently affects their choice of the business sector (World Bank Group, 2018). Equality of opportunity allows women to make the choices that are best for them, their families, and their communities (World Bank Group, 2018).

Historically, women entrepreneurs in Indonesia have remained a largely untapped potential source of economic growth (Lubis, 2016). Following the keynote addresses at the international conference for female entrepreneurs in the Southern African Development Community (SADC) region in 1996, it was suggested that governments in the SADC target females as beneficiaries of government programs for entrepreneurial development. The International

Labour Organization (ILO) also emphasized the importance of the government in creating a favorable environment for the promotion of sustainable entrepreneurship among women.

Previous studies have also shown that women are less likely to have degrees in the Science, Technology, Engineering, and Mathematics (STEM) fields, which provide fertile ground for both revenue generation and employment opportunities and tend to be a "nesting ground" for growth-oriented industries in areas such as technology, biosciences, and health care. Despite the ostensible benefits of women entrepreneurs to an economy, the full potential of the women entrepreneurial sector has not been realized (Aidis, 2006). The previous findings correlate with the Indonesian and Zimbabwean women's entrepreneurship literature in various aspects. It must be mentioned that Zimbabwe and Indonesia make an interesting comparison because of the existence of vital similarities and differences as shown in Table 1.

Bandung city and Harare city were chosen as the contexts of this study because both cities have the highest entrepreneurial activity in both countries, with a total of 5,141 and 4,954 SMEs in 2015 respectively. More so, the cities have higher numbers of women entrepreneurs whose businesses are in traditionally female-dominated business sectors such as beauty, food, retail, and wholesale trading. Only a few comparative research studies have been done in developing countries. In addition, the landscape of women's entrepreneurship is gendered terrain.

It is in this vein that this study explored the determinants of sector preferences in women's entrepreneurship in Bandung city, Indonesia and Harare city, Zimbabwe. Ahl (2006) suggested that future research would be an in-depth and qualitative study to explore internal and external factors that precipitate women entrepreneurs' underrepresentation in many sectors and yet there is high sector crowding in trade and services, for example, retail, wholesale, education, hotel, and catering. Most of the microenterprises are owned by women. Following this line of thought, the research questions (RQs) are thus as follows:

Table 1. General comparison of Indonesia and Zimbabwe.

No	Characteristics of Indonesia	Characteristics of Zimbabwe
1	Long-term president serving 34 years	Long-term president serving 34 years
2	Patriarchal society (gender inequality)	Patriarchal society (gender inequality)
3	Women population is slightly lower than Men, 49.7% of 254.9 million	Women population is higher than, men 52% of 13 million people
4	Women empowerment: Underrepresentation of women in high decision-making positions in most entities at the local and national level	Women empowerment (top positions): 48% parliamentarians, 40% provincial ministers, 26% public sector, 47% judiciary, 47% media
5	Religion is mainly Islam: Does not necessarily support women to have businesses and come in contact with men	Religion is mainly Christian: Women not necessarily prohibited to have businesses and socially coming in contact with men
6	Underrepresentation of women's education in engineering, science, and technology fields	Underrepresentation of women's education in engineering, science, and technology fields
7	Women own 60% of the 57.9 million SMEs	Women own more than 50% of the total SMEs
8	Most women are entrepreneurs by necessity: 47.4% necessity-driven, 37.2% stable, and 15.4% are growth-oriented entrepreneurs, with an average of 2.4 employees	About 80% of the women are necessity entrepreneurs because of high unemployment rate, having 0–4 employees
9	Women's business sector type:trade, manufacturing, beauty, food, and other services	Women's business sector type: food preparation, health and beauty, trading, education services
10	National policy, not for profit organizations, women have entrepreneurial policies favoring women including NGO presence, such as PNPM, IWAPI, and ASSPUK	Various entrepreneurial policies favoring women handled by MWAGCD ministry at the national level though marginalized

Source: Compiled by the authors from various sources.

RQ 1. What intrinsic and extrinsic motivating factors drive female entrepreneurs to choose the sector/industry they venture into in Bandung city of Indonesia and Harare city in Zimbabwe?

RQ 2. Which factors discourage or hinder women entrepreneurs from entering male-dominated sectors in Bandung city of Indonesia and Harare city in Zimbabwe?

RQ 3. Which factors have similar and different influences on women entrepreneurs' sector choices in Bandung city of Indonesia and Harare city in Zimbabwe?

2 METHODOLOGY

Data collection was done through the conduction of unstructured interviews because it allows an uninterrupted conversation or narrative of lived experiences to take place (Neergaard & Leitch, 2015). An interview is a specialized pattern of interaction, a two-person face-to-face dialogue, where one person asks the questions and the other answers them to the best of his or her ability. Interviews are used when the objective of the research concerns human experience, perceptions, and beliefs, and the purpose of the qualitative research interview is to explore the beliefs, experiences, and views connected to this experience (Kvale, 1995). The unstructured interview focuses on eliciting a direct description of a particular situation or event as it is lived through without offering causal explanations or interpretive generalizations. The unstructured interview is frequently guided by very few open-ended questions, maybe only one or two, with follow-up questions concerned with clarification or elaboration on what is said (Adams & van Manen, 2008; Neergaard & Leitch, 2015).

The research context was two locations in different cities: Harare city of Zimbabwe in Southern Africa and Bandung city of Indonesia in Southeast Asia. The rationale for choosing these locations was based on the fact that the authors have lived in both cities and the experience revealed high entrepreneurial activity among women. Five respondents were interviewed from each city, which means the findings are based on ten interviewees.

3 RESULTS

3.1 *RQ 1. What intrinsic and extrinsic motivating factors drive female entrepreneurs to choose the sector/industry they venture into in Bandung city of Indonesia and Harare city in Zimbabwe?*

With regard to whether intrinsic or extrinsic motivation played a significant role in choosing their businesses, most of the respondents indicated that their motivation was more extrinsic than intrinsic. Only two out of six respondents were attracted to the fashion industry by the human capital concept, as they were educated and had experience in the fashion industry. The rest of the respondents were motivated extrinsically by market availability, social networks and exposure, family influence, work–life balance issues, poor job security, unemployment, and poverty. The Zimbabwean respondents cited unemployment and poverty as the main push factors whereas Indonesian respondents highlighted that they wanted to spend more time with their families or their family was in this business so they decided to start their own in the same sector. The Zimbabwean economy has taught people to be resilient, and buying and selling have been a refuge for many because the capital injection required is very little. In this study, it was found that most women started their businesses from their savings.

3.2 *RQ 2. Which factors discourage or hinder women entrepreneurs from entering male-dominated sectors in Bandung city of Indonesia and Harare city in Zimbabwe?*

Respondents gave similar responses outlining that the traditionally male sectors normally require a huge capital injection; for example, in mining acquiring a mining claim is expensive, and in transport one needs to have the buses or trucks, which are expensive to acquire.

Moreover, businesses require experience and greater attention. There is a tiresome bureaucracy in acquiring licenses and lack of experience in the area. These findings do not fall far from the results obtained by Bianco et al. (2017), who found that gender ideologies were manifested in the forms of interrelated structural barriers that restricted women entrepreneurs' access to resources. Only two respondents from both countries said that they never thought of it because their opportunity was in the fashion arena and they lacked exposure in any other industrial sectors, which shows high-risk aversion. The data show that much needs to be done in order get more females into the male-dominated sectors, for example, communicating about and giving access to the opportunities, injecting funding for those businesses, educating women in science and technical fields so as to lay their foundation for entrepreneurship in future sectors, as well as employing the women in those gendered institutions. If women entrepreneurs do not seek, or if they are not able to obtain, external capital, their prospects for growing their firms are considerably diminished.

3.3 *RQ 3. Which factors have similar and different influences on women entrepreneurs' sector choices in Bandung city of Indonesia and Harare city in Zimbabwe?*

The responses show that family plays a fundamental role in deciding someone's choice of business, especially when the person is young and unmarried. The findings from this study further cement findings from previous research that established that family commitment is an enabler of female entrepreneurship. The family may supply various types of support to assist the business context, for example, in the decision-making process, and comprises emotional and financial support from spouse, parents, friends, and relatives. In addition, respondents outlined that religion supports women entrepreneurship irrespective of the sector as long it is a Godly business. Discrepancies are noted on cultural issues, as in Zimbabwe women in business are viewed as people of loose morals and thus most of the women entrepreneurs are single or divorced.

4 CONCLUSION

Many factors drive women to start their own businesses, mainly family influence, human capital, work–life balance, funding or capital, poverty, and unemployment. Most of the women believe that factors that discourage women from venturing in the male-dominated sectors are capital, lack of opportunities, license acquisitions, and lack of experience and knowledge in those areas. The individual characteristics and acquired a knowledge, coupled with the sociocultural environment, are factors that were found to directly influence women entrepreneurial sector choices, indicating some social environment values, for example, tolerance, cooperation, and cultural kinship. The practical implications of the results are as follows: (1) governments should find ways to communicate to a greater and wider extent with women that leads to countries having more women entrepreneurs; and (2) the ministries of education from both countries need to communicate with female students and encourage them to study STEM fields in order to enhance the presence of women in growth-oriented and male-dominated sectors.

REFERENCES

Adams, C., & van Manen, M. 2008. *Phenomenology: The SAGE encyclopedia of qualitative research methods.* Thousand Oaks, CA: SAGE.
Ahl, H. 2006. Why research on women entrepreneurs needs new directions. *Entrepreneurship Theory and Practice, 30*(2), 595–621.
Aidis, R. 2006. *Laws and Customs: Entrepreneurship, Institutions and Gender during Economic Transition.* London: School of Slavonic and East European Studies, University College London.
Anna, A., Chandler, G., Jansen, E., & Mero, N. 2000. Women business owners in traditional and non-traditional industries. *Journal of Business Venturing, 15*(3), 279–303.

Bianco, E. M., Lombe, M., & Bolis, M. 2017. Challenging gender norms and practices through women's entrepreneurship. *International Journal of Gender and Entrepreneurship*, 9(4), 338–358.

Butler, J. E. 2003. *New Perspectives on Women Entrepreneurs*. Greenwich, CT: Information Age Publishers.

Coleman, S. 1999. Sources of small business capital: A comparison of men and women-owned small businesses. *Journal of Applied Management and Entrepreneurship*, 4(2), 138–151.

Du Rietz, A., & Henrekson, M. 2000. Testing the female underperformance hypothesis. *Small Business Economics*, 14(1), 1–10.

Global Entrepreneurship Monitor (GEM). 2017. Special Report: Women's Entrepreneurship.

Kvale, S. 1995. *The Social Construct of Validity: Qualitative Inquiry, Qualitative Researching*, 2nd. London: SAGE.

Lubis, R. L. 2016. Building an entrepreneurial thinking of women graduate students: What else beyond learning and dreaming? *Journal of Teaching and Education*, 5(1), 469–504.

Neergaard, H., & Leitch, C. 2015. *Handbook of Qualitative Research Techniques and Analysis in Entrepreneurship*. Cheltenham: Edward Elgar Publishing.

Singh, G., & Belwal, R. 2008. Entrepreneurship and SMEs in Ethiopia: Evaluating the role, prospects and problems faced by women in this emergent sector. *Gender in Management: An International Journal*, 23(2), 120–136.

World Bank Group. 2018. Women, Business and the Law. Washington, DC: World Bank.

The compliance to the statement of risk management and internal control in Malaysia

H. Johari & N. Jaffar
Multimedia University, Cyberjaya, Selangor, Malaysia

ABSTRACT: This study explores the compliance of Malaysian listed companies to the Statement of Risk Management and Internal Control (SRMIC). Risk management and internal control (RMIC) are vital for an effective corporate governance. Effective risk management system helps companies achieve targeted performance. Internal control system provides reasonable assurance that an adverse impact arising from future event is at an acceptable level. In 2015, SRMIC commissions the need to examine the compliance to its requirements. Annual reports of 746 companies listed on the Bursa Malaysia for 2015 and 2016 were examined. Findings show that only 2.5% companies disclose all the mandatory risk management items and only 0.01% of companies disclose all internal control information. Companies voluntarily disclose information on external auditors, internal audit function, internal control opinion, risk appetite, and soundness of the RMIC system. The findings evidence the companies' engagements towards the demands for a more effective corporate governance.

1 INTRODUCTION

Endless occurrences of corporate scandals have raised debates on the effectiveness of corporate governance. Corporate governance is the system of rules, practices and processes by which a company is directed and controlled. It not only provides a framework for attaining a company's objectives, but also encompasses practically every sphere of management, from action plans and internal controls to performance measurement and corporate disclosure. A company cannot achieve its objectives and sustain success without effective governance, risk management and internal control processes. Risk management and internal control are embedded in the governance framework (Bursa Malaysia, 2015).Place the cursor immediately before the T of Title at the top of your newly named file and type the title of the paper in lower case (no caps except for proper names). The title should not be longer than 75 characters). Delete the word Title (do not delete the paragraph end).

Risk management is an important part of planning for businesses. The process of risk management is designed to identify, assess and respond to risk to achieve the company's objectives. An effective risk management process helps a company to achieve its performance and profitability targets by providing risk information to enable better decisions. The system of internal control, on the other hand, is defined as "the actions taken by the board and management to manage risk and increase the likelihood that established goals will be achieved" (Bursa Malaysia, 2015). The internal control system should provide reasonable assurance that the likelihood of a significant adverse impact on objective arising from a future event or situation is at a level acceptable to the business.

In line with the need for companies to manage risk appropriately, the Statement of Internal Control – Guidance for Directors of Public Listed Companies (SIC) had been revised to incorporate risk management in the annual report. As such in 2015, Statement of Risk Management and Internal Control – Guidelines for Directors of Listed Issuers (SRMIC) was issued to guide directors of listed companies in making mandatory disclosures concerning risk management and internal control in their companies' annual reports. Prior studies on SIC found low level of mandatory disclosures on internal control. Haron, Ibrahim, Jeyaraman, & Chye (2010) found less than a quarter (22.3%) of companies fully complying with all eight mandatory disclosures of internal control as stated in the SIC. Nonetheless, the study is based on internal control disclosures as specified by SIC. It is thus timely to assess the compliance with the SRMIC by assessing the level of disclosure of risk management and internal control information. Hence, this study aims to investigate the compliance with the SRMIC's requirements by all companies listed on the Bursa Malaysia.

2 LITERATURE REVIEW

Corporate scandals have placed the nation in the limelight for all the wrong reasons. Transmile Berhad and Technology Resources Industries (TRI) Berhad are among a few of the many companies embroiled in corporate scandals in Malaysia. This results in the demands for improved corporate governance, specifically risk management and internal control system. As a consequence, Malaysian regulatory bodies are proactively putting their efforts to strengthen corporate reporting practice of listed companies. Securities Commission has updated its Malaysian Code on Corporate Governance (MCCG) in 2016 which includes improvements in risk management and internal control system (Oh, 2016). In addition to this, Bursa Malaysia requires public listed companies to furnish disclosures on the state of risk management and internal control in accordance with the revised Statement on Risk Management and Internal Control- Guidelines for Directors of Listed Issuers 2015. Such revision is necessary as concerns have been raised over the dubious utility of internal control reports that contain vague disclosures of unclear meaning or too general and confusing statements (Fadzil, Haron, & Jantan, 2005).

2.1 *Risk management and internal control*

The risk management and internal control system encompasses the policies, culture, organisation, behaviours, processes, systems and other aspects of a company that facilitate its effective and efficient operation by enabling it to assess current and emerging risks, respond appropriately to risks and significant control failures and to safeguard its assets. The risk management and internal control systems should be embedded in the operations of the company and be capable of responding quickly to evolving business risks, whether they arise from factors within the company or from changes in the business environment. Prior studies have found positive correlation between risk management, internal control, and listed companies' performance (Mohamad & Ibrahim, 2012; Brown, Pott, & Wömpener, 2014; Mandzila & Zeghal, 2016). Apart from that, these systems should not be seen as a periodic compliance exercise, but instead as an integral part of the company's day to day business processes (FRC, 2014). The guidelines provided in the SRMIC on the disclosure of risk management and internal control system set out the obligations of management and the board of directors with respect to the disclosures on risk management and internal control.

2.2 *Risk management and internal control disclosure compliance for different sectors*

Companies in different industrial sectors have different compliance levels when disclosing information to the shareholders (Haron, Ibrahim, Jeyaraman, and Chye, 2010; Jo & Na, 2012; Mohamad and Ibrahim, 2012; Dumay & Linlin, 2015; Khlif & Hussainey, 2016) due to their unique characteristics. Haron et al. (2010) discovered only 27 out of 121 companies (22.3%) across industrial sectors fully complied with all eight mandatory disclosures of internal control information. The eight mandatory internal control information to be disclosed were based on SIC, the predecessor of SRMIC, which treated risk management disclosure as voluntary. The SRMIC, on the other hand, identifies disclosure on risk management and internal control as mandatory. Wong, Alashwal, & Mohd Rahim (2019) studies the disclosure of risk management information among Malaysian public listed companies in the construction sector based on the SRMIC and found improved disclosure level of risk management information over time.

3 RESEARCH METHOD

3.1 Research design and sample

This study uses annual reports of all companies that are listed on the Main Market of Bursa Malaysia for the year 2015 as source of data for the compliance and disclosure practices of risk management and internal control information by Malaysian listed companies. From 814 annual reports obtained from the Bursa Malaysia website only data of 746 companies are used. There are 68 companies being excluded due to incomplete data.

3.2 Disclosure index

Disclosure index is developed based on the SRMIC guidelines and Haron et al. (2010). The disclosure of risk management and internal control information are measured by gathering the level of disclosure of eight mandatory disclosure category and six general voluntary disclosure category as presented in the companies' annual reports using an information index.

4 DATA ANALYSIS

4.1 Descriptive analysis

The population of the study comprises of 746 public listed companies from 14 industrial sectors. The population is strongly represented by four sectors which are industrial product (26.9%), trading or services (22.1%), consumer product (16.1%), and properties (12.2%). The other ten sectors in the order of size are construction, plantation, finance, technology, real estate investment trust, hotel, infrastructure project corporation, special purpose acquisition company, close end fund and mining.

4.2 Mandatory disclosures

SRMIC requires companies to disclose three types of information which are risk management, internal control, and assurance from the CEO and CFO on whether risk management and internal control system are operating adequately and effectively. There are nineteen companies (2.5%) fully disclosing information on risk management while the remaining companies disclose the information at 88%. The sector with the highest percentage of companies fully disclosing risk management information is the construction sector which is consistent with the findings of Wong et al. (2019). Only one company (0.01%) from trading or services sector fully disclosed all of internal control information. Majority of the companies (96%) disclose internal control information at 70%. Additionally, majority of the companies comprising of 622 companies (83.4%), disclose receiving assurance from the CEO and CFO on whether risk management and internal control system are operating adequately and effectively while the remaining 124 companies (16.6%) did not disclose this information. All companies in the close end fund, infrastructure project corporation and special purpose acquisition company sectors disclosed the assurance received from the CEO and CFO, and conversely, there is no disclosure on this information in the mining sector.

4.3 Voluntary disclosures

Majority of the companies, 558 (74.8%), fully disclose information on external auditors and conversely only one company from the industrial product sector disclosed information on internal control opinion. Disclosure level in the highest quartile from 76% to 100% were obtained for internal audit function (six companies), risk appetite (four companies), and sound risk management (ten companies). No company disclosed sound internal control system in the highest quartile. The sector with highest number of companies fully disclosing information on external auditors is the plantation sector. On the other hand, the finance sector has the highest number of companies disclosing information in the highest quartile of 76% to 100% on internal audit function, risk appetite, and sound risk management.

5 CONCLUSION

In summary, in terms of mandatory disclosure, only a small number of companies fully disclosed risk management (2.5%) and internal control information (0.01%). Conversely, majority (83.4%) of the companies disclosed CEO and CFO assurance on whether risk management and internal control system are operating adequately and effectively. For voluntary disclosures, majority (74.8%) of the companies disclosed information on external auditors and just a minority of the companies disclosed high level of information on internal audit function, risk appetite, and sound risk management. The findings of this study may offer some insights to the regulators namely Companies Commission of Malaysia, Securities Commission and Bursa Malaysia on the needs to impose stricter rules on the disclosure of risk management and internal control information in the companies' annual reports.

Tables of the descriptive analysis are available upon request. They are not provided in this article due to the quantity and size.

REFERENCES

Brown, N. C., Pott, C., & Wömpener, A. 2014. The effect of internal control and risk management regulation on earnings quality: Evidence from Germany. *Journal of Accounting and Public Policy*, *33*(1): 1–31.

Bursa Malaysia. 2015. *Statement on Risk Management and Internal Control: Guidelines for Directors of Listed Issuers*.

Dumay, J., & Linlin, C. 2015. Using content analysis as a research methodology for investigating intellectual capital disclosure. *Journal of Intellectual Capital*, *16*(1).

Fadzil, F. H., Haron, H., & Jantan, M. 2005. Internal auditing practices and internal control system. *Managerial Auditing Journal*, *20*(8): 844–866.

Haron, H., Ibrahim, D. D. N., Jeyaraman, K., & Chye, O. H. 2010. Determinants of internal control characteristics influencing voluntary and mandatory disclosures: A Malaysian perspective. *Managerial Auditing Journal*, *25*(2): 140–159.

Jo, H., & Na, H. 2012. Does CSR Reduce Firm Risk? Evidence from Controversial Industry Sectors. *Journal of Business Ethics*, *110*(4): 441–456.

Khlif, H., & Hussainey, K. 2016. The association between risk disclosure and firm characteristics: a meta-analysis. *Journal of Risk Research*, *19*(2): 181–211.

Mandzila, E., & Zeghal, D. 2016. Content Analysis Of Board Reports On Corporate Governance, Internal Controls And Risk Management: Evidence From France. *Journal of Applied Business Research*, *32*(3): 637–648.

Mohamad, M., & Ibrahim, N. A. 2012. Governance oversight roles on the voluntary disclosure of internal control systems and its impact on firms performance of Malaysian initial public offerings (IPOs). *ISBEIA 2012 - IEEE Symposium on Business, Engineering and Industrial Applications*, 279–284.

Oh, E. 2016. Corporate governance code gets a sweet 16 makeover - Business News _ The Star Online. *The Star*. Retrieved from http://www.thestar.com.my/business/business-news/2016/04/30/corporate-governance-code-gets-a-sweet-16-makeover/.

Wong, C. C., Alashwal, A. M., & Mohd Rahim, F. A. 2019. Risk Disclosure and Performance of Malaysian Construction Public Listed Companies. *MATEC Web of Conferences*, *266*, 03012.

Toward a model of social entrepreneurship using the soft system methodology approach

G. Anggadwita & G.C.W. Pratami
Telkom University, Bandung, Indonesia

ABSTRACT: Indonesia has several communities with special needs. As a result, a social rehabilitation center was built for people with visual disabilities called Wyata Guna Bandung, with most of the graduates working in massage parlors. This study used qualitative data collected via purposive sampling obtained from the results of in-depth interviews with stakeholders that included society, academics, companies, and students of Wyata Guna. This study also reports the application of Soft System Methodology (SSM) in the Wyata Guna Bandung. Using the theory of social entrepreneurship, this study applied the SSM at each stage that analyzes and explores how each stakeholder thinks. This study only summarizes the preliminary results because further study is required to analyze the application of the system in the Wyata Guna Bandung.

1 INTRODUCTION

Today, the difficulty of finding a job is felt not only by ordinary people, but also by Indonesian residents who have special needs. Therefore, the government needs to pay attention to the disability community in Indonesia. According to Tohari (2014), one form of measuring a democracy is to weigh the country's ability to fulfill and guarantee the rights of its citizens. This means that the state is the provider and protector of the rights of all its citizens. People with disabilities are no exception and have the same rights as ordinary people, for example, to have a job.

About 15% of Indonesian people have special needs; one group is blind people. It is known that at the Wyata Guna Bandung Social Rehabilitation Center for People with Eye Sensory Disability, most of the students have a greater interest in massage because they think that blind people can only do massage, as evidenced by the majority of graduates of the Wyata Guna Bandung who work at a massage parlor. However, from 2016 to 2018, the number of graduates of Wyata Guna who have become entrepreneurs increased. From interviews conducted at the Wyata Guna, it was found that being an entrepreneur can be achieved by only a few graduates who meet certain criteria. This often happens because of increased competition, making it difficult for students at the Wyata Guna Bandung to find work.

The purpose of this study was to identify the forms of social entrepreneurship models for blind people with a Soft System Methodology (SSM) approach. This study provides an overview of the process of social entrepreneurship at the Social Rehabilitation Center and focuses on identifying models of social entrepreneurship that are appropriate for blind people.

2 RESEARCH METHODOLOGY

This study used a qualitative method and was very much tied to the population and sample. However, qualitative research does not use populations but rather a social situation. The sample in qualitative research is not called a respondent, but an interviewee or participant. The determination of data sources in the interviewees is conducted purposively, with specific considerations and objectives. Therefore, the Quadruple Helix in this study had the role of knowing key informants, namely society, academics, companies, and the students of Wyata Guna. This study used the technique of triangulation or a combination of observation, interviews, and documentation. The data collected were analyzed using SSM, which has seven stages, but this research only reached stage 4 because the next stage is the application of the model produced.

3 FINDINGS AND ARGUMENT SSM MODELING STEPS

3.1 The first stage of SSM

The develop a rich picture, a number of sources of information need to be used to see views from all perspectives (Novani et al., 2014). The results of this study found several problems as follows:

- Students of BRSPDSN Wyata Guna Bandung have several traits that must be developed for each individual to become a social entrepreneur.
- College students must reeducate the BRSPDSN Wyata Guna Bandung students.
- The production installation offers several programs such as capacity building and advanced guidance. However, some of these programs have been stopped in recent years.
- The government does not provide capital in the form of money but in the form of skills and tools for massage only.
- There is a lack of teaching practices in entrepreneurship for the BRSPDSN Wyata Guna Bandung students.

3.2 The second stage of SSM

After finding several problems in the first stage, a rich picture may be described as shown in Figure 1.

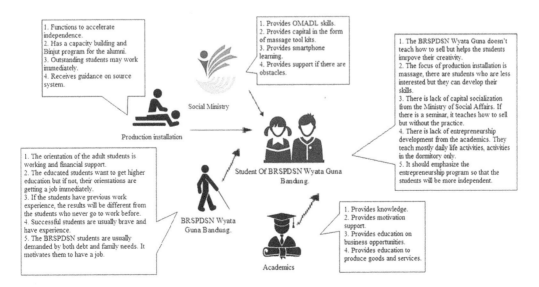

Figure 1. Rich picture.

3.3 The third stage of SSM

The root definition of this study is to develop the potential of the BRSPDSN Wyata Guna Bandung students to learn social entrepreneurship taught by the stakeholders. To check whether the root definition has been described correctly, CATWOE which stands for a customer, actor, transformation, *weltanschauung*, owner, and the environment, is used (Table 1).

3.4 The fourth stage of SSM

At this stage, this study compiled a conceptual model generally obtained from the results of the study (Figure 2).

Table 1. CATWOE analysis.

Item	Description
Customer	The students of BRSPDSN Wyata Guna Bandung
Actor	BRSPDSN Wyata Guna Bandung, massage parlors (production installation), academics, and social ministry
Transformation	From low entrepreneurship knowledge, to being able to develop their own businesses
Weltanschauung	Providing knowledge of the latest innovations as well as providing motivation to move forward and be able to develop is one of the good ways to develop knowledge of social entrepreneurship.
Owner	BRSPDSN Wyata Guna Bandung
The environment	Knowledge of social entrepreneurship

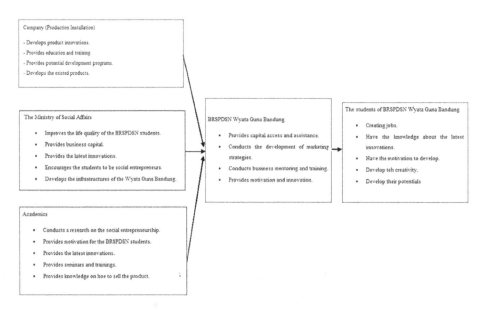

Figure 2. Conceptual model.

4 CONCLUSIONS AND RECOMMENDATIONS

By using SSM, this study analyzed social entrepreneurship using the authors' perspective. To be able to apply the social entrepreneurship model generated in this study, all stakeholders must work together to achieve the main goal of providing an outlet for the BPRSPDSN Wyata Guna Bandung students who are able to develop social entrepreneurship such as establishing their own businesses.

All stakeholders at the BRSPDSN Wyata Guna Bandung, such as companies, can develop product innovations, provide education and training support to the students, provide potential development programs, and develop the existing products. The Ministry of Social Affairs is expected to improve the quality of life for students, provide business capital, provide knowledge of new innovations such as the latest technologies, encourage BRSPDSN students to become socialcial entrepreneurs, and develop the infrastructure such as a special place for buying and selling products. Academics may conduct research on social entrepreneurship; provide motivation for the students; provide innovations, seminars, and training on social entrepreneurship development; and provide knowledge on how to market their products. BRSPDSN Wyata Guna Bandung may create a business incubator that may provide capital access and assistance, develop marketing strategies, conduct business mentoring and training as well as provide motivation, innovation ideas, and creations. The last outcome of the social entrepreneurship model in this study is that BRSPDSN Wyata Guba Bandung students may create jobs, have knowledge of new innovations, become motivated, can develop creativity, and develop their potential. It is emphasized that the results presented in this study are preliminary. Analysis needs to be conducted in the application at the next stage in the SSM.

REFERENCES

Arnkil, R., Järvensivu, A., Koski, P., & Piirainen, T. 2010. Exploring Quadruple Helix: Outlining user-oriented innovation models. Työraportteja 85/2010 Working Papers.

BRSPDSN Wyata Guna Bandung. 2019. *Data Penyaluran Penerima Manfaat PSBN Wyata Guna yang Disalurkan Tahun 2014–2018*. Bandung: BRSPDSN Wyata Guna Bandung.

Carayannis, E. G., et al. 2017. The ecosystem as helix: An exploratory theory-building study of regional competitive entrepreneurial ecosystems as Quadruple/Quintuple Helix innovation models. *R&D Management*, 148–162.

Firdaus, A., & Maarif, M. S. 2015. Aplikasi Soft System Methodology (SSM) untuk Perencanaan Terintergrasi Biofuel dalam Sektor Pertanian dan Sektor Energi. *Jurnal PASTI, IX*, 1–9.

Indrawati. 2015. *Metode Penelitian Manajemen dan Bisnis (Konvergensi Teknologi Komunikasi dan Informasi)*.. Bandung: PT Refika Aditama.

Irwanto, et al., 2010. *Analisis Situasi Penyandang Disabilitas Di Indonesia: Sebuah Desk-Review*. Jakarta: ilo.org.

Littlewood, D., & Holt, D. 2018. Social entrepreneurship in South Africa: Exploring the influence of environment. *Business & Society 2018*, 57(3), 525–561.

Novani, S., Putro, U. S., & Hermawan, P. 2014. An application of Soft System Methodology in batik industrial cluster solo by using service system science perspective. *Procedia: Social and Behavioral Sciences*, 115, 324–331.

Patel, N. V. 1995. Application of soft systems methodology to the real world process of teaching and learning. *International Journal of Educational Management*, 9(1), 13–23.

Sofia, I. P. 2015. Konstruksi model kewirausahaan sosial (social entrepreneur) sebagai gagasan inovasi sosial bagi pembangunan perekonomian. *Jurnal Universitas Pembangunan Jaya, 2*, 2–23.

Sugiyono. 2017. *Metode Penelitian Kuantitatif (untuk penelitian yang bersifat: eksploratif, enterpretid, interaktif dan konstruktif)*. Bandung: Alfabeta.

Tohari, S. 2014. Pandangan Disabilitas dan Aksesibilitas Fasilitas Publik bagi Penyandang Disabilitas di Kota Malang. *Indonesian Journal of Disability Studies*, 27.

Motivation to become agropreneurs among youths in Malaysia

M.F.A. Rahim, K.W. Chew & M.N. Zainuddin
Multimedia University, Malaysia

A.S. Bujang
Malaysian Agricultural Research and Development Institute, Malaysia

ABSTRACT: While numerous government initiatives have been implemented to encourage entrepreneurship activities among Malaysian youths, agricultural-based entrepreneurship (agropreneurship) seems to be an unfavoured choice despite its strategic economic importance. Young entrepreneurs prefer to choose sectors that are trendier and happening, such as food and beverages, technology, and services. Therefore, the motivation to become agropreneurs seems to be rather exclusive and is influenced by unknown factors. This study sought to examine the factors that influence the motivation of Malaysian youth to become agropreneurs. A survey method via self-administered questionnaires was used in this study. Data were collected from young agropreneurs from various states in Malaysia, who were asked about their motivations to start a business in the agriculture sector. Among the key motivators for the respondents to start their agro-based businesses were pursuing their dream, elevating their socio-economic status, and challenging themselves.

1 INTRODUCTION

The Malaysian government's emphasis on modernising the agriculture industry is evidenced by the allocation of more than RM4 billion for the Ministry of Agriculture and Agro-based Industry Malaysia (MOA) budget in 2017 (MOA, 2017). In addition to being an important source of food, agriculture can serve as a medium to overcome poverty and mitigate unemployment in Malaysia. This is in line with one of the strategic thrusts of the Eleventh Malaysia Plan (RMK11), which is to raise the average income and share of total income of the bottom 40% household income group (B40 household) (RMK11, 2017). Under the Government Transformation Programme 2.0 (GTP2.0), agricultural-based entrepreneurship (agropreneurship) has the potential to help the country achieve two of the National Key Result Areas (NKRAs), including raising living standards of low-income households and addressing the cost of living (GTP, 2017).

Furthermore, agriculture is also one of the focal points of the National Key Economic Area (NKEA), evidenced by the identification of 16 Entry Point Projects (EPP) and 11 Business Opportunities (BO). This is part of the Economic Transformation Plan (ETP) aimed at contributing to the increase of the country's gross national income (GNI) amounting to RM21.44 billion by the year 2020, which would result in the creation of 74,000 additional jobs in Malaysia (MOA, 2017). The creation of more agriculture-based entrepreneurs would also support the New Economic Model's (NEM) Strategic Reform Initiatives (SRI), specifically by "Re-energising the private sector to drive growth" (SRI 1) and "Creating a competitive domestic economy" (SRI 3).

Despite numerous government initiatives towards encouraging entrepreneurship activities across all economic sectors, agriculture-based entrepreneurship (agropreneurship) seems to be an unfavoured choice among Malaysian youths, albeit its strategic economic importance. Young entrepreneurs prefer to choose sectors that are trendier and happening, such as food and beverages, technology, and services. Hence, the motivation to become agropreneurs seems to be rather exclusive and is influenced by unknown factors. Thus, the current study aimed to investigate the factors that influence the motivation of Malaysian young agropreneurs.

2 LITERATURE REVIEW

2.1 Agropreneur and entrepreneur

Schumpeter (1934) describes entrepreneurs as innovators who disrupt the existing order to introduce new products or services. Meanwhile, the term agropreneur or agripreneur was coined as the combination of two words: "agriculture" and "entrepreneur" to signify entrepreneurs who are based in the agricultural sector (Nagalakshmi & Sudhakar, 2013; Waribugo, 2016). The Malaysian government's focus on nurturing entrepreneurship is evidenced in the various initiatives that were allocated in the national budget of 2019 (MOF, 2019). This saw the reestablishment of the Ministry of Entrepreneur Development (MED) on 2 July 2018 and the launching of the National Entrepreneurship Policy 2030, with a specific focus on empowering the B40 group by inculcating a culture of entrepreneurship (MED, 2019).

2.2 MARDI Young Agropreneur Programme

The Malaysia Agricultural Research and Development Institute (MARDI) is an agency under the Ministry of Agriculture and Agro-Based Industries (MOA). MARDI is one of the government agencies that was tasked to spearhead the implementation of MOA's Young Agropreneur Programme. The programme focuses on instilling and encouraging young (aged 18–40 years old) Malaysians' interest in agriculture, which in turn would create entrepreneurs who will prosper in agro-based businesses. With income generation a focal point in this initiative, the program also targets lesser dependency on foreign imports of food commodities to safeguard the nation's food security. Since its inception in 2014, MARDI's Young Agropreneur Programme has developed many young agropreneurs in various agricultural sectors including vegetable fertigation systems, mushroom cultivation, stingless bee farming, as well as food processing and agro-based processing. MARDI is tasked to organise trainings, short courses, technical consultancies, as well as facilitating application of grants to achieve these goals. This programme also provides platforms and avenues for agropreneurs to promote their products and expand their marketing network. This is done through many activities including roadshows, exhibitions, and product showcases nationwide (MARDI, 2016).

2.3 Entrepreneurial motivation

Entrepreneurial motivation explains the deep connection between the mind and behaviour that is triggered when individuals (entrepreneurs) respond to an opportunity that they had previously recognised and decide to pursue (Shane et al., 2003; Carsrud et al., 2017). Accordingly, several theoretical models describe how entrepreneurial motivation operates. These models assume that human action is the result of both motivational and cognitive factors that include skills, ability, intelligence, and heuristics (Locke, 2000; Bryant, 2007; Chaston & Sadler, 2012) that can lead to recognition and exploitation of opportunities. In addition, individuals' action is also influenced by a series of external factors such as the state of the economy, the availability and accessibility of venture capital or funding, and the local government regulation and initiatives.

3 METHODOLOGY

Following a positivist approach to research, this study employed the quantitative approach to data collection by distributing self-administered survey questionnaires. The questionnaire items, particularly for motivators to start a business, were adapted from previous literature (Volery et al., 1997; Choo & Wong, 2006). The questionnaire underwent back-translation to translate the original English language questions into Bahasa Malaysia. The focus of the translation was to ensure that the questionnaire maintained the meanings of the questions and was not merely a literal translation (Dillman et al., 2009). Pretesting was conducted by having the questionnaire reviewed by entrepreneurship experts, including actual entrepreneurs and academics, to provide content validity. A sample of 233 responses was collected from participants who attended MARDI's AGORA Young Agropreneur training programme. Data analysis was conducted by using the Statistical Package for Social Sciences (SPSS) software.

4 RESULTS

Table 1 illustrates the demographic profiles of respondents in this study. The majority of the respondents were male (53.2%) and most of them were 26–30 years old (29.2%). Most respondents had only Sijil Pelajaran Malaysia (SPM) as their highest academic qualification (29.6%), followed by a bachelor's degree (23.6%) and diploma (24.5%). Table 2 shows the descriptive statistics for motivators to start a business. Based on the mean scores, it was found that the main motivators for the respondents to start a business were to realise their dream ($\mu = 6.53$), to increase their status ($\mu = 6.50$), and to challenge themselves ($\mu = 6.46$). All three items fall under the intrinsic motivators.

Table 1. Demographic profiles of respondents.

Characteristic		Frequency	%	Characteristic		Frequency	%
Gender	Male	124	53.2	Education	Doctorate	1	0.4
	Female	109	46.8		Master's degree	9	3.9
					Bachelor's degree	55	23.6
Age	18–20 years old	12	5.2		Professional	1	0.4
	21–25 years old	45	19.3		Diploma	57	24.5
	26–30 years old	68	29.2		Certificate	24	10.3
	31–35 years old	64	27.5		STPM	7	3.0
	36–40 years old	44	18.8		SPM	69	29.6
					Others	10	4.3

Table 2. Descriptive statistics of motivators to start a business.

Item	Mean	Std. dev.
To provide a comfortable retirement	6.02	1.370
To earn more money	6.17	1.271
To follow the example of a person I admire	5.88	1.421
To maintain a family tradition	5.00	3.369
To take advantage of a market opportunity	6.43	0.906
To take advantage of my own creative talents	6.31	1.078
To challenge myself	6.46	0.906
To realise my dream	6.53	0.871
To increase my status	6.50	0.881

5 CONCLUSION

Among the key motivators for the respondents to start their agro-based businesses are to pursue their dream, to elevate their socio-economic status, and to challenge themselves. This shows that young agropreneurs are mostly motivated by intrinsic factors. Thus, in order to achieve higher participation from young entrepreneurs in their entrepreneurship programs, government agencies such as MARDI should focus more on highlighting the intrinsic benefits of becoming agropreneurs and participating in the programmes.

ACKNOWLEDGMENTS

This work was supported by the Fundamental Research Grant Scheme (FRGS) 2017 (FRGS/1/2017/SS03/MMU/02/6).

REFERENCES

Bryant, P. 2007. Self-regulation and decision heuristics in entrepreneurial opportunity evaluation and exploitation. *Management Decision*, 45(4), 732–748.

Carsrud, A., Brännback, M., Elfving, J., & Brandt, K. 2017. Motivations: The entrepreneurial mind and behaviour. In M. Brännback & A. Carsrud (eds.), *Revisiting the Entrepreneurial Mind* (pp. 185–209). Berlin and Heidelberg: Springer.

Chaston, I., & Sadler, S. E. 2012. Entrepreneurial cognition, entrepreneurial orientation and firm capability in the creative industries. *British Journal of Management*, 233), 415–432.

Choo, S., & Wong, M. 2006. Entrepreneurial intention. *Singapore Management Review*, 28 (2), 47–64.

Dillman, D. A., Smyth, J. D., & Christian, L. M. 2009. *Internet, Mail and Mixed-Mode Surveys: The Tailored Design Method*, 3rd ed. Hoboken, NJ: John Wiley & Sons.

GTP (2017). *Government Transformation Programme*. Retrieved from http://gtp.pemandu.gov.my/gtp/What_Are_NKRAs%5E-@-NKRAs_Overview.aspx

Locke, E. 2000. Motivation, cognition, and action: An analysis of studies of task goals and knowledge. *Applied Psychology*, 49(3), 408–429.

MARDI. 2016. Malaysian Agriculture Research and Development Institute. Retrieved from http://www.mardi.gov.my/(accessed 28 September, 2016).

MED. 2019. Ministry of Entrepreneur Development, Malaysia. Retrieved from http://www.med.gov.my/portal/index.php

MOA. 2017. Ministry of Agriculture and Agro-based Industry Malaysia. Retrieved from http://www.moa.gov.my/bajet-moa-2017

MOF. 2019. Ministry of Finance, Malaysia. Retrieved from http://www.treasury.gov.my/

RMK11. 2017. The Eleventh Malaysia Plan. Retrieved from http://rmk11.epu.gov.my/index.php/en/

Schumpeter, J. A. 1934. *The Theory of Economic Development: An Inquiry into Profits, Capital, Credit, Interest, and the Business Sycle*. Piscataway, NJ: Transaction Publishers.

Shane, S., Locke, E. A., & Collins, C. J. 2003. Entrepreneurial motivation. *Human Resource Management Review*, 13(2), 257–279.

Volery, T., Doss, N., Mazzarol, T., & Thein, V. 1997. Triggers and barriers affecting entrepreneurial intentionality. In *42nd ICSB World Conference*, 21–24 June, San Francisco.

Customer journey maps of Muslim young agropreneurs

M.F.A. Rahim, J.W. Ong & N.M. Yatim
Multimedia University, Cyberjaya, Malaysia

H.A. Yanan & M.N.M. Nizat
Malaysian Agricultural Research and Development Institute, Kuala Lumpur, Malaysia

ABSTRACT: The Malaysian government has targeted to encourage more youths to be agriculture-related entrepreneurs (agropreneur) with the the Malaysian Agricultural Research and Development Institution (MARDI) being tasked to groom young agopeneurs in Malaysia via the Young Agropreneur Programme. This study aims to develop a template of customer journey map to investigate the experience of Muslim Young Agropreneurs under MARDI's Young Agropreneur Programme through review of literature. The expected output of this article is a template of customer journey map for MARDI's Young Agropreneur Programme. The journey map would capture the channel, customer expectation, experience and emotion throughout their contact with the MARDI's Young Agropreneur Programme. This article reviews the literature of customer journey map to be used to study the experience of Muslim Young Agropreneurs in dealing with MARDI. The comprehensiveness of customer journey map can help to more comprehensively capture their experience in dealing with the service provider.

1 INTRODUCTION

The agricultural sector is important to the Malaysian economy as it contributes approximately 8.2 percent or equivalent to RM96 billion to the Gross Domestic Product (GDP) of Malaysia in 2017 (Department of Statistics [DOS] 2018). Agricultural-related entrepreneurship (agropreneurship) provides opportunities for the 2.7 million people in the bottom 40 percent of Malaysian households with monthly income of RM3,900 and below (B40 group) to be uplifted to middle-class society (RMK11 2017). In Budget 2017, RM100 million was allocated to implement the Young Agropreneur Programme to produce 3,000 young entrepreneurs, particularly from the B40 group. Participation in agropreneurship activities would help Malaysians to increase their income and improve their living standards. However, the current agropreneurship participation rate among Malaysian youth is at 15 percent, which is very low (Hussin 2016). Agencies under the Ministry of Agriculture (MOA), including MARDI were tasked to alleviate this issue by conducting programmes to nurture agrorpreneurship among the younger generation. However, there is a need to study the effectiveness of the said programmes to justify further investment in the future.

Thus, this study aims to investigate the customer experience journey of Muslim Young Agropreneurs who have been involved in programmes conducted by MARDI. The customer journey map will reveal the experience of the Muslim Young Agropreneur while engaging with MARDI. The journey will be captured from the awareness stage until the post-event services. This will assist MARDI in understanding their strengths and weaknesses in the complete customer engagement process. Subsequently, the information would be utilised to guide MARDI to improve their processes.

2 LITERATURE REVIEW

2.1 *Entrepreneurship and agropreneurship*

Timmons and Spinelli (2003) describe an entrepreneur as someone who is an innovator or developer that recognizes and seizes opportunities, converts those opportunities into a workable or marketable idea, adds value through time, effort, money or skills, assumes the risks of the competitive marketplace to implement these ideas and realizes the rewards from these efforts. The term Agropreneur or Agripreneur was coined as the combination of two words: 'agriculture' and 'entrepreneur' to signify entrepreneurs that are based in the agricultural sector (Nagalakshmi & Sudhakar 2013; Waribugo 2016). In Malaysia, various agencies including the Malaysia Agricultural Research and Development Institute (MARDI), and Federal Agricultural Marketing Agency (FAMA) were tasked to spearhead the development entrepreneurship and agropreneurship, with a specific focus on youths. For example, the Ninth Malaysian Plan was aimed at producing a total 260,928 agropreneurs (Mohamed, Rezai, Shamsudin, & Mahmud 2012).

2.2 *MARDI young agropreneur program*

Unit Agropreneur Muda (UAM) or Young Agropreneur Unit was established in September 2013 by the Ministry of Agriculture and Agro-Based Industries (MOA) as aspired by the Honorable Minister of Agriculture and Agro-based Industries (MOA). The main objective of this Unit is to instill and encourage interest of young Malaysians (aged 18-40 years old) in agriculture that would in turn create entrepreneurs that will prosper in agro-based businesses. This is done through various structured and packaged programs covering the aspects of trainings, business development and entrepreneurship, preparation of project's site/business premise, promotion and marketing as well as funding. The vision of this Unit is to create young agropreneurs who are progressive, competitive, creative, innovative and will ultimately generate high income from their ventures. In addition, the Unit's mission is to promote and inculcate agriculture as a modern and dynamic sector as well as the chosen career. In light of this, MARDI's Young Agropreneur Programme (Unit) was established in September 2014 to coordinate and facilitate technology and tacit knowledge transfer activities in supporting the government's mission of creating innovative and progressive young agropreneurs. Since its establishment, more than 5,000 youths have registered and received various benefits in the forms of technical advice, training courses, in-kind contribution grants and marketing which in turn resulted in the development of around 400 progressive young agropreneurs throughout the nation so far (Mohd Nizam 2019).

2.3 *Customer journey map*

Customers form a critical part for success of businesses (Johnston & Kong 2011). A greater customer experience will no doubt improve the customer's satisfaction and subsequently the businesses (Jonston & Kong 2011). On top of that, Jonston and Kong (2011) and Rawson, Duncan and Jones (2013) found improvement in customer experience does improve the business performance and benefit the employees as well. While businesses try to model their provision of services or value to the customers through their service blueprint, it is important to also examine the customers' side of experience through the journey map (Halvorsrud, Kvale & Folstad 2016). The customer journey map focuses on empathizing the customers through the service delivery process by focusing on their experience in each possible touchpoint in chronological order, including the emotion of the customers (Halvorsrud et al. 2016; Haugstveit, Halvorsrud & Karahasanovic 2016). It is important for both to be studied as any discrepancy could indicate a deviation or a service gap that leads to customer's dissatisfaction (Halvorsrd et al. 2016; Haugstveit et al. 2016). There are four possible forms of deviations including "missing", "failing", "ad-hoc" and "irregularities" (Halvorsrd et al. 2016; Haugstveit et al. 2016).

The customer journey, according to Halvorsrud et al. (2016, p. 846) is defined as "customer's interactions with one or more service providers to achieve a specific goal" while the touchpoint is defined as "instance of communication between a customer and a service provider". Lemon and Verhoef (2016) explained the customer journey as the customer experience with a business over time. According to them, the journey can be broadly divided into prepurchase, purchase and postpurchase stage. In addition, their previous and future experience should also be considered as part of their journey with the business (Lemon & Verhoef 2016). The customer journey map, on the other hand represents the visual presentation of the customer journey (Marquez, Downy & Clement 2015). Rawson et al. (2013) suggested that focusing on the complete journey is important to evaluate the overall satisfaction of the customers toward a business, rather than overly focus on a specific touchpoint within the process. In terms of the touchpoints, they categorised the touchpoints according to brand-owned, partner-owned, customer-owned and social/external/independent touchpoints indicating the ownership and initiator of the touchpoints (Lemon & Verhoef 2016). Since the customer experiences are personal and unique, it is recommended that the customer journey map should be kept at individual level instead of aggregating multiple maps of various customers (Halvorsrud et al. 2016).

3 METHODOLOGY

The study develops the customer journey map for MARDI's Young Agropreneur Programme through review of literature and standard operating procedure of MARDI. Based on the review, the important touch points for the Muslim young agropreneurs' journey in participating in the MARDI's Young Agropreneur Programme are identified.

4 FINDINGS

Based on the review, the drafted general journey map for MARDI's Young Agropreneur Programme is shown in Figure 1. In general, the Muslim young agropreneurs are expected to go through five general stages from exposure or awareness with the programme, officially engage with the programme through registration, participating in the training and consultation, opportunities for grant, funding or technical assistant, and lastly the continuous monitoring.

With the general journey map, the researchers will proceed to interviews the targeted Muslim Young involved in the process throughout the journey maps. The respondents will be asked to reveal their experience with MARDI. The experience includes their tasks, their feelings and their perception towards whether MARDI has delivered. Based on that, the Muslim Young Agropreneurs journey map will be developed.

5 CONCLUSION

This article reviews the literatures surrounding the importance of the agricultural sector and agropreneurship to the Malaysian economy, and the customer journey map to be used to study the experience of Muslim Young Agropreneurs in dealing with MARDI, a governmental agency. The comprehensiveness of the customer journey map can help MARDI to more comprehensively capture the customers' experience in dealing with the service provider. The findings of this study would help to improve the experience of Muslim

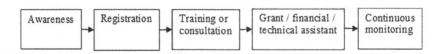

Figure 1. Proposed general journey map of MARDI's Young Agropreneur Programme.

young agropreneurs in dealing with government agencies like MARDI. Cultivating a more pleasant experience would encourage greater working relationships amongst stakeholders and will ultimately lead to greater income and standard of living of agropreneurs.

ACKNOWLEDGMENTS

This work was supported by the Fisabilillah R&D Grant Scheme (FRDGS) 2018 (FRDGS2018/30).

REFERENCES

DOS 2018, *Department of Statistics*. Retrieved from https://www.statistics.gov.my/index.php

GTP 2017, *Government Transformation Programme*. Retrieved from http://gtp.pemandu.gov.my/gtp/What_Are_NKRAs%5E-@-NKRAs_Overview.aspx

Lemon, KN and Verhoef, PC 2016, Understanding Customer Experience throughout the Customer Journey. *Journal of Marketing*, vol. 80 (November 2016), pp. 69–96.

Halim, MASA & Hamid, AC 2011, The Relationship of Persionality Traits and Enterepreneurial Commitment among Agropreneurs in Pasar Tani. *Canadian Social Science*, vol. 7, no. 2, pp. 52–59. http://doi.org/http://dx.doi.org/10.3968/g1281

Halvorsrud, R, Kvale, K & Folstad, A 2016, Improving Service Quality through Customer Journey Analysis. *Journal of Service Theory and Practice*, vol. 26, no. 6, pp. 840–867.

Haugstveit, IM, Halvorsrud, R & Karahasanovic, A 2016, Supporting redesign of C2C services through customer journey mapping. In *Service Design Geographies. Proceedings of the ServDes. 2016 Conference*, no. 125, pp. 215–227, Linköping University Electronic Press.

Hussin, M 2016, March, *Unit Agropreneur Muda pelapis*. Retrieved from http://www.hmetro.com.my/node/119266

MARDI 2016 *Malaysian Agriculture Research and Development Institute*. Retrieved September 28, 2016, from http://www.mardi.gov.my/

Marquez, JJ, Downey, A & Clement, R 2015, Walking a Mile in the User's Shoes: Customer Journey Mapping as a Method to Understanding the User Experience. *Internet reference Services Quarterly*, vol. 20, pp. 135–150.

Ministry of Youth and Sports 2016, *Ministry of Youth and Sports, Malaysia*. Retrieved from http://www.kbs.gov.my/en/

Mohamed, Z, Rezai, G, Shamsudin, MN & Mahmud, MM 2012, Enhancing young graduates' intention towards entrepreneurship development in Malaysia. *Education + Training*, vol. 54, no. 7, pp. 605–618. http://doi.org/10.1108/00400911211265648

MOA 2016, *Ministry of Agriculture and Agro-based Industry Malaysia*. Retrieved from http://www.moa.gov.my/bajet-moa-2017

Mohammad Nor, NAA, Nik Mohd Masdek, NR & Maidin, MKH 2015, Youth inclination towards agricultural entrepreneurship. *Economic and Technology Management Review*, vol. 10a, pp. 47–55.

Mohd Nizam, M.N. (2019), Implementation of an Innovative Structured Training Program to Encourage.

Participation of Malaysian Youth in Agribusiness. *Country Paper for Workshop on Innovations in Agribusiness for Young Entrepreneurs, 1 – 5 July 2019, Taipei, Republic of China*.

Rawson, A, Duncan, E & Jones, C 2013, The truth about customer experience. *Harvard Business Review*, vol. 91, no. 9, pp. 90–98.

RMK11(2017), *The Eleventh Malaysia Plan*. Retrieved from http://rmk11.epu.gov.my/index.php/en/

Timmons, J.A & Spinelli, S 2003, New Venture Creation: Entrepreneurship for the 21st Century. New York: Irwin, McGraw Hill.

New culinary trends based on the most popular Instagram accounts

G. Anggadwita & E. Yuliana
Telkom University, Bandung, Indonesia

D.T. Alamanda & A. Ramdhani
Garut University, Garut, Indonesia

A. Permatasari
President University, Bekasi, Indonesia

ABSTRACT: Besides being known for its mountains and beaches, Garut district is known for its culinary diversity. The purpose of this study was to identify Garut's culinary products that receive the attention of citizens while being able to be used as a new icon of Garut's typical culinary business. The @jajanangarut Instagram account, as the center of information and review of Garut culinary types and centers, was chosen as the research subject. This research primarily used qualitative content analysis. By paying attention to 9,400 posts as of June 2018, the data were then filtered into 3,624 posts in the observation period January 2018–June 2019. The results of the study found there are 10 categories of Garut culinary types, where snacks are the most popular, with 1,178 posts, followed bynoodles and meatballs.

1 INTRODUCTION

Garut is one of the districts in Indonesia that has a tremendous culinary business. Those culinary businesses were introduced by several popular Instagram accounts, namely @jajanangarut, @kulinergarut, @kulinerkotagarut, @garutdelicious etc. Instagram is one of the social media platforms that provides photos, videos, and other visual content to share with the public. Instagram also can display advertisements and marketing and other interesting messages online (Hashim, 2017). Also, with Instagram, marketers can easily get feedback from customers and make closer connections with potential customers (Hashim, 2017).

The advantages of Instagram are captured by one of the Instagram accounts, @jajanangarut, which positions itself as an information center for snacks and culinary review in the Garut district. However, the richness of its data is yet to be utilized for a more profitable digital platform. Therefore, this study aimed to describe the culinary information provided by the @jajanangarut account and its impact on the online delivery media available in Garut Regency.

2 LITERATURE REVIEW

Chatzinakos (2016) explored the potential of culinary tourism in an area, concluding that culinary tourism can strengthen the relationship between the food and the places where food is produced, and it triggered the development of brand image and purpose occurs (Sotiriadis, 2012). Almost all businesses have adopted some form of online promotion through social media (Permatasari & Kuswadi, 2017; Ghoshal, 2019); therefore the role of social media is very important for the purchasing decision-makers. Ghoshal (2019) found that social media is one of the most powerful, profitable, and effective platforms for a company to increase its visibility among its target customers. In line with this, Ghoshal et al. (2018) stated that the promotional activities of e-WOM will increase the brand image of an Instagram account, which could influence consumer purchase intentions. Fahy and Jobber (2012) described marketing communication consisting of 4C's, namely, clarity, credibility, consistency, and competitiveness.

Instagram invites users to be able to see the world in a fun way by sharing photos (Instagram, 2019). In addition to sharing photos, Instagram also utilizes images as a business medium, as well as using sponsored advertisements that in fact create positive feelings from Instagram users (Rochman & Iskandar, 2015). Rizqia and Hudrasyah (2015) further found that Instagram is a text-based technology that is used by businesses as Word-of-Mouth (eWOM) electronic devices to attract customers and then lead to purchase intentions.

The use of Instagram as a promotional medium has been studied by Azwar and Sulthonah (2018) at libraries in Indonesia. Instagram is not only used as a promotional medium but also considered as an effective medium for sharing information and communicating. However, Lestari and Aldianto (2016) stated that promotional efforts through Instagram did not significantly affect the promotion objectives, which occurs if the business actor cannot understand effective promotional tools that can influence potential customers to get to know the brand. The brand of the products will affect buying decisions for a new customer or a potential customer.

Hashim (2017) found that more than 100 nascent entrepreneurs stated that Instagram is a must-have platform to share products and services with potential customers. Specifically, in the culinary business, Stone et al. (2017) mentioned in greater detail that culinary experiences become very memorable because of, among other factors, food/drinks consumed, location, friends, opportunities, and authenticity or novelty. Instagram also helps them as new entrepreneurs to make vital and important business decisions. The connection with this research was shown by Smith and Sanderson (2015), who stated that Instagram is a platform that provides a framework for research using content analysis.

3 RESEARCH METHOD

3.1 Research characteristics

This study was a type of exploratory descriptive research that aims to access and understand the interaction of all stakeholders relating to the enhancement of typical culinary tourism in Garut through the Instagram account @jajanangarut. The initial stages of this study used a content analysis approach. The content analysis describes or explains a problem whose results can be generalized and do not concern the depth of the data (Ward, 2016). The data analyzed 9,400 @jajanangarut posts. Online content analysis was chosen to prioritize the breadth of data from the Instagram account @jajanangarut so that the data presented would represent the entire population, shown by systematically identifying visible communication.

3.2 Construction category

Construction category is a tool used to explore research problems. Its function is to sort the contents of the written message into a picture (in the form of data) that can be analyzed to answer the problems raised. According to Seals et al. (2000) three things that need to be considered in making categories: (1) the categories must be relevant to the objectives of the study, (2) the categories must be functional, and (3) the system of the categories must be applied.

3.3 Research stages

The study began by determining the research period based on information from admin @jajanangarut. Crawling using data mining was carried out on a number of posts that comprised the research data population. The next step was to categorize the posts with consideration of the popularity of culinary stalls and snacks. Therefore, the triangulation process enlisted three experts to validate the category findings. Further posts were grouped by type, price range, and location of culinary stalls and snacks.

4 RESULTS AND DISCUSSION

From 9,400 @jajanangarut photo uploads, data were taken in the period January 2018–June 2019, so that as many as 3,624 uploads were taken as research samples. This period was chosen because according to admin @jajanangarut, several businesses that posted before January 2018 had closed for reasons of seeking other opportunities or bankrupcy. Based on crawling results with data mining, posts are categorized by culinary type, price range, and culinary stall location.

4.1 *Culinary type*

Based on @jajanangarut posts, there are 10 categories that are considered to represent all posts. The categories were validated by three speakers, where two people are restaurant chefs and one food blogger. The category findings in the study are slightly different from the 21 food categories that exist in the Go-Food application (2019). The 10 categories were snacks, meatballs and noodles, rice meal packages, beverages, fast food, chicken meal packages, Western food, bread and cakes, Indonesian pancakes, and miscellaneous. The difference in categories is because Go-food in the Go-jek application is a car-based application platform whose features are applicable nationally while the @jajanangarut information media comes from snacks and culinary centers that exist only in Garut District. Even though the number of categories in Go-Food is far greater, the number of business actors involved is far more at @jajanangarut, because not all culinary and snacks businesses in Garut District are partnering with Go-Food.

Snacks are the most dominant type, with 1,178 posts (32.5%), followed by meatballs and noodles with 830 posts (22.9%) and rice meal packages with 450 posts (12.4%); the number of posts in the other 7 categories is below 10%. The miscellaneous are the least plentiful types (0.4%), which means they are generally represented by the previous categories.

4.2 *Price range*

Based on the price range, there are eight categories starting from the price range of 1 K but less than 5 K (category 1) to a price range more than or equal to 50 K (category 8). The price range of 15 K but less than 20 K (category 4) dominates, with 2,768 posts, where the culinary type that is mostly in the price range is meatballs and noodles (1,034 posts). The lowest price range is in category 6 (25 K but less than 30 K) and even more have a price range more than or equal to 50 K, with the dominance of bread and cakes products.

4.3 *Culinary center location*

Three sub-districts in Garut district became culinary centers with a total of 945, 734, and 876 posts, respectively: Garut Kota, Tarogong Kidul, and Tarogong Kaler. Garut Kota sub-district dominates the type of culinary snacks, namely 620 posts (17%). Meatball and noodle snacks are the most common culinary type from Tarogong Kidul district (245 posts) and Garut Kota sub-district (220 posts). Rice meal packages are most commonly found in posts originating from Tarogong Kaler sub-district (219 posts) and Tarogong Kidul sub-district (177 posts). Tarogong Kidul sub-district dominates the beverage culinary type, with 117 posts or 42% of the total posts regarding beverages.

The online transportation application Jajap Garut is a Go-Food competitor that appeared only in May 2019 and is still continuing to develop applications. The purpose of Garut Jap is to help the small and medium-sized enterprises in Garut Regency to market food products or souvenirs. The Garut Jajap application itself has been downloaded by more than 1,000 smartphone users, competing with Go-Food, which is the most popular app in Indonesian society. Therefore Garut Jajap has the potential market to develop local-based features of the Garut community by collaborating with other information media such as @jajanangarut.

5 CONCLUSION

This study found a gap between the culinary information provided by the @jajanangarut account and the online delivery media available in Garut District. Deficiencies of @jajanangarut are the inconsistency of information between posts and the availabilityof only three categories in almost all posts, namely the culinary type, price range, and culinary center location. However, the @jajanangarut account has a very significant number of followers and an extraordinary number of posts. The other potential that needs to be encouraged is an enormous amount of business interest to market products at @jajanangarut. On the other hand, there are three food ordering applications, one of which is the Garut Jajap app. The disadvantage of Garut Jajap is that it is still in the development stage so that it still lacks features and minimal data compared to its two more popular competitors. Therefore, the proposed blueprint for the development of the data-driven Jajap Garut application @jajanangarut is a collaboration between mutually beneficial platforms.

REFERENCES

Azwar, M., & Sulthonah, S. 2018. The utilization of Instagram as a media promotion: The case study of library in Indonesia. *Journal of Islam and Humanities*, 2(2).

Chatzinakos, G. 2016. Exploring potentials for culinary tourism through a food festival: The case of Thessaloniki Food Festival. *Transnational Marketing Journal*, 4(2), 110–124.

Fahy, J., & Jobber, D. 2012. *Foundations of Marketing*. New York: McGraw-Hill Education.

Ghoshal, M. 2019. Social media as an effective tool to promote business: An empirical study. S *Global Journal of Management and Business Research*, 19(1), 15–25.

Hashim, N. A. 2017. Embracing the Instagram waves: The new business episode to the potential entrepreneurs. *Journal of Entrepreneurship and Business Innovation*, 4(2), 13–29.

Instagram. 2019. *Instagram definition*. Retrieved from Instagram.com: https://www.instagram.com/about/us/(accessed June 21, 2019).

Khalisa, N., & Kesuma, T.M. 2018. The impact of electronic word of mouth Instagram as recommendations halal culinary tours in Banda Aceh. *International Journal of Academic Research in Business and Social Sciences*, 8(5), 1042–1060.

Lestari, S., & Aldianto, L. 2016. Effect of using hashtag, celebrity endorsement, and paid promote to achieve promotion objective in Instagram; case study: Women fashion brand. *The Journal of Innovation and Entrepreneurship*, 1(1), 1–7.

Permatasari, A., & Kuswadi, E. 2017. The impact of social media on consumers' purchase intention: A study of ecommerce sites in Jakarta, Indonesia. *Review of Integrative Business and Economics*, 6(S1), 321–335.

Rizqia, C.D., & Hudrasyah, H. 2015. The effect of electronic word-of-mouth on customer purchase intention, case study: Bandung culinary Instagram account. *International Journal of Humanities and Management Sciences*, 3(3), 155–160.

Rochman, E.A., & Iskandar, B. P. 2015. User's engagement toward the brand accounts in Instagram based on AISAS MODEL. *Journal of Business and Management*, 4(8), 890–900.

Smith, L.R., & Sanderson, J. 2015. I'm going to Instagram it! An analysis of athlete self-presentation on Instagram. *Journal of Broadcasting & Electronic Media*, 59(2), 342–358.

Sotiriadis, M. 2015. Culinary tourism assets and events: Suggesting a strategic planning tool. . *International Journal of Contemporary Hospitality Management*, 27(6), 1214–1232. doi: http://doi.org/10.1108/IJCHM-11-2013-0519

Stone, J.E., Soulard, J., Migacz, S., & Wolf, E. 2018. Elements of memorable food, drink, and culinary tourism experiences. *Journal of Travel Research*, 57(8), 1121–1132.

Ward, J. 2016. A content analysis of celebrity Instagram posts and parasocial interaction. *Elon Journal of Undergraduate Research in Communications*, 7(1).

Sentiment analysis of social media engagement to purchasing intention

A. Wiliam, Sasmoko, W. Kosasih & Y. Indrianti
Bina Nusantara University, Jakarta, Indonesia

ABSTRACT: M-commerce is currently growing rapidly in Indonesia with the reinforcement of social media that encourages customers to be more involved in the products in the market. Businesses and organizations always want to find consumer or public opinions about their products and services. When a business needed public or consumer opinions, it conducted surveys, opinion polls, and focus group. This research aimed to investigate customer engagement with the mobile commerce platform. This study specifically analyzed sentiment analysis trends of different mobile commerce brands using the Brand24 sentiment analysis website application on the popular social media platform Twitter. The results show that consumers in Indonesia can be encouraged through social media engagement in driving m-commerce purchasing intentions.

1 INTRODUCTION

1.1 Research background

Social spending in Indonesia is a driver for consumer behavior, so that social media engagement becomes an important new concept in m-commerce. Social media engagement is realized through the strengthening of social media that encourages customers to be more involved in products on the market (Ibrahim et al., 2017). Engagement is related to a positive psychological condition that comprises both cognitive perception and physical statements that energize one's dedication and absorption of something (Bakker et al., 2008).

Social media such as forums, blogs, and microblogs is now becoming increasingly used for public information sharing and opinions exchange. It has changed the way that the online community interacts and has led to a new trend of engagement for online retailers, especially on microblogging websites such as Twitter. In this study, we investigated the impact of online retailers' engagement with the online brand communities on users' perceptions of brand image and service (Ibrahim et al., 2017). A study found that customers in m-commerce can increase engagement when the needs for efficiency, convenience, personalized service, and convenient operation process are fulfilled (Parker & Wang, 2016).

2 LITERATURE REVIEW

The rapid development of the cellular world has given birth to a new platform in the field of trade and technology, especially for the people of Indonesia. One of them is the birth of m-commerce. Just like the internet and web browser, cellular phones that have data readiness and are connected with digital communication networks will be able to fulfill the basic requirements for m-commerce (Dholakia & Dholakia, 2004). Customer engagement can be seen from two different perspectives, the customer side and the company side (Alexander & Abdellatif, 2018). Conditions in which digital trade interactions are carried out through social media and include customer engagement are interpreted as social media engagement. Purchase intentions and word of mouth are measured with items provided by Zeithaml et al. (1996).

Sentiment analysis is the computational detection and investigation of opinion, sentiments, emotions, and subjectivities in texts (He et al., 2015). As a special application of text mining, sentiment analysis is concerned with the automatic extraction of positive or negative opinions from texts (Ibrahim et al., 2017). Sentiment analysis has been widely used in linguistics and machine learning studies with various classifier and language models (Prager, 2006). Sentiments are closely related to how people behave and different contexts of relationships emerge from them (Ibrahim et al., 2017). In this study, sentiment analysis was chosen because of the strong correlation between m-commerce and social media.

3 RESEARCH METHOD

The research design for this study is displayed in Figure 1. The dataset was obtained by scraping from Twitter. Five leading brands were selected based on brands ranked in the top 10 on the mobile commerce retailer. Tweets were extracted using the Brand24 sentiment analysis website application and were cleaned to remove unnecessary noise and duplications. Sentiment analysis was executed to obtain research insights.

Figure 1. Research design.

3.1 *Data collection*

Tweets were collected using NodeXL Pro (version 1.0.1.362) through Twitter search API for a one-month period from July 26 to 26 August 26, 2019. NodeXL is a program that enables researchers to perform keyword searches to download social media comments. The Twitter Engagement Calculator collected data using the Phlanx web application.

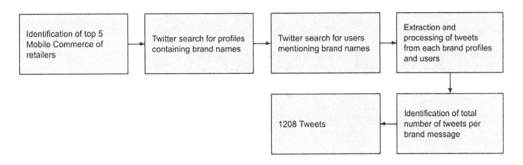

Figure 2. Data collection summary.

4 RESULTS AND DISCUSSION

4.1 Statistical approach to brand sentiment

From an analysis of 1,208 tweets collected for five brands of Indonesian online retailers, sentiment analysis was conducted for each brand, as shown in the detailed analysis in Figure 3.

4.2 Twitter m-commerce brand engagement rate

An analysis was undertaken to examine the overall sentiment of each brand in the existing industry. Expectedly, customers' sentiment on microblogging social media affects the engagement rate of the brand. The highest engagement rate goes to Shopee, with 8.52%. This research also found that tweets from customers can help firms to develop services based on customers' needs. This research also found that negative sentiments can be motivating and

Figure 3. Summary of mobile commerce brand results.

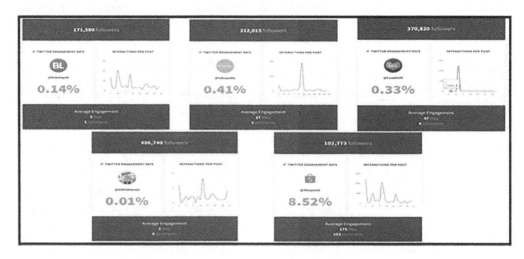

Figure 4. Summary of results of Twitter m-commerce brand engagement.

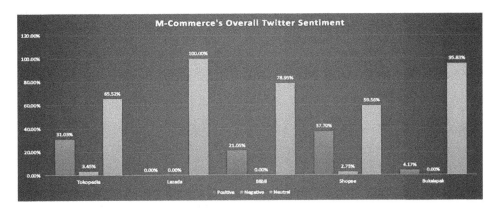

Figure 5. Brand sentiment comparison.

encouraging. This type of tweet helps firms to grow and provide much room for improvement. Social media is at the early stages of development as a social commerce tool; therefore, guidance to brands may be needed as to what drives brand engagement, trust, and purchasing intention in social commerce (Erdoğmuş & Tatar, 2015). The results of social media engagement analysis on m-commerce are developed from the concept of customer engagement (Brodie et al., 2011).

REFERENCES

Alexander, M., & Abdellatif, E. K. M. 2018. Exploring customer engagement marketing (CEM) and its impact on customer engagement behavior (CEB) stimulation. In *10th SERVSIG Conference*, 1–8.

Bakker, A. B., Schaufeli, W. B., Leiter, M. P., & Taris, T. W. 2008. Work engagement: An emerging concept in occupational health psychology. *Work and Stress*, 22(3), 187–200. https://doi.org/10.1080/02678370802393649

Brodie, R. J., Hollebeek, L. D., Jurić, B., & Ilić, A. 2011. Customer engagement: Conceptual domain, fundamental propositions, and implications for research. *Journal of Service Research*, 14(3), 252–271. https://doi.org/10.1177/1094670511411703

Dholakia, R. R., & Dholakia, N. 2004. Mobility and markets: Emerging outlines of m-commerce. *Journal of Business Research*, 57 (12SPEC.ISS.), 1391–1396. https://doi.org/10.1016/S0148-2963(02)00427-7

Erdoğmuş, İ. E., & Tatar, Ş. B. 2015. Drivers of social commerce through brand engagement. *Procedia: Social and Behavioral Sciences*, 207(212), 189–195. https://doi.org/10.1016/j.sbspro.2015.10.087

He, W., Wu, H., Yan, G., Akula, V., & Shen, J. 2015. A novel social media competitive analytics framework with sentiment benchmarks. *Information and Management*, 52(7), 801–812. https://doi.org/10.1016/j.im.2015.04.006

Ibrahim, N. F., Wang, X., & Bourne, H. 2017. Exploring the effect of user engagement in online brand communities: Evidence from Twitter. *Computers in Human Behavior*, 72, 321–338. https://doi.org/10.1016/j.chb.2017.03.005

Parasuraman, A., Zeithaml, V. A., & Malhotra, A. 2005. E-S-QUAL a multiple-item scale for assessing electronic service quality. *Journal of Service Research*, 7(3), 213–233. https://doi.org/10.1177/1094670504271156

Parker, C.J., & Wang, H. 2016. Examining hedonic and utilitarian motivations for m-commerce fashion retail app engagement. *Journal of Fashion Marketing and Management*, 20(4), 487–506. https://doi.org/10.1108/JFMM-02-2016-0015

Prager, J. 2006. Open-domain question-answering. *Foundations and Trends in Information Retrieval*, 1(2),91–233. https://doi.org/10.1561/1500000001

User satisfaction among Malaysian music streamers

A.H. Kaur & S. Gopinathan
Multimedia University, Cyberjaya, Selangor, Malaysia

ABSTRACT: On-demand music streaming is gaining popularity among users worldwide and this is mainly caused by a shift in the music-consumption landscape. The number of users continues to increase yearly and Malaysia is no exception to the trend. This study aims to investigate the role of information quality, system quality, service quality and trustability on user satisfaction among Malaysian music streamers. Trustability was added to the existing information systems success model in order to enhance research findings. A non-probability sampling method was applied and data was collected from 250 respondents via questionnaire. Respondents for this research are Malaysians who stream music on platforms such as Apple Music, Joox and Spotify. The relationships in the research model were then examined and analysed using the Partial Least Squares (PLS) approach. Results revealed that all four variables have a significant positive effect on user satisfaction.

1 INTRODUCTION

Digital music is currently consumed by users in the form of Music as a Service (MaaS) (Dörr & Benlian, 2013), or better known as online music streaming or on-demand music streaming. A shift in the music consumption landscape from physical (cassettes and compact discs) to digital ultimately reflects a shift in information systems as well as a development in the music industry. The increase in number of users for online music streaming platforms has led to studies being conducted by third party organisations to determine user-level satisfaction (J.D. Power, 2016).

2 LITERATURE REVIEW

DeLone & McLean (1992) emphasised that user satisfaction is one of the most significant factors used in measuring the success of information systems. User satisfaction in this context is described as the degree to which users believe the system will satisfy their information needs (Ives, et al., 1983). Amin, et al. (2014) believes that user satisfaction is the user's positive experience or feelings as a whole towards the service provided. On that premise, DeLone & McLean (2003) introduced a widely accepted information systems success model which is popular among researchers in this field. The information systems success model consists of three information systems qualities which have a significant impact on user satisfaction, namely information quality, system quality and service quality.

For this study, trustability was also considered as a potential variable which could affect user satisfaction. Currently, there are no definitive dimensions and factors of trust which researchers could work on, making it open to interpretations (Vance, et al., 2008). From a marketing stand-point, trust can be seen as a gesture whereby consumers are willing to continue purchasing the product or service offered as they are confident and satisfied with the brand. Consumers with a high level of trust are more inclined to pay a premium price for a product, continue to remain loyal to the brand and are willing to share information pertaining to their preferences (Chaudhuri & Holbrook, 2001; Busacca & Castaldo, 2003). That being said, this particular concept of trust was adapted for this research.

At present, no known published journals have developed frameworks on user satisfaction in the context of online music streaming. Therefore, a research model was developed based on past literatures to examine the significance of information quality, system quality, service quality and trustability on user satisfaction towards music streaming platforms.

The following hypotheses were also derived for this study:

H$_1$: *There is a significant positive relationship between information quality and user satisfaction among Malaysian music streamers.*

H$_2$: *There is a significant positive relationship between system quality and user satisfaction among Malaysian music streamers.*

H$_3$: *There is a significant positive relationship between service quality and user satisfaction among Malaysian music streamers.*

H$_4$: *There is a significant positive relationship between trustability and user satisfaction among Malaysian music streamers.*

3 METHODOLOGY

The working population for this study constitutes of Malaysians who are actively streaming music on platforms such as Apple Music, Joox and Spotify. A non-probability purposive sampling method was used for this research. This method does not necessitate underlying theories and does not require a strict number of participants (Ilker, et al., 2016). A questionnaire using the five-point Likert scale was developed using Google Forms and was then deployed via social networking sites such as Facebook, Twitter, Instagram and Reddit. The questionnaire was mainly designed based on past studies conducted by Bailey & Pearson (1983) as well as DeLone & McLean (2003). Data collection was conducted for a time period of one month. A total 250 individuals responded to the questionnaire. Data was then analysed using the Partial Least Squares (PLS) approach. Hair et al. (2017) explains that PLS is capable of evaluating the measurement model and the structural model, all while trying to minimise error variance (Chin, 1998; Gil-Garcia, 2008, Marcoulides, 2013).

4 DATA ANALYSIS AND RESULTS

4.1 *Measurement model evaluation*

A measurement model is generally derived to specify the relationship between the indicators and latent variable (Henseler, et al., 2009). Indicator reliability, internal consistency, convergent validity and discriminant validity are used to test the reliability and validity of the model (Hair, et al., 2011).

Indicator reliability evaluates the extent to which a set of variables is consistent with what it proposes to measure (Urbach & Ahlemann, 2010) and is assessed by referring to factor loadings. Factor loadings higher than 0.70 (as well as loadings which fall in the lower threshold of 0.50 to 0.69) are deemed to be acceptable (Chin, 1998; Urbach & Ahlemann, 2010; Vinzi, et al., 2010). SQ5, US4, US5 and US6 were deleted from the model due to low factor loadings. The deletion resulted in an increase in Average Variance Extracted (AVE) ratings. The factor loadings for this study range from 0.632 to 0.879, and this demonstrates that there is indicator reliability.

Composite reliability (CR) is used to measure internal consistency reliability as it takes into account the different outer loadings of each indicator. Internal consistency reliability is present in the model as the value of CR for each variable is greater than 0.70 (IQ = 0.874, SQ = 0.836, SvQ = 0.851, Trust = 0.893, User Satis = 0.865). Convergent validity, on the other hand, is usually assessed using AVE and takes into account the factor loadings and composite reliabilities as well. Constructs with AVE values of 0.50 or higher signify that the variable in a model is able to explain more than half of its respective indicators on average (Gotz, et al., 2009). Convergent validity is present in this model (AVE IQ = 0.581, AVE SQ = 0.563, AVE SvQ = 0.535, AVE Trust = 0.676, AVE User Satis = 0.682).

Discriminant validity is observed to ensure that the items within a construct do not measure something else by mistake (Urbach & Ahlemann, 2010). Items should illustrate stronger loadings on their own constructs compared to other constructs within the model (Fornell & Larcker, 1981). With each item showing a stronger loading on their own construct (IQ = 0.762, SQ = 0.750, SvQ = 0.731, Trust = 0.822 and User Satis = 0.826), discriminant validity is present.

4.2 *Structural model evaluation*

Structural model allows researchers to systematically evaluate whether the hypotheses are supported by the data attained (Henseler, et al., 2009). A structural model may only be assessed when the measurement model is deemed to be reliable and valid. Table 1 depicts the path coefficient (β), t-statistics and p-values for each path in the model. The path coefficient value has to be at least 0.1 in order to account for an impact in the research model. The t-statistic has to be equal to or greater than 1.645 at a 0.05 alpha level.

Table 1. Beta, T-Statistic, P-Value and hypothesis decision.

Relationship	Path Coefficient (β)	T-Statistic	P-Value	Hypothesis
IQ →User Satis	0.295	4.519	0.000	Supported
SQ →User Satis	0.182	3.152	0.002	Supported
SvQ →User Satis	0.175	2.556	0.011	Supported
Trust →User Satis	0.266	4.237	0.000	Supported

5 DISCUSSION

H_1: *There is a significant positive relationship between information quality and user satisfaction among Malaysian music streamers.*

Easy to understand: Information in this context refers to navigation pane labels, name of playlists and description of content. Timeliness: Users expect the service providers to conduct regular software and application updates in order to constantly receive up-to-date information of the content in the platforms. Accuracy: Retrieval of saved content or playlists. Completeness: Users expect the platform to provide them with complete information of their account (personal details and payment details). Relevancy: Information in the platform is curated based on users' listening and browsing habits.

H_2: *There is a significant positive relationship between system quality and user satisfaction among Malaysian music streamers.*

Availability (eliminated): Users agreed that all three music streaming platforms had a weakness in terms of the type of content made available. Ease of use: Users believe that they are able to seamlessly navigate between the content of each platform. Response time: The loading time and request time (for content) were minimal or non-existent. Performance: Lags and buffering times should be minimal or non-existent for these three platforms. Adaptability: Users are able to navigate within the platform and save the selected content without much hassle.

H_3: *There is a significant positive relationship between service quality and user satisfaction among Malaysian music streamers.*

Customisation: Users prefer to have control over the content which they choose to see in their respective profiles. Reliability: The platforms provide users with the exact service promised in websites, social media and advertisements. Design of system: Each streaming platform has its own aesthetic which pleases different groups of people. Empathy: Users believe that the system takes into account their best interests by providing them with personal offers (Spotify) and notification messages (Joox). Assurance: Regardless of their subscription status, users expect the content to still be made available to them. The only exception in this case is Apple Music whereby non-subscribers are not given access to the music catalogue.

H_4: *There is a significant positive relationship between trustability and user satisfaction among Malaysian music streamers.*

Personal information safety: Apple Music, Joox and Spotify continue to ensure and affirm that the information stored will not be distributed to third parties without the consent of their users. Payment security: Users believe that the payment system is secure as these platforms are trusted names in the industry. Presence of privacy policy: All three platforms are transparent with their privacy policies and these policies are made available on their respective websites. Design quality: Users trust these platforms as the design quality is deemed to be impeccable.

Previously, a similar research on this topic of interest was conducted by the authors. With a sample size of 130 respondents, the results revealed that service quality does not have a significant positive impact on user satisfaction. There was an impression whereby users assumed that service quality meant more of a 'customer service' aspect instead of a more IS-themed aspect. They believed that service quality was not truly important in terms of achieving user satisfaction. For this research, the questionnaire was refined after an extended study, especially in terms of the constructs used in service quality where the SERVQUAL model was heavily applied. This was done with hopes that the constructs reflect more towards the information systems aspect of service quality.

6 CONCLUSION

Significant positive relationships between the independent variables (information quality, system quality, service quality and trustability) and dependent variable (user satisfaction) were established after conducting a thorough analysis using PLS. An additional variable which was not originally from the information systems success model was added to enhance the model and to make the research relevant to the topic in hand. The need to conduct this research was certain as no study on user satisfaction towards music streaming platforms has been conducted, be it globally or in Malaysia. The music streaming landscape might evolve into something better, just as how music consumption formats have changed over the century.

REFERENCES

Amin, M., Rezaei, S. & Abolghasemi, M., 2014. User satisfaction with mobile websites: the impact of perceived usefulness (PU), perceived ease of use (PEOU) and trust. *Nankai Business Review International*, 5(3), pp. 258–274.

Chaudhuri, A. & Holbrook, M. B., 2001. The chain of effects from brand trust and brand affect to brand performance: the role of brand loyalty. *Journal of Marketing*, 65(2), pp. 81–93.

Dörr, J. & Benlian, A., 2013. Music as a Service as an Alternative to Music Piracy? An Empirical Investigation of the Intention to Use Music Streaming Services. *Business & Information Systems Engineering*, 17 May.pp. 383–396.

Giletti, T., 2012. *Why Pay If It's Free? Streaming, Downloading, and Digital Music Consumption in the "iTunes Era"*. London: London School of Economics and Political Science.

Ives, B., Olson, M. H. & Baroudi, J. J., 1983. The measurement of user information satisfaction. *Communications of the ACM*, 26(10), pp. 785–793.

Urbach, N. & Ahlemann, F., 2010. Structural Equation Modeling in Information Systems Research Using Partial Least Squares. *Journal of Information Technology Theory and Application (JITTA)*, 11 (2),p. Article 2.

Vance, A., Elie-Dit-Cosaque, C. & Straub, D. W., 2008. Examining Trust in Information Technology Artifacts: The Effects of System Quality and Culture. *Journal of Management Information Systems*.

Effective tax rate and reporting quality for Malaysian manufacturing companies

S.M. Ali, M. Norhashim & N. Jaafar
Multimedia University, Malaysia

ABSTRACT: Tax incentives are used to reduce overall tax burden, and the excess cash flows can be used to finance growth in capital expenditures and many other investments. Unfortunately, determining a tax incentive's costs and benefits is inherently difficult because it is unknown if the firm would make the investment without the incentives. Review of recent studies has made progress measuring tax incentive effectiveness; however, the findings are inconclusive. This suggests that another explanation is needed for a firm's decisions to take up the tax incentive. This study aims to fill the gap by investigating if the utilization of tax incentive by small and medium-sized enterprises within the manufacturing sector reduces the overall tax burden measured by effective tax rate (ETR) and significantly reduces the exercise of earnings management.

1 INTRODUCTION

1.1 Tax incentive effectiveness

The need to assess the effectiveness of tax incentives, encouraging various scopes of (intended) development, arises from the lack of theoretical soundness and low practical effectiveness of such benefits (Rumina et al., 2015). Without a detailed assessment of the effectiveness of tax incentives, the unsystematic withdrawal of some benefits and the emergence of others, quite often not effective ones, takes place. At present there is no single methodology for assessing the effectiveness of tax incentives in general (Rumina et al., 2015). Therefore, in order to further coordination, currently a great deal of attention should be paid to monitoring of the effectiveness of tax incentive programs.

In Malaysia, the Malaysian Industrial Development Authority (MIDA) will help enhance Malaysia's ability to attract foreign investment and stimulate domestic investment. One of the government initiatives to facilitate this desired growth is to continue introducing tax incentives. More recently, for example, in the 2017 budget proposals, the government introduced a new scheme specifically for assessment years 2017 and 2018 which will benefit small and medium-sized enterprises (SMEs). This scheme will provide a reduction by stages based on a percentage increase in income compared to the previous assessment year. Despite an understanding of tax incentive objectives, determining a tax incentive's costs and benefits is inherently difficult—partly because it is impossible to know the level and mix of economic activity that would occur without the credit although there is large and diverse research.

2 LITERATURE REVIEW

Previous studies show the target economic activities are tax sensitive. For example, Brown and Krull (2008) argued that investment in research and development (R & D) is tax incentive sensitive, suggesting that tax incentives play a role in influencing investment in R&D decisions. These findings suggest that tax incentives are one of the significant determinants of company spending on certain economic activities. Nonetheless, there are also other incentives that may influence these economic activities. For example, managers may tend to cut on R & D spending to boost reported income to a certain level or benchmarks to satisfy certain personal objectives rather than increase the R& D spending to enjoy tax incentive benefits. This notion is consistent with a study by Brown and Krull (2008) suggesting that there are two mainstream studies that examine the relationship between the increase and decrease in R & D investment. First, studies show that firms cut spending on R & D to achieve earnings benchmarks (see, e.g., Bushee, 1998), while the second stream shows tax credits (tax incentives) have an important role in promoting R & D spending. Interestingly, whether these other incentives may jointly influence or trade-off the motivation of managers in their spending decisions with the presence of these tax incentives remains an unanswered empirical question. The answer to this question gives rise to an opportunity to provide alternative measures of tax incentive effectiveness by examining the plausible trade-off and/or joint effects of tax incentives and other incentives, particularly managers' reporting incentives. In other words, tax incentives may not be deemed as effective mechanisms if they fail to mitigate managers' opportunistic reporting incentives and thus fail to meet the national objectives of striving to bring certain economic activities to a certain level and help achieve target growth by private sectors.

There are many efforts to measure the effectiveness of tax incentives in the literature. Most of the studies investigate specific tax incentives and how they encourage specific intended activities. For example, Parys and James (2010) investigated tax incentives in order to encourage private investment in research, development, and innovation (R&D+i) in Argentina. The results imply that the intervention has been effective in increasing firms' innovation efforts. Crespi et al. (2016) found that effects of tax incentives vary depending on the type of investment being subsidized, the industrial sector, and the size of the firm. Nonetheless, if the intended activities may still occur without such incentives is unknown. Further, Rumina et al. (2015) argue that the formation of the modern tax mechanism is not possible without a thorough evaluation of the effectiveness of those tools that the government uses to promote the tax incentives. The problem of developing a system of evaluation criteria of the efficiency of tax benefits—the method of assessment of the effectiveness of tax incentives—should include a list of universal evaluation criteria, such as budget and economic and social efficiency. This study, however, focused on general tax incentive effects, captured by lowered tax burden due to various tax incentives and their association with managerial opportunistic behaviors. This is consistent with Hanlon and Heitzman (2010), who argue that one of the factors that is still under research when it comes to the determination of joint and/or trade-off factors of taxation in influencing optimum investment decisions is that from the financial accounting and reporting stream.

The earliest study that attempted to link accounting and taxation was initiated by Berger (1993), who studied the effectiveness of tax incentives for R&D expenditures, and evidence suggests that R & D spending is sensitive to the rate of tax and credit incentives (see also Klassen et al., 2004) However, under certain conditions, financial reporting incentives have the potential to reduce the response of investment to tax incentives. For example, in the accounting literature there is evidence that companies reduce R&D expenses to increase revenue (e.g., Bushee, 1998).

Further, Hanlon and Heitzman (2010) argue that the effectiveness of the tax incentives should be reflected in business decisions such as business investment, corporate capital structure, and other decision. Under certain conditions, financial reporting incentives are argued to have the potential to reduce the response of investment to tax incentives. In the established accounting literature, there is evidence that companies reduce R & D expenses to increase revenue (e.g., Bushee, 1998). In addition, a recent study by Brown and Krull (2008) intersected financial incentives and tax reporting in investment decisions in R & D. Based on this evidence, it is suggested that tax incentives and the impact on income tax expense for financial accounting reduce real earnings management through R & D and vice versa, that the incentive of financial statements can reduce the effectiveness of tax incentives on real investment.

Because the study of tax incentives involves multidisciplinary streams of study, this study focuses on the perspective of the financial reporting literature. It is argued that divergent reporting incentives for tax and financial reporting purposes lead to empirical studies of trade-offs between tax costs and financial accounting earnings. This study is expected to be able to put bounds on managers' value of incremental accounting earnings because we can measure the cash tax cost incurred to alter the measure of the earnings. Conversely, by understanding how managers balance tax incentives with external reporting incentives, this study can provide evidence on the effectiveness of tax policy. This study aims to investigate if tax incentives influence financial reporting incentives (i.e., managing reported accounting income). Specifically, this study focuses on whether tax incentives reduce both the real earnings management and accrual earnings management and if the presence of the earnings management will negatively affect the investment decisions on capital expenditures. The outcome will provide an alternative basis to evaluate the tax policy framework, specifically within SMEs of the manufacturing sector.

3 DEVELOPMENT OF HYPOTHESES AND TESTING MODEL

3.1 *Hypotheses*

It is argued that in Malaysia, the corporate tax system is used as a one of mechanisms to achieve economic growth (Mahenthiran & Kasipillai, 2012). The changes in tax laws together with the provision of various tax incentives in many forms, such as reduction in statutory tax rate, tax exemptions, deductions, and abatements, are reflected in a company's effective tax rate (ETR) (Mahenthiran & Kasipillai, 2012). ETRs provide a summary statistic of tax burden, describing the amount of tax paid by a firm relative to its gross profit (for example)—a measure that incorporates the use of both tax shelters and incentives (Harris & Feeny, 2003; Rohaya et al., 2010). As the Malaysia government will be absolutely focused on implementing the tax incentives given to various industries and targeted economic activities, it is expected that ETR is lowered as the result. Among others, SMEs and manufacturing sectors are given substantially high incentives packages, which lead to the development of the following hypothesis:

H1: The effective tax rates of SMEs in manufacturing sectors are lower than of SMEs in nonmanufacturing sectors.

The tax incentives will reduce the cost for investment. Therefore, it is also expected that SMEs in the manufacturing sectors will have a significant increase in capital expenditure as compared to SMEs in nonmanufacturing sectors. Therefore hypothesis 2 is developed as follows:

H2: The capital expenditure of SMEs in manufacturing sectors is higher than of SMEs in nonmanufacturing sectors.

Owing to the influences of various tax incentives granted by the government for SMEs in the manufacturing sectors, it is expected that these tax incentives will align management incentives to owners' incentives and therefore reduce the opportunistic incentives by managers. This research question leads to the following testable hypothesis,stated in the alternative:

H3: The discretionary accruals earnings management of SMEs in the manufacturing sectors is different than of SMEs in nonmanufacturing sectors.

3.2 *The overall model*

SMEs = $\int$ (ETR) + (DACC) + (CapEX) + Control Variables (SIZE, LEVERAGE, CORPORATE GOVERNANCE, AUDIT QUALITY)

where
SMEs = Dummy variables for SMEs of (1) and (0) otherwise
DACC = Discretionary accruals detection modified from Dechow and Dichev (2002)
CapEX = The cash flow to capital expenditure ratio, or CF/CapEX ratio, relates to a company's ability to acquire long-term assets using free cash flow. The cash flow to

capital expenditures ratio will often fluctuate as businesses go through cycles of large and small capital expenditures. A ratio greater than 1 could mean that the company's operations are generating the cash needed to fund its asset acquisitions. On the other hand, a low ratio may indicate that the company is having issues with cash inflows and hence its purchase of capital assets. A company with a ratio less than 1 may need to borrow money to fund its purchase of capital assets.

ETR = ETR summarizes various tax incentives (Rohaya et al., 2010). Defined as the ratio of current income tax expense divided by income before interest and taxes.

4 CONCLUSIONS

The results suggest that if tax incentives motivate managers to align the resources accordingly, there will be limits to personal (opportunistic) purposes and therefore lead to a high quality of reporting. Further, they will also give rise to alternate measures of tax incentive effectiveness (in the perspective of financial reporting and internal decision) to evaluate current tax policy and suggest new tax reforms if necessary.

REFERENCES

Berger, P.G. 1993. Explicit and implicit tax effects of the R&D tax credit. *Journal of Accounting Research*, *31*(2), 131–171.

Brown, J.L., & Krull, L.K. 2008. Stock options, R&D, and the R&D Tax Credit. (May).

Bushee, B.J. 1998. The influence of institutional investors on myopic R&D behavior investment. *The Accounting Review*, *73*(3), 305–333.

Crespi, G., Giuliodori, D., Giuliodori, R., & Rodriguez, A. 2016. The effectiveness of tax incentives for R & D + i in developing countries: The case of Argentina. *Research Policy*, *45*(10), 2023–2035.

Dechow, P.M., & Dichev, I.D. 2002. The qualityof accruals and earnings: The role of accruals estimation errors. *The Accounting Review*, *77*, 35–59.

Hanlon, M., & Heitzman, S. 2010. A review of tax research. *Journal of Accounting and Economics*, *50* (2–3), 127–178.

Harris, M.N., & Feeny, S. (2003). Habit persistence in effective tax rates, 951–958.

Klassen, K. J., Pittman, J. A., & Reed, M. P. 2004. A cross-national comparison of R&D expenditure decisions: Tax incentives and financial constraints. *Contemporary Accounting Research*, *21*(3), 639–680.

Mahenthiran, S., & Kasipillai, J. 2012. Influence of ownership structure and corporate governance on effective tax rates and tax planning: Malaysian evidence. *Australian Tax Forum*, *27*(4), 941–970.

Parys, S. Van, & James, S. 2010. The effectiveness of tax incentives in attracting investment: Panel data evidence from the CFA Franc zone. 400–429.

Rohaya, M.N., Nur Syazwani, M.F., & Nor'Azam, M. 2010. Corporate tax planning: A study on corporate effective tax rates of Malaysian listed companies, *1*(2), 189–194.

Rumina, U.A., Balandina, A.S., & Bannova, K.A. 2015. Evaluating the effectiveness of tax incentives in order to create a modern tax mechanism innovation development. *Procedia: Social and Behavioral Sciences*, *166*, 156–160.

The effects of viral marketing on users' attitudes toward JOOX Indonesia

I. Nilasari
Faculty of Business & Management, Universitas Widyatama, Bandung, Indonesia

D. Tricahyono
School of Economics and Business, Telkom University, Bandung, Indonesia

ABSTRACT: This research focused on the Indonesian key elements of viral marketing that drive music streaming users's attitudes toward JOOX Indonesia (a popular music streaming provider in Indonesia). This study collected data by using Google Forms and 373 valid responses as respondents. The factor analysis discovered that from four dimensions of viral marketing (i.e., informativeness, entertainment, irritation, and source credibility) only two dimensions emerged: entertaining message and trusted information as merged factors from the original dimensions of viral marketing. Multiple regression analysis found that both variables significantly affected the JOOX Indonesia users' attitudes.

1 INTRODUCTION

The popularity of music streaming in Indonesia began in 2015, when JOOX (which belongs to Tencent-China) was launched in Indonesia. Before 2015, there were some streaming music services on the market: Langitmusik, Deezer, Guvera, and Apple Music. Since then, the growth of music streaming has been increasing by CAGR of 1.2% between 2013 and 2018 and is expected to grow by 2.9% for the next five years until 2023 (Statista.com, 2019). Today, according to the survey conducted by DailySocial.id (2018), JOOX and Spotify are the top two popular music streaming services in Indonesia, subscribed to by 70.37% and 47.70% of respondents, followed by Langitmusik (28.51%), SoundCloud (19.75%), Apple Music (16.5%), and others.

This study focused on viral marketing for three reasons. First, viral marketing is low cost, efficient, and has a widespread capability for online peer-to-peer communications (Wei, 2014). Second, the targeted customers, which are mostly young people, are the most digitally savvy segments (Gamble & Gilmore, 2013). Finally, viral marketing is a relatively new knowledge (Zernigah & Sohail, 2012). JOOX Indonesia was chosen as a case study because of the popularity of this application among Indonesians. The present study aimed to understand how viral marketing elements will affect user intention to use JOOX.

2 LITERATURE REVIEW

It was Steve Jurvetson, a venture capitalist, who in 1997 used the term "viral marketing" to describe advertisement content attached to the outgoing user mail (Agam, 2017). Since then, viral marketing has been used to describe electronic-WOM (e-WOM) as a new generation of WOM marketing (Gamble & Gilmore, 2013). Subramahi and Rajagopalan (2003, p. 1) noted that "viral marketing or word-of-mouth (WOM) or buzz marketing is the tactic of creating a process where interested people can market to each other." In other words, viral marketing is a type of marketing that works by using networks of anyone who is involved in voluntarily sharing messages (Wei, 2014). In this case, companies do not need to control the entire process (because it automatically spreads among networks' members), except (perhaps) for the content.

Some researchers found a relationship between viral marketing and consumers' intentions to use a product (Zernigah & Sohail, 2012; Wei, 2014; Rukuni et al., 2017; Tricahyono et al., 2018). Ferguson (2008) asserted that viral marketing builds awareness. It means that viral marketing may influence consumers' attitudes toward a product, specifically a consumer's intention to use a product. Among the reasons for the success of viral marketing success are (1) the effectiveness of spontaneous electronic referrals that create awareness, trigger interest, generate sales or product adoption (De Bruyn & Lilien, 2008); (2) driving imagination, fun, and intrigue, encouraging ease of use and visibility, targeting credible sources, and leveraging combinations of technology (Dobele et al., 2016); and (3) low cost and higher profit (Wei, 2014).

3 RESEARCH METHODS

3.1 Research approach

The present study used quantitative research implementing cross-sectional data. The questionnaires were distributed and collected through Google Forms spread to JOOX users in Indonesia. In this case, the snowballing effect could occur. Since there is no exact information about the population, the present study decided to use the rule of thumb of sample size calculation and distribute 500 questionnaires (i.e. "Sample sizes larger than 30 and less than 500 are appropriate for most research" [Sekaran & Bougie, 2016, p. 264]) in the period of January–February 2019 and found 373 valid responses were returned and became a basis for further analysis. The questionnaire was adapted from Rukuni et al. (2017).

The present study conducted face validity to ensure that all questions were valid and also conducted construct validity by calculating Cronbach's alpha. As all the scores are above .70, all dimensions are acceptable, and based on the SPSS version 23 results, all the items are reliable as well (Sekaran & Bougie, 2016). The factor analysis found that out of four original dimensions, only two dimensions were confirmed: entertaining message (contain four items) and trusted information (contain six items). Table 1 shows the result of factor analysis.

Table 1. Factor analysis results.

Items of viral marketing (dimensions): Status	F1	F2
I find e-mail marketing messages of JOOX informative (Infor.): dropped	—	—
Marketing messages displayed on social networking sites for JOOX are informative (Infor.): dropped	—	—
Fan pages on social networking sites for JOOX are informative (Infor.): dropped	—	—
Internet blogs of JOOX are an important source of information (Infor.): on		.767
I find marketing messages received from JOOX via e-mails entertaining (Entert.): on		.779
Joining fan pages of JOOX on social networking sites is entertaining and exciting (Entert.): on		.723
I find the blogs of JOOX entertaining as people share their views about products and services that make them interesting and worth reading (Entert.): on		.759
E-mail marketing messages of JOOX do not irritate me (Irrit.): on	.608	
I believe that unsolicited e-mails from JOOX contain no viruses (Irrit.): on	.751	
Marketing messages on social media sites of JOOX are not irritating (Irrit.): on	.682	
I do not find blogs of JOOX irritating as the information provided is misleading and unreliable (Irrit.)	.587	
I trust the information provided by marketers of JOOX through e-mails (SCredible): on	.746	
Fan pages on social media sites of JOOX are trustworthy (SCredible): on	.684	
I trust the information provided by bloggers of JOOX (SCredible): dropped	—	—

KMO: .788; Bartlett's test of sphericity: approx. Chi square: 1049.577, df: 45, sig.: .000; Cumm variance: 53.119%.

3.2 Framework of thinking and hypotheses

Following Rukuni et al. (2017) and Tricahyono et al. (2018), the present study constructed viral marketing (as IV) into four dimensions. First, informativeness is the usefulness of information that is provided either by the brand owner (firm) or by the users. Second, the entertainment dimension is related to marketing messages that can generate positive attitudes among consumers. Third, irritation of a marketing message affects people because it is offensive, annoying, or manipulative. Marketers use tactics that irritate consumers to grab customers' attention and it creates a positive consumer attitude toward viral marketing. However, if consumers become confuse because of too many messages (i.e., spams), a negative reaction could arise. Finally, source credibility refers to consumers' perceptions of the openness and reliability of marketing messages. People tend to believe the messages that come from a credible source (Wei, 2014).

The present study tried to examine how all four dimensions of viral marketing will influence users' attitudes toward a product. As Dependent Variable, a user's attitude is defined as a user's tendency to use a JOOX music streaming application. Figure 1 shows the original and modified framework of thinking. Two hypotheses were:

H1: Entertaining message will positively influence a user's intention to use JOOX.
H2: Trusted information will positively influence a user's intention to use JOOX.

4 RESULTS AND DISCUSSION

4.1 Characteristics of respondents

Based on 373 responses, some characteristics of the respondents could be discovered. The respondents comprised 57.3% women and 42.7% men. They were predominantly millennials between 18 and 30 years of age (90.2%). They were mostly students (46.1%) and private organization employees (30.6%). In terms of educational background, they were predominantly high school graduates (39.3%) and bachelor's degree holders (48.73%). Interestingly, almost half of the respondents (59.1%) claimed they listen to JOOX every day.

4.2 Hypotheses testing results

The multiple regression analysis found that an entertaining message and trusted information dimensions significantly influenced users' intentions to use JOOX. The R^2 indicates that viral marketing may explain 45.6% of users' intentions to use and the remaining 54.4% is explained by other variables. Table 2 shows the results of multiple regression using SPSS version 23.

The present study supports previous research, for example, that by Rukuni et al. (2017), Tricahyono et al. (2018), Wei (2014), and Zernigah and Sohail (2012). However, the present study discovered that in the Indonesian JOOX case, users tend to focus on trusted information and entertaining message for expressing viral marketing. As a modern tool, viral marketing is

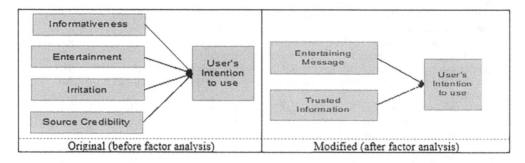

Figure 1. Framework of thinking.

Table 2. Multiple regression results.

	Dependent variable: User's intention to use
Independent variable:	
Entertaining message	.134**
Trusted information	.616**
R^2	.456
Adjusted R^2	.453
F	154.931**

Significance levels: ** $p < .01$; * $p < .05$.

working in getting the attention of new users. If we compare between two dimensions, Indonesian JOOX users tend to choose trusted information more than entertaining message. Indonesian JOOX users react positively if they trust the information and are entertained at the same time. Thus, JOOX should consider using prominent influencers such as YouTubers or famous celebrities as their brand ambassadors to influence more users.

5 CONCLUSION

The present study demonstrates that viral marketing is an effective and powerful modern tool for marketing to the millennial generation that listens to music via streaming. In order to implement viral marketing, the present study suggests hiring prominent influencers such as YouTubers or famous celebrities or perhaps those from a trusted source such as radio or TV programs/actors. The users will tend to use an application when first, they trust the information and second, they feel entertained by the message they receive.

Future research should dig deeper into the role of viral marketing in influencing the attitudes of users, for instance, in applications other than music. Do the users of other applications react the same way as music application users? Another possibility for future research is evaluating the role of communities in conducting viral marketing. As a community is a closed user group, perhaps some unique behaviors could be revealed.

REFERENCES

Agam, D. N. L. A. 2017. The impact of viral marketing through Instagram. *Australasian Journal of Business, Social Science and Information Technology*, 4(1), 40–45.
DailySocial.id. 2018. *Online Music Streaming Survey 2018* [online]. Available at https://dailysocial.id/report/post/online-music-streaming-in-indonesia-survey-2018 (Accessed January 25, 2019).
De Bruyn, A., & Lilien, G. L. 2018. A multi-stage model of word-of-mouth influence through viral marketing. *International Journal of Research in Marketing*, 25, 151–163.
Dobele, A., Beverland, M., Lindgreen, A., & Wijk, R.V. 2016. *Forwarding Viral Messages: What Part Does Emotion Play?* Brisbane, Qld.: Queensland University of Technology.
Ferguson, R. 2008. Word of mouth and viral marketing: taking the temperature of the hottest trends in marketing. *Journal of Consumer Marketing*, 25(3), 179–182.
Gamble, J., & Gilmore, A. 2013. A new era of consumer marketing? An application of co-creational marketing in the music industry. *European Journal of Marketing*, 47(11/12), 1859–1888.
Rukuni, F. T., Shaw, G., Chetty, Y., Kgama, P., & Kekana, P. 2017. Viral marketing strategies and customer buying behavioural intentions at retail store in Johannesburg. *Business Management and Strategy*, 8(1), 58–83.
Sekaran, U., & Bougie, R. 2016. *Research Methods for Business: A Skill Building Approach*, 7th ed. Chichester, UK: John Wiley & Sons.
Statista.com. 2019. *Music Streaming* [online]. Indonesia. Available at: https://www.statista.com/outlook/209/120/music-streaming/indonesia. (Accessed January 28, 2019).

Subramani, M. R., & Rajagopalan, B. 2003. Knowledge-sharing and influence in online social networks via viral marketing. *Communications of the ACM, 46*(12), 300–307.

Tricahyono D., Utami, L. W., & Safitri, W. 2018. The impact of viral marketing on consumer's intention to use. Case Study: Spotify Indonesia. In *Proceeding of the First International Conference on Economics, Business, Entrepreneurship, and Finance (ICEBEF)*. Bandung: UPI.

Wei, S. L. 2014. *The Attitudes of Consumers towards Viral Marketing in Malaysia*. Doctoral dissertation, Universitas Tunku Abdul Rahman Database, Malaysia.

Zernigah, K. I., & Sohail, K. 2012. Consumers' attitude towards viral marketing in Pakistan. *Management & Marketing Challenges for the Knowledge Society, 7*(4), 645–662.

Political connection, internal audit and audit fees in Malaysia

A.A. Saprudin
Multimedia University, Cyberjaya, Malaysia

ABSTRACT: Accounting firms are increasingly under a great deal of pressure to minimize audit costs and deliver services efficiently while improving the quality of their audit services. This study examines the association between political connection (POLCON), internal audit (IA) investment and arrangement with audit fees in Malaysia. Multiple regression was used to test the models and different tests were employed to enhance the rigor of the results produced. The sample comprised of 2209 firms observations listed in the Main Board of Bursa Malaysia, encompassing data from 2011 to 2013. The results show that the majority of the variables exhibit significant association with audit fees. POLCON displays positive significant association with audit fees while both IAs have negative significant association with audit fees. The result of this study contributes to the industry and existing literature by suggesting that the agency problems in POLCON firms can be minimized through effective internal control mechanisms.

1 AUDIT FEES, POLITICAL CONNECTION AND INTERNAL AUDIT

1.1 Audit fees

The recent series of corporate collapses in stiffer competition environment have triggered attention on the viability of financial statement audit and excessive audit fees. A survey conducted by the Financial Executives International in 2015 reported an increased median of 3.1% over the fees paid by the publicly-held companies to their external auditors in the previous year, primarily due to an increase in the amount of work related to mergers and acquisitions and manual controls reviews by Public Company Accounting Oversight Board (PCAOB) (Financial Executives International (FEI), 2015). The increasing pressures are on the accounting firms to provide services in more efficient ways and minimize audit costs while improving the quality of their audit offerings.

Over the years a number of researches have been conducted in examining the determinants of audit fees. Audit fees tend to vary with the size, complexity, riskiness (Gul, 2006; Che-Ahmad & Houghton, 1996; Collier & Gregory, 1996; Craswell, Francis & Taylor, 1995) and other characteristics of the audited entity such as internal control. The relationship between internal control risks and audit fees have been tested in past researches (Munsif, Raghunandan, Rama, & Singhvi, 2011; Hoag & Hollingsworth, 2011; Raghunandan & Rama, 2006). These researches posit that audit fees are significantly higher for organizations that unveil material weakness in internal control. These results also conclude that auditors manage and control risk by exerting more effort in audits of firms with weak internal controls, resulting in increased audit fees.

1.2 Political connection

Past studies have discovered the benefits that a firm may obtain from POLCON as well as from the ruling parties, including lucrative government contracts, larger long-term loans with lower financing costs, tariff protection, lower taxation, higher likelihood of been bailed out by the government in an event of a financial distress and relaxed regulatory oversight (Su &

Fung, 2013; Faccio, 2010; Claessens, Feijen & Laeven, 2008; Faccio, Masulis, & McConnell, 2006). These economic benefits enhance the market valuation of POLCON firms, thus increasing the wealth of the shareholders. However, studies also suggest POLCON firms are engulfed with issues of poor firm governance and transparency (Chen, Li, Su, & Sun, 2011; Chen, Ding, & Kim, 2010; Shumin, 2010; Leuz & Oberholzer-Gee, 2006). The agency problems are deepened when a firm, having close ties with the ruling government is known to have easier access to bank financing.

1.3 Internal audit

Internal audit has become an essential function in assisting organizations accomplish their goals and secure their assets. IA's main role is to independently provide assurance on the operation effectiveness in organization's governance, internal control and risk management processes. External auditors and their audit clients may have a potential avenue to reduce audit costs that is by relying on the works undertaken by the IA function (Felix, Gramling & Maletta, 2001). IA function is integral in governance element with duties to attest and enhance internal control, risk management and other governance mechanisms in an organization. In addition, the scope of job done by IA could possibly cover spaces usually done by the external auditors during their audit engagement. Furthermore, Arena and Azzone (2009) and Coram, Ferguson and Moroney (2008) indicate that there is an upward trend on the awareness of the function and worth of IA and how they can add value in modern organizations. For example, in preparing for the audit engagement, the risk of material misstatement in the client's financial statements is being assessed by the external auditors. They need to comprehend the client's internal controls as it is part of the assessment. Felix et al. (2001) highlight that IA function is in ideal position as an in-house monitoring mechanism of organizational internal controls and risk in assisting external auditors to perform such duties.

2 HYPOTHESES FORMULATION

This study predicts that POLCON firms have higher likelihood to pay higher audit fees as compared to non-POLCON firms. Next, the involvement of IA in heightening controls within the firm could potentially lead to a lower audit risk assessment, hence, implying a negative relationship between IA and external audit fees (Felix et al., 2001; Turpen, 1990). In addition, the involvement of IA in heightening controls within the firm could potentially lead to a lower audit risk assessment, hence, implying a negative relationship between IA and external audit fees (Felix et al., 2001; Turpen, 1990). Hence, the following hypotheses are therefore proposed:

H1: Politically connected firms are more likely to pay higher audit fees.
H2: Firms with higher internal audit investment are more likely to be associated with lower audit fees.
H3: Firms that outsource their internal audit function are associated with lower audit fees.

3 METHODOLOGY

The sample of this study consists of listed firms in Bursa Malaysia Main Board from the years 2011 to 2013 encompassing ten industries. Data on audit fees, internal audit (IA) cost and arrangement, and some control variables such as board size, number of subsidiaries, foreign subsidiaries, audit opinion, big 4 auditor and financial year end are hand-collected from the annual reports based on the list in the Bursa Malaysia's (Malaysian Stock Exchange) database. Other financial data namely size (total asset), return on asset (ROA), leverage and Altman Z-Score are extracted from the Bloomberg and Datastream databases. The data for POLCON firms for the period of 2011 to 2013 in this study follows the definition and list used in the past studies (Fung,

Gul, Radakrishnan, 2015; Bliss & Gul, 2012; Johnson & Mitton, 2003) which define POLCON firms as listed companies that comprise ministers, retired ministers, or any member in the board of directors who have close contact with the government. The main panel regression is based on the unbalanced pooled dataset of 2,772 firm year observations.

4 RESULTS

First, the results indicate that POLCON firms are more likely to pay higher audit fees. This is evident when the results show that POLCON firms are positively and significantly associated with AUDITFEE, which supports H1. This first finding therefore supports the agency theory argument that POLCON firms face severe agency problems due to their opaque financial reporting. The second finding is that internal audit investment is negatively associated with audit fees. The negative relationship implies that higher internal audit investment through the dollar value (RM) of cost invested in internal audit is associated with lower external audit fee paid by the firm. This is consistent with the substitution perspective which suggests that the audit fee is reduced when there is an increase in governance mechanisms. Therefore, H2 is supported. The rationale behind this argument is that, while audit clients/companies are generally concerned with controlling external audit costs, they also have an interest in obtaining the benefit from their IA investment (Aldhizer & Cashell, 1996). Third, this study documents a negative association between internal audit arrangement and audit fees, which supports H3. The significant negative relationship implies that internal audit arrangement through outsourcing is associated with lower audit fees. This suggests that firms that outsource their internal audit function to external parties, which are mainly experts and experienced, can expect to have lesser complex external audit works, ultimately lowering their statutory audit fees. This is in line with the substitution perspective and consistent with prior outsourcing studies such as by Carey, Subramaniam and Ching (2006) which posit that the external provider of internal audit services is more likely to have appropriate and relevant technical competence.

5 CONCLUSION

This study documents strong evidence that political connection has a major impact on the level of expenditure of a public listed company, particularly in audit fees. It provides an indication to firms that even though such connection would invite certain lucrative benefits, it comes with a price; highly prone to agency problems that could potentially affect their audit assessment (i.e. fees and opinion). In addition, findings indicate that the level of internal audit costs can have an influence on the level of audit works to be conducted by the external auditors. In line with the substitution perspective, firms that have significant level of IA investment can expect lesser audit works to be carried out by their external auditors. Therefore, this study should assist organization to allocate a considerable amount of budget for IA because this could lead to some operational cost savings, particularly audit fees. Due to Malaysia's special institutional setting, this study's main findings may not be generalizable for all countries. However, this study's findings may be relevant to certain countries such as Singapore and China, where their governments have strong influence in the financial markets. These countries share some similarity with Malaysia in the sense that funds owned or controlled by the government operate actively in the capital markets.

REFERENCES

Aldhizer III, G. R. & Cashell, J. D. (1996). Internal audit outsourcing. *The CPA Journal*, 66(10), 40.
Arena, M. & Azzone, G. (2009). Identifying organizational drivers of internal audit effectiveness. *International Journal of Auditing*, 13(1), 43–60.

Bliss, M.A. & Gul, F.A. (2012a). Political connection and cost of debt: Some Malaysian evidence. *Journal of Banking and Finance, 36(5)*, 1520–1527.

Carey, P., Subramaniam, N. & Ching, K. C. W. (2006). Internal audit outsourcing in Australia. *Accounting & Finance, 46 (1)*,11–30. Public Company Accounting Oversight Board (PCAOB)2012). An Audit of Internal Control over Financial Reporting Performed in Conjunction with an Audit of Financial Statements. *Auditing Standard No. 2*.

Chen, C.J.P., Ding, Y. & Kim, C. (2010). High-level politically connected firms, corruption, and analyst forecast accuracy around the world. *Journal of International Business Studies, 41(9)*, 1505–1524.

Chen, C.J.P., Li, Z., Su, X. & Sun, Z. (2011). Rent-seeking incentives, corporate political connections and the control structure of private firm: Chinese evidence. *Journal of Corporate Finance, 17(2)*, 229–243.

Che-Ahmad, A. & Houghton, K. A. (1996). Audit fee premiums of big eight firms: Evidence from the market for medium-size UK auditees. *Journal of International Accounting, Auditing and Taxation, 5(1)*, 53–72.

Claessens, S., Feijen, E. & Laeven, L. (2008). Political connections and preferential access to finance. *Journal of Financial Economics, 88(3)*, 554–580.

Collier, P. & Gregory, A. (1996). Audit committee effectiveness and the audit fee. *European Accounting Review, 5(2)*, 177–198.

Coram, P., Ferguson, C. & Moroney, R. (2008). Internal audit, alternative internal audit structures and the level of misappropriation of assets fraud. *Accounting& Finance, 48(4)*, 543–559.

Craswell, A. T., Francis, J. R., & Taylor, S. L. (1995). Auditor brand name reputations and industry specializations. *Journal of accounting and economics, 20(3)*, 297–322.

Faccio, M. (2006). Politically connected firms. *American Economic Review, 96(1)*, 369–386.

Felix, W. L., Jr., A. A. Gramling & M. J. Maletta (2001). The contribution of internal audit as a determinant of external audit fees and factors influencing this contribution. *Journal of Accounting Research 39(3)*, 513–534.

Financial Executives International (FEI) (2015). 2015 Audit Fee Report. Available online at https://www.financialexecutives.org/Research/Publications/2015/2014-Audit-Fee-Survey.aspx.

Fung, S.Y.K., Gul, F.A. & Radhakrishnan, S. (2015). Corporate political connections and the 2008 Malaysian election. *Accounting, Organisations and Society, 43(1)*, 67–86.

Gul, F.A. (2006). Auditor's response to political connections and cronyism in Malaysia. *Journal of Accounting Research, 44(5)*, 931–963.

Hoag, M. L. & Hollingsworth, C. W. (2011). An intertemporal analysis of audit fees and Section 404 material weaknesses. *Auditing: A Journal of Practice & Theory, 30(2)*, 173–200.

Johnson, S. & Mitton, T. (2003). Cronyism and capital controls: Evidence from Malaysia. *Journal of Financial Economics, 67(2)*, 351–382.

Leuz., C. & Oberholzer-Gee, F. (2006). Political relationships, global financing, and corporate transparency: Evidence from Indonesia. *Journal of Financial Economics, 81(3)*, 411–439.

Munsif, V., Raghunandan, K., Rama, D. V. & Singhvi, M. (2011). Audit fees after remediation of internal control weaknesses. *Accounting Horizons, 25(1)*, 87–105.

Raghunandan, K. & Rama, D. V. (2006). SOX Section 404 material weakness disclosures and audit fees. *Auditing: A Journal of Practice & Theory, 25(1)*, 99–114.

Su, Z. Q. & Fung, H. G. (2013). Political connections and firm performance in Chinese companies. *Pacific economic review, 18(3)*, 283–317.

Turpen, R. A. (1990). Differential pricing on auditors initial engagements-further evidence. *Auditing-A Journal of Practice & Theory, 9(2)*, 60–76.

Cyber-entrepreneurial intentions of the Malaysian university students

N. Ismail, N. Jaffar & T.S. Hooi
Multimedia University, Cyberjaya, Malaysia

ABSTRACT: Cyber-entrepreneur has been well regarded as an important career option since the number of unemployed in Malaysia keeps increasing. The youth contributes to the highest number of unemployed. Despite many studies explored on self-entrepreneur, there are only few studies that focused on the readiness and willingness to become an internet-based self-entrepreneur and none using Entrepreneurial Attitude Orientation (EAO) model. Hence, focus of this study is to explore the cyber-entrepreneurial intentions of the Malaysian University students using the EAO scale to measure the students' entrepreneurial attitude. A survey approach is adopted by distributing questionnaires to 2000 students in Malaysia. The results confirmed that self-esteem, innovation and personal control have significant and positive relationships with cyber-entrepreneurial intention. Achievement, on the other hand, is significant but negative relationship. These findings provide an important insight to relevant parties to embed more cyber-entrepreneurial workshop to educate universities students in building their self-confidence and skills.

1 INTRODUCTION

Every year the number of graduates entering the labour market surpassed the number of job opportunities created for them. According to Bank Negara Malaysia's annual report 2018, on average between 2010 and 2017, 173,457 diploma and degree graduates entered the labour market annually but, only 98,514 high-skilled jobs were created over that time. This scenario will give a huge negative impact on the unemployment rate in Malaysia. Based on the Department of Statistics in Malaysia, the number of unemployed workers in general increased from 503.3 thousands in 2017 to 504.3 thousands in 2018. In addition, the duration of unemployed for more than one year increased from 8.7% in 2017 to 9.7% in 2018. Hence, many studies start to explore the possibility to make self-entrepreneur as an important career option. However, few studies focused on the readiness and willingness to become an internet-based self-entrepreneur or cyber-entrepreneur as a career option among students and none using Entrepreneurial Attitude Orientation (EAO) model.

According to Robinson, Stimpson, Heufner, and Hunt (1991), attitude is a good approach to describe entrepreneurship. They have developed EAO scale which is tested to be high in validity and reliability. Ismail, Jaffar and Tan (2013), reported that based on the EAO model, entrepreneurial attitudes such as personal control, self-esteem and innovation have statistically significant impact on self-entrepreneurial intention. However, achievement is found to have no significant impacts on self-entrepreneurial intention. Other study done by Ismail, Jaffar, Khan and Tan (2012) has shown that strong levels of self-entrepreneurial intention are related to low levels of cyber entrepreneurial intention. A common attitude definition for entrepreneurs may also be applied to cyber-entrepreneurs. Individual's intention can be predicted with high possibility. The intention is important compared to interest for commitment and acceptance using of advanced technology (Mamat, Abdullah, Wan Ismail, Muhammad & Samsudin, 2018). Hence, this study is carried out to explore cyber-entrepreneurial intention factors among Malaysian university students.

2 CONCEPTUAL FRAMEWORK AND HYPOTHESIS DEVELOPMENT

2.1 *Entrepreneurial Attitude Orientation (EAO) Model*

This study adopts the Entrepreneurial Attitude Orientation Model to predict the Malaysian university students' cyber-entrepreneurial intention. The EAO Model is developed by Robinson et al. (1991) with the purpose to offer an alternative approach to study entrepreneurship. Many studies have adopted this model but their main focus is on self-entrepreneurial intention (Aloulou 2016, Ismail, et al. 2013). It contains four subscales which are:

i. Achievement in business, referring to concrete results associated with the start-up and growth of a business venture.
ii. Innovation in business, relating to perceiving and acting upon business activities in new and unique ways.
iii. Perceived personal control of business outcomes, concerning the individual's perception of control and influence over his or her business.
iv. Perceived self-esteem in business, pertaining to the self-confidence and perceived competency of an individual in conjuction with his or her business affairs.

2.2 *Hypothesis development and research questions*

H1: Higher levels of achievement in business are associated with higher levels of cyber-entrepreneurial intention
H2: Higher levels of personal control of business outcomes are associated with higher levels of cyber- entrepreneurial intention
H3: Higher levels of self-esteem in business are associated with higher levels of cyber- entrepreneurial intention
H4: Higher levels of innovation in business are associated with higher levels of cyber- entrepreneurial intention

RQ1: How will the levels of achievement in business affects the levels of cyber-entrepreneurial intention
RQ2: How will the levels of personal control of business affects the levels of cyber-entrepreneurial intention
RQ3: How will the levels of self-esteem in business affects the levels of cyber-entrepreneurial intention
RQ4: How will the levels of innovation in business affects the levels of cyber-entrepreneurial intention

3 RESEARCH METHOD

A survey approach is utilized in this study by distributing randomly the questionnaires to 2000 business and non-business degree holders from seven public universities and 11 private universities with each university has the same number of respondents. The questionnaire is developed by modifiying those of Robinson et al. (1991) to suit the focus of this study and the number of items for each variable as stated in Table 1. All the variables in the questionnaire are measured on a 5-point Likert scaling ranging from strongly disagree to strongly agree.

Table 1. Results of factor analysis.

Independent variables	No. of items	KMO	Eigenvalues
Achievement	6	0.834	2.927
Personal Control	5	0.811	2.694
Self-Esteem	5	0.724	2.069
Innovation	3	0.625	1.708
Dependent variable			
Cyber-entrepreneurship Intention	6	0.886	3.971

4 RESULT AND DISCUSSION

In addressing the research questions of this study, only respondents with high self-employment intention will be included. From the total of 1380 usable questionnaires, factor analysis is first performed on the 15 items relevant to the self-employment intention of the university undergraduates and factor score is computed. 664 out of the 1380 respondents are found to have factor score greater than zero and hence are used in testing the relationship between the entrepreneurial attitude of the universities' undergraduates and the cyber-entrepreneurial intention.

4.1 *Factor analysis*

Principal component factor analysis was performed on the four dimensions of entrepreneurial attitudes (i.e. Achievement, Personal Control, Self-esteem and Innovation) to define the underlying structure among the variables in the analysis (Table 1). The values of the Kaiser-Meyer-Olkin (KMO) measure of sampling adequacy for each factor were all above 0.50 while each of the Bartlett's test of sphericity was significant at 5 percent. Hence, the four dimensions in this model were adequate to represent the data. In addition, all the items on every construct were shown to have high factor loadings (> 0.50).

Factor analysis was also conducted to confirm the validity of cyber-entrepreneurship intention (6 items). The result of factor analysis for this intention is also summarized in Table 1. A single factor solution emerged with an eigenvalue greater than 1. The KMO measure of sampling adequacy was 0.886 indicating a sufficient inter-correlation, while the Bartlett's test of sphericity was significant at 5 percent. Most importantly, all items of cyber-entrepreneurship intention were shown to have factor loadings greater than 0.50.

4.2 *Regression analysis*

Table 2 presents the regression estimation of cyber-entrepreneurial intention. With the significant F statistic at 5 percent level (F-statistic = 11.685), the overall model provides a statistically significant relationship between entrepreneurial attitudes and cyber-entrepreneurial intention. Specifically, self-esteem and innovation were positively significant with cyber-entrepreneurial intention at 1 percent significant level. Personal Control, on the other hand, was positively significant at 5 percent significant level. Since a common attitude definition for entrepreneurs may also be applied to cyber-entrepreneurs (Ismail, et al. 2012), the findings are in line with Aloulou (2016) and Ismail, et al. (2013) who studied on self-employment intention. Surprisingly, achievement was found to have a negative relationship with cyber-entrepreneurial intention at 5 percent significant level. Aloulou (2016) found achievement to have a significant and positive relationship on self-entrepreneurial intention whereas Ismail, et al. (2013) found no significant for achievement variable. In summary, hypothesis H2, H3 and H4 are supported but not for the hypothesis H1.

Table 2. Results of regression analysis.

Model	Unstandardized Coefficients		Standardized Coefficients		
	B	SE	β	t	Sig
Constant	-0.001	0.038		-0.029	0.977
Achievement	-0.138	0.065	-0.137	-2.129	0.034**
Personal Control	0.131	0.059	0.130	2.236	0.026**
Self-Esteem	0.180	0.047	0.179	3.813	0.000***
Innovation	0.128	0.047	0.128	2.748	0.006***
R^2	0.067				
Adjusted R^2	0.061				
F	11.685				
Sig. F	0.000				

*** and ** indicate significant at 1% and 5% respectively.

4.3 Conclusions

The results of this study to some extent conform to the past study that entrepreneurial attitudes do have a positive impact on cyber-entrepreneurial intention (Ismail, et al. 2012). Personal control, self-esteem and innovation were positively significant with cyber-entrepreneurial intention whereas achievement was negatively significant with cyber-entrepreneurial intention. These findings indicate that the universities' students who are perceived to have high achievement in business do not intent to be cyber-entrepreneur or to operate own cyber-business. Although achievement has a negative significant relationship on cyber-entrepreneurial intention, it still plays an important factor for universities to educate their students to be self-employed.

REFERENCES

Aloulou, W.J. 2016. Predicting entrepreneurial intentions of freshmen students from EAO modeling and personal background: A Saudi perspective. *Journal of Entrepreneurship in Emerging Economies* 8(2): 180–203.

Ismail, N., Jaffar, N., Khan, S., & Tan, S.L. 2012. Tracking the cyber entrepreneurial intention of private universities students in Malaysia. *International Journal Entrepreneurship and Small Business* 17(4): 538–546.

Ismail, N., Jaffar, N., & Tan, S.H. 2013. Self-employment intentions among the universities' students in Malaysia. *Corporate Ownership & Control* 10(3): 456–463.

Mamat, S.A., Abdullah, Z., Wan Ismail, W. A.A., Muhammad, S., & Samsudin, M. R. 2018. A conceptual development of cyberpreneurs intention among higher education students. *International Journal of Digital Information and Wireless Communications* 8(3): 203.

Robinson, B.P., Stimpson, V.D., Heufner, S.J. & Hunt, K.H. 1991. An attitude approach to the prediction of entrepreneurship. *Entrepreneurship Theory and Practice* 15(4): 13–30.

Financial literacy of the younger generation in Malaysia

Z. Selamat, N. Jaffar, H. Hamzah & I.S. Awaludin
Multimedia University, Malaysia

ABSTRACT: In recent years, there has been a rising interest in the financial literacy of the younger generation among the academic community, financial institutions, organization, consumers, government agencies, and policymakers. This is because financial matters are an important part of everyday life for individuals and an ability to manage finances is essential to success in life. In this context, the purpose of the present study was to investigate the level of financial literacy of the younger generation in Malaysia. The results of the study indicate that the younger generation was severely lacking in financial knowledge. The findings also reveal that females were less financially literate compared to male respondents.

1 INTRODUCTION

Financial literacy means an understanding of personal financial matters. In simple terms, financial literacy is a basic concept in understanding money and its use in daily life. It is a combination of financial knowledge, skills, attitudes, and behaviors necessary to make good financial decisions. In recent years, there has been a rising interest in the financial literacy of the younger generation among the academic community, financial institutions, organization, consumers, government agencies, and policymakers. This is because financial matters are an important part of everyday life for individuals and an ability to manage finances is essential to success in life. Financial literacy enables individuals to improve their level of understanding of financial matters, make informed investment decisions, and minimize chances of being misled.

Unfortunately, according to the Malaysian Insolvency Department (MDL) survey between 2013 and 2017, a total of 100,610 Malaysians were declared bankrupt, 60% of whom were between 18 and 44 years old. In addition, various newspapers (*The Star*, August 4, 2018; *New Straits Times*, February 18, 2013; *Utusan Malaysia*, May 9, 2012; *Berita Harian*, October 28, 2017; *Kosmo*, August 22, 2017) have highlighted the seriousness of the problem and the need to address the causes of personal financial difficulties among members of the younger generation. Thus, the above scenario can be worrying and the government views bankruptcy among the younger generation as a serious problem not only for the individual but also with regard to a loss to the country.

In this context, the purpose of this study was to examine the financial literacy levels of the younger generation and its differences between the genders. It was essential to conduct this study, as it is claimed that financial literacy can strengthen an individual's self-confidence, the capability to manage his or her personal finances, self-control, and independence (Remund, 2010). This study also intended to contribute to the existing literature, providing additional evidence on the levels of financial literacy of the Malaysian younger generation. Results from the study provide some information for financial educators and policymakers seeking to enhance the younger generation's financial literacy and well-being.

2 LITERATURE REVIEWS

Financial literacy can broadly be defined as the capacity to have familiarity with and understanding of financial market products, especially rewards and risks, to make informed choices in economic life. Different researchers and organizations have defined and measured financial literacy in many different ways. The most common definition of financial literacy is the ability to understand finance. More specifically, it refers to an individual's ability to read, interpret, and analyze; manage money; communicate about the personal financial conditions that affect material well-being; compute; develop independent judgments; and take actions resulting from those processes in order to thrive in our complex financial world (Vitt et al., 2000). It includes the ability to discern financial choices, discuss money and financial issues without (or despite) discomfort, plan for the future, and respond competently to life events that affect everyday financial decisions, including events in the general economy. The Organisation for Economic Co-operation and Development (OECD, 2012) defines financial literacy as a combination of financial awareness, knowledge, skills, attitude, and behavior necessary to make sound financial decisions and ultimately achieve individual and societal financial well-being.

Financial literacy skills enable individuals to navigate the financial world, make informed decisions about their money, and minimize their chances of being misled on financial matters (Beal & Delpachitra, 2003). The need for financial literacy has become significant with the deregulation of financial markets and the easier access to credit as financial institutions compete strongly with each other for market share, the rapid growth in development and marketing of financial products, and the government's encouragement for people to take more responsibility for their retirement incomes.

Many studies investigated financial literacy among young people. Among them; Chen and Volpe (1998) examined the relationship between personal literacy and college students' characteristics and the impact of the literacy on students' opinions and decisions. Results showed that participants answered about 53% of the questions correctly. Young women with non-business majors and with little work experience have lower levels of financial literacy. Income and race were not important factors in determining financial literacy. The authors concluded that college students are not knowledgeable about personal finance and this incompetence will limit their ability to make informed financial decisions. This could be an issue because the financial habits students have while in college tend to carry on into adult life. The lack of financial literacy when they leave college may lead to making poor financial choices that can have negative consequences on their financial well-being, whereas the better their financial literacy, the fewer financial hardships they may have in life (Grable & Joo, 1998).

Later on, the study by Marzieh et al. (2013) revealed that age and education are positively correlated with financial literacy and financial well-being. Married people and men are more financially literate. Higher financial literacy leads to greater financial well-being and fewer financial concerns. Finally, financial well-being leads to fewer financial concerns.

Despite the importance of university or college students as a customer segment for financial products and services in Malaysia, research on their financial behaviors is very limited. A study by Sabri and Donald (2010) analyzed the relationship between savings behavior and financial problems and financial literacy among college students in Malaysia. The results of their study found that students with higher financial knowledge were more likely to report savings behavior and also reported fewer financial problems, while students with greater influence from socialization agents and late exposure in their childhood consumer experience were less likely to engage in savings behavior. In another study, Sabri et al. (2012) found that savings habits, financial literacy, and financial socialization agent factors have a significant influence on students' financial well-being levels including their current financial situation and financial management skills. The study also verified that there were important differences between the Malay and Chinese ethnic groups in Malaysia.

Mahdzan and Tabiani (2013) examined the relationship between financial literacy and individual saving in Malaysia. The results indicated that the level of financial literacy had a significant, positive impact on individual saving. In addition, saving regularity, gender, income, and educational level positively influenced the probability of saving.

3 METHODOLOGY

This study focused on measuring the financial literacy of the younger generation in Malaysia. A questionnaire was developed to gather the required data. The questionnaire was replicated from that adopted by Lusardi et al. (2010) with some amendments made to suit the purpose of the study. The first section of the questionnaire comprised questions relating to the financial literacy of the respondents, which was assessed by examining their understanding of the basic principles of interest rate, inflation, and the concept of diversification. The second section sought sociodemographic information, including information on gender, age, and number of years of study.

The targeted population of this study was undergraduate students between 18 and 30 years old studying in bachelor programs of the Faculty of Management Multimedia University (MMU), Malaysia. Questionnaires were distributed personally to randomly approached students. From 150 questionnaires distributed, only 115 could be collected and 9 were incomplete. Therefore, the response rate for this study was 71%.

4 FINDINGS

Data were analyzed using descriptive statistics. Of the 106 students who responded to the survey, 43.4% (46) were males and 56.6% (50) were females. A larger percentage of female respondents is consistent with the national overall population averages. Most of the respondents were Malay (45.3%), followed by Chinese (34%), other ethnicities (11.3%), and Indian (9.4%). The highest percentage of respondents were in the age group of 22–25 years (59.43%), followed by 18–21 years (38.69%) and 26–30 years (1.88%). The majority of the students, 37.7%, were from the first year, 33% from the second year, 25.5% from the third year, and 3.8% from the fourth year of study.

Table 1 reports result from the three questions that measured respondent levels of financial literacy. Concerning knowledge of the interest rate, 60.4% of respondents answered the question correctly. As for inflation, 34.9% answered correctly, 42.5% answered incorrectly, and 22.6% responded that they did not know the answer to the inflation question. Only 33.33% answered the risk diversification question correctly, and 52.37% respondents answered incorrectly. Considering that the questions were basic and simple, the low correct response rates, particularly to the inflation and risk diversification questions, suggest that university students' financial literacy level is inadequate. According to Lusardi et al. (2010), the high "don't know" response rate was particularly troubling, as it reflects that the respondents with "don't know" answers have a very low level of financial knowledge.

While the overall level of financial knowledge was low among the young, there were significant differences in financial literacy between males and females. The study found that females were less likely to respond correctly to the interest and risk diversification questions. This result is consistent with the study of Chen and Volpe (1998), who found that male students are more knowledgeable in financial matters than female students.

Table 1. Levels of financial literacy.

Responses	Correct	Incorrect	Don't know
Interest rate	60.4	27.3	12.3
Inflation	34.9	42.5	22.6
Risk diversification	33.33	52.37	14.3

Table 2. Differences between genders.

Gender	Interest rate	Inflation	Risk diversification
Male	67.39	30.44	34.78
Female	55	38.33	31.67

5 CONCLUSION

The main purpose of this study was to examine the level of financial literacy among the younger generation in Malaysia. The results of the study indicate that members of the younger generation were severely lacking in their financial knowledge and therefore financial education is vital to improving their knowledge of personal finance. Further, females were found to be less financially literate compared to male respondents. These results reveal the need for universities to pursue a sound financial literacy program that is tailored according to the characteristics, individual preferences, and learning styles of the students. The training format should be kept simple and relevant to the participants' day-to-day lives. Emphasis should be on learning by doing and linking to the benefits and knowing how to use available financial products and services. In addition, financial education should be made mandatory in continuous professional development and personal growth programs to empower and equip the younger generation with the knowledge and skills that are necessary to make informed financial decisions in their lives and build a more secure future for themselves and their families.

The findings of this study have to be seen in the light of some limitations. The study was limited to only MMU undergraduate students aged between 20 and 30 years. Undergraduate students are differing from other youths who have already started their working lives and have certain attitudes and behaviors that may vary from those of other people as well; therefore the results cannot be generalized. Moreover, further research needs to be conducted using survey research on other institutes of higher learning, both public and private universities in Malaysia.

REFERENCES

Beal, D.J., & Delpachitra, S.B. 2003. Financial literacy among Australian university students. *Economic Papers: A Journal of Applied Economics and Policy*, 22, 65–78.

Chen, H., & Volpe, R. P. 1998. An analysis of personal financial literacy among college students / *Financial Services Review*, 7(2), 107–128.

Grable, J. E., & Joo, S. 1998. Does financial education affect knowledge, attitudes, and behavior? An empirical analysis. *Personal Finances and Worker Productivity*, 2(2), 213–220.

Lusardi, A., Mitchell, O.S., & Curto, V. 2010. Financial literacy among the young. *Journal of Consumer Affairs*, 44(2), 358–380.

Mahdzan, N.S., & Tabiani, S. 2013. The impact of financial literacy on individual saving: An exploratory study in the Malaysian. *Transformations in Business & Economics*, 12(1[28]), 41–55.

Marzieh, K.T., Zare Z.H., Seyyed, M.T.M., & Abdoreza, R. 2013. The relation between financial Literacy, financial wellbeing and financial concerns. *International Journal of Business and Management*, 8(11), 63–75.

OECD INFE 2012. *OECD/INFE High-Level Principles on National Strategies for Financial Education*. Paris: OECD Publishing.

Remund, D. L. 2010. Financial literacy explicated: the case for a clearer definition in an increasingly complex economy. *The Journal of Consumer Affairs*, 44(2), 276.

Sabri M.F., Cook C.C., & Gudmunson, C.G. 2012. Financial well-being of Malaysian college students. *Asian Education and Development Studies*, 1(2), 153–170.

Sabri, M. F, & Donald, M. M. 2010. Savings behavior and financial problems among college students: The role of financial literacy in Malaysia. *Cross-Cultural Communication*, 6(3), 103–110.

Vitt, L. A., Anderson, C., Kent, J., Lyter, D. M., Siegenthaler, J. K., & Ward, J. 2000. *Personal Finance and the Rush to Competence: Financial Literacy Education in the U.S.* Middleburg, VA: Fannie Mae Foundation.

Financial performance evaluation for network facility and service providers

S. Segaran & L.T.P. Nguyen
Multimedia University, Cyberjaya, Selangor, Malaysia

ABSTRACT: Growing trend of broadband penetration in the past few years has led to significant increase in the number of licensees in the Malaysian telecommunication industry. However, the cancellation rate of the Individual licenses are equally on the rise since 2007. This put the country's telecommunication industry at a risk as Malaysia may not achieve its national digital economy agenda by 2020. Thus, this study examines: (1) the extent to which the telecommunication industry has contributed to the Malaysian economy, and (2) whether poor financial performance is the main reason that led to the rise of the license cancellation. Using financial statements of 6 firms from 2012 to 2016, regression analysis was carried out. Results obtained show that the firms contributes significantly to GDP of this particular industry. This implies that the overall GDP is contributed by the exit of the firms' holding the Individual licenses.

1 INTRODUCTION

The telecommunication (Telco) industry has contributed 6.1% of revenue to Malaysia's Gross Domestic Product (GDP). There was another 1% increase in the GDP in 2010 where a total of 135,000 jobs were newly created in the information, communications and technology (ICT) sector following the successful deployment of High Speed Broadband (HSBB) and Broadband to the General Population (BBGP) (Annual Report 2008, MCMC). The overall contribution of the ICT to the national economy in 2017 is 18.3% with a registered a value of RM247.1 billion. ICT GDP accounted for 13.2 % in 2017 with a growth of 8.4%. (ICTSA 2017, 2018).

The trend of jobs creation by the ICT sector appear to be quite stable and has seen an increase of approximately 28.57% since the year 2008 (ICTSA 2016 (2017) & ICTSA 2005-2013 (2014), Department of Statistics Malaysia). In 2017, the employment of the ICT industry increased to 1.09 million and contributed to 7.6% of total employment in Malaysia (ICTSA 2017, 2018).

Despite the growing number of licensees in the Telco sector, not much firms have grown as steadily as the incumbent fixed and mobile operators to provide service provisioning to the end users. The growth of the industry is reflected in the number of issuance of Individual licenses i.e. the Network Facilities Provider, NFP(I) and Network Service Provider, NSP(I) by the Malaysian Communications Multimedia Commission (MCMC) (Register of Individual Licenses, MCMC). The issuance of NFP(I) and NSP(I) licenses are on the rise and popular as the licenses are essential to deploy the relevant network facilities and services in the Telco industry.

Notwithstanding the increase in the issuance of Individual licenses, it is observed that the number of licensees whose Individual licenses are being cancelled by the MCMC have also been correspondingly increasing as shown in Figure 1. Firms mostly fail to sustain their presence in the industry due to non-payment of annual license fees to the MCMC, and/or inability to meet the forecasted deployment plan. The upward trends of cancellation of Individual licenses and the reasons behind the cancellation strongly implies that there are some challenges

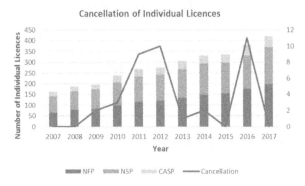

Figure 1. Cancellation of individual licenses by MCMC.

in the Telco industry which must be acknowledged to ensure the sector's continuous growth and contribution to GDP of the nation.

1.1 *Research questions*

The study raises the following research questions: (1) How do the NFP(I) and NSP(I) licensees perform in terms of their financial status?; (2) To what extent does the financial performance influence the cancellation of the NFP(I) and NSP(I) licenses by the MCMC?; and (3) What is the impact of the cancellation of NFP(I) and NSP(I) licenses towards the growth of the Telco industry in Malaysia?

2 LITERATURE REVIEW

2.1 *Licensing framework*

The NFP(I) license is required by a company that owns and/or deploys facilities in respect to network which spreads from submarine cables to towers, earth stations and other communication elements such as optical fibre cables and transmitters for radio communication purposes. It does also include satellite communication elements. The licensable services via the NSP(I) license encompasses bandwidth management, broadcasting distribution, cellular mobile services, access applications, space, switching and gateway related. The licensee will be able to provide connectivity and bandwidth to provision a range of applications. The NSP(I) licensee usually owns or deploys the network facilities or may provide service using the network facilities owned by a different licensee who holds the NFP (I) (Licensing Guidebook, 2015). The research focuses on the NFP(I) and NSP(I) license because the licenses are required to achieve the present government initiatives to drive the digital economy era, namely, the proliferation of Internet of Things to establish the Smart Sustainable City (SSC) initiative in Malaysia.

2.2 *Past models*

Kiew (2017) examined the corporate governance for Digi Telecommunications Berhad by analyzing the firm's risk and financial performance. An analysis on the firm's financial ratios, i.e. return on assets, return on equity, current ratio, profitability, liquidity and leverage was undertaken for 5 years (2011 - 2015). The study provides substantiation of the important relationship between profitability and liquidity and how these ratios will give a great weight to Digi group of companies and allow the financial performance of the company to become more stable (Kiew, 2017). Kiew (2017) supported the suggestion that a research based on the profitability and liquidity ratios ought to be carried out before concluding liquidity positions and the relationship between the firms' performances and business hazard of Digi companies. The result with the highest t-value was GDP, hence this variable impacts much on profitability measurements i.e. return on common stockholders' equity (ROE). The results show that an effective of ROE have significant effect towards the GDP, hence the firm's focus into profitability should take precedence starting 2015 besides the GDP and liquidity. The research achieved a significant relationship between the profitability and the GDP of the company. The implication of the research is that in principle, the indicators of firm's performance provides an overview of liquidity and profitability position. Kiew (2017)'s approach was also embraced by Indian researchers, Khan and Safiuddin (2016), where both studied the liquidity and profitability ratios to measure the financial performance and identify the financial health status of two leading telecom companies in India, Bharti Airtel and Vodafone India from 2011 to 2015 and 2010 to 2015.

3 METHODOLOGY

3.1 Research framework

The study measures the firm's performances based on the financial results obtained from the annual report of the selected NFP(I) and NSP(I) licensees. Similar to the study conducted by Kiew (2017), this research suggests the dependent variable to be the firm's financial performance which will be calculated via the profitability ratios, i.e. return on assets (ROA) and return on common stockholder's equity (ROE). The independent variables established for testing will be the other key financial ratios associated with liquidity and solvency. Additionally, cash and cash equivalents and the firm's total equities influence towards monetary position of firms will be included to monitor the effect of a firm's financial performance. The research aims to expand Kiew (2017)'s work by including profitability ratios, i.e. profit margin, ROA and ROE as the dependent variables and four new solvency ratios, namely, the long-term debt to total capitalization ratio, debt to equity ratio, long term debt to equity ratio and the total debt to asset ratio as the independent variables. The measure for liquidity ratio for this research is current ratio.

3.2 Data and sample selection

The list of NFP(I) and NSP(I) licensees in the respective registers maintained by the MCMC is used as the secondary data. The research will focus on six firms namely, Telekom Malaysia Berhad ('TM'), Maxis Berhad ('Maxis'), Digi. Com Berhad ('Digi'), Celcom Axiata Berhad ('Celcom'), TT dotCom Sdn Bhd ('TT dotCom') and U Mobile Sdn Bhd ('U Mobile') from the years 2012 to 2016 for research question 1 and 1 other firm, Jaring Communications Sdn Bhd ('Jaring') to address research question 2. For Research Question 3, the data of the number of firms that exited from the industry (Register of Individual Licenses, MCMC) from the year 2008 till 2016 and the GDP data acquired from the report published by Department of Statistics Malaysia were used. The Multiple Linear Regression (MLR) analysis were conducted via SPSS to test the impact of financial performance for the selected licensees.

4 RESULTS AND DISCUSSION

4.1 Multiple Linear Regression ('MLR') analysis

The hypothesis for significance of models for the profit margin, ROA and ROE is provided as Ho: The model is not significant; and H1: The model is significant. The research indicated that the model is significant for all 3 independent variables tested and that there are no multicollinearity issues since all of the VIF values are below 10 for all variables. The significance of the model for profit margin, ROA and ROE is provided by p-value = 0.000 < 0.05. The research therefore, reject null hypothesis, Ho, and conclude that the models are significant indicating at least one of the independent variables entered can be used to model the profit margin, ROA and ROE variables. The Adjusted R-Square for profit margin = 0.551. 55.1% of the variation in profit margin can be described by the variation in the total debt to asset ratio, debt to equity ratio and cash and cash equivalents variables. For profit margin, the cash and cash equivalents is the most significant independent variable for profit margin with p-value = 0.000 < 0.05. The regression model is given as Profit Margin = 0.37 + 5.125E-09 (Cash and cash equivalents) + 0.117 (Debt to equity ratio) − 0.754 (Total debt to asset ratio). The Adjusted R-Square for ROA = 0.545. 54.5% of the variation in ROA can be described by the variation in the total debt to asset ratio and total equity variables. The model consisting of total debt to asset ratio and total equity variables are the most significant independent variables for ROA with p-value = 0.000 < 0.05 and p-value = 0.002 < 0.05. The regression model is given as ROA = 1.129 − 1.908 (Total debt to asset ratio) − 3.918E-11 (Total equity). Since the total equity's increase is extremely minimal, the research concludes that the total debt to asset ratio is the most significant independent variable to model the ROA variable. The Adjusted R-Square for ROE is 0.932. 93.2% of the variation in ROE can be described by the variation in the total debt to asset ratio, debt to equity ratio and cash and cash equivalents variables. The total debt to asset ratio and debt to equity ratio variables are the most significant independent variables for ROE with p-value = 0.000 < 0.05 and p-value = 0.000 < 0.05. The regression model is given as ROE = − 1.411 + 18.141 (Total debt to asset ratio) − 4.332 (Debt to equity ratio) + 3.047E -09 (Cash and cash equivalents).

4.2 Ratio analysis: Jaring

The ratio analysis for Jaring that exited the industry via cancellation is provided in Table 1.

4.3 Correlation analysis & MLR to test impact of the cancellation

The independent variable for this analysis is the number of NFP(I) and NSP(I) licensees that were cancelled whereas the dependent variable is the GDP by Kind of Economic Activity (Communications).The correlation test results showed that debt to equity has a strong correlation of 0.762 indicating that the number of NFP(I) and NSP(I) licensees that exited the market have a significant relationship with the GDP contribution by the Telco industry. The Adjusted R-Square is 0.520. The hypothesis is provided as H_o: The model is not significant and H_1: The model is significant. The significance of the model is shown by p-value = 0.017 < 0.05, thus, the research reject null hypothesis, H_o, and conclude that the model is significant. The research also found that the number of cancellation of the NFP(I) and NSP(I) licensees as a significant independent variable for the GDP by kind of economic activity (communications) with p-value = 0.017 < 0.05.

Table 1. Profitability and solvency ratios of Jaring.

Year	Profit Margin	Return on Assets	Return on Common Stockholders' Equity	Total Debt-Asset Ratio	Debt-Equity
2009_Jaring	(0.08)	(0.07)	(0.14)	0.83	4.84
2010_Jaring	(0.07)	(0.09)	(0.12)	0.72	2.54
2011_Jaring	(0.30)	(0.45)	(0.37)	0.97	32.11
2012_Jaring	(0.30)	(0.41)	(0.27)	0.92	10.84
2013_Jaring	(0.23)	(0.32)	(0.15)	0.90	9.48

5 CONCLUSION

Overall, the results indicated that there is no relationship between profitability ratios with the liquidity ratios although a strong relationship is observed between profitability and solvency ratios. The research concludes that the firms' ability to meet its relative claim of creditors in the short term does not necessarily affect the profitability measures of the firms. The financial performance of firms impacts cancellation of firms' licenses. The key financial ratios for Jaring showed continuous losses for its profitability and high solvency ratios (close to 1 and > than 1). The role of finance is thus, a significant contributing factor behind the sluggish deployment by the firms that failed in the industry. Finally, it can be observed that there is an adverse impact to the overall GDP of the country should there is a decrease of percentage share by the Telco industry caused by the exit of the firms' holding NFP(I) and NSP(I) licenses.

REFERENCES

Annual Report 2008. *Malaysian Communications Multimedia Commission (MCMC)*.
Information and Communication Technology Satellite Account (ICTSA) 2005 – 2013 (2014), ICTSA 2016 (2017) and ICTSA 2017 (2018). *Dept. of Statistics, Malaysia*.
Khan, M.M. and Safiuddin, S.K., (2016). Liquidity and Profitability Performance Analysis of Selected Telecoms Companies. *International Journal of Research in Regional Studies, Law, Social Sciences, Journalism and Management Practices (AIJRRLSJM), Vol. 1 (8).p.*365–376.
Kiew, S.Y., (2017). Firm Risk and Performance: The Role of Corporate Governance of Digi Telecommunication Berhad. *MPRA Paper No.78313 (2017)*. p.1–18.
Licensing Guidebook (2015). *MCMC*. Register of Individual Licenses. *MCMC*. Retrieved on 20th December 2017 from https://www.mcmc.gov.my/legal/registers/cma-registers.

Nonlinear impact of institutional quality on economic performance within the comprehensive and progressive agreement for trans-pacific partnership

C.Y. Chong
Faculty of Management, Multimedia University, Cyberjaya, Selangor, Malaysia

ABSTRACT: Institutions are the fundamental determinants of economic growth. Therefore, a high institutional quality of among the Comprehensive and Progressive Agreement for Trans-Pacific Partnership (CPTPP) countries will be important to ensure continuing economic performance after the withdrawal of the United States from the TPP Agreement. This study aimed to investigate how institutional quality can improve the economic performance of member countries within the CPTPP. Unbalanced panel data for 10 members of CPTPP (without Brunei) covering the years 2002–2015 were analyzed using the Least Square Dummy Variable (LSDV) model. An inverted-U shape of the relationship was detected, implying a nonlinear impact of institutional quality on economic performance. Better overall quality of institutions helps to boost economic performance. However, a reverse relationship was found when institutional quality surpasses the level of 0.638. Therefore, CPTPP countries can enhance their economic performance if their institutional quality is kept below the threshold value.

1 INTRODUCTION

The Trans-Pacific Partnership Agreement (TPPA) was originally signed by 12 different countries in 2016 with the intention of deepening trade integration through reduced tariffs. After the withdrawal of the United States from the TPPA, the remaining partners continued with the Comprehensive and Progressive Agreement for Trans-Pacific Partnership (CPTPP) in March 2018. The sustainability of CPTPP depends on continuing economic growth, and institutions are one crucial determinant of economic growth.

Institutional quality (IQ) covers performance in many aspects, and it should not be measured only in one aspect. Worldwide Governance Indicators (WGIs) from the World Bank measure governance performance in six key dimensions including voice and accountability, political stability and absence of violence, government effectiveness, regulatory quality, rule of law, and control of corruption. However, IQ indicators found in most of the empirical studies focused on a single indicator. With a more comprehensive composite index, we can investigate the relationship between economic growth and IQ, which includes multiple aspects. In this definition, a high IQ can promote the country's stability, reduce corruption, and ensure efficient allocation of resources in productive economic activities.

Theoretically, there is a positive impact of IQ on economic growth. However, some empirical studies that focused on one aspect of IQ reported evidence of a nonlinear impact on economic growth (Barro, 1996; Aidt et al., 2008; Marakbi & Turcu, 2016). As a result, this study aimed to examine the relationship between economic performance and IQ using a more comprehensive measure. In addition, this study investigated the presence of nonlinearity in the relationship between these two variables.

The rest of this article is organized as follows: Section 2 discusses the relevant literature reviews. The research methodology and data adopted in this study are explained in Section 3. Section 4 reports the findings, followed by some concluding remarks in the last section.

2 LITERATURE REVIEWS

Most of the available empirical studies revealed a positive impact of IQ on economic growth. Gwartney et al. (2004) disclosed that economic freedom positively influenced economic growth. Butkiewicz and Yanikkaya (2006) explained that economic growth can be improved by the maintenance of the rule of law. However, they did not find a significant impact of democracy on economic growth. Positive impacts can be found only when certain estimation techniques were adopted. Studies conducted by Acemoglu et al. (2001) (using 3 IQ indicators) Valeriani and Peluso (2011) (using 3 IQ indicators), and Vergil and Teyyare (2017) (using 12 IQ indicators) revealed the same conclusion, that better institutions promote economic growth.

On the other end, Chong and Calderon (2000), who used the BERI Index as an IQ indicator along with Williams (2006), using a newly developed composite index, revealed a two-way causal effect between IQ and economic growth. In addition, Nawaz et al. (2014) reported a positive impact of IQ on economic growth when the World Bank WGI was adopted. They further explained that IQ is more effective among the developed Asian countries compared to developing Asian countries.

Barro (1996) reported a nonlinear relationship between democracy and economic growth. Democracy can boost economic growth, but an adverse impact was observed once democracy surpassed a moderate level. Some studies reported nonlinearities in the corruption–growth relationship as well (Aidt et al., 2008; Marakbi and Turcu, 2016). Corruption showed a negative impact on economic growth among countries with higher IQ whereas corruption had no impact among countries with low IQ.

3 METHODOLOGY AND DATA

In order to control for significant country effect and time effect in addition to a small sample size of CPTPP in this study, the Least Square Dummy Variable (LSDV) model was adopted. The linkage of IQ and economic performance was first examined using the following linear growth equation:

$$y_{it} = \alpha + \beta_1 IQ_{it} + \beta_2 X_{it} + \varepsilon_{it} \qquad (1)$$

where y_{it} refers to the logarithm of real gross domestic product (GDP) per capita while IQ refers to the institutional quality indicator. The vector of standard control variables is represented by X_{it}. Lastly, the error term was captured by ε_{it}. A significant positive β_1 implies a positive impact of IQ on economic performance.

On the other hand, a nonlinear relationship between IQ and economic performance can be examined by including a squared term of IQ into equation (1) as follows:

$$y_{it} = \alpha + \beta_1 IQ_{it} + \beta_2 IQ_{it}^2 + \beta_3 X_{it} + \varepsilon_{it} \qquad (2)$$

A U-shaped relationship between IQ and economic performance is found if both the IQ and IQ^2 are significant, with a negative and positive sign respectively. On the contrary, an inverted U-shaped of relationship is found when both the IQ and IQ^2 are significant with a positive and negative sign respectively. Nonlinearity can also be tested using the U-test proposed by Lind and Mehlum (2010).

An unbalanced panel of data from 2002 to 2015 was collected from the database of the International Monetary Fund, United Nations Statistics Division (UNSD), and the World Bank. Of the remaining 12 member countries of CPTPP, Brunei was excluded in this study because of the unavailability of data. Economic performance is measured by real GDP per capita in US dollars, at the constant rate of year 2010. Institutional quality is measured by the average of six indicators from the World Bank Worldwide Governance Indicators (WGIs).

Each indicator ranges from the value of –2.5 (weak governance performance) to 2.5 (strong governance performance). The control variables in the growth equation included initial real GDP, population growth rate, market capitalization of listed domestic companies (financial development indicator), GDP deflator (inflation indicator), life expectancy (human capital indicator), and trade in percentage of GDP (trade openness indicator). All these variables were logarithmically transformed in except for the inflation indicator because of possible negative values in the sample data.

4 FINDINGS

An overview of overall IQ is presented in Figure 1. A decreasing trend can be observed until 2008 and remained at a lower level below 1. These figures indicate that the IQ measured by governance performance among the CPTPP countries was at a moderate level throughout these years.

The impact of IQ on economic performance among these countries is reported in Table 1. Within the linear model specification, IQ was found to show a negative impact on economic performance among CPTPP countries, but it is not significant. However, in the nonlinear model specification, both the IQ and IQ^2 are significant, showing a positive and negative sign

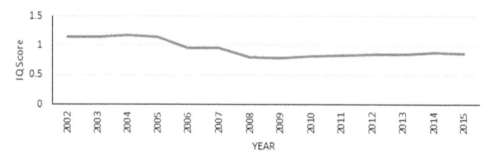

Figure 1. Overall institutional quality within the CPTPP.
Source: World Bank

Table 1. Results of the LSDV model for both linear and nonlinear specifications.

	Model specification	
Variables	Linear model	Nonlinear model
IQ	–0.0019 (0.0077)	0.0213* (0.012)
IQ^2	—	–0.0167** (0.0082)
Initial income	0.9055*** (0.0312)	0.8762*** (0.0372)
Population growth	–0.0030* (0.0017)	–0.003* (0.0016)
Financial development	0.0232*** (0.0086)	0.0311*** (0.0092)
Inflation	0.000002 (0.0003)	0.0001 (0.0002)
Human capital	–0.1075 (0.8555)	–0.0781 (0.8222)
Trade openness	–0.0079 (0.0165)	–0.0123 (0.0164)
Constants	0.6252 (1.5561)	0.7016 (1.4674)
R^2	0.9998	0.9998
Lind–Mehlum U-test statistics	—	1.85**
Threshold IQ	—	0.638

Significance level at *10%, **5%,***1%. Standard errors are reported in the parentheses.

respectively. These findings implied an inverted U-shape of the relationship between IQ and economic performance. The Lind–Mehlum U-test asserted the nonlinearity, with the threshold level of IQ identified at a score of 0.638. This threshold level explains that increasing overall IQ of the CPTPP helps to improve their overall economic performance before reaching the IQ score of 0.638. An adverse impact of IQ will be observed if overall IQ surpasses the threshold level.

5 CONCLUSIONS

This study aimed to examine how IQ helps improve economic performance among CPTPP countries. Most of the previous empirical findings supported that better IQ promotes economic growth. However, this study differs from others by uncovering a nonlinear impact of overall IQ on economic performance within the CPTPP. Continuing to improve overall IQ will not continuously enhance economic performance within the CTCPP. An adverse effect will be experienced if the overall IQ goes beyond a moderate level, which had been identified at the score of 0.638. With these findings, CPTPP countries can have their economic performance improved as long as their overall IQ is kept below this threshold level.

ACKNOWLEDGMENT

The author would like to express gratitude to the Malaysian Ministry of Education for providing financial support for this study through an FRGS grant.

REFERENCES

Acemoglu, D., Johnson, S., & Robinson, J. A. 2001. The colonial origins of comparative development: An empirical investigation. *The American Economic Review*, *91*(5), 1369–1401.

Aidt, T., Dutta, J., & Sena, V. 2008. Governance regimes, corruption and growth: Theory and evidence. *Journal of Comparative Economics*, *36*(2), 195–220.

Barro, R. J. 1996. Determinants of economic growth: A cross-country empirical study. (No. 5698). National Bureau of Economic Research.

Butkiewicz, J. L., & Yanikkaya, H. 2006. Institutional quality and economic growth: Maintenance of the rule of law or democratic institutions, or both? *Economic Modelling*, *23*(4), 648–661.

Chong, A., & Calderon, C. 2000. Causality and feedback between institutional measures and economic growth. *Economics and Politics*, *12*(1), 69–81.

Gwartney, J., Holcombe, R. G., & Lawson, R. A. 2004. Economic freedom, institutional quality, and cross-country differences in income and growth. *The Cato Journal*, *24*(3), 205–233.

Lind, J. T., & Mehlum, H. 2010. With or without u? The appropriate test for a U-shaped relationship. *Oxford Bulletin of Economics and Statistics*, *72*(1), 109–118.

Marakbi, R., & Turcu, C. 2016. Corruption, institutional quality and growth: a panel smooth transition regression approach. Available at: http://www.leo-univ-orleans.fr/mbFiles/documents/site-du-leo/seminaires-2016/seminaires-doctorants-2016/marakbi-turcu.pdf

Nawaz, S., Iqbal, N., & Khan, M. A. 2014. The impact of institutional quality on economic growth: The case of developing countries. *The Pakistan Development Review*, *53*(1), 15–31.

United Nations Statistics Division (UNSD) 2014. Available at: http://unstats.un.org/unsd/snaama/dnlList.asp.

Valeriani, E., & Peluso, S. 2011. The impact of institutional quality on economic growth and development: An empirical study. *Journal of Knowledge Management, Economics and Information Technology*, *1*(6), 1–25.

Vergil, H., & Teyyare, E. 2017. Crisis, institutional quality and economic growth, *Bogazici Journal: Review of Social, Economic & Administrative Studies*, *31*(2), 1–20.

Williams, A. 2006. *The link between institutional quality and economic growth: Evidence from a panel of countries*. University of Western Australia.

World Bank. 2018. Available at: https://data.worldbank.org/indicator.

External knowledge acquisition and innovation in small and medium-sized enterprises

I.S. Rosdi, A.M. Noor & N. Fauzi
Multimedia University, Cyberjaya, Selangor, Malaysia

ABSTRACT: Continuous renewal of organizational knowledge is crucial for firms intending to produce innovative outputs. For small and medium-sized enterprises (SMEs), external sources of knowledge becomes especially important, as their internal knowledge is relatively limited compared to that of large organizations. Based on a model of knowledge-driven innovation in SMEs, this study examined the perception and behavior of 221 employees in innovative Malaysian SMEs with regard to how their choices of using specific types of external knowledge sources subsequently influence the types of innovation produced by their firms. Results show that the employees' use of either market-oriented or scientific-oriented knowledge sources determined the type of innovative outputs for their firm, i.e., incremental or radical. The main contribution of this study is the emphasis on how employees' knowledge acquisition behavior significantly impacts innovation at the firm level.

1 INTRODUCTION

Small and medium-sized enterprises (SMEs) are important for the social and economic development of Malaysia, contributing greatly to the country's progress toward becoming a fully developed country in a knowledge economy (SME Corporation Malaysia, 2018). There is a positive relationship between organizational innovation and performance of SMEs (Salim & Sulaiman, 2011). Highly competitive business environments and fierce competition from both international and local companies result in increasing pressure on Malaysian SMEs to improve their innovation performance (Government of Malaysia, 2010; Jala, 2013).

1.1 Innovation and organizational innovative capability

Innovation refers to the drastic changes in organizations or their products and processes to gain a sustainable competitive advantage over other firms (De Leede & Looise, 2005). Innovative capability refers to the ability of a firm to manage its key competences and resources to develop innovative activities (Fleury et al., 2013). Incremental innovative capability refers to a firm's ability to produce incremental innovations, which refine and improve existing products and services (Chandy & Tellis, 2000), leading to minor changes in the firm as a result of minimal new knowledge (Hansen & Ockwell, 2014). On the other hand, radical innovative capability creates major transformations of existing products, services, or technologies, and makes existing technologies obsolete (Chandy & Tellis, 2000).

1.2 Knowledge sources

Knowledge is key in the fast-changing business environment, and the continuous acquisition of new knowledge is critical for the process of innovation in organizations (Fores & Camison, 2016; Nonaka & Takeuchi, 1995). Knowledge sources being used by SMEs can come from both the firm's internal and external environment (Henttonen, 2012). Internal sources are within the firm, such as employee experience or outcomes of a firm's research and development (R&D) activities while external knowledge comes from a firm's external environment such as customers, competitors, alliance partners, and universities (Berchicci, 2013; Henttonen, 2012).

External knowledge is of special important to SMEs, as their internal knowledge bank is relatively smaller compared to that of bigger firms (Berchicci, 2013; Henttonen, 2012). Furthermore, external knowledge sources are categorized as either scientific oriented or market oriented, with scientific-oriented sources being research centers or universities while market-oriented sources include customers, competitors, and suppliers (Henttonen, 2012).

1.3 Research background and objective

An empirical study using a survey as a data collection tool by Lee et al. (2019) had looked into the influences of external knowledge sources utilized by innovative Malaysian SMEs on their innovative capability. It involved 221 employees in Malaysian SMEs certified innovative by the Malaysian government, and the study reveals that the innovative Malaysian SMEs:

1. Rely more on market-oriented compared to scientific-oriented knowledge sources.
2. Produce innovative capability that is more incremental in nature compared to radical innovative capability.

Hence the objective of this article is to present a detailed analysis of the behavior of the employee respondents based on average ratings of the survey items, which will provide a relevant explanation of the research findings.

2 FINDINGS

2.1 Respondent demographics

Most of the respondents (62.09%) were made up of young workers between 25 and 35 years old. Among the respondents, 13.27% belonged to the "less than 25 years old" category, while the rest of respondents were distributed mostly between the "36–45 years old" (11.37%), "46–55 years old" (9.48%) categories and only eight employees (3.79%) were more than 55 years old. In terms of gender, the respondents were quite equally distributed between male (55.92%) and female (44.08%). All respondents in this study were Malaysians.

As for education, most had a bachelor's degree (50.71%), followed by diploma (20.38%), SPM or equivalent (13.74%), certificate (7.58%), and master's degree (6.64%). There was only one respondent from each respective STPM/Foundation/pre-U courses and PHD/DBA category. In terms of career profile, most respondents were at "Executive" level (62.56%), followed by "Non-Executive" (17.54%), "Managerial" (14.69%), and "Top Management" (5.21%) levels. Most were still relatively new to the job, as most of them had served only "1–3 years" (36.49%) and "4–6 years" (36.02%) with the current organization. The rest of the employees served "7–9 years" (16.59%) and only about 10.90% of the respondents had been with the current organizations for more than 9 years. In terms of the total years of employment, the majority of respondents were distributed equally among "4–6 years" (28.91%), "More than 9 years" (27.49%), and "1–3 years" (26.07%). The lowest number of respondents (17.54%) had "7–9 years" of employment.

2.2 Choice of knowledge sources and impact on innovative capability

Average ratings of individual survey items show that employees' relationship with customers is an important source of external knowledge (mean score = 3.93). Employees also depend highly on alliance partners (mean score = 3.88), suppliers (mean score = 3.88), and even competitors (mean score = 3.88) to gain new market knowledge. As for scientific-oriented sources, employees depend on universities and/or polytechnics (mean score = 3.69), consulting agencies, commercial laboratories, and private research institutes (mean score = 3.62) and on public and/or private nonprofit research institutes (mean score = 3.61). It seems that even though utilization of all types of scientific-oriented knowledge sources is above average, it is still significantly and relatively lower than utilization of market-oriented knowledge sources. This situation may be due to the easier access and lower costs of engaging with market-oriented sources. As for the impact on innovative capability, the companies are perceived by employees as producing more incremental innovative capability (mean score = 4.02) as compared to radical innovative capability (mean score = 3.73). This suggests that although the companies are innovating, they are mostly doing so in an incremental capacity.

3 DISCUSSION AND RECOMMENDATIONS

From the findings, it can be concluded that there needs to be properly strategized intervention to encourage SMEs to reach out further to the scientific actors. Perhaps with more systematic intervention by policymakers and SME governing entities, SMEs can be encouraged to utilize relevant external knowledge sources, especially the scientific-oriented ones, to help build their radical innovative capability.

Future research could be done on innovative SMEs in specific sectors and the results compared to see if there are differences between sectors. This could help in government policy development, as there is a benefit to understanding the perception of employees in different sectors. The use of cross-industry or cross-country study would allow a comparison of the potentially different outcomes from the use of external knowledge sources.

On the industry front, governmental bodies monitoring the progress of SMEs across all sectors in Malaysia must play an important role in stressing the importance of external knowledge sources to enhance the innovative capability of Malaysian SMEs. They could provide incentives or financial assistance for participation in interorganizational collaborative projects leading to knowledge sharing and also encourage SMEs to work with universities or other research institutes to spur radical innovative capability. The continuous pursuit of new knowledge should be encouraged in Malaysian SMEs.

4 CONCLUSION

Results show that the employees' use of either market-oriented or scientific-oriented knowledge sources determines the type of innovative outputs for their firm, i.e., incremental or radical innovative outputs. The main contribution of this study is the emphasis on how employees' knowledge acquisition behavior significantly impacts innovation at the firm level. Findings from the study show that employees rely more on market-oriented knowledge sources compared to scientific-oriented knowledge sources, and because of this practice, their firms develop innovative capability that is more incremental in nature compared to radical innovative capability. If the SMEs can be encouraged to utilize more scientific-oriented knowledge sources in the future, it may enable them to acquire the capability of producing more radical innovations. Hence, data from the study have important implications for both future researchers and industry players, as they show the different roles of both market-oriented and scientific-oriented knowledge sources in driving organizational innovation in SMEs toward a more sustainable future.

REFERENCES

Berchicci, L. 2013. Towards an open R&D system: Internal R&D investment, external knowledge acquisition and innovative performance. *Research Policy*, *42*(1), 117–127.

Chandy, R. K., & Tellis, G. J. 2000. The incumbent's curse? Incumbency, size, and radical product innovation. *Journal of Marketing*, *64*(3), 1–17.

De Leede, J., & Looise, J. K. 2005. Innovation and HRM: Towards an integrated framework. *Creativity and Innovation Management*, *14*(2), 108–117.

Fleury, A., Fleury, M. T. L., & Borini, F. M. 2013. The Brazilian multinationals' approaches to innovation. *Journal of International Management*, *19*(3), 260–275.

Fores, B., & Camison, C. 2016. Does incremental and radical innovation performance depend on different types of knowledge accumulation capabilities and organizational size? *Journal of Business Research*, *69*(2), 831–848.

Government of Malaysia. 2010. Keynote address by the Prime Minister of Malaysia at Invest Malaysia 2010 [Online]. Available at: http://nitc.kkmm.gov.my/index.php/key-ict-initiatives/new-economic-model-nem-for-malaysia-2010/keynote-address-by-the-prime-minister-of-malaysia-at-invest-malaysia-2010-and-the-new-economic-model-2010 (accessed September 10, 2016).

Hansen, U. E., & Ockwell, D. 2014. Learning and technological capability building in emerging economies: The case of the biomass power equipment industry in Malaysia. *Technovation*, *34*(10), 617–630.

Henttonen, K. 2012. The relation between external sources of knowledge and different types of innovation in SMEs. In *Proceedings of the International Society for Professional Innovation Management (ISPIM)*.

Jala, I. 2013. Innovation is not dead in Malaysia. *[Online]*. Available at: http://www.thestar.com.my/business/business-news/2013/08/19/innovation-is-not-dead-in-malaysia/(accessed August 22, 2016).

Nonaka, I., & Takeuchi, H. 1995. *The Knowledge-Creating Company: How Japanese Companies Create the Dynamics of Innovation*. New York: Oxford University Press.

Salim, I. M., & Sulaiman, M. 2011. Organizational learning, innovation and performance: A study of Malaysian small and medium sized enterprises. *International Journal of Business and Management*, *6*(12), 118.

SME Corporation Malaysia. 2018. SME Statistics [Online]. Available: at: http://www.smecorp.gov.my/index.php/en/policies/2015-12-21-09-09-49/sme-statistics (accessed July 13, 2018).

Overcoming math anxiety and developing mathematical resilience via e-learning

R.A. Razak
Multimedia University, Malaysia

ABSTRACT: Math anxiety is one of the psychological factors that affect students' math skills and achievement. Math-anxious individuals often stay away from or skip mathematics classes, take up fewer mathematics courses, and have lower confidence in the subject compared to their non-anxious peers. These behaviours are often displayed by business and management students at the university. With the increasing demand for STEM-skilled graduates, math anxiety in business and management students should be overcome so that they can become relevant in the future industry. This paper discusses how interactive e-learning can reduce students' math anxiety and develops their mathematical resilience for effective learning.

1 INTRODUCTION

The urgent need to prepare a skilled workforce for Industry 4.0 has increased the demand for science, technology, engineering and math (STEM) graduates in Malaysia. For the purpose of ensuring graduates that are Industry 4.0 ready, many higher education institutions have begun incorporating big data and data analytics in their programmes, including business and management programmes. These changes have posed greater challenges for academics in the business and management faculties to increase students' higher-order thinking and mathematical skills. The students who enrolled in the business and management programmes are usually students with little or no mathematical background. Recent studies have shown that students who came from non-science background and those having high stats anxiety were more likely to underperform in Mathematics (Syed Wahid et al. 2018). This issue is really concerning and has to be addressed so that students from the non-science background can remain relevant in and beyond the Industry 4.0 era.

The use of e-learning for teaching mathematics could be effective in developing students' mathematical resilience and reduce math anxiety. An excellent e-learning program should suit all type of learners. In general, the e-learning environment promotes active learning, where educational videos and tools are created to get students engage with the subject such as solving problems through a diverse range of methods and forum discussions. Learning at own pace instead of following the pace of the whole batch motivates one to learn more. However, many mathematics academics felt that the traditional way of teaching math is much more effective. Therefore, this paper aims to discuss how interactive e-learning can reduce students' math anxiety and develops their mathematical resilience for effective learning.

2 LITERATURE REVIEW

2.1 *Math anxiety*

Math anxiety is one of the psychological factors that greatly affect students' performance in mathematics and statistics (Paechter et al. 2017). This anxiety is a pervasive issue in education that requires attention from both educators and researchers to help students reach their full academic potential (Ramirez et al 2018). Math anxiety is defined as a negative response to participation in any activities that are related to mathematics (Hembree 1990, Suárez-Pellicioni et al. 2016). For example, for a student with math anxiety, the act of opening a mathematics textbook or entering a mathematics class can trigger a negative emotional response. Simple daily activities such as reading a cash register receipt can send math-anxious individuals into a panic state. People with math anxiety performs poorly when numerical information is involved despite having good comprehension, critical thinking and reasoning skills.

Low mathematical abilities are often associated with high mathematics anxiety. Mathematics anxious individuals often stay away from or skip mathematics and classes, take up less mathematics courses, rate themselves as less skilled in the subject and have lower confidence in the subject compared to their non-anxious peers (Hembree 1990). These behaviors are often displayed by business and management students at the university, especially students who came from non-science background. Past studies found that students' math anxiety is impediment to their mathematics achievements (Maloney & Beilock 2012, Syed Wahid et al. 2018).

To date, studies on math anxiety and mathematics performance are continuously being studied in order to find out the potential factors or mediators that influence the relationship between the two variables (Suárez-Pellicioni et al. 2016, Paechter et al. 2017, Ramirez et al. 2018). In a recent study, Suárez-Pellicioni et al. (2016) provide three potential causes for the relationship between math anxiety and performance, namely, weakness in fundamental cognitive and spatial skills, low working memory and negative learning experience.

2.2 *Mathematical resilience*

The concept of mathematical resilience was proposed by Johnston-Wilder & Lee (2010) to describe a positive stance towards mathematics. A learner with mathematical resilience has a growth belief related to their mathematical ability. Such learner has confidence to overcome mathematical difficulties by utilizing the resources and support provided. Williams (2014) defined mathematical resilience as what it takes to be safely challenged while being in the growth zone, where the learner develops confidence, persistence and perseverance. This characteristic allows the learner to overcome math anxiety.

2.3 *The growth model*

Marshall et al. (2017) applied the growth zone model to develop strategies for students to overcome their math anxiety. The growth zone model has three zones, namely, the comfort zone, the growth zone and the anxiety zone. In the comfort zone, the learner feels safe, can build self-confidence, able to learn independently and do not require any help. The growth zone is where new learning happens. Learners in the growth zone learn from their mistakes in their calculations and use available resource to solidify their understanding of the knowledge learnt. The anxiety zone has students who often feels helpless with math. New learning will never take place in this danger zone even with good support.

2.4 *Interactive e-learning for effective math learning*

Technological advancement has influenced the characteristics of millennial students and the way they learn. Among the characteristics of millennial students are high reliance on technology, and short-term attention span. Through interactive e-learning, students can learn at their own pace, anywhere and anytime. They can also get peer support online instantly and instructor support online conveniently. They can also assess their understanding anytime via self-tests.

Effective mathematics learning requires understanding rather than memorization, sufficient practice, peer-learning and good support or resources. The e-learning modules prepared by the instructor should be illustrative and interactive so that they suit millennial learners. When a math anxious student learns a topic via e-learning and attempts the interactive math exercise, it can be said that the student has entered the growth learning zone. New learning keeps occurring whenever the student attempts more exercises or learn another topic though videos. Overall, the e-learning should also be effective such that it helps students overcome their math anxiety and develop their mathematical resilience.

3 CONCLUSION

This paper lays the foundation of a research work by discussing how interactive e-learning can reduce students' math anxiety and develops their mathematical resilience for effective learning. The effectiveness of e-learning in reducing student math anxiety and developing mathematical resilience will be carried out in the next study. In addition, interactive e-learning is inferred to be a moderating factor of the learning zones.

REFERENCES

Hembree, R. 1990. The nature, effects, and relief of mathematics anxiety. *J. Res. Math. Educ.*, 21: 33–46.
Johnston-Wilder, S. & Lee, C. 2010. Mathematical Resilience. *Mathematics Teaching*, 218: 38–41.
Johnston-Wilder, S., Pardoe, S., Almehrz, H., Evans, B., Marsh, J. & Richards, S. 2016. Developing teaching for mathematical resilience in further education. *9th International Conference of Education, Research and Innovation* (ICERI2016), Seville, 14 – 16 November, 2016.
Maloney, E.A. & Beilock, S. L. 2012. Math anxiety: who has it, why it develops, and how to guard against it. *Trends in Cognitive Sciences*, 16(8): 404–406.
Marshall, E., Mann, V., Wilson, D. & Staddon, R. 2017. Learning and teaching toolkit: Maths anxiety. https://www.sheffield.ac.uk/polopoly_fs/1.753619!/file/Maths_anxiety_strategies.pdf
Paechter, M., Macher, D., Martskvishvili, K., Wimmer, S. & Papousek, I. 2017. Mathematics anxiety and statistics anxiety. Shared but also unshared components and antagonistic contribution to performance in statistics. *Frontiers in Psychology*, July (8): 1–13.
Ramirez, G., Shaw, S. T. & Maloney, E. A. 2018. Math anxiety: Past research, promising interventions and a new interpretation framework. *Educational Psychologist*, 53(3): 145–164.
Suárez-Pellicioni, M., Núñez-Peña, M. I. & Colomé, A. 2016. Math anxiety: A review of its cognitive consequences, psychophysiological correlates and brain bases. *Cognitive, Affective and Behavioral Neuroscience*, 16(1): 3–22.
Syed Wahid, S. N., Yusof, Y. & Mohamed Nor, A. H. 2018. Effect of mathematics anxiety on student's performance in higher education level: A comparative study on gender. *AIP Conference Proceedings* 1974, 050010(2018).
Williams, G. 2014. Optimistic problem-solving activity: enacting confidence, persistence and perseverance. *ZDM*, 46(3): 407–422.

Relational social capital, innovation capability, and small and medium-sized enterprise performance

Rochmi Widayanti & Ratna Damayanti
Universitas Islam Batik, Surakarta, Indonesia

Nuryakin & Susanto
Master of Management Department, Universitas Muhammadiyah Yogyakarta, Indonesia

ABSTRACT: This research aimed to analyze the effect of relational social capital on performance of small and medium-sized enterprises with innovation capability as the mediating variables. The sample of this research was the cluster pattern of SMEs in the export-based brass industry in the Central Java region. The unit of analysis that was carried out was the SMEs owners. Furthermore, the sample of this research comprised 200 respondents. The results of hypothesis testing used the Structural Equation Modeling (SEM) approach with the AMOS program. The results of this research showed that the social-relational capital significantly affected SME performance and innovation capability. This research also empirically found that innovation capability can mediate the relationship between social-relational capital and SME performance.

1 INTRODUCTION

The meaningful existence of small and medium-sized enterprises (SMEs) in improving the economic activities of a region becomes an essential factor that needs to be considered. The presence of SMEs also has significant challenges, especially in developing countries, including Indonesia. Daou et al. (2013) explain that SMEs in developing countries have a limited capability to establish networking and contribute a suggestion for government policy. However, it is challenging to meet the aspect of capital and development of relational capital, and challenging to communicate with the network in the international market.

SME performance is one of the variables in this research that refers to the terminology of financial performance (Chen et al., 2012; Malhotra, 2005; Wulandari et al., 2017; Zohdi et al., 2013) and nonfinancial performance (Low et al., 2007; Nätti et al., 2014; Osakwe et al., 2016). The benchmark for nonfinancial performance is the activities that relate to the three dimensions of efficiency, quality, and time. The dynamic change of the company's environment triggers SMEs to be modern business organization and learning organization-based business with a sustainable transformation process (Eris & Ozmen, 2012; Tohidi et al., 2012).

This study aimed to analyze the effect of social-relational capital on SME performance with innovation capability as the mediating variable. This study also tried to answer the research problem on the factors that support the improvement of SME performance through the critical role of innovation capability.

2 THEORETICAL REVIEW AND HYPOTHESIS DEVELOPMENT

2.1 *The relationship between relational social capital and SME performance*

The concept of social capital is mostly related to SMEs. The studies of Chen et al.,(2007) and Felicio. et al. (2014) showed that social capital could improve the performance of SMEs significantly. The relation mentioned before leads to the direction of the information diffusion acceleration, opportunity for new technology and products, niche markets, and financial resources. These products are obtained through the expansion of business networks externally. Furthermore, efforts to maintain the level of trust and interdependence among the partners in business networks can improve performance. According to Reagan and Zuckerman (2001), a team with high social diversity will have bound or interconnected relationships that will enhance performance, and there will be an exchange of information from the social interactions that occur. Based on the above description, the research hypothesis is as follows:

H1: Social-relational capital has a significant positive effect on the performance of SMEs.

2.2 The relationship between relational social capital and innovation capability

Social capital and community characteristics can affect business behavior, for example, risk taking and innovation that will ultimately have an impact on improving the company (Ismail et al., 2014). The basis of innovation capability is the resources and competencies. The resources obtained are physical, economic, intellectual, and social. Innovation capability is the combination of these resources with the capability to use and implement them. Therefore, social capital is one of the primary elements in innovation capability. The existence of a social relation will provide an opportunity in creating innovation (Tura & Harmaakorpi, 2005). Tsai dan Ghoshal, (1998) examined the relationship of structural and relational dimensions with cognitive social capital on the pattern of resources and product innovation exchange in a company. The social interaction is the manifestation of the social structural capital dimension. Based on the above description, the research hypothesis is as follows:

H2: Social-relational capital has a significant positive effect on innovation capability.

2.3 The relationship between innovation capability and SME performance

Innovation capability is one of the critical factors that enable SMEs to reach the level of both national and international market competition. Therefore, the promotion and capability to maintain the innovation are becoming the main focus for the managers of SMEs in the top management ranks (Erturk, 2010). At the same time, to prove the existence of a relationship between a company's innovation and performance, Tidd (2001) constructs two categories. The first group is related to the accounting and financial performance such as profitability, stock prices, and return on investment. The second group is associated with market share and market development. Based on the above description, the research hypothesis is as follows:

H3: Innovation capability has a significant positive effect on SME performance.

3 RESEARCH METHODS

Data collected in this study came from SMEs of the brass exporter group in the region of Central Java, in Boyolali regency. The respondents of this study were the owners of brass SMEs based on export. The data sampling process consisted of respondents filling out the questionnaire and participating in an interview. The researcher used a purposive sampling method and took a sample of 200 respondents. The questionnaires with complete answers were then tested with advanced testing to predict the research constructs.

4 RESULTS

Testing of the hypotheses in this study used Structural Equation Modeling (SEM) analysis of the AMOS program. The result of the confirmatory full model test shows a good result that fulfilled the goodness of fit criteria. The model structure was used to describe the causality models of the research with a tiered relationship. The test results fulfilled the goodness of fit criteria, with a chi-square of 103.566. The probability value is 0.001. Those two assumptions have been fulfilled. TLI value is 0.764, GFI value is 0.812, AGFI value is 0.872, and RMSEA value is 0.061. These values indicate that they follow the specified cut-off, indicating that the research model is accepted and meets the criteria (standards) decided in the testing of matrix correlation between the constructs of social-relational capital.

Table 1 indicates the relationship of SEM among social-relational capital, innovation capabilities, and performance of SMEs. The results of this research are shown in Table 1 with three hypotheses. These three hypotheses are the relationships among social-relational capital, innovation capabilities, and performance of SMEs. The statistical calculation using AMOS 21 is shown in Table 1. The t-value and probability value portray a positive relationship and the significance of each variable.

A discussion of the relationship of each variable to SMEs follows.

Social-relational capital significantly influences the performance of SMEs. Table 1 shows the structural path model that explains the relationship between social-relational capital and SME performance. The result of the statistical test shows the value of ($t = 3.582 > 1.96$), significance value of ($0.000 < 0.05$). So, hypothesis 1 stating that social-relational capital positively affects SME performance is proved.

Social-relational capital significantly influences innovation capability. Table 1 shows the structural path model that explains the relationship between social-relational capital and innovation capability. The result of the statistical test shows the value of ($t = 3.293 > 1.96$), significance value of ($0.000 < 0.05$). So, hypothesis 2 stating that the social-relational capital positively and significantly affects innovation capability is proved.

Innovation capability significantly influences SME performance. Table 1 shows the structural path model that explains the relationship between innovation capability and SME performance. The result of the statistical test shows the value of ($t = 4.790 > 1.96$), significance value of ($0.000 < 0.05$). So, hypothesis 3 stating that innovation capability positively and significantly affects SME performance is proved.

5 DISCUSSION, IMPLICATIONS, AND LIMITATIONS

This study found that social-relational capital directly and significantly influences SME performance. This study follows the studies by Chenet al. (2007) and Felicio et al. (2014), which showed that social capital could significantly improve SME performance. Reagan and Zuckerman (2001) found that high social diversity would have an interconnection and enhance the performance. Also, there would be an exchange of information from the social interactions that occur (Schoonhoven &Romanelli, 2001).

Social-relational capital directly and significantly influences innovation capability, and innovation capability directly and substantially affect SME performance. The result of this research is in line with the previous study, which found that social capital could be observed as a business network. Thus, it facilitates the business in achieving excellent performance and a competitive advantage (Batjargal, 2003). Another research result that is in line with this study also states that the ability to maintain innovation is the main focus of the SME managers of the top management ranks (Ertürk, 2010).

Table 1. Summary of the results from the SEM models.

Relationship	Path coeff.	CR	Prob	Hypotheses
Social-relational capital × SME performance	0.382	3.582	0.000	H1: Accepted
Social-relational capital × innovation capability	0.317	3.293	0.000	H2: Accepted
Innovation capability SME performance	0.410	4.790	0.000	H3: Accepted

**$p < 0.01$; *$p < 0.05$.

This article tried to answer the aim of the research regarding the influence of social-relational capital toward innovation capability that improves SME performance. This study has limitations in the sample. It allows the results of testing models detecting problematic data in the testing of data normality and outliers. Thus, it requires accuracy on the part of the researcher.

ACKNOWLEDGMENTS

KEMENRISTEK DIKTI Indonesia supported this research.

REFERENCES

Chen, Y.-Y., Yeh, S.-P., & Huang, H.-L. 2012. Does knowledge management "fit" matter to business performance? *Journal of Knowledge Management*, 16(5), 671–687. doi: 10.1108/13673271211262745

Daou, A., Karuranga, E., & Su, Z. 2013. Intellectual capital in Mexican SMEs from the perspective of the resources-based and dynamic capabilities views. *The Journal of Applied Business Research*, 29(6), 1673.

Eris, E. D., & Ozmen, O. N. T. 2012. The effect of market orientation, learning orientation and innovativeness. *International Journal of Economic Sciences and Applied Research*, 5(1),77–108.

Low, D. R., Chapman, R. L., & Sloan, T. R. 2007. Inter-relationships between innovation and market orientation in SMEs. *Management Research News*, 30(12), 878–891. doi: 10.1108/01409170710833321

Malhotra, Y. 2005. Integrating knowledge management technologies in organizational business processes: Getting real time enterprises to deliver real business performance. *Journal of Knowledge Management*, 9(1), 7.

Nätti, S., Hurmelinna-Laukkanen, P., & Johnston, W. J. 2014. Absorptive capacity and network orchestration in innovation communities: Promoting service innovation. *Journal of Business & Industrial Marketing*, 29(2), 173–184. doi: 10.1108/jbim-08-2013-0167

Osakwe, C. N., Chovancova, M., & Ogbonna, B. U. 2016. Linking SMEs profitability to brand orientation and market-sensing capability: A service sector evidence. *Periodica Polytechnica Social and Management Sciences*, 24(1), 34–40. doi: 10.3311/PPso.8069

Tohidi, H., Seyedaliakbar, S. M., & Mandegari, M. (2012). Organizational learning measurement and the effect on firm innovation. *Journal of Enterprise Information Management Decision*, 25(3,2), 219–245. doi: 10.1108/17410391211224390

Wulandari, F., Djastuti, I., & Nuryakin. 2017). Reassessment of the entrepreneurial motivation among female business owners to enhance SMEs business performance in Indonesia. *European Research Studies Journal*, XX((4A), 18–34.

Zohdi, M., Shafeai, R., & Hashemi, R. 2013. Influence of relational capabilities on business performance: Case of: Kermanshah industrial city SMEs. *International Research Journal of Applied and Basic Sciences*, 4 (3), 589–596.

Author Index

Adityawarman, 90
Ahmed, W. 192
Akter, H. 192
Alam, M.F.Q. 227
Alamanda, D.T. 358
Alamsyah, A. 280, 285
Alexandra, M.O. 172
Ali, S.M. 370
Alif, M. G. 163
Amini, A. 112
Andriawan, F. 201
Anggadwita, G. 346, 358
Anthonysamy, L. 108, 133
Ariyanti, M. 27
Arnaz, R. 71
Arumsari, A. 258
Arviansyah, 99, 214
Atamtajani, A.S.M. 150
Aurachman, R. 180, 184, 188
Awaludin, I.S. 387

Bachtiar, R. 85
Baranti, L.A. 141
Banjarnahor, R.E. 99
Baskoro, H.A. 112
Bionda, A.R. 289
Bujang, A.S. 350

Cardiah, T. 267
Cariawan, U. 99
Chalid, D.A. 271
Chew, K.W. 350
Chong, C.Y. 396
Choo, K.A. 108
Christian, D. 54
Ciptandi, F. 302

Damayanti, R. 407
Dalimunthe, Z. 289
Desa, M.A.B.M. 328
Dharmiko, A. 145
Digpasari, D.A. 285
Dudija, N. 145

Edbert, 49
Endriawan, D. 137

Fauzi, N. 400
Feisal, F. 18
Fitri, I. 240

Gandakusuma, I. 289
Geni, S.L. 316
Gitasari, R.A. 58
Gopinathan, S. 366

Hamsal, M. 210
Hamzah, H. 387
Hana, G.S. 307
Hanafi, H.M. 324
Hasana, S.A. 76
Hendrawan, R. 158
Hidayat, D. 328
Hidayat, M.I. 248
Hin, H.S. 108
Hizam, S.M. 192
Hooi, T.S. 383
Husodo, Z.A 63
Hussin, A.A.A. 219

Indrawati, 5, 67, 76, 81, 103, 129, 196, 201
Indrianti, Y. 362
Irawan, H. 280, 285
Ismail, N. 383
Ismawaty, 40
Ismiraini, M.A.S. 125
Izzuddin, F.D. 324

Jaafar, N. 370
Jaffar, N. 342, 383, 387
Jahja, J. 244
Jalil, A. 219
Januarizka, C. 196
Johari, H. 342

Kaur, A.H. 366
Khan, N. 333

Khan, S. 333
Koesoemadinata, M.I.P. 328
Kosasih, W. 362
Kristanti, F.T. 223, 248, 262, 275
Kristian, T. 223
Kurnia, F.R. 20
Kurniawan, R. 67
Kusdinar, P. 27
Kusmara, A.R. 258

Lubis, L.M. 232
Lubis, M. 205, 297
Lutfi, A. 141

Mapfumo, S. 337
Maryati, T. 13
Maulida, N. 94
Mengko, T.L.R. 307
Miala, M.I. 262
Mon, C.S. 176
Muafah, A.R. 297
Mukdamanee, V. 9
Muthaiyah, S. 1, 5, 196, 201

Nadeak, L. 154
Nguyen, L.T.P. 391
Nidya, P.A. 81
Nilasari, I. 374
Nizat, M.N.M. 354
Noor, A.M. 400
Nopendri, 180
Norhashim, M. 370
Nugroho, R.L. 337
Nuryakin, 407
Nuryakin, S. 13

Octaviani, M.B. 63
Ong, J.W. 354

Pasaribu, R.D. 17
Permatasari, A. 358
Pillai, S.K.B. 76, 103

Prabowo, H. 210
Pradhina, N.P. 103
Pratami, G.C.W. 346
Princes, E. 117, 121
Puspitasari, W. 232
Putri, S.A. 150

Rachmawati, R. 90
Rahardjo, B. 307
Rahardjo, S. 311
Rahim, M.F.A. 350, 354
Ramantoko, G. 40, 71
Ramdhani, A. 358
Rapiz, K. 214
Razak, R.A. 404
Retno, S.L. 24
Rikumahu, B. 227
Rokhim, R. 58
Rosby, K. 271
Rosdi, I.S. 400

Sachari, A. 258
Sadeli, J. 293
Said, A. 210
Sandyopasana, T.N. 163
Saprudin, A.A. 379
Sarihati, T. 267

Sarman, 253
Sasmoko, 121, 362
Satriawana, P.W. 158
Segaran, S. 391
Selamat, Z. 387
Sembiring, P.E. 244
Sembung, N.F. 275
Sentosa, I. 192
Setiawan, J. 117
Sham, R. 219
Shigeno, Y. 45
Simatupang, B. 210
Sitorus, P.M.T. 20, 172, 320
Sudiana, K. 168
Sugiat, M.A. 168
Sugiyanto, 236
Sulthoni, A. 293
Suryani, R. 205
Suryawardani, B. 35
Susanto, 407
Suyanto, A.M.A. 253
Susanty, A.I. 85
Syafrizal, S. 316

Tan, O. 333
Tanudjaya, E. 54

Tricahyono, D. 154, 196, 201, 374
Trihanondo, D. 137

Wahyudi, E. 31
Wahyuningtyas, R. 94
Widaryanti, E. 129
Widayanti, R. 407
Widhaningrat, S.K. 31, 45, 49
Widodo, T. 240
Widyawati, R.S. 280
Wiliam, A. 362
Witjaksono, R.W. 205, 297
Wulandari, A. 35
Wulandari, R. 267

Yanan, H.A. 354
Yatim, N.M. 354
Yuldinawati, L. 125
Yuliana, E. 358
Yuliansyah, Y. 320

Zainuddin, M.N. 350
Zaw, T.O.K. 1, 5